THE HYDROGEN ECONOMY

Opportunities, Costs, Barriers, and R&D Needs

Committee on Alternatives and Strategies
for Future Hydrogen Production and Use

Board on Energy and Environmental Systems

Division on Engineering and Physical Sciences

NATIONAL RESEARCH COUNCIL *AND*
NATIONAL ACADEMY OF ENGINEERING
OF THE NATIONAL ACADEMIES

THE NATIONAL ACADEMIES PRESS
Washington, D.C.
www.nap.edu

NATIONAL ACADEMIES PRESS • 500 Fifth Street, N.W. • Washington, DC 20001

This report and the study on which it is based were supported by Grant No. DE-FG36-02GO12114 from the U.S. Department of Energy. Any opinions, findings, conclusions, or recommendations expressed in this publication are those of the authors and do not necessarily reflect the views of the organizations or agencies that provided support for the project.

International Standard Book Number 0-309-09163-2 (Book)
International Standard Book Number 0-309-53068-7 (PDF)
Library of Congress Control Number 2004108605

Available in limited supply from:
Board on Energy and Environmental
Systems
National Research Council
500 Fifth Street, N.W.
KECK-W934
Washington, DC 20001
202-334-3344

Additional copies available for sale from:
National Academies Press
2101 Constitution Avenue, N.W.
Box 285
Washington, DC 20055
800-624-6242 or 202-334-3313 (in the Washington metropolitan area)
http://www.nap.edu

Printed in the United States of America

THE NATIONAL ACADEMIES

Advisers to the Nation on Science, Engineering, and Medicine

The **National Academy of Sciences** is a private, nonprofit, self-perpetuating society of distinguished scholars engaged in scientific and engineering research, dedicated to the furtherance of science and technology and to their use for the general welfare. Upon the authority of the charter granted to it by the Congress in 1863, the Academy has a mandate that requires it to advise the federal government on scientific and technical matters. Dr. Bruce M. Alberts is president of the National Academy of Sciences.

The **National Academy of Engineering** was established in 1964, under the charter of the National Academy of Sciences, as a parallel organization of outstanding engineers. It is autonomous in its administration and in the selection of its members, sharing with the National Academy of Sciences the responsibility for advising the federal government. The National Academy of Engineering also sponsors engineering programs aimed at meeting national needs, encourages education and research, and recognizes the superior achievements of engineers. Dr. Wm. A. Wulf is president of the National Academy of Engineering.

The **Institute of Medicine** was established in 1970 by the National Academy of Sciences to secure the services of eminent members of appropriate professions in the examination of policy matters pertaining to the health of the public. The Institute acts under the responsibility given to the National Academy of Sciences by its congressional charter to be an adviser to the federal government and, upon its own initiative, to identify issues of medical care, research, and education. Dr. Harvey V. Fineberg is president of the Institute of Medicine.

The **National Research Council** was organized by the National Academy of Sciences in 1916 to associate the broad community of science and technology with the Academy's purposes of furthering knowledge and advising the federal government. Functioning in accordance with general policies determined by the Academy, the Council has become the principal operating agency of both the National Academy of Sciences and the National Academy of Engineering in providing services to the government, the public, and the scientific and engineering communities. The Council is administered jointly by both Academies and the Institute of Medicine. Dr. Bruce M. Alberts and Dr. Wm. A. Wulf are chair and vice chair, respectively, of the National Research Council.

www.national-academies.org

COMMITTEE ON ALTERNATIVES AND STRATEGIES FOR FUTURE HYDROGEN PRODUCTION AND USE

MICHAEL P. RAMAGE, NAE,[1] *Chair*, ExxonMobil Research and Engineering Company (retired), Moorestown, New Jersey
RAKESH AGRAWAL, NAE, Air Products and Chemicals, Inc., Allentown, Pennsylvania
DAVID L. BODDE, University of Missouri, Kansas City
ROBERT EPPERLY, Consultant, Mountain View, California
ANTONIA V. HERZOG, Natural Resources Defense Council, Washington, D.C.
ROBERT L. HIRSCH, Science Applications International Corporation, Alexandria, Virginia
MUJID S. KAZIMI, Massachusetts Institute of Technology, Cambridge
ALEXANDER MACLACHLAN, NAE, E.I. du Pont de Nemours & Company (retired), Wilmington, Delaware
GENE NEMANICH, Independent Consultant, Sugar Land, Texas
WILLIAM F. POWERS, NAE, Ford Motor Company (retired), Ann Arbor, Michigan
MAXINE L. SAVITZ, NAE, Consultant (retired, Honeywell), Los Angeles, California
WALTER W. (CHIP) SCHROEDER, Proton Energy Systems, Inc., Wallingford, Connecticut
ROBERT H. SOCOLOW, Princeton University, Princeton, New Jersey
DANIEL SPERLING, University of California, Davis
ALFRED M. SPORMANN, Stanford University, Stanford, California
JAMES L. SWEENEY, Stanford University, Stanford, California

Project Staff

Board on Energy and Environmental Systems (BEES)

MARTIN OFFUTT, Study Director
ALAN CRANE, Senior Program Officer
JAMES J. ZUCCHETTO, Director, BEES
PANOLA GOLSON, Senior Project Assistant

NAE Program Office

JACK FRITZ, Senior Program Officer

Consultants

Dale Simbeck, SFA Pacific, Inc.
Elaine Chang, SFA Pacific, Inc.

[1]NAE = member, National Academy of Engineering.

BOARD ON ENERGY AND ENVIRONMENTAL SYSTEMS

DOUGLAS M. CHAPIN, NAE,[1] *Chair*, MPR Associates, Alexandria, Virginia
ROBERT W. FRI, *Vice Chair,* Resources for the Future, Washington, D.C.
ALLEN J. BARD, NAS,[2] University of Texas, Austin
DAVID L. BODDE, University of Missouri, Kansas City
PHILIP R. CLARK, NAE, GPU Nuclear Corporation (retired), Boonton, New Jersey
CHARLES GOODMAN, Southern Company Services, Birmingham, Alabama
DAVID G. HAWKINS, Natural Resources Defense Council, Washington, D.C.
MARTHA A. KREBS, California Nanosystems Institute (retired), Los Angeles, California
GERALD L. KULCINSKI, NAE, University of Wisconsin, Madison
JAMES J. MARKOWSKY, NAE, American Electric Power (retired), North Falmouth, Massachusetts
DAVID K. OWENS, Edison Electric Institute, Washington, D.C.
WILLIAM F. POWERS, NAE, Ford Motor Company (retired), Ann Arbor, Michigan
EDWARD S. RUBIN, Carnegie Mellon University, Pittsburgh, Pennsylvania
MAXINE L. SAVITZ, NAE, Honeywell, Inc. (retired), Los Angeles, California
PHILIP R. SHARP, Harvard University, Cambridge, Massachusetts
ROBERT W. SHAW, JR., Aretê Corporation, Center Harbor, New Hampshire
SCOTT W. TINKER, University of Texas, Austin
JOHN J. WISE, NAE, Mobil Research and Development Company (retired), Princeton, New Jersey

Staff

JAMES J. ZUCCHETTO, Director
ALAN CRANE, Senior Program Officer
MARTIN OFFUTT, Program Officer
DANA CAINES, Financial Associate
PANOLA GOLSON, Project Assistant

[1]NAE = member, National Academy of Engineering.
[2]NAS = member, National Academy of Sciences.

Acknowledgments

The Committee on Alternatives and Strategies for Future Hydrogen Production and Use wishes to acknowledge and thank the many individuals who contributed significantly of their time and effort to this National Academies' National Research Council (NRC) study, which was done jointly with the National Academy of Engineering (NAE) Program Office. The presentations at committee meetings provided valuable information and insight on advanced technologies and development initiatives that assisted the committee in formulating the recommendations included in this report.

The committee expresses its thanks to the following individuals who briefed the committee: Alex Bell (University of California, Berkeley); Larry Burns (General Motors); John Cassidy (UTC, Inc.); Steve Chalk (U.S. Department of Energy [DOE]); Elaine Chang (SFA Pacific); Roxanne Danz (DOE); Pete Devlin (DOE); Jon Ebacher (GE Power Systems); Charles Forsberg (Oak Ridge National Laboratory [ORNL]); David Friedman (Union of Concerned Scientists); David Garman (DOE); David Gray (Mitretek); Cathy Gregoire-Padro (National Renewable Energy Laboratory [NREL]); Dave Henderson (DOE); Gardiner Hill (BP); Bill Innes (ExxonMobil Research and Engineering); Scott Jorgensen (General Motors); Nathan Lewis (California Institute of Technology); Margaret Mann (NREL); Lowell Miller (DOE); JoAnn Milliken (DOE); Joan Ogden (Princeton University); Lynn Orr, Jr. (Stanford University); Ralph Overend (NREL); Mark Pastor (DOE); David Pimentel (Cornell University); Dan Reicher (Northern Power Systems and New Energy Capital); Neal Richter (ChevronTexaco); Jens Rostrup-Nielsen (Haldor Topsoe); Dale Simbeck (SFA Pacific); and Joseph Strakey (DOE National Energy Technology Laboratory).

The committee offers special thanks to Steve Chalk, DOE Office of Hydrogen, Fuel Cells and Infrastructure Technologies, and to Roxanne Danz, DOE Office of Energy Efficiency and Renewable Energy, for being responsive to its needs for information. In addition, the committee wishes to acknowledge Dale Simbeck and Elaine Chang, both of SFA Pacific, Inc., for providing support as consultants to the committee.

Finally, the chair gratefully recognizes the committee members and the staffs of the NRC's Board on Energy and Environmental Systems and the NAE Program Office for their hard work in organizing and planning committee meetings and their individual efforts in gathering information and writing sections of the report.

This report has been reviewed in draft form by individuals chosen for their diverse perspectives and technical expertise, in accordance with procedures approved by the NRC's Report Review Committee. The purpose of this independent review is to provide candid and critical comments that will assist the institution in making its published report as sound as possible and to ensure that the report meets institutional standards for objectivity, evidence, and responsiveness to the study charge. The review comments and draft manuscript remain confidential to protect the integrity of the deliberative process. We wish to thank the following individuals for their review of this report:

Allen Bard (NAS), University of Texas, Austin;
Seymour Baron (NAE), retired, Medical University of South Carolina;
Douglas Chapin (NAE), MPR Associates, Inc.;
James Corman, Energy Alternative Systems;
Francis J. DiSalvo (NAS), Cornell University;
Mildred Dresselhaus (NAE, NAS), Massachusetts Institute of Technology;
Seth Dunn, Yale School of Management, and School of Forestry & Environmental Studies;
David Friedman, Union of Concerned Scientists;
Robert Friedman, The Center for the Advancement of Genomics;
Robert D. Hall, CDG Management, Inc.;
James G. Hansel, Air Products and Chemicals, Inc.;
H.M. (Hub) Hubbard, retired, Pacific International Center for High Technology Research;
Trevor Jones (NAE), Biomec;
James R. Katzer (NAE), ExxonMobil Research and Engineering Company;
Alan Lloyd, California Air Resources Board;
John P. Longwell (NAE), retired, Massachusetts Institute of Technology;
Alden Meyer, Union of Concerned Scientists;
Robert W. Shaw, Jr., Aretê Corporation; and
Richard S. Stein, (NAS, NAE) retired, University of Massachusetts.

Although the reviewers listed above have provided many constructive comments and suggestions, they were not asked to endorse the conclusions or recommendations, nor did they see the final draft of the report before its release. The review of this report was overseen by William G. Agnew (NAE), General Motors Corporation (retired). Appointed by the National Research Council, he was responsible for making certain that an independent examination of this report was carried out in accordance with institutional procedures and that all review comments were carefully considered. Responsibility for the final content of this report rests entirely with the authoring committee and the institution.

Contents

Tables and Figures

TABLES

FIGURES

Executive Summary

The National Academies' National Research Council appointed the Committee on Alternatives and Strategies for Future Hydrogen Production and Use in the fall of 2002 to address the complex subject of the "hydrogen economy." In particular, the committee carried out these tasks:

- Assessed the current state of technology for producing hydrogen from a variety of energy sources;
- Made estimates on a consistent basis of current and future projected costs, carbon dioxide (CO_2) emissions, and energy efficiencies for hydrogen technologies;
- Considered scenarios for the potential penetration of hydrogen into the economy and associated impacts on oil imports and CO_2 gas emissions;
- Addressed the problem of how hydrogen might be distributed, stored, and dispensed to end uses—together with associated infrastructure issues—with particular emphasis on light-duty vehicles in the transportation sector;
- Reviewed the U.S. Department of Energy's (DOE's) research, development, and demonstration (RD&D) plan for hydrogen; and
- Made recommendations to the DOE on RD&D, including directions, priorities, and strategies.

The vision of the hydrogen economy is based on two expectations: (1) that hydrogen can be produced from domestic energy sources in a manner that is affordable and environmentally benign, and (2) that applications using hydrogen—fuel cell vehicles, for example—can gain market share in competition with the alternatives. To the extent that these expectations can be met, the United States, and indeed the world, would benefit from reduced vulnerability to energy disruptions and improved environmental quality, especially through lower carbon emissions. However, before this vision can become a reality, many technical, social, and policy challenges must be overcome. This report focuses on the steps that should be taken to move toward the hydrogen vision and to achieve the sought-after benefits. The report focuses exclusively on hydrogen, although it notes that alternative or complementary strategies might also serve these same goals well.

The Executive Summary presents the basic conclusions of the report and the major recommendations of the committee. The report's chapters present additional findings and recommendations related to specific technologies and issues that the committee considered.

BASIC CONCLUSIONS

As described below, the committee's basic conclusions address four topics: implications for national goals, priorities for research and development (R&D), the challenge of transition, and the impacts of hydrogen-fueled light-duty vehicles on energy security and CO_2 emissions.

Implications for National Goals

A transition to hydrogen as a major fuel in the next 50 years could fundamentally transform the U.S. energy system, creating opportunities to increase energy security through the use of a variety of domestic energy sources for hydrogen production while reducing environmental impacts, including atmospheric CO_2 emissions and criteria pollutants.[1] In his State of the Union address of January 28, 2003, President Bush moved energy, and especially hydrogen for vehicles, to the forefront of the U.S. political and technical debate. The President noted: "A simple chemical reaction between hydrogen and oxygen generates energy, which can be used to power a car producing only water, not exhaust fumes. With a new national commitment, our scientists and engineers will overcome obstacles to taking these cars from

[1]Criteria pollutants are air pollutants (e.g., lead, sulfur dioxide, and so on) emitted from numerous or diverse stationary or mobile sources for which National Ambient Air Quality Standards have been set to protect human health and public welfare.

laboratory to showroom so that the first car driven by a child born today could be powered by hydrogen, and pollution-free."[2] This committee believes that investigating and conducting RD&D activities to determine whether a hydrogen economy might be realized are important to the nation. There is a potential for replacing essentially all gasoline with hydrogen over the next half century using only domestic resources. And there is a potential for eliminating almost all CO_2 and criteria pollutants from vehicular emissions. However, there are currently many barriers to be overcome before that potential can be realized.

Of course there are other strategies for reducing oil imports and CO_2 emissions, and thus the DOE should keep a balanced portfolio of R&D efforts and continue to explore supply-and-demand alternatives that do not depend upon hydrogen. If battery technology improved dramatically, for example, all-electric vehicles might become the preferred alternative. Furthermore, hybrid electric vehicle technology is commercially available today, and benefits from this technology can therefore be realized immediately. Fossil-fuel-based or biomass-based synthetic fuels could also be used in place of gasoline.

Research and Development Priorities

There are major hurdles on the path to achieving the vision of the hydrogen economy; the path will not be simple or straightforward. Many of the committee's observations generalize across the entire hydrogen economy: the hydrogen system must be cost-competitive, it must be safe and appealing to the consumer, and it would preferably offer advantages from the perspectives of energy security and CO_2 emissions. Specifically for the transportation sector, dramatic progress in the development of fuel cells, storage devices, and distribution systems is especially critical. Widespread success is not certain.

The committee believes that for hydrogen-fueled transportation, the four most fundamental technological and economic challenges are these:

1. *To develop and introduce cost-effective, durable, safe, and environmentally desirable fuel cell systems and hydrogen storage systems.* Current fuel cell lifetimes are much too short and fuel cell costs are at least an order of magnitude too high. An on-board vehicular hydrogen storage system that has an energy density approaching that of gasoline systems has not been developed. Thus, the resulting range of vehicles with existing hydrogen storage systems is much too short.

2. *To develop the infrastructure to provide hydrogen for the light-duty-vehicle user.* Hydrogen is currently produced in large quantities at reasonable costs for industrial purposes. The committee's analysis indicates that at a future, mature stage of development, hydrogen (H_2) can be produced and used in fuel cell vehicles at reasonable cost. The challenge, with today's industrial hydrogen as well as tomorrow's hydrogen, is the high cost of distributing H_2 to dispersed locations. This challenge is especially severe during the early years of a transition, when demand is even more dispersed. The costs of a mature hydrogen pipeline system would be spread over many users, as the cost of the natural gas system is today. But the transition is difficult to imagine in detail. It requires many technological innovations related to the development of small-scale production units. Also, nontechnical factors such as financing, siting, security, environmental impact, and the perceived safety of hydrogen pipelines and dispensing systems will play a significant role. All of these hurdles must be overcome before there can be widespread use. An initial stage during which hydrogen is produced at small scale near the small user seems likely. In this case, production costs for small production units must be sharply reduced, which may be possible with expanded research.

3. *To reduce sharply the costs of hydrogen production from renewable energy sources, over a time frame of decades.* Tremendous progress has been made in reducing the cost of making electricity from renewable energy sources. But making hydrogen from renewable energy through the intermediate step of making electricity, a premium energy source, requires further breakthroughs in order to be competitive. Basically, these technology pathways for hydrogen production make electricity, which is converted to hydrogen, which is later converted by a fuel cell back to electricity. These steps add costs and energy losses that are particularly significant when the hydrogen competes as a commodity transportation fuel—leading the committee to believe that most current approaches—except possibly that of wind energy—need to be redirected. The committee believes that the required cost reductions can be achieved only by targeted fundamental and exploratory research on hydrogen production by photobiological, photochemical, and thin-film solar processes.

4. *To capture and store ("sequester") the carbon dioxide by-product of hydrogen production from coal.* Coal is a massive domestic U.S. energy resource that has the potential for producing cost-competitive hydrogen. However, coal processing generates large amounts of CO_2. In order to reduce CO_2 emissions from coal processing in a carbon-constrained future, massive amounts of CO_2 would have to be captured and safely and reliably sequestered for hundreds of years. Key to the commercialization of a large-scale, coal-based hydrogen production option (and also for natural-gas-based options) is achieving broad public acceptance, along with additional technical development, for CO_2 sequestration.

For a viable hydrogen transportation system to emerge, all four of these challenges must be addressed.

[2] *Weekly Compilation of Presidential Documents.* Monday, February 3, 2003. Vol. 39, No. 5, p. 111. Washington, D.C.: Government Printing Office.

The Challenge of Transition

There will likely be a lengthy transition period during which fuel cell vehicles and hydrogen are not competitive with internal combustion engine vehicles, including conventional gasoline and diesel fuel vehicles, and hybrid gasoline electric vehicles. The committee believes that the transition to a hydrogen fuel system will best be accomplished initially through distributed production of hydrogen, because distributed generation avoids many of the substantial infrastructure barriers faced by centralized generation. Small hydrogen-production units located at dispensing stations can produce hydrogen through natural gas reforming or electrolysis. Natural gas pipelines and electricity transmission and distribution systems already exist; for distributed generation of hydrogen, these systems would need to be expanded only moderately in the early years of the transition. During this transition period, distributed renewable energy (e.g., wind or solar energy) might provide electricity to onsite hydrogen production systems, particularly in areas of the country where electricity costs from wind or solar energy are particularly low. A transition emphasizing distributed production allows time for the development of new technologies and concepts capable of potentially overcoming the challenges facing the widespread use of hydrogen. The distributed transition approach allows time for the market to develop before too much fixed investment is set in place. While this approach allows time for the ultimate hydrogen infrastructure to emerge, the committee believes that it cannot yet be fully identified and defined.

Impacts of Hydrogen-Fueled Light-Duty Vehicles

Several findings from the committee's analysis (see Chapter 6) show the impact on the U.S. energy system if successful market penetration of hydrogen fuel cell vehicles is achieved. In order to analyze these impacts, the committee posited that fuel cell vehicle technology would be developed successfully and that hydrogen would be available to fuel light-duty vehicles (cars and light trucks). These findings are as follows:

- The committee's upper-bound market penetration case for fuel cell vehicles, premised on hybrid vehicle experience, assumes that fuel cell vehicles enter the U.S. light-duty vehicle market in 2015 in competition with conventional and hybrid electric vehicles, reaching 25 percent of light-duty vehicle sales around 2027. The demand for hydrogen in about 2027 would be about equal to the current production of 9 million short tons (tons) per year, which would be only a small fraction of the 110 million tons required for full replacement of gasoline light-duty vehicles with hydrogen vehicles, posited to take place in 2050.
- If coal, renewable energy, or nuclear energy is used to produce hydrogen, a transition to a light-duty fleet of vehicles fueled entirely by hydrogen would reduce total energy imports by the amount of oil consumption displaced. However, if natural gas is used to produce hydrogen, and if, on the margin, natural gas is imported, there would be little if any reduction in total energy imports, because natural gas for hydrogen would displace petroleum for gasoline.
- CO_2 emissions from vehicles can be cut significantly if the hydrogen is produced entirely from renewables or nuclear energy, or from fossil fuels with sequestration of CO_2. The use of a combination of natural gas without sequestration and renewable energy can also significantly reduce CO_2 emissions. However, emissions of CO_2 associated with light-duty vehicles contribute only a portion of projected CO_2 emissions; thus, sharply reducing overall CO_2 releases will require carbon reductions in other parts of the economy, particularly in electricity production.
- Overall, although a transition to hydrogen could greatly transform the U.S. energy system in the long run, the impacts on oil imports and CO_2 emissions are likely to be minor during the next 25 years. However, thereafter, if R&D is successful and large investments are made in hydrogen and fuel cells, the impact on the U.S. energy system could be great.

MAJOR RECOMMENDATIONS

Systems Analysis of U.S. Energy Options

The U.S. energy system will change in many ways over the next 50 years. Some of the drivers for such change are already recognized, including at present the geology and geopolitics of fossil fuels and, perhaps eventually, the rising CO_2 concentration in the atmosphere. Other drivers will emerge from options made available by new technologies. The U.S. energy system can be expected to continue to have substantial diversity; one should expect the emergence of neither a single primary energy source nor a single energy carrier. Moreover, more-energy-efficient technologies for the household, office, factory, and vehicle will continue to be developed and introduced into the energy system. The role of the DOE hydrogen program[3] in the restructuring of the overall national energy system will evolve with time.

To help shape the DOE hydrogen program, the committee sees a critical role for systems analysis. Systems analysis will be needed both to coordinate the multiple parallel efforts within the hydrogen program and to integrate the program within a balanced, overall DOE national energy R&D effort. Internal coordination must address the many primary sources from which hydrogen can be produced, the various

[3]The words "hydrogen program" refer collectively to the programs concerned with hydrogen production, distribution, and use within DOE's Office of Energy Efficiency and Renewable Energy, Office of Fossil Energy, Office of Science, and Office of Nuclear Energy, Science, and Technology. There is no single program with this title.

scales of production, the options for hydrogen distribution, the crosscutting challenges of storage and safety, and the hydrogen-using devices. Integration within the overall DOE effort must address the place of hydrogen relative to other secondary energy sources—helping, in particular, to clarify the competition between electricity-based, liquid-fuel-based (e.g., cellulosic ethanol), and hydrogen-based transportation. This is particularly important as clean alternative fuel internal combustion engines, fuel cells, and batteries evolve. Integration within the overall DOE effort must also address interactions with end-use energy efficiency, as represented, for example, by high-fuel-economy options such as hybrid vehicles. Implications of safety, security, and environmental concerns will need to be better understood. So will issues of timing and sequencing: depending on the details of system design, a hydrogen transportation system initially based on distributed hydrogen production, for example, might or might not easily evolve into a centralized system as density of use increases.

Recommendation ES-1. The Department of Energy should continue to develop its hydrogen initiative as a potential long-term contributor to improving U.S. energy security and environmental protection. The program plan should be reviewed and updated regularly to reflect progress, potential synergisms within the program, and interactions with other energy programs and partnerships (e.g., the California Fuel Cell Partnership). In order to achieve this objective, the committee recommends that the DOE develop and employ a systems analysis approach to understanding full costs, defining options, evaluating research results, and helping balance its hydrogen program for the short, medium, and long term. Such an approach should be implemented for all U.S. energy options, not only for hydrogen.

As part of its systems analysis, the DOE should map out and evaluate a transition plan consistent with developing the infrastructure and hydrogen resources necessary to support the committee's hydrogen vehicle penetration scenario or another similar demand scenario. The DOE should estimate what levels of investment over time are required—and in which program and project areas—in order to achieve a significant reduction in carbon dioxide emissions from passenger vehicles by midcentury.

Fuel Cell Vehicle Technology

The committee observes that the federal government has been active in fuel cell research for roughly 40 years, while proton exchange membrane (PEM) fuel cells applied to hydrogen vehicle systems are a relatively recent development (as of the late 1980s). In spite of substantial R&D spending by the DOE and industry, costs are still a factor of 10 to 20 times too expensive, these fuel cells are short of required durability, and their energy efficiency is still too low for light-duty-vehicle applications. Accordingly, the challenges of developing PEM fuel cells for automotive applications are large, and the solutions to overcoming these challenges are uncertain.

The committee estimates that the fuel cell system, including on-board storage of hydrogen, will have to decrease in cost to less than \$100 per kilowatt (kW)[4] before fuel cell vehicles (FCVs) become a plausible commercial option, and that it will take at least a decade for this to happen. In particular, if the cost of the fuel cell system for light-duty vehicles does not eventually decrease to the \$50/kW range, fuel cells will not propel the hydrogen economy without some regulatory mandate or incentive.

Automakers have demonstrated FCVs in which hydrogen is stored on board in different ways, primarily as high-pressure compressed gas or as a cryogenic liquid. At the current state of development, both of these options have serious shortcomings that are likely to preclude their long-term commercial viability. New solutions are needed in order to lead to vehicles that have at least a 300 mile driving range; that are compact, lightweight, and inexpensive; and that meet future safety standards.

Given the current state of knowledge with respect to fuel cell durability, on-board storage systems, and existing component costs, the committee believes that the near-term DOE milestones for FCVs are unrealistically aggressive.

Recommendation ES-2. Given that large improvements are still needed in fuel cell technology and given that industry is investing considerable funding in technology development, increased government funding on research and development should be dedicated to the research on breakthroughs in on-board storage systems, in fuel cell costs, and in materials for durability in order to attack known inhibitors of the high-volume production of fuel cell vehicles.

Infrastructure

A nationwide, high-quality, safe, and efficient hydrogen infrastructure will be required in order for hydrogen to be used widely in the consumer sector. While it will be many years before hydrogen use is significant enough to justify an integrated national infrastructure—as much as two decades in the scenario posited by the committee—regional infrastructures could evolve sooner. The relationship between hydrogen production, delivery, and dispensing is very complex, even for regional infrastructures, as it depends on many variables associated with logistics systems and on many public and private entities. Codes and standards for infrastructure development could be a significant deterrent to hydrogen advancement if not established well ahead of the hydrogen market. Similarly, since resilience to terrorist at-

[4]The cost includes the fuel cell module, precious metals, the fuel processor, compressed hydrogen storage, balance of plant, and assembly, labor, and depreciation.

tack has become a major performance criterion for any infrastructure system, the design of future hydrogen infrastructure systems may need to consider protection against such risks.

In the area of infrastructure and delivery there seem to be significant opportunities for making major improvements. The DOE does not yet have a strong program on hydrogen infrastructures. DOE leadership is critical, because the current incentives for companies to make early investments in hydrogen infrastructure are relatively weak.

Recommendation ES-3a. The Department of Energy program in infrastructure requires greater emphasis and support. The Department of Energy should strive to create better linkages between its seemingly disconnected programs in large-scale and small-scale hydrogen production. The hydrogen infrastructure program should address issues such as storage requirements, hydrogen purity, pipeline materials, compressors, leak detection, and permitting, with the objective of clarifying the conditions under which large-scale and small-scale hydrogen production will become competitive, complementary, or independent. The logistics of interconnecting hydrogen production and end use are daunting, and all current methods of hydrogen delivery have poor energy-efficiency characteristics and difficult logistics. Accordingly, the committee believes that exploratory research focused on new concepts for hydrogen delivery requires additional funding. The committee recognizes that there is little understanding of future logistics systems and new concepts for hydrogen delivery—thus making a systems approach very important.

Recommendation ES-3b. The Department of Energy should accelerate work on codes and standards and on permitting, addressing head-on the difficulties of working across existing and emerging hydrogen standards in cities, counties, states, and the nation.

Transition

The transition to a hydrogen economy involves challenges that cannot be overcome by research and development and demonstrations alone. Unresolved issues of policy development, infrastructure development, and safety will slow the penetration of hydrogen into the market even if the technical hurdles of production cost and energy efficiency are overcome. Significant industry investments in advance of market forces will not be made unless government creates a business environment that reflects societal priorities with respect to greenhouse gas emissions and oil imports.

Recommendation ES-4. The policy analysis capability of the Department of Energy with respect to the hydrogen economy should be strengthened, and the role of government in supporting and facilitating industry investments to help bring about a transition to a hydrogen economy needs to be better understood.

The committee believes that a hydrogen economy will not result from a straightforward replacement of the present fossil-fuel-based economy. There are great uncertainties surrounding a transition period, because many innovations and technological breakthroughs will be required to address the costs and energy-efficiency, distribution, and nontechnical issues. The hydrogen fuel for the very early transitional period, before distributed generation takes hold, would probably be supplied in the form of pressurized or liquefied molecular hydrogen, trucked from existing, centralized production facilities. But, as volume grows, such an approach may be judged too expensive and/or too hazardous. It seems likely that, in the next 10 to 30 years, hydrogen produced in distributed rather than centralized facilities will dominate. Distributed production of hydrogen seems most likely to be done with small-scale natural gas reformers or by electrolysis of water; however, new concepts in distributed production could be developed over this time period.

Recommendation ES-5. Distributed hydrogen production systems deserve increased research and development investments by the Department of Energy. Increased R&D efforts and accelerated program timing could decrease the cost and increase the energy efficiency of small-scale natural gas reformers and water electrolysis systems. In addition, a program should be initiated to develop new concepts in distributed hydrogen production systems that have the potential to compete—in cost, energy efficiency, and safety—with centralized systems. As this program develops new concepts bearing on the safety of local hydrogen storage and delivery systems, it may be possible to apply these concepts in large-scale hydrogen generation systems as well.

Safety

Safety will be a major issue from the standpoint of commercialization of hydrogen-powered vehicles. Much evidence suggests that hydrogen can be manufactured and used in professionally managed systems with acceptable safety, but experts differ markedly in their views of the safety of hydrogen in a consumer-centered transportation system. A particularly salient and underexplored issue is that of leakage in enclosed structures, such as garages in homes and commercial establishments. Hydrogen safety, from both a technological and a societal perspective, will be one of the major hurdles that must be overcome in order to achieve the hydrogen economy.

Recommendation ES-6. The committee believes that the Department of Energy program in safety is well planned and should be a priority. However, the committee emphasizes the following:

• Safety policy goals should be proposed and discussed by the Department of Energy with stakeholder groups early in the hydrogen technology development process.

• The Department of Energy should continue its work with standards development organizations and ensure increased emphasis on distributed production of hydrogen.

• Department of Energy systems analysis should specifically include safety, and it should be understood to be an overriding criterion.

• The goal of the physical testing program should be to resolve safety issues in advance of commercial use.

• The Department of Energy's public education program should continue to focus on hydrogen safety, particularly the safe use of hydrogen in distributed production and in consumer environments.

Carbon Dioxide-Free Hydrogen

The long timescale associated with the development of viable hydrogen fuel cells and hydrogen storage provides a time window for a more intensive DOE program to develop hydrogen from electrolysis, which, if economic, has the potential to lead to major reductions in CO_2 emissions and enhanced energy security. The committee believes that if the cost of fuel cells can be reduced to $50 per kilowatt, with focused research a corresponding dramatic drop in the cost of electrolytic cells to electrolyze water can be expected (to ~$125/kW). If such a low electrolyzer cost is achieved, the cost of hydrogen produced by electrolysis will be dominated by the cost of the electricity, not by the cost of the electrolyzer. Thus, in conjunction with research to lower the cost of electrolyzers, research focused on reducing electricity costs from renewable energy and nuclear energy has the potential to reduce overall hydrogen production costs substantially.

Recommendation ES-7. The Department of Energy should increase emphasis on electrolyzer development, with a target of $125 per kilowatt and a significant increase in efficiency toward a goal of over 70 percent (lower heating value basis). In such a program, care must be taken to properly account for the inherent intermittency of wind and solar energy, which can be a major limitation to their wide-scale use. In parallel, more aggressive electricity cost targets should be set for unsubsidized nuclear and renewable energy that might be used directly to generate electricity. Success in these areas would greatly increase the potential for carbon dioxide-free hydrogen production.

Carbon Capture and Storage

The DOE's various efforts with respect to hydrogen and fuel cell technology will benefit from close integration with carbon capture and storage (sequestration) activities and programs in the Office of Fossil Energy. If there is an expanded role for hydrogen produced from fossil fuels in providing energy services, the probability of achieving substantial reductions in net CO_2 emissions through sequestration will be greatly enhanced through close program integration. Integration will enable the DOE to identify critical technologies and research areas that can enable hydrogen production from fossil fuels with CO_2 capture and storage. Close integration will promote the analysis of overlapping issues such as the co-capture and co-storage with CO_2 of pollutants such as sulfur produced during hydrogen production.

Many early carbon capture and storage projects will not involve hydrogen, but rather will involve the capture of the CO_2 impurity in natural gas, the capture of CO_2 produced at electric plants, or the capture of CO_2 at ammonia and synfuels plants. All of these routes to capture, however, share carbon storage as a common component, and carbon storage is the area in which the most difficult institutional issues and the challenges related to public acceptance arise.

Recommendation ES-8. The Department of Energy should tighten the coupling of its efforts on hydrogen and fuel cell technology with the DOE Office of Fossil Energy's programs on carbon capture and storage (sequestration). Because of the hydrogen program's large stake in the successful launching of carbon capture and storage activity, the hydrogen program should participate in all of the early carbon capture and storage projects, even those that do not directly involve carbon capture during hydrogen production. These projects will address the most difficult institutional issues and the challenges related to issues of public acceptance, which have the potential of delaying the introduction of hydrogen in the marketplace.

The Department of Energy's Hydrogen Research, Development, and Demonstration Plan

As part of its effort, the committee reviewed the DOE's draft "Hydrogen, Fuel Cells & Infrastructure Technologies Program: Multi-Year Research, Development and Demonstration Plan," dated June 3, 2003 (DOE, 2003b). The committee's deliberations focused only on the hydrogen production and demand portion of the overall DOE plan. For example, while the committee makes recommendations on the use of renewable energy for hydrogen production, it did not review the entire DOE renewables program in depth. The committee is impressed by how well the hydrogen program has progressed. From its analysis, the committee makes two overall observations about the program:

• First, the plan is focused primarily on the activities in the Office of Hydrogen, Fuel Cells, and Infrastructure Technologies Program within the Office of Energy Efficiency and Renewable Energy, and on some activities in the Office of Fossil Energy. The activities related to hydrogen in the Office of Nuclear Energy, Science, and Technology, and in the Office of Science, as well as activities related to carbon capture and storage in the Office of Fossil Energy, are important, but they are mentioned only casually in the plan. The development of an overall DOE program will require better integration across all DOE programs.

• Second, the plan's priorities are unclear, as they are lost within the myriad of activities that are proposed. The general budget for DOE's hydrogen program is contained in the appendix of the plan, but the plan provides no dollar numbers at the project level, even for existing projects and programs. The committee found it difficult to judge the priorities and the go/no-go decision points for each of the R&D areas.

Recommendation ES-9. The Department of Energy should continue to develop its hydrogen research, development, and demonstration (RD&D) plan to improve the integration and balance of activities within the Office of Energy Efficiency and Renewable Energy; the Office of Fossil Energy (including programs related to carbon sequestration); the Office of Nuclear Energy, Science, and Technology; and the Office of Science. The committee believes that, overall, the production, distribution, and dispensing portion of the program is probably underfunded, particularly because a significant fraction of appropriated funds is already earmarked. The committee understands that of the $78 million appropriated for hydrogen technology for FY 2004 in the Energy and Water appropriations bill (Public Law 108-137), $37 million is earmarked for activities that will not particularly advance the hydrogen initiative. The committee also believes that the hydrogen program, in an attempt to meet the extreme challenges set by senior government and DOE leaders, has tried to establish RD&D activities in too many areas, creating a very diverse, somewhat unfocused program. Thus, prioritizing the efforts both within and across program areas, establishing milestones and go/no-go decisions, and adjusting the program on the basis of results are all extremely important in a program with so many challenges. This approach will also help determine when it is appropriate to take a program to the demonstration stage. And finally, the committee believes that the probability of success in bringing the United States to a hydrogen economy will be greatly increased by partnering with a broader range of academic and industrial organizations—possibly including an international focus[5]—and by establishing an independent program review process and board.

Recommendation ES-10. There should be a shift in the hydrogen program away from some development areas and toward exploratory work—as has been done in the area of hydrogen storage. A hydrogen economy will require a number of technological and conceptual breakthroughs. The Department of Energy program calls for increased funding in some important exploratory research areas such as hydrogen storage and photoelectrochemical hydrogen production. However, the committee believes that much more exploratory research is needed. Other areas likely to benefit from an increased emphasis on exploratory research include delivery systems, pipeline materials, electrolysis, and materials science for many applications. The execution of such changes in emphasis would be facilitated by the establishment of DOE-sponsored academic energy research centers. These centers should focus on interdisciplinary areas of new science and engineering—such as materials research into nanostructures, and modeling for materials design—in which there are opportunities for breakthrough solutions to energy issues.

Recommendation ES-11. As a framework for recommending and prioritizing the Department of Energy program, the committee considered the following:

• Technologies that could significantly impact U.S. energy security and carbon dioxide emissions,
• The timescale for the evolution of the hydrogen economy,
• Technology developments needed for both the transition period and the steady state,
• Externalities that would decelerate technology implementation, and
• The comparative advantage of the DOE in research and development of technologies at the pre-competitive stage.

The committee recommends that the following areas receive increased emphasis:

• *Fuel cell vehicle development.* Increase research and development (R&D) to facilitate breakthroughs in fuel cell costs and in durability of fuel cell materials, as well as breakthroughs in on-board hydrogen storage systems;
• *Distributed hydrogen generation.* Increase R&D in small-scale natural gas reforming, electrolysis, and new concepts for distributed hydrogen production systems;
• *Infrastructure analysis.* Accelerate and increase efforts in systems modeling and analysis for hydrogen delivery, with the objective of developing options and helping guide R&D in large-scale infrastructure development;
• *Carbon sequestration and FutureGen.* Accelerate development and early evaluation of the viability of carbon capture and storage (sequestration) on a large scale because of its implications for the long-term use of coal for hydrogen production. Continue the FutureGen Project as a high-priority task; and
• *Carbon dioxide-free energy technologies.* Increase emphasis on the development of wind-energy-to-hydrogen as an important technology for the hydrogen transition period and potentially for the longer term. Increase exploratory and fundamental research on hydrogen production by photobiological, photoelectrochemical, thin-film solar, and nuclear heat processes.

[5]Secretary of Energy Spencer Abraham, joined by ministers representing 14 nations and the European Commission, signed an agreement on November 20, 2003, to formally establish the International Partnership for the Hydrogen Economy.

1

Introduction

The January 2003 announcement by President Bush of the Hydrogen Fuel Initiative stimulated the interest of both the technical community and the broader public in the "hydrogen economy." As it is frequently envisioned, the hydrogen economy comprises the production of molecular hydrogen using coal, natural gas, nuclear energy, or renewable energy (e.g., biomass, wind, solar);[1] the transport and storage of hydrogen in some fashion; and the end use of hydrogen in fuel cells, which combine oxygen with the hydrogen to produce electricity (and some heat).[2] Fuel cells are under development for powering vehicles or to produce electricity and heat for residential, commercial, and industrial buildings. Many of the technologies for realizing such extensive use of hydrogen in the economy face significant barriers to development and successful commercialization. The challenges range from fundamental research and development (R&D) needs to overcoming infrastructure barriers and achieving social acceptance.

ORIGIN OF THE STUDY

In response to a request from the U.S. Department of Energy (DOE), the National Research Council (NRC) formed the Committee on Alternatives and Strategies for Future Hydrogen Production and Use (see Appendix A for biographical information). Formed by the NRC's Board on Energy and Environmental Systems and the National Academy of Engineering Program Office, the committee evaluated the cost and status of technologies for the production, transportation, storage, and end use of hydrogen and reviewed DOE's hydrogen research, development, and demonstration (RD&D) strategy.

In April 2003, the committee submitted an interim letter report to the Department of Energy. The letter report was prepared to provide early feedback and recommendations for assisting the DOE in preparations for its Fiscal Year (FY) 2005 hydrogen R&D programs. (The complete text of the letter report is presented in Appendix B.) In the present report, the committee expands on the four recommendations in the letter report and further develops its views.

DEPARTMENT OF ENERGY OFFICES INVOLVED IN WORK ON HYDROGEN

Within the DOE, and reporting to the Undersecretary for Energy, Science, and Environment, are three applied energy offices: the Office of Energy Efficiency and Renewable Energy (EERE), the Office of Fossil Energy (FE), and the Office of Nuclear Energy, Science, and Technology (NE). The Office of Science (SC) also has a role to play in that its support of basic science, especially in areas such as fundamental materials science, could lead to key breakthroughs needed for widespread use of hydrogen in the U.S. economy. All four of these offices are involved to one degree or another in hydrogen-related work, although their respective overall missions are much broader and total budgets larger than the segments focused on hydrogen-related work. Summed across all four offices (EERE, FE, NE, SC), the President's budget request for FY 2004 for the hydrogen program[3] was \$181 million for direct programs and \$301 million for associated programs (DOE, 2003a; see Appendix C regarding the hy-

[1]Hydrogen in the lithosphere is, with few exceptions, bound to other elements (e.g., as in water) and must be separated by using other sources of energy to produce molecular hydrogen. Properly considered, hydrogen fuel is not a primary energy source in the context of a hydrogen economy.

[2]Hydrogen can also be burned in internal combustion engines or in turbines, but fuel cells have the advantage of high efficiencies and virtually zero emissions except for water.

[3] The words "hydrogen program" refer collectively to the programs concerned with hydrogen production, distribution, and use within DOE's Office of Energy Efficiency and Renewable Energy, Office of Fossil Energy, Office of Science, and Office of Nuclear Energy, Science, and Technology. There is no single program with this title.

drogen program budget).[4] The funding level for direct programs would represent a near doubling of budget authority (appropriated funds) over funding for FY 2003, during which direct programs received $96.6 million.

SCOPE, ORGANIZATION, AND FOCUS OF THIS REPORT

Statement of Task

The committee assessed the current state of technology for producing hydrogen from a variety of energy sources; made estimates on a consistent basis of current and future projected costs for hydrogen; considered potential scenarios for the penetration of hydrogen technologies into the economy and the associated impacts on oil imports and carbon dioxide (CO_2) gas emissions; addressed the problems and associated infrastructure issues of how hydrogen might be distributed, stored, and dispensed to end uses, such as cars; reviewed the DOE's RD&D plan for hydrogen; and made recommendations to the DOE on RD&D, including directions, priorities, and strategies.

The current study is modeled after an NRC study that resulted in the 1990 report *Fuels to Drive Our Future* (NRC, 1990), which analyzed the status of technologies for producing liquid transportation fuels from domestic resources, such as biomass, coal, natural gas, oil shale, and tar sands. That study evaluated the cost of producing various liquid transportation fuels from these resources on a consistent basis, estimated opportunities for reducing costs, and identified R&D needs to improve technologies and reduce costs. *Fuels to Drive Our Future* did not include the production and use of hydrogen, which is the subject of this committee's report.

The statement of task for the committee was as follows:

> This study is similar in intent to a 1990 report by the National Research Council (NRC), *Fuels to Drive Our Future*, which evaluated the options for producing liquid fuels for transportation use. The use of that comprehensive study was proposed by DOE as the model for this one on hydrogen. With revisions to account for the different end use applications, process technologies, and current concerns about climate change and energy security, it will be used as a general guide for the report to be produced in this work. In particular, the NRC will appoint a committee that will address the following tasks:
>
> 1. Identify and evaluate the current status of the major alternative technologies and sources for producing hydrogen, for transmitting and storing hydrogen, and for using hydrogen to provide energy services especially in the transportation, but also the utility, residential, industrial and commercial sectors of the economy.
> 2. Assess the feasibility of operating each of these conversion technologies both at a small scale appropriate for a building or vehicle and at a large scale typical of current centralized energy conversion systems such as refineries or power plants. This question is important because it is not currently known whether it will be better to produce hydrogen at a central facility for distribution or to produce it locally near the points of end-use. This assessment will include factors such as societal acceptability (the NIMBY problem), operating difficulties, environmental issues including CO_2 emission, security concerns, and the possible advantages of each technology in special markets such as remote locations or particularly hot or cold climates.
> 3. Estimate current costs of the identified technologies and the cost reductions that the committee judges would be required to make the technologies competitive in the market place. As part of this assessment, the committee will consider the future prospects for hydrogen production and end-use technologies (e.g., in the 2010 to 2020, 2020–2050, and beyond 2050 time frames). This assessment may include scenarios for the introduction and subsequent commercial development of a hydrogen economy based on the use of predominantly domestic resources (e.g., natural gas, coal, biomass, renewables [e.g., solar, geothermal, wind], nuclear, municipal and industrial wastes, petroleum coke, and other potential resources), and consider constraints to their use.
> 4. Based on the technical and cost assessments, and considering potential problems with making the "chicken and egg" transition to a widespread hydrogen economy using each technology, review DOE's current RD&D programs and plans, and suggest an RD&D strategy with recommendations to DOE on the R&D priority needs within each technology area and on the priority for work in each area.
> 5. Provide a letter report on the committee's interim findings no later than February 2003 so this information can be used in DOE's budget and program planning for Fiscal Year 2005.
> 6. Publish a written final report on its work, approximately 13 months from contract initiation.
>
> The committee's interim letter report and final report will be reviewed in accordance with National Research Council (NRC) report review procedures before release to the sponsor and the public.

Structure of This Report

Chapter 2 describes the U.S. energy system as it exists today and explains how energy infrastructure is built up and how production technologies mature. The chapter also describes key, overarching issues that will be treated in later chapters. Chapter 3 discusses the demand side—describing the categories of technologies, such as automotive and stationary fuel cells, that use hydrogen and postulating the future demand for these units should hydrogen become a com-

[4]"Direct funding" is defined by the DOE as funding that would not be requested if there were no hydrogen-related activities. "Associated" efforts are those necessary for a hydrogen pathway, such as hybrid electric components in the DOE's budget within the FreedomCAR Partnership, a cooperative research effort between the DOE and the United States Council for Automotive Research (USCAR).

mercial fuel. Chapter 4 explains the barriers to be overcome in establishing an economic and reliable infrastructure for the transmission and storage of hydrogen, including onboard vehicle storage in the discussion.

Chapter 5 presents the committee's analysis of the total supply chain costs of hydrogen involved in the methods for producing hydrogen using various feedstocks at different scales. From a baseline of the cost to produce hydrogen using currently available technology, the analysis postulates future cases for the various technologies on the basis of the committee's judgment about possible cost reduction. Chapter 6 builds on the results presented in the previous chapter to consider potential scenarios for the penetration of hydrogen technologies into the economy and associated impacts on oil imports and CO_2 gas emissions. Chapter 7 addresses the issue of capture and storage of CO_2 from fossil-fuel-based hydrogen production processes.

Chapter 8 discusses the supply side—treating in greater detail the hydrogen feedstock technologies that were analyzed in Chapters 5 and 6. (Appendix G presents extensive additional discussion of these technologies.) Chapter 9 discusses several crosscutting issues, such as systems analysis, hydrogen safety, and environmental issues. Lastly, Chapter 10 includes the committee's major findings and recommendations on the programs of the DOE applied energy offices (EERE, FE, NE) on hydrogen.

Sources of Information

The committee held four meetings with sessions that were open to the public, hearing presentations from more than 30 outside speakers—including persons from industry (involved with both hydrogen production and use), nongovernmental organizations, and academia. Appendix D provides a listing of all of the committee's meetings and the speakers and topics at the open sessions.

The committee reviewed several documents in connection with this study. First (see item 4 of the statement of task, above) was the Office of Energy Efficiency and Renewable Energy's "Hydrogen, Fuel Cells & Infrastructure Technologies Program: Multi-Year Research, Development and Demonstration Plan" (DOE, 2003b), or multi-year program plan (MYPP). This plan identifies "critical path" barriers that the DOE believes must be overcome if a hydrogen economy is to be realized. The MYPP includes milestones and measures of progress with respect to these barriers, all leading to a commercialization decision in 2015. Most of the focus of the MYPP is on replacing gasoline use in light-duty vehicles (automobiles and light trucks) with hydrogen; some attention is directed to stationary applications of hydrogen.

The committee also reviewed the Office of Fossil Energy's *Hydrogen Program Plan, Hydrogen from Natural Gas and Coal: The Road to a Sustainable Energy Future* (DOE, 2003c), which concentrates on stationary applications of hydrogen (e.g., distributed power, industry, buildings). (The Office of Fossil Energy does not necessarily address the use of fuel cells for industry or building applications. These applications are mostly addressed in EERE.)

Other documents reviewed by the committee include the *Hydrogen Posture Plan: An Integrated Research, Development, and Demonstration Plan* (DOE, 2003a). This plan integrates program activities across EERE, FE, NE, and SC that relate to hydrogen, in accordance with the *National Hydrogen Energy Roadmap* (DOE, 2002a), also reviewed.

Two strategic goals common to the DOE plans referred to above are energy security and environmental quality—the latter including reduction of CO_2 from the combustion of fossil fuels with the implications of such reductions for climate change. This report includes discussion and analysis of these two strategic goals, in particular in Chapters 5 and 6, in which the results of the committee's analysis of current and future hydrogen technologies are presented.

Focus of This Report

This report does not offer a prediction of whether the transition to a hydrogen-fueled transportation system will be attempted or whether the hydrogen economy will be realized. Instead, the committee offers an assessment of the current status of technologies for the production, storage, distribution, and use of hydrogen and, with that as a baseline, posits potential future cases for the cost of the hydrogen supply chain and its implications for oil dependence, CO_2 emissions, and market penetration of fuel cell vehicles. In presenting these future cost reductions, the committee also estimates what might be achieved with concerted research and development. The committee is not predicting that this research will occur, nor is it predicting that such research would necessarily bring the posited cost reductions. Finally, liquid carriers of hydrogen such as methanol and ethanol were not considered in this study.

2

A Framework for Thinking About the Hydrogen Economy

This report concerns research and development (R&D) to advance the hydrogen economy, a transition to a national energy system envisioned to rely on hydrogen as the commercial fuel that would deliver a substantial fraction of the nation's energy-based goods and services. While the focus of the report is on technology recommendations, the committee also recognizes that any technological change must take place within a larger economic and societal context. Therefore, this analysis begins with a perspective on the context in which the R&D programs of the Department of Energy (DOE) are embedded—a framework for thinking about a hydrogen economy.

OVERVIEW OF NATIONAL ENERGY SUPPLY AND USE

The transition to a hydrogen economy would begin in the context of a mature and reasonably efficient energy system; indeed, hydrogen technologies must compete effectively with that system if the transition is to occur at all. As shown in Figure 2-1, U.S. primary energy consumption has risen over recent decades, and is likely to continue increasing. To the consumers who contribute to this demand, energy is valuable not in its own right but rather as a source of products and services that are highly valued. In the United States, these services are customarily organized into sectors—residential, commercial, industrial, and transportation sectors—as shown in Figure 2-2. Fossil fuels overwhelmingly drive this consumption, as shown in Figure 2-3. Domestic production of energy, especially petroleum, has not kept pace with consumption (see Figure 2-4), resulting in increasing imports.

The national energy system contains great inertia, and several persistent trends will influence the energy economy well into the future. Most fundamentally, the Energy Information Administration's *Annual Energy Outlook 2003* (EIA, 2003) projects total energy consumption to increase at an annual average rate of 1.5 percent out to 2025, as shown in Figure 2-1. This increase is more rapid than projected growth in domestic energy production, leading to increasing dependence on imported fuels. For example, natural gas imports from Canada are projected by the EIA (2003) to provide 15 percent of the total U.S. natural gas supply in 2025, and liquefied natural gas (LNG) imports from overseas are expected to grow dramatically to 6 percent of the total from near zero today. While the Canadian imports can be presumed stable, the same cannot be said of the LNG imports that increasingly come from the most politically volatile regions of the globe. Import dependence for energy products is growing too. Refining capacity in the United States is projected to increase to nearly 20 million barrels per day in 2025, but this country will still depend on foreign refineries for roughly 33 percent of its petroleum products.

Over the same 2003–2025 time period, the EIA (2003) projects that CO_2 emissions from energy use will rise in step with energy use, an average of 1.5 percent per year under current policies and practices. Atmospheric concentrations of CO_2 are likely to increase. And though the environmental implications cannot be specified with precision, it seems reasonable to believe that as human activity continues to change the chemical content of the atmosphere, some kind of negative consequence will result.

ENERGY TRANSITIONS

The earliest transition to a modern energy system coincided with the Industrial Revolution. New ways to produce goods and services demanded large quantities of fuels with predictable burning characteristics. Fuels were tailored to the devices that burned them (steam engines, lamps, furnaces, and so forth), and these devices were designed around assumptions about fuels, a pattern that continues to the present day.

Over time, the fuels sector has undergone two kinds of transition. The first is a general trend toward greater efficiency in the use of energy to produce the goods and services desired by the world's economy, coupled with structural

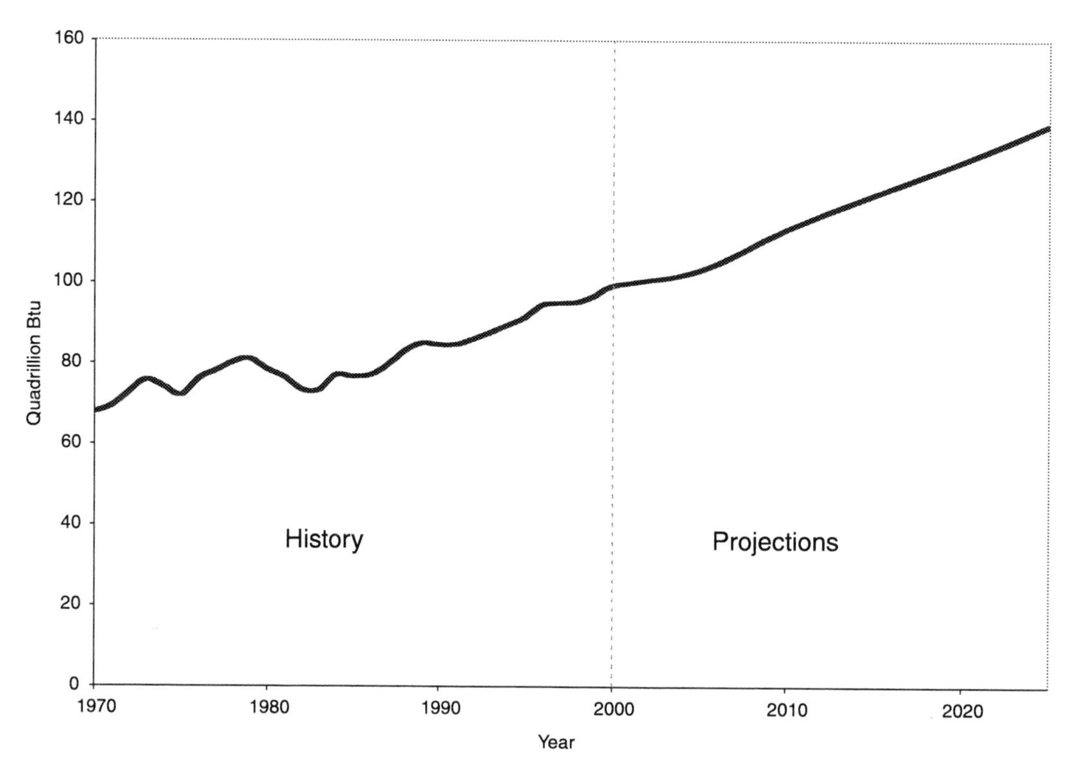

FIGURE 2-1 U.S. primary energy consumption, historical and projected, 1970 to 2025. SOURCE: EIA (2003).

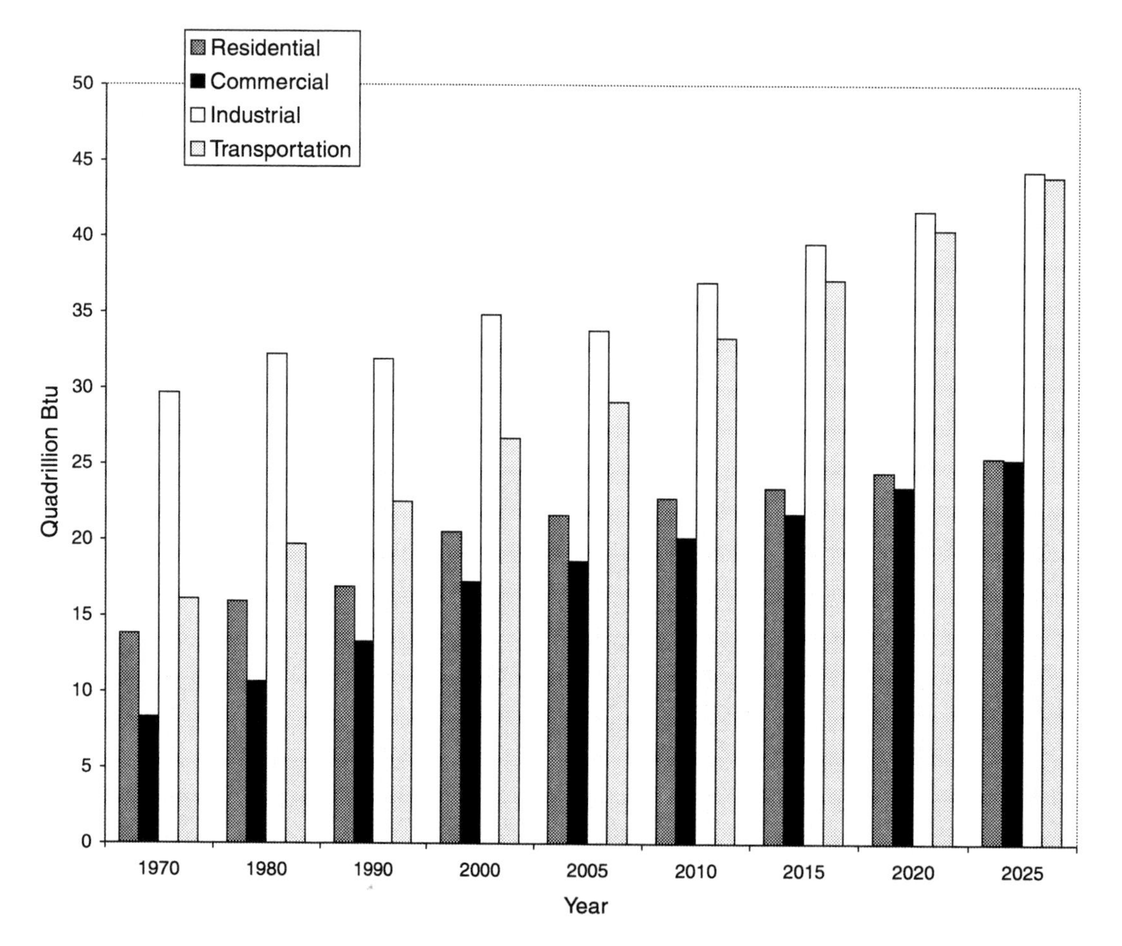

FIGURE 2-2 U.S. primary energy consumption, by sector, historical and projected, 1970 to 2025. SOURCE: EIA (2003).

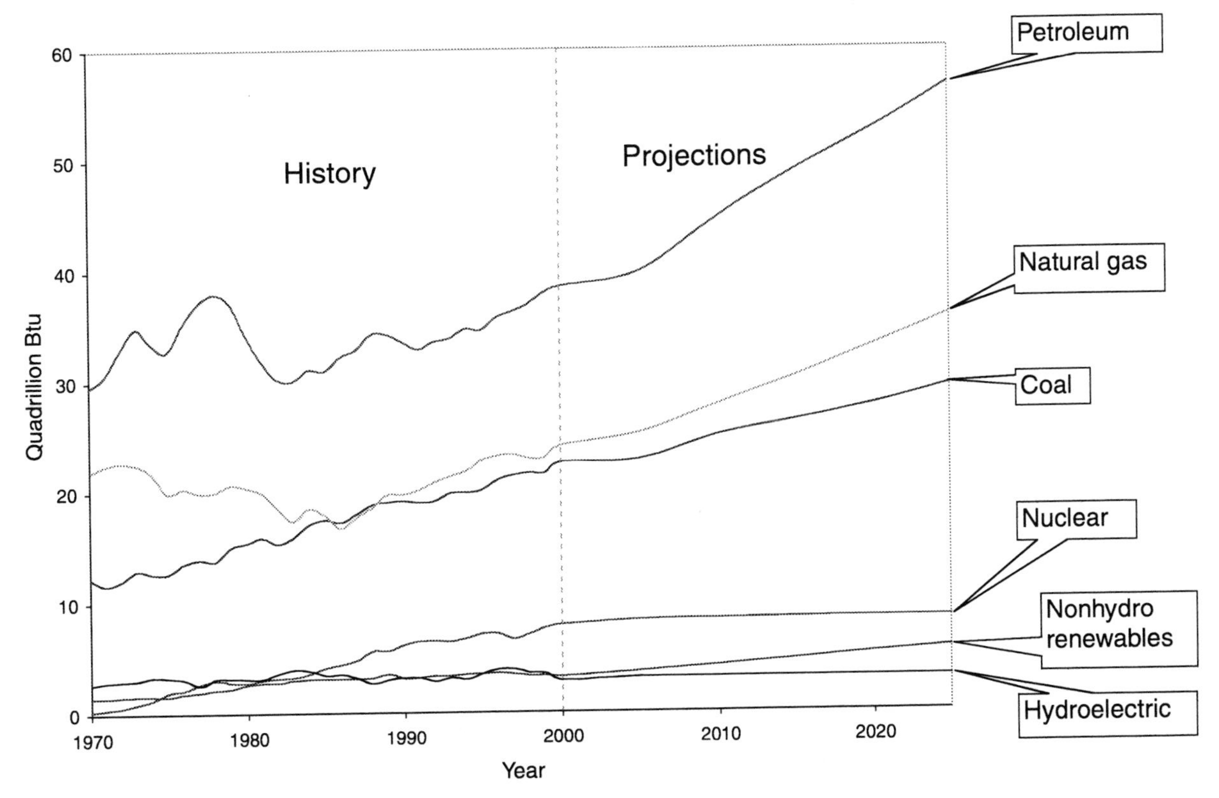

FIGURE 2-3 U.S. primary energy consumption, by fuel type, historical and projected, 1970 to 2025. SOURCE: EIA (2003).

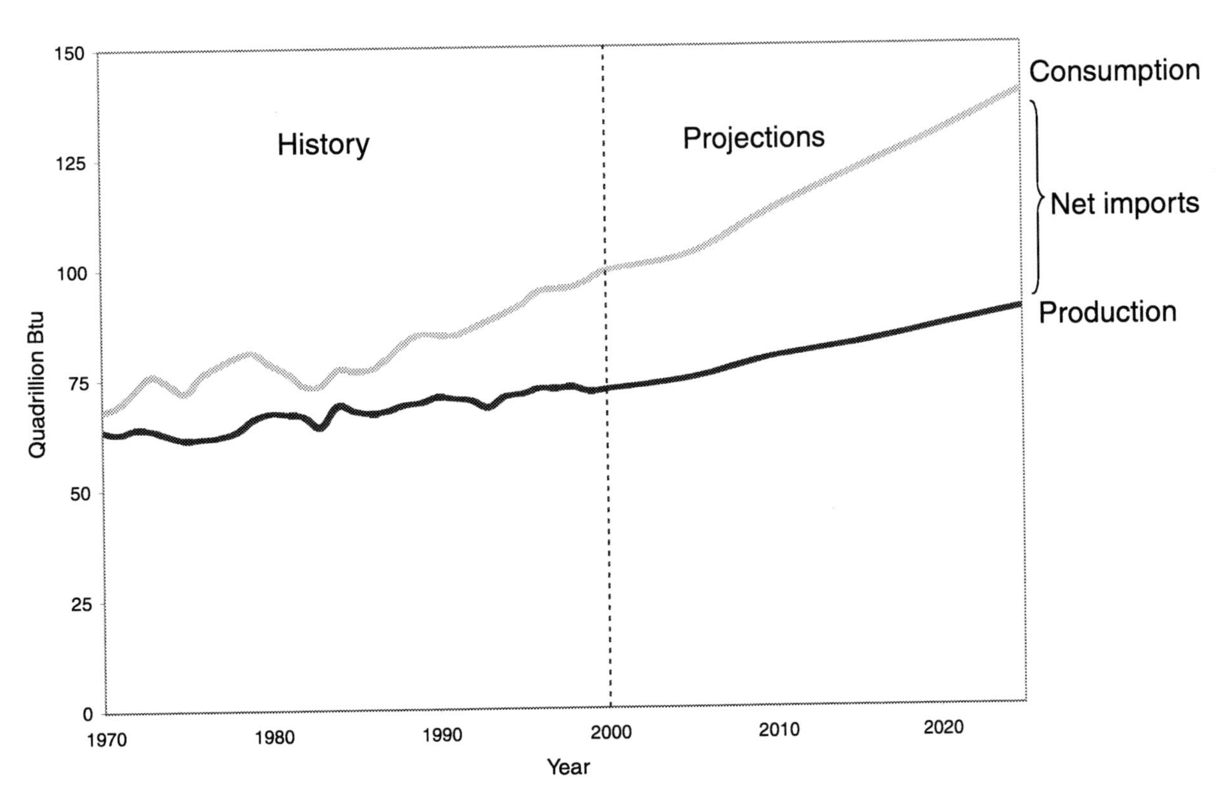

FIGURE 2-4 Total U.S. primary energy production and consumption, historical and projected, 1970 to 2025. SOURCE: EIA (2003).

changes in developed economies away from manufacturing toward services. This tendency has been most pronounced in the United States, in which the energy intensity of the economy fell from about 70 megajoules (MJ) per constant dollar of gross domestic product (GDP) in the mid-19th century to about 20 MJ today (Schrattenholzer, 1998).

The second transition comprises a change in market share among the various commercial fuels; this change has favored fuels with lower ratios of carbon to hydrogen. In general, solid fuel has lost market share to liquid fuel, especially in transportation, where the greater energy density (energy per unit of volume) of the liquids offers significant advantages. More recently, the share of natural gas has grown steadily, though chiefly in stationary applications in which the lower energy density of natural gas presents no disadvantage. As an unintended consequence of this interfuel competition, the more carbonaceous fuels such as wood and coal have been superseded by less carbonaceous fuels such as oil and methane.

This substitution, together with the rise of knowledge-based industries, has caused a general reduction in the carbon intensity of the global economy—the amount of carbon released to the atmosphere per unit of primary energy—as shown in Figure 2-5. Even if no changes are made to the current energy infrastructure, this decline will probably continue into the future, driven by continued interfuel substitution and by the ongoing shift in the balance of value creation from heavy industry to a knowledge-based economy. Nevertheless, world carbon emissions continue to rise, despite this drop in carbon intensity, as economic growth outpaces business-as-usual improvements in both energy efficiency and carbon intensity (see Figure 2-6; EIA, 2003). The amount of carbon emitted varies widely around the globe, but its survival time in the lower atmosphere is sufficiently long that it is spread around by wind and becomes evenly mixed spatially across latitudes and longitudes (NRC, 2001b). The remainder of this chapter and the rest of the report, however, concentrate on hydrogen technology policies specifically for the United States.

MOTIVATION AND POLICY CONTEXT: PUBLIC BENEFITS OF A HYDROGEN ENERGY SYSTEM

Two public goals—environmental quality, especially the reduction of greenhouse gas emissions, and energy security—provide the policy foundation for the hydrogen programs of the DOE (DOE, 2003a). The first of these goals

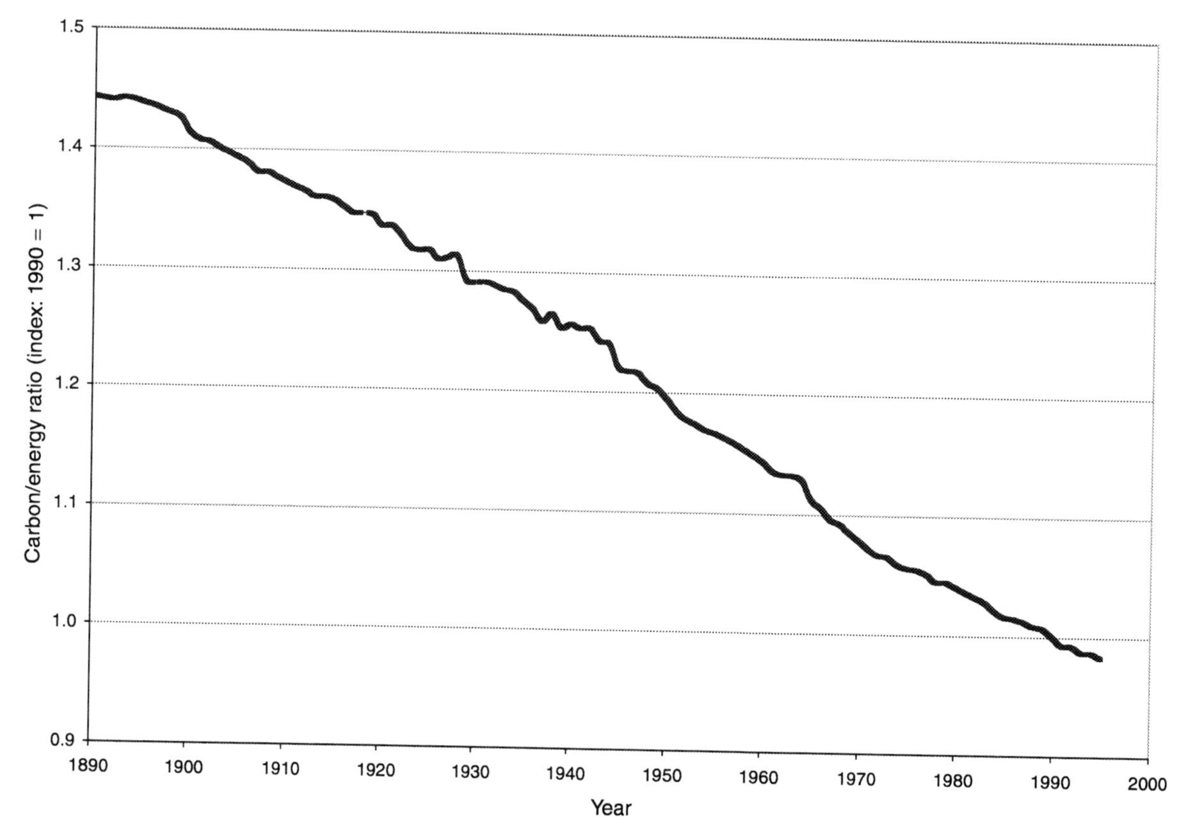

FIGURE 2-5 Carbon intensity of global primary energy consumption, 1890 to 1995. SOURCE: Adapted from Arnulf Grübler, data available online at http://www.iiasa.ac.at/~gruebler/Data/TechnologyAndGlobalChange/. Accessed November 15, 2003.

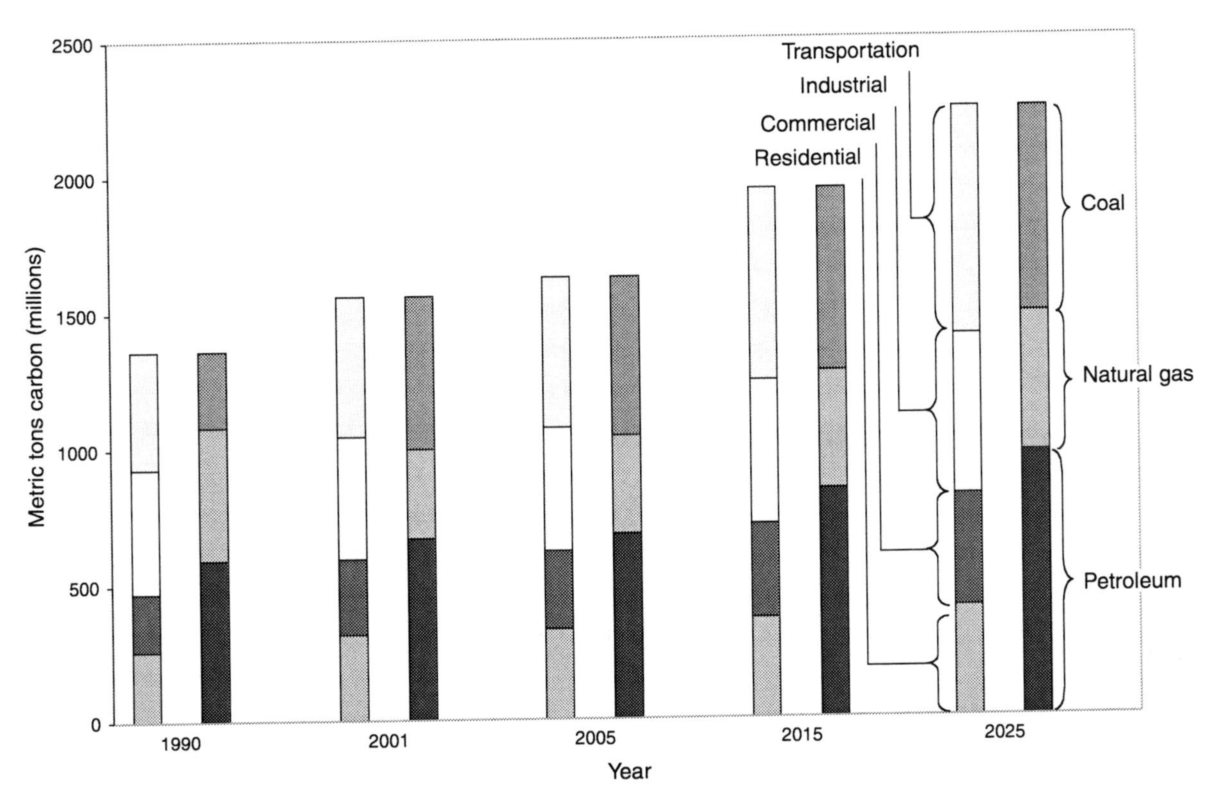

FIGURE 2-6 Trends and projections in U.S. carbon emissions, by sector and by fuel, 1990 to 2025. SOURCE: EIA (2003).

seeks to reduce emissions of criteria pollutants[1] and the anticipated releases of carbon dioxide (and other greenhouse gases) into the atmosphere. In the United States, two intermediate demand sectors stand out as the source of much of the energy-related carbon: those involving (1) the burning of coal to produce electricity and (2) the burning of petroleum in transportation fuels (see Figure 2-7). Any hydrogen-based energy system must address these sectors in order to achieve the full environmental benefit of hydrogen energy. The second policy goal seeks to enhance national security by reducing the nation's dependence on fuels imported from insecure regions of the world and on increasingly imported liquefied natural gas. These policy goals set two of the criteria that the committee used to weigh competing energy systems and technologies.

The dual policy goals described above intersect in the transportation sector, which has become the focus of much of the DOE hydrogen program (DOE, 2003a). Present-day transportation in the United States relies almost exclusively on petroleum and contributes an amount of carbon to the atmosphere nearly equal to that from coal used in electric power production (see Figure 2-7). Thus, in principle, the substitution of hydrogen for petroleum in ground transportation would benefit both goals. The benefits, however, accrue to the respective goals quite differently.

Consider, for example, a kilogram of hydrogen, produced in a way that does not emit carbon, displacing about 1.67 gallons of gasoline[2] at some future time when hydrogen gains a meaningful share of the motor fuel market (in the committee's scenarios presented in Chapter 6, sometime in the period 2025 to 2050). With regard to CO_2 emissions, the benefit would be direct: the carbon that would otherwise have been emitted from the displaced gasoline is kept from the atmosphere. But with regard to energy security, the situation becomes more complex. This is so because the first petroleum displaced is as likely to come from high-cost foreign and domestic producers as from the low-cost Persian Gulf producers. Indeed, the market share of the Persian Gulf producers might actually rise as their higher-cost competitors are displaced. Thus, the most meaningful security gains could be achieved only if hydrogen were to displace essentially all petroleum used in ground transportation—around

[1]Criteria pollutants are air pollutants (e.g., lead, sulfur dioxide, and so forth) emitted from numerous or diverse stationary or mobile sources for which National Ambient Air Quality Standards have been set to protect human health and public welfare.

[2]A gasoline hybrid electric vehicle having fuel economy of 45 miles per gallon would travel as far on 1.67 gallons of gasoline as would a fuel cell vehicle on 1 kilogram of hydrogen, assuming that the efficiency of the latter is 75 miles per kilogram of hydrogen. The committee's assumptions about efficiencies for the different vehicle and power plant types are discussed further in Chapter 3.

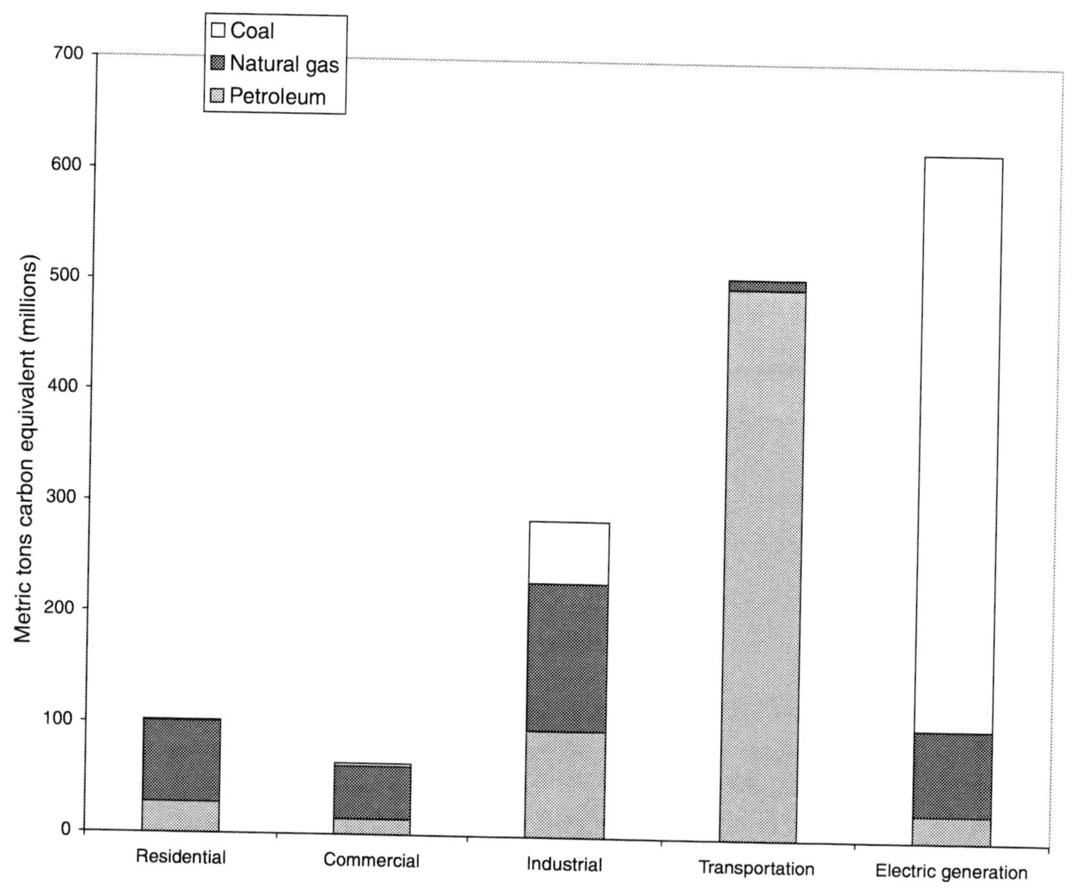

FIGURE 2-7 U.S. emissions of carbon dioxide, by sector and fuels, 2000. SOURCE: EIA (2002).

2040 to 2050 as depicted in the scenarios in Chapter 6. Offsetting this possibility somewhat, the economic effects of an oil supply disruption could diminish in direct proportion to the share of the world economy dependent on oil.

These dual policy objectives also carry broader implications for hydrogen development strategies. With respect to environmental quality, for example, using natural gas in preference to coal without carbon sequestration as a feedstock for hydrogen production would result in lower carbon emissions. This advantage of natural gas can be made greater at large production scale,[3] at which carbon capture is likely to be most economic—a proposition that may not be true of natural gas reformers at distributed scale. But long-term use of natural gas as a hydrogen-producing feedstock does not solve the security concern if that gas is imported from unstable regions.

Like electricity, hydrogen is not a primary energy source, although it is a high-quality energy carrier. Large-scale manufacturing of hydrogen from a primary energy source such as coal would imply, for example, a resurgence of coal production with increased carbon emissions unless the co-produced CO_2 were captured and sequestered. In effect, capture and sequestration could separate carbon intensity from carbon release (see Chapter 7).

SCOPE OF THE TRANSITION TO A HYDROGEN ENERGY SYSTEM

The scope of change that would be required poses some of the largest challenges to the transition to a hydrogen energy system. Both the supply side (the technologies and resources that produce hydrogen) and the demand side (the technologies and devices that convert hydrogen to services desired in the marketplace) must undergo a fundamental transformation. The one will not work without the other. This has not been the case in previous energy transitions. In promoting nuclear power, for example, the government simply sought to add a potentially attractive new power source. The rest of the electric power system remained the same, and customers' use of electricity went unaffected. Similarly, government intervention has become significant in protecting some industry segments (tax concessions for

[3]The committee considered three illustrative scales of facilities that produce hydrogen. The first two scales—large (central station) and midsize—require distribution infrastructure for produced hydrogen. The third and smallest, the distributed scale, comprises small facilities at the point of the dispensing of hydrogen.

domestic oil production, for example), promoting others (wind subsidies, for example), or shaping the performance of others (regulations on the mining and burning of coal, for example). But in no prior case has the government attempted to promote the replacement of an entire, mature, networked energy infrastructure before market forces did the job. The magnitude of change required if a meaningful fraction of the U.S. energy system is to shift to hydrogen exceeds by a wide margin that of previous transitions in which the government has intervened. This raises the question of whether research, development, and demonstration programs will be sufficient or whether additional policy measures might be required.

The interlocked nature of the current energy infrastructure—the systems that produce and distribute energy and the devices to convert that energy into useful services—presents a challenge to policy makers seeking to promote a complete fuel change. The components of this challenge include these:

- Both the new hydrogen production systems and the devices to convert that hydrogen into services that consumers will freely purchase must be developed in parallel. Neither serves any purpose without the other.
- The incumbent technologies do not stand still, but continue to improve in performance, albeit within the envelope of the other components of the energy system—for example, more fuel-efficient internal combustion engine (ICE) vehicles and hybrid propulsion systems that make better use of the existing fueling infrastructure.
- The cost of the current energy infrastructure is already sunk, which increases the barrier to new technologies that require new infrastructure. In addition, selected components of the current energy structure benefit from economic subsidies and favorable regulation.
- New hydrogen-based technologies will require a transition period during which old and new systems must operate simultaneously. During this transition, neither system is likely to function at peak efficiency.

These factors all tend to lock in the current energy infrastructure and pose severe competitive challenges for a society that would rely on markets to allocate economic resources.

COMPETITIVE CHALLENGES

Any future hydrogen energy system will be subject to market preferences and to competition from other energy carriers and among hydrogen feedstocks. The choices that a market economy makes about its energy services will influence the utilization of hydrogen and hydrogen feedstocks and the attributes of the hydrogen end-use technologies. As discussed in the subsections below, the issues that frame the competitive challenge in using hydrogen include the following:

- *Energy demand.* In what situations would the use of hydrogen offer the greatest economic advantage? The greatest environmental and security advantage?
- *Energy supply.* How should hydrogen be produced from primary resources, such as coal, methane, nuclear, and renewable energy (solar, wind, and so forth)? What environmental consequences and trade-offs arise from its production from each resource?
- *Logistics and infrastructure.* How can a storage-and-delivery infrastructure best connect the demand for hydrogen with its supply and ensure the public safety?
- *Transition.* How can the mature, highly integrated energy system of the United States make the transition to a hydrogen economy?

Energy Demand

The world economy currently consumes about 42 million tons of hydrogen per year. About 60 percent of this becomes feedstock for ammonia production and subsequent use in fertilizer (ORNL, 2003). Petroleum refining consumes another 23 percent,[4] chiefly to remove sulfur and to upgrade the heavier fractions into more valuable products. Another 9 percent is used to manufacture methanol (ORNL, 2003), and the remainder goes for chemical, metallurgical, and space purposes (Holt, 2003). The United States produces about 9 million tons of hydrogen per year, 7.5 million tons of which are consumed at the place of manufacture. The remaining 1.5 million tons are considered "merchant" hydrogen.[5]

If a transition from the use of hydrogen in industrial markets to a broader hydrogen economy is to occur, devices that use hydrogen (e.g., fuel cells) must compete successfully with devices that use competing fuels (e.g., hybrid propulsion systems). Equally important, hydrogen must compete successfully with electricity and secondary fuels (e.g., gasoline, diesel fuel, and methanol). The following discussion of energy demand considers both of these issues—market preferences and energy competition.

Market Preferences in Energy

The nature of the competition in which hydrogen would be engaged is shaped by the unique role of energy in the economy: the demand for energy is not a final demand, but rather derives from the demand for other goods and services. Both the *amount of primary energy used* and the *physical characteristics of the final energy carrier* (e.g., gasoline, methane, electricity, or possibly hydrogen) depend on the devices that convert energy into products (e.g., cars, furnaces, air conditioners, telephones, and computers) or ser-

[4]Refers to consumption only, not net production. Petroleum refineries are roughly in balance between hydrogen produced and consumed onsite.

[5]Jim Hansel, Air Products and Chemicals, Inc., personal communication to Martin Offutt, National Research Council, October 3, 2003.

vices (e.g., transportation, heating, cooling, communications, and computing).

In a market economy, the amount of energy used depends on trade-offs among desirable attributes such as the following:

- The cost of building greater efficiency into the device, relative to the subsequent (and discounted) benefits in fuel saving;
- The value of time versus the cost of the energy needed to save time—for example, motor trips take longer when people drive at a relatively fuel-efficient 55 mph rather than at the less efficient 70 mph, but the lower speed costs drivers a valuable resource, their time; and
- The price of the energy input seen by the particular consumer as distinct from its cost to produce—for example, electric energy consumed during peak hours costs more to produce than that consumed at other times of day, yet the price is the same at all times.

The physical characteristics of the final energy carrier depend on the nature of the service that the market demands. In transportation, for example, the need for fuels with high energy density and rapid refueling strongly favors liquid hydrocarbons, mostly derived from petroleum. By contrast, devices such as computers operate with electric energy, which can be made from a variety of fuels (e.g., coal, natural gas, nuclear, and petroleum) including less-energy-dense fuels, as well as gaseous and solid fuels.

Various preferential interventions in the form of taxes, subsidies, and regulations also influence consumer prices, and hence consumer behavior. At the same time, however, the cost of important external effects, such as the stress on the global climatic system or lower national security, are also excluded from the prices that influence consumer trade-offs. And if the full cost of the mine-to-waste cycle needed to provide an energy-based service does not appear in the price of that service, then it will be consumed inefficiently.

Competition and Synergy

If large quantities of hydrogen can be produced at competitive costs and without undue carbon release, the use of hydrogen would offer marked advantages in the competition with other secondary fuels. First, hydrogen is likely to burn more cleanly in combustion engines. Second, hydrogen is better matched to fuel cell use than competing fuels are; and the fuel cell could become the disruptive technology that will transform the energy system and enable hydrogen to displace petroleum and carbon-releasing fuel cycles. If cost-effective and durable fuel cell vehicles can be developed, they could prove attractive to manufacturers, marketers, and consumers insofar as they can achieve the following:

- Replace mechanical/hydraulic subsystems with electric energy delivered by wire, potentially improving efficiency and opening up the design envelope;
- Reduce manufacturing costs as manufacturers are able to use fewer vehicle platforms; and
- Enable the vehicle to offer mobile, high-power electricity, which could provide accessories and on-vehicle services more effectively than could alternatives.

However, gasoline hybrid electric vehicles (GHEVs) can offer many of these attractive features while at the same time retaining the current fuel infrastructure. Even though GHEVs cannot achieve the fuel efficiency envisioned for fuel cell vehicles (FCVs) and despite the significant cost of battery replacement, some consumers might find that the convenience of the familiar "gas station" offsets these disadvantages well into any hydrogen transition. This suggests that fuel cell vehicles will face stiff market competition from hybrids for many years into the future.

In a fuel cell vehicle, hydrogen produces electricity, which is converted electromechanically into torque in the wheels which drives the vehicle; in effect, hydrogen fuel powers a mobile electric generator. In a mature hydrogen infrastructure, new synergies might be found in large-scale production and distribution. One visionary concept is the national Energy Supergrid, advanced by Chauncey Starr, founder and emeritus president of the Electric Power Research Institute. This supergrid would combine hydrogen and electric energy in two components: (1) a network of superconducting, high-voltage, direct current cables for power transmission, with (2) liquid hydrogen as the coolant required to maintain superconductivity in the cables. The electric power and hydrogen would be supplied from nuclear and renewable energy power plants spaced along the grid. Electric energy would exit the system at various taps, connecting into the existing power grid. The hydrogen would also be tapped to provide a readily available fuel for automotive or other use (*National Energy Supergrid Workshop Report*, 2002). On a smaller scale, others have proposed similar hydrogen-electric projects as a way to move renewable energy from remote sources to markets—for example, from wind farms in North Dakota to load centers like Chicago.

Hydrogen might also enjoy a synergistic relationship with renewable energy. The chief difficulty with many renewable technologies is the intermittency of the resource itself—the Sun doesn't always shine or the wind always blow, and when they do they are variable. But if sufficiently low-cost hydrogen storage could be developed, hydrogen might provide a pathway to market for renewable energy because it could be manufactured whenever sufficient energy was available. The problem of intermittency would be mitigated, because the stored hydrogen could be used to produce electricity during times when sunlight or wind was not available.

Finally, hydrogen might compete directly with electricity as an energy carrier, with each using a separate production and distribution system. This competition can be analyzed in terms of a specific application—for example, energy storage on board an automobile. Here, hydrogen enjoys a distinct advantage over electricity, even if grid electricity might be less expensive than hydrogen. This advantage derives from energy storage—in its current state of development, the battery technology needed to make grid electricity applicable to mobile uses is unable to provide vehicles with the range, power, and convenience that consumers require. If, however, battery technology were to achieve a major breakthrough, then the availability of relatively inexpensive energy from the grid would put hydrogen at a competitive disadvantage. Even without improved batteries, electricity from an onboard generator is available in several hybrid vehicles now on the market. The resulting fuel economy of these hybrid vehicles is substantially higher than that achievable with conventional vehicles. As this technology gains manufacturing scale, it will prove a formidable competitor for hydrogen, especially at the beginning of any transition. However, hybrid vehicle technology seems unlikely to match the ultimate performance of the hydrogen fuel cell vehicle, if all of the relevant technologies are successfully developed.

Energy Supply

The U.S. energy system has evolved over the past century into a massive infrastructure involving extraction, processing, transportation, and end-use equipment. The replacement value of the current system and related end-use equipment would be in the multi-trillion-dollar range.[6] Major changes to the system have typically taken decades. If hydrogen is to succeed as a fuel, it must be in the context of this energy system. For example, insofar as hydrogen may compete with petroleum, it faces an established infrastructure of 161 oil refineries, 2,000 oil storage terminals, roughly 220,000 miles of crude oil and oil products lines, and more than 175,000 gasoline service stations (NRC, 2002). Much of this infrastructure would have to be replaced or heavily modified if hydrogen is to become the dominant fuel for the highway transportation sector. (A description of the U.S. energy system is presented in Appendix F.)

Hydrogen production technologies based on various primary energy resources—renewable energy resources,[7] carbonaceous fuel resources, and nuclear energy—would compete for market share in an envisioned hydrogen economy. Each promises advantages, involves uncertainties, and raises currently unresolved issues. The technologies for producing hydrogen from these various primary resources can be deployed at varying scales of production, and in Chapter 5 the committee presents its analysis of total supply chain costs for hydrogen generation at three illustrative scales of production—central station, midsize, and distributed.[8] The following subsections present an overview of the attributes associated with the various production scales, and primary energy sources and associated technologies for hydrogen generation at each scale are discussed.

Central Station (Very Large Scale)

At very large scale, around a gigawatt and above, the principal supply options include carbonaceous fuels and nuclear energy. About 100 such plants would be able to supply the current world demand for hydrogen, about 42 million tons per year (ORNL, 2003), and about 20 such plants would be able to supply the current U.S. demand for hydrogen of about 9 million tons per year.

With regard to a carbonaceous feedstock, hydrogen could be manufactured from natural gas or coal. The carbon would be converted into synthesis gas (syngas—$CO + H_2$)—used either for combustion for electricity generation or for further chemical processing into hydrogen and CO_2, which can be captured for sequestration. The chief advantage of this approach is the abundance of domestic coal: the United States has the world's largest recoverable coal reserves, sufficient to manufacture hydrogen for a *very* long time. The large scale of operation would yield attractive economies of scale. In contrast, natural gas will increasingly have to be imported, raising new energy security concerns.

Two salient issues would arise from the use of carbonaceous fuels as a major source of hydrogen. The first is concerned with whether the carbon really can be captured and sequestered in a manner that is both environmentally acceptable[9] and cost-effective. If this cannot be achieved, hydrogen production from carbonaceous fuel resources, particularly coal, offers none of the sought-after large reductions in (net) carbon emissions. The second issue derives from the scale of operation. Demand for hydrogen must be sufficient to justify investment in a large-scale plant, and a matching distribution infrastructure would be required. In addition, a satisfactory means for bulk storage of hydrogen would have

---

[6]For example, replacing existing electric generators with new units averaging $1000 per kilowatt (electric) would cost about $800 billion. A new transmission system, at $1 million per mile, would cost $160 billion. Oil refineries and pipelines would be several hundred billion dollars more. The natural gas transmission and distribution systems would also cost hundreds of billions. Then add the cost of replacing all of the factories, buildings, and vehicles that are designed for a specific type of fuel. Clearly, a detailed calculation would show a total value of multi-trillion dollars (NRC, 2002).

[7]Strictly speaking, the primary energy resource is the Sun for solar renewable energy (e.g., photovoltaic) and wind energy. Renewable energy is a primary resource for hydrogen in the sense that hydrogen is the product of chemical processes using renewable feedstocks (e.g., biomass) or of electrolysis of water powered by renewable electricity sources.

[8]In the committee's analysis, central station plants are assumed to produce hydrogen on average 1,080,000 kilograms per day (kg/d); midsize plants, 21,600 kg/d; and distributed facilities, 432 kg/d. (See Chapter 5.)

[9]As used in this report, the term "environmentally acceptable" implies a high probability that the carbon will not leak into the atmosphere during processing and handling, that it will remain sequestered from the atmosphere essentially in perpetuity, and that it will not cause adverse side effects, such as harmful chemical reactions, while so sequestered.

to be found. A transitional strategy to address these requirements must precede the move to producing hydrogen fuel in very large scale plants.

Nuclear energy could produce hydrogen in one of three ways: (1) through electrolysis, the splitting of water molecules with electricity generated by dedicated nuclear power plants; (2) through process heat provided by advanced high-temperature reactors for the steam reforming of methane; or (3) through a thermochemical cycle, such as the sulfur-iodine process. Among the three, the issue of carbon capture and storage arises only for steam reforming; otherwise, the nuclear option is carbon-free. Scale, however, remains an issue, as it does for the large coal plants. In addition, delays in the development and deployment cycle for nuclear plants might arise from concerns with the storage and disposal of nuclear fuels, the security of nuclear facilities against terrorist attack, and the siting and licensing of nuclear facilities. These issues could prolong the time to realization of a full-scale hydrogen economy.

Midsize Scale

At midsize scale, a few tens of megawatts, both natural gas and renewable energy technologies offer production possibilities. Megawatt-scale production is especially attractive for biomass-based energy sources. Natural gas production at this scale could provide an efficient response to early market demand for hydrogen, but could not offer sufficient scale economies to compete effectively in mature hydrogen markets.

Distributed Scale

At the distributed end of the size range, large-scale pipeline systems would not be required because hydrogen production could be colocated with hydrogen dispensing and/or use. Distributed production might rely on primary energy from renewable resources, to the extent that those could be located reasonably near the point of use. Alternatively, grid electricity, possibly used during off-peak hours, might serve as the energy source. A distributed approach offers clear advantages during a transition from the current energy infrastructure, although it might not be sustainable in a mature hydrogen economy.

The advantages of distributed production during a transition are economic. The costs of a large-scale hydrogen logistic system, which many analysts believe will dominate a mature hydrogen economy, could be deferred until the demand for hydrogen increased sufficiently. This would mitigate the problem of "lumpy" investment—large production and distribution facilities that provide economies of scale but lead to underused capital while the demand for their output catches up. In contrast, distributed production systems could be installed rapidly as the demand for hydrogen increased, thus allowing hydrogen production to grow at a pace reasonably matched with hydrogen demand. Instead of static economies of scale, distributed production would rely on dynamic economies of scale in the manufacture of small hydrogen conversion and storage devices. Nevertheless, the cost of hydrogen compared with that of gasoline would likely be more expensive during this transition phase (see Chapters 5 and 6).

One major disadvantage of distributed production is environmental. If the hydrogen were produced by small-scale electrolysis and if the energy inputs to the electrolyzer were to come from the grid, the carbon consequences would be the same as for any other use of electric energy on a per kilowatt basis. If the hydrogen were produced by small-scale reformers, the collection of the carbon and its shipment to a sequestration site might prove an insurmountable challenge. Indeed, distributed-scale production in a mature hydrogen economy might require a costly reverse-logistic system to move the carbon captured from the dispersed production sites to the places of sequestration if the environmental benefits are to be achieved. The cost of a dispersed capture and disposal system might make distributed production unattractive in a mature hydrogen economy. During a transition period, however, the carbon from distributed production could simply be vented while the economic advantages of scalability and demand-following investment served to start the hydrogen economy.

Logistics and Infrastructure Issues

Between the production of hydrogen at any scale and the use of hydrogen in an energy device, the following series of logistic operations will exist:

- *Packaging.* The hydrogen must be put into a form suitable for shipping. This form might be a compressed gas, a liquid, some form of hydride, or some chemical compound.
- *Distribution.* The hydrogen must be moved to the point of use. Pipelines, pipes, roads, and railroads are typical shipping modes.
- *Dispensing.* The hydrogen must be transferred from the care of retailers into the care of consumers.
- *Storage.* In the interval between production and use, the hydrogen must be stored. Pressurized containers or cryogenic containers typify current practices.

With the technologies now available, many of these logistic steps themselves become significant consumers of energy; some analyses suggest that logistic costs will dominate the economics of any hydrogen energy system (Boessel et al., 2003). This consideration emphasizes the importance of viewing R&D objectives in the context of complete prototypical hydrogen energy systems rather than in isolation (NRC, 2003b).

Transition Issues

The transition to a hydrogen economy is unlikely to be achieved through the linear substitution of hydrogen components for their counterparts in the current energy infrastructure. Consider refueling, for example. It might emerge that refueling systems for hydrogen vehicles would become entirely modular, so that refueling would be more like purchasing and loading a videocassette into a recorder than filling a present-day automobile with gasoline. That could result in the flourishing of customer advantages and business models quite distinct from those common to the current fuels infrastructure.

Indeed, the ultimate timing and configuration of a mature hydrogen economy cannot be known, because they turn on resolution of the four pivotal questions discussed at the end of the chapter. Thus, the DOE might have its greatest impact by leading the private economy toward transition strategies rather than to ultimate visions of an energy infrastructure markedly different from the one now in place.

Developing Strategies for the Transition

The set of technologies and business models capable of beginning a transition to the hydrogen economy might be very different from those that would be most desirable in a mature energy system. This possibility challenges the DOE to maintain its focus on the goals to be achieved by the hydrogen economy, but also to cultivate flexibility, learning, and responsiveness in assisting the transition pathways leading to it.

Subsidies

As part of a transition strategy, some form of buy-down of the cost of technology might be required in order to initiate and accelerate the pace of transition. An example might be a set of temporary subsidies to encourage the early adoption of hydrogen technology; they could be phased out once scale economies had been achieved and mainstream markets opened. The societal benefits of promoting a more rapid transition to hydrogen might justify this use of subsidies. The challenge for any subsidization strategy would be to support the kind of "game-changing" technologies that can actually deliver public benefits. Otherwise, buy-down tends to become an entitlement, entrenching the subsidized rather than accelerating systemic change.

Regulatory and Social Issues

Public apprehensions regarding hydrogen must be addressed early in a transition—otherwise the hydrogen economy might never reach the steady state. Of these concerns, safety appears to be foremost. To be sure, hazards exist with the current fuels infrastructure—there can be natural gas explosions in homes, or auto fires, for example. However, the public has grown accustomed to the possibility of these hazards, and the relevant safety precautions are widely known. By contrast, hydrogen's distinct properties lead to distinct safety issues (see Chapter 9).

Safety issues cut across all segments of the hydrogen economy and become operational in two forms: concern with loss of human life and property, and codes and standards that shape the configuration and location of hydrogen facilities and vehicles. Much evidence demonstrates that hydrogen can be manufactured and used in professionally managed systems with acceptable safety. The concerns arise from prospects of its widespread use in the consumer economy, where careful handling and proper maintenance cannot be fully ensured.

Technology demonstrations might mitigate public skepticism, both by displaying the merits of the technology and by educating local officials regarding emergency response procedures and effective zoning codes. Beyond that remains the issue of how DOE R&D programs can best inform, and in turn be informed by, state and local authorities.

None of these precautions, however, can compensate for the casual approach that some consumers will inevitably take to their own safety. Engineering aimed at reducing the possibilities for mishandling can help lower the number of accidents but can never preclude them all. Some hydrogen logistic systems will prove superior in allowing a more benign consumer interface, and the issue for the DOE will be to identify and promote these systems.

Finally, the successful sequestration of massive quantities of carbon may be essential for any hydrogen economy that makes more than transitional use of carbonaceous fuels. The history of radioactive waste disposal suggests that dedicated opposition can overcome general public acceptance of a technology and its waste disposal plan. Thus, even energy systems that now appear to enjoy widespread acceptance can become vulnerable to delays and costly false starts. The carbon sequestration issue falls into that category (see Chapter 7).

Technology Development for the Transition

Much of the policy analysis now performed on the subject speaks to hydrogen supply and demand under steady-state conditions. But if an effective transition cannot be achieved, neither can the benefits of the steady state. Thus, technologies and policies developed explicitly for a transition remain important, even if they do not carry over into the mature hydrogen economy. This issue of how to effect the transition has several dimensions:

- Should the DOE seek to guide the transition into the pathways it selects, or should it let development be guided principally by the industrial stakeholders?
- In either case, how can the DOE know which transitional technologies to develop?
- What assumptions should be made regarding the success of pivotal technologies such as carbon capture and sequestration?
- What incentives will entrepreneurs and investors in the interim technologies need before they commit their capital resources?

ENERGY USE IN THE TRANSPORTATION SECTOR

In order to examine the potential demand for hydrogen, it is necessary to examine the ways in which hydrogen would be used in the economy. Two generic uses were considered by the committee—those of hydrogen as a fuel for transportation vehicles and hydrogen as a fuel for electricity generation. The committee's analysis focused on the first of these two potential uses of hydrogen. In particular, the committee examined the use of hydrogen as a fuel for light-duty vehicles (i.e., passenger cars, pickup trucks, vans, and sport utility vehicles), as this is where most of the DOE's hydrogen research is focused. With respect to the use of hydrogen for electricity generation, the committee notes the difficulty that such use would have competing with natural gas turbines. (See the discussion earlier in this chapter, in the section entitled "Competitive Challenges," as well as in Chapter 3.)

In order for hydrogen to compete successfully as a fuel for light-duty vehicles, vehicle manufacturers and purchasers must believe that hydrogen-fueled vehicles offer advantages over the available light-duty-vehicle alternatives. Those alternatives could involve diverse possibilities of energy carriers and the particular vehicle technologies that utilize them.[10] Figure 2-8 illustrates the possible combinations of energy carriers and vehicle technologies that could conceivably characterize the future vehicle stock for personal transportation in the United States.

In successive columns, Figure 2-8 shows three distinctions among the possible combinations of energy carriers and technologies. Storage on board the vehicle, with periodic refueling, has been the norm for personal passenger vehicles, trucks, buses, and aircraft, and that is the committee's approach to light-duty vehicles. Various gaseous, liquid, or solid fuels could be supplied to the vehicle. In the first column, "on-board energy carriers" distinguish the various forms of energy that could be supplied to the vehicle.

Currently, most light-duty vehicles are fueled by petroleum products, primarily gasoline and secondarily diesel fuel, although some vehicles are fueled by nonpetroleum hydrocarbons and alcohol fuels. Compressed natural gas and propane are routinely used to fuel light-duty vehicles. Among alcohol fuels, ethanol is used in light-duty vehicles, and methanol has been widely discussed as an alternative. Hydrocarbons can be used in combination with alcohol fuels, such as gasoline with ethanol. Bio-based diesel fuel currently exists in the marketplace. Another generic alternative is electricity supplied to the vehicle. That electricity is then converted and stored in the form of electrochemical energy in a battery, or mechanical energy in a flywheel. The last energy carrier in the column is the alternative that the committee examined, molecular hydrogen.

The last two columns in Figure 2-8 denote the conversion process (second column) applied to the energy carrier by the motor (third column). Fuels such as petroleum products, nonpetroleum hydrocarbons, alcohols, or molecular hydrogen could be converted to mechanical power through a combustion cycle. The current generation of internal combustion engines could be used, or advanced combustion techniques could conceivably transform such engines. (Hydrogen internal combustion engines were not analyzed, since the committee determined that in North America the demand for hydrogen was more likely to be due to fuel cell vehicles.[11]) Alternatively, each of these fuels could be used to generate on-board electricity, most likely through an electrochemical conversion device, such as a fuel cell. Within the realm of imagination would be microturbines that use the fuels to generate electricity that would be used directly in electric motors to propel the vehicle.

Hybrids of electric and combustion processes could also be used. Currently, hybrid electric vehicle technology combines the combustion of petroleum products (gasoline or diesel), over a wide range of degrees of hybridization, with electric motors for propulsion. Hybrids could be created for any of the other fuels. Hybrids of fuel cells and batteries are under consideration today.

The locus of competition, therefore, could be both among fuels supplied to the vehicles and among vehicle technologies that use those fuels. Thus, if molecular hydrogen were widely available as a fuel source for light-duty vehicles, the competition would be between fuel cell vehicles and internal combustion vehicles using hydrogen, and perhaps other technologies that use hydrogen as a fuel. And molecular hydrogen in these vehicles would compete with the direct use of electricity, and with the use of petroleum products, nonpetroleum hydrocarbons, and alcohols, either combusted or electrochemically converted to electricity.

Some of the technologies discussed above have been well developed already, some need significant developmental work, some require technological breakthroughs for success, and presumably some require initial conceptualization. Just as there is a high degree of uncertainty about the success of hydrogen technologies, there is a high degree of uncertainty about the success of those alternative technologies that require technological breakthroughs, and even more for technologies that have yet to be conceptualized! For example, possible future reductions in the cost and increases in the range of batteries could ultimately make dedicated electric vehicles, with batteries charged from grid-supplied electricity, much less expensive and more practical than they are

[10]The term "energy carrier" refers to electricity as well as to gas and liquid (or solid) fuels. When the term "fuels" is used in an unqualified sense, it refers to all of these energy carriers, but not to electricity.

[11]Larry Burns, General Motors Corporation, "Fuel Cell Vehicles and the Hydrogen Economy," presentation to the committee, June 11, 2003.

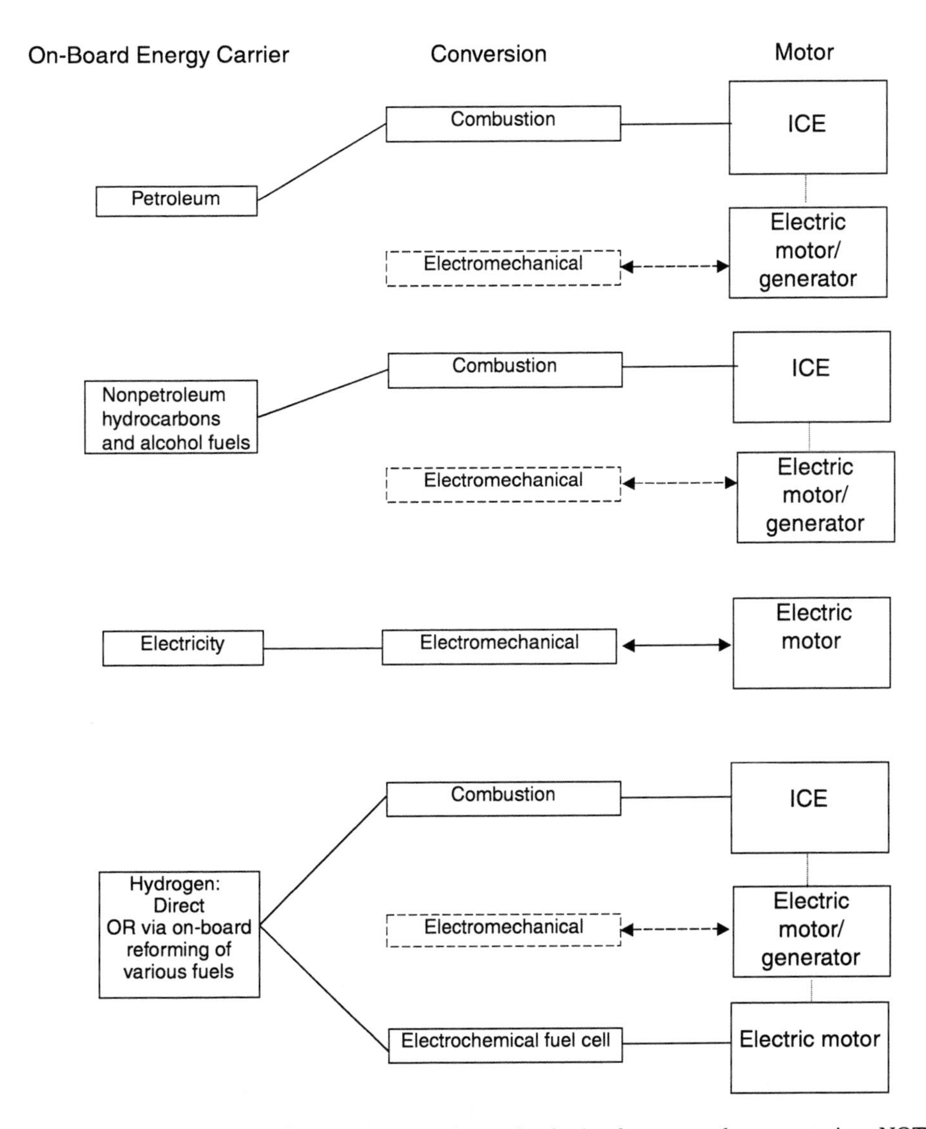

FIGURE 2-8 Possible combinations of on-board fuels and conversion technologies for personal transportation. NOTE: ICE = internal combustion engine.

currently. There is much uncertainty about whether such technologies would ultimately lead to vehicles that are less costly and more convenient than fuel cell vehicles.

For this study, the committee was not able to examine all of the options that may shape the future competition. Figure 2-9 illustrates the comparisons that were developed within this study. In particular, the committee focused on the competition between vehicles with on-board storage: fuel cell vehicles supplied by molecular hydrogen in competition with internal combustion, gasoline-fueled vehicles, either as conventional vehicles or as gasoline hybrid electric vehicles.

FOUR PIVOTAL QUESTIONS

From the foregoing analysis, the following four pivotal questions emerge as decisive:

- When will vehicular fuel cells achieve the durability, efficiency, cost, and performance needed to gain a meaningful share of the automotive market? The future demand for hydrogen depends on the answer.
- Can carbon be captured and sequestered in a manner that provides adequate environmental protection but allows hydrogen to remain cost-competitive? The entire future of carbonaceous fuels in a hydrogen economy may depend on the answer.
- Can vehicular hydrogen storage systems be developed that offer cost and safety equivalent to that of fuels in use today? The future of transportation uses depends on the answer.
- Can an economic transition to an entirely new energy infrastructure, both the supply and the demand side, be achieved in the face of competition from the accustomed benefits of the current infrastructure? The future of the hydrogen economy depends on the answer.

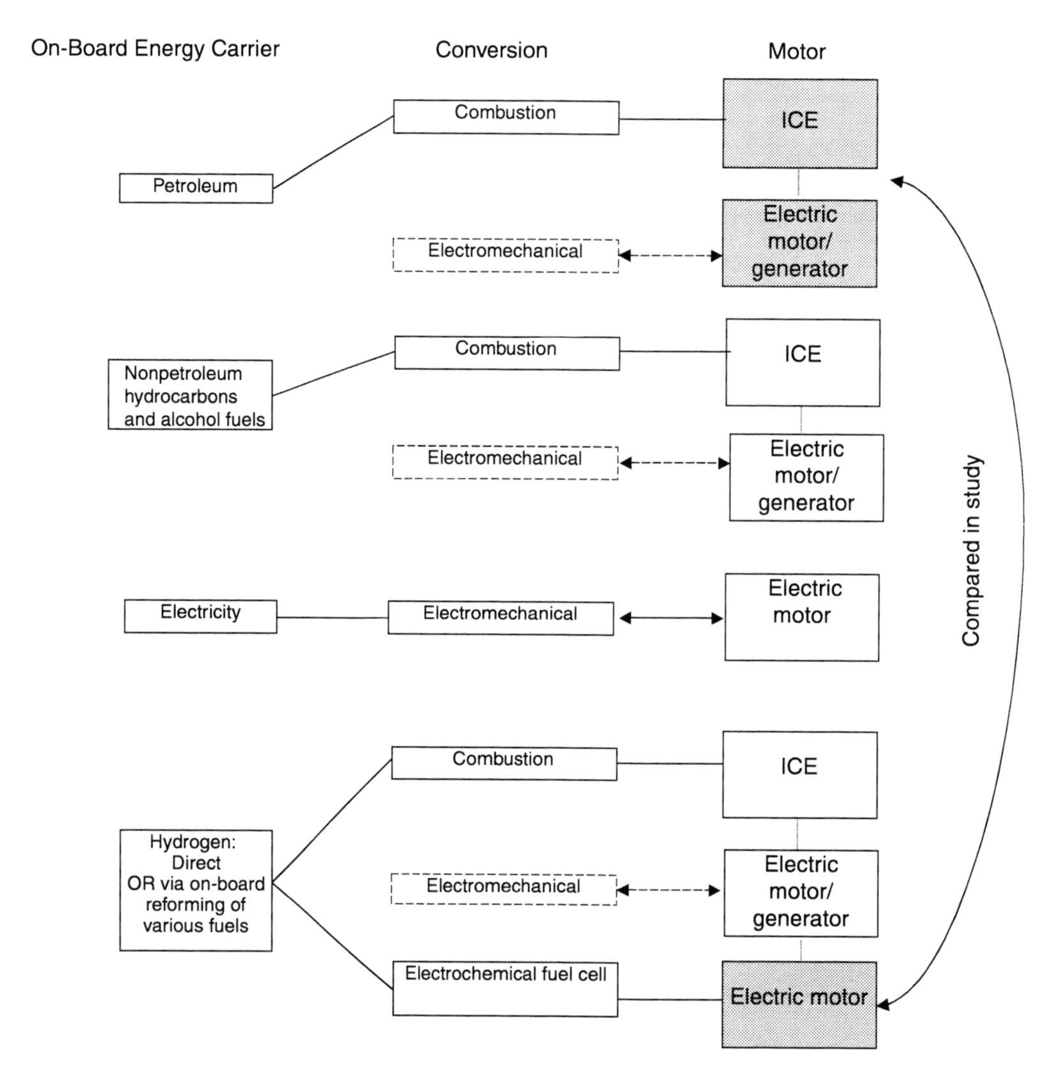

FIGURE 2-9 Combinations of fuels and conversion technologies analyzed in this report. The committee conducted cost analyses of hydrogen fuel converted electrochemically in fuel cells versus gasoline use in internal combustion engines (ICEs) in standard and hybrid configurations. Other combinations of fuels and energy conversion technology are discussed in the report.

3

The Demand Side: Hydrogen End-Use Technologies

The transition to a new energy carrier requires a series of investments and enhancements not only in energy supply and distribution, but also in vehicles and other end-use technologies. This chapter addresses the demand side in three major categories, namely, transportation, stationary power, and industrial uses. Transportation demand scenarios are postulated for the present (i.e., 2002), the near term (2020), and the long term (2050).

In the 1970s a number of studies, driven by energy-independence considerations, predicted that a hydrogen economy might emerge as early as the year 2000. Today the interest in a transition to a hydrogen economy is driven not only by concerns about energy security but also by those about global climate change and air quality. Rapid improvements in the proton exchange membrane fuel cell (PEMFC) during the past decade have been a catalyst for this renewed interest in a hydrogen economy because of the fuel cell's potential in transportation applications. In this chapter, the nature and magnitude of demand for hydrogen (H_2) are examined in a number of categories, with special attention focused on customer and regulatory attributes. (See Chapter 5 and Appendix E for estimates of well-to-wheels energy use.) On the basis of these analyses, technology barriers are identified that will need to be addressed in the DOE's research, development, and demonstration (RD&D) activities.

The focus of most of this report and this chapter is on light-duty passenger vehicles, the largest segment of the vehicle market. Stationary power systems to produce electric power from hydrogen may be an important part of a possible future H_2 energy system, both in a transition to a hydrogen economy and also in the steady state. The committee did not do an extensive analysis of the future stationary electric power system in the United States and the role that H_2 may play, but the section below entitled "Stationary Power: Utilities and Residential Uses" delineates some of the developments and opportunities in fuel cells and turbines for stationary power.

TRANSPORTATION

Background and Barriers

The transition to new fuels and/or energy carriers is especially problematic in the transportation sector because of the diffuse nature of the system and its complex public-private composition. Considering land vehicles only, there are more than 750 million passenger cars and commercial vehicles worldwide, with an annual production rate of 56 million units in 2001 (Ward's Communication, 2002). The geographically diffuse distribution of vehicles favors fuels that are easy to transport and store—that is, fuels that are liquid at room temperature. Consider, for instance, that natural gas fuels and electricity are generally less expensive (on a per unit of energy basis) and tend to be "cleaner" than liquid fuels are, but they are much more difficult to transport, in the case of natural gas, and much more difficult to store, in the case of electricity. Alcohol fuels are easy to transport and store, but they tend to be more expensive than are petroleum fuels, natural gas, and electricity. Most vehicular fuels continue to be gasoline and diesel fuels. The convenience of the petroleum-based fuel distribution system is a key factor in the continuing dominance of vehicles running on liquid fossil fuels. It explains in large part why gasoline hybrid electric vehicles[1] (GHEVs) have been successful in penetrating the consumer market, while grid-connected electric vehicles (including grid-connected hybrid electric vehicles) have not.

[1]Hybrid electric vehicles incorporate an energy storage device (e.g., a battery) along with the primary energy converter (the engine, which can be a gasoline internal combustion engine [ICE], a diesel engine, or a fuel cell, and so on) and a traction electric motor. The energy storage device can allow the possibility of recovering a significant portion of a vehicle's kinetic energy as the vehicle decelerates during braking. It also allows the primary energy converter (i.e., the engine) to be smaller and to operate under load and speed conditions that are independent of the vehicle's immediate needs, permitting the efficiency of the engine, for example, to be optimized.

Because of the large number of vehicles on the road and their relatively slow turnover, a change in fuel and/or energy carrier must be transitional—that is, sufficient fuel must be available for the large existing fleet while the new fuel is introduced in parallel. The success of energy-efficient GHEVs is instructive in two ways: (1) it demonstrates the huge challenge in moving beyond the relatively simple gasoline system now in widespread use, and (2) it creates an even greater barrier to newer technologies, such as the hydrogen fuel cell vehicle (FCV), by enhancing the fuel economy of "conventional" vehicles. The major "demand parameters" for a light-duty vehicle are shown in Table 3-1.

Transportation applications of fuel cell technology and hydrogen fuels not discussed in this report include urban buses, heavy-duty truck auxiliary power units (APUs) (Lutsey et al., 2003; Winter and Kelly, 2003), delivery vehicles, forklifts, airport baggage-handling vehicles, mining vehicles, golf carts, scooters, boats, and even airplanes. Of these, the hydrogen-fueled urban bus market segment has received the most attention.

Fuel Cell Vehicle Technology

The success of hydrogen in the transportation sector will be dependent on the development and commercialization of competitive FCVs. The challenge is to develop automotive fuel cell systems that are lightweight and compact (i.e., have high power densities by both mass and volume), tolerant to rapid cycling and on-road vibration, reliable for 4000 to 5000 hours or so of noncontinuous use in cold and hot weather, and able to respond rapidly to transient demands for power (perhaps by being hybridized with a battery or ultracapacitor for electrical storage on the vehicle), and able to use hydrogen of varying purity.

TABLE 3-1 Key "Demand Parameters" for a Light-Duty Vehicle

Demand Category	Parameter
Customer	Initial cost
	Operational and maintenance costs
	Quality
	Range (between refueling) and refueling convenience
	Passenger/cargo space
	Performance (acceleration, speed, ride quality, acceptably low levels of noise, vibration, and harshness)
	Safety
Regulatory	Emissions of pollutants (carbon monoxide [CO], oxides of nitrogen [NO_x], hydrocarbons [HC], particulates)
	Fuel efficiency
	Greenhouse gas emissions
	Safety

One of the most important attributes for FCVs is fuel efficiency, since less fuel means lower fuel costs, less expensive and bulky on-board hydrogen storage, and less upstream environmental impact. Wang (2002) summarizes the numerous studies comparing the fuel efficiency and life-cycle impacts of FCVs, hybrid electric vehicles (including GHEVs), and potential "transition vehicles" with baseline gasoline and diesel vehicles. Ignoring life-cycle impacts, fuel cells operating on hydrogen are much more energy-efficient than are internal combustion engine (ICE) systems. It is impossible to specify accurately how much more efficient they are, since fuel cells have very different efficiency characteristics (e.g., they are many times more efficient at low speeds and loads, but are less efficient at higher speeds and loads) and because automotive fuel cell systems are in their technological infancy and so their future performance cannot be accurately predicted.

For the purposes of quantitative comparisons, after extensive deliberation and literature review, the committee selected a fuel-efficiency improvement factor of 2.40 for FCVs versus a baseline gasoline vehicle—that is, today's gasoline vehicles are assumed to use two-and-a-half times as much energy as a comparable FCV. This comparison, an average for all light-duty vehicles, is based on average U.S. driving conditions. (For detailed assumptions, see Wang [2002].) The committee selected a fuel-efficiency factor of 1.45 for GHEVs versus a baseline gasoline vehicle. (See the discussion of hybrid technology in the following subsection, "Market Acceptance and Demand Trajectories.") Fuel-efficiency factors for diesel-powered hybrid electric vehicles would fall between 1.45 and 2.40. These assumptions of fuel economy are based on averages from Wang's (2002) review of other studies. In practice, actual differences in fuel economy may vary considerably. For instance, automakers might take advantage of the on-board electricity capability of FCVs and introduce a range of high-energy-consuming appliances and services, which would dramatically increase fuel consumption. Alternatively, FCVs might have relatively higher fuel economy because they disproportionately replace gasoline vehicles in urban settings or because traffic congestion results in slower driving speeds—in both cases taking advantage of FCVs' better fuel efficiency at lower speeds.

Given these requirements, hybrid and nonhybrid PEMFC systems are the leading contenders for automotive fuel cell power, with additional attention focusing on the direct-methanol fuel cell (DMFC) version of the technology and the possibility of using solid oxide fuel cell (SOFC) systems as auxiliary power units for cars and trucks.

An important attraction of all of these fuel cell systems, both as main vehicle power systems and as APUs, is their ability to support the new wave of vehicle electronics that is being introduced. New or planned electronic gadgetry on vehicles includes navigation systems; extensive on-board communications; voice-actuated controls; exterior alternating current (ac) power supplies; computer-controlled, power-assisted active suspension; collision-avoidance systems; electric air-conditioning compressors; "drive-by-wire" steering; side and rear-view bumper cameras; electronic tire pressure control; and generally greater computer power for increasing control of the various vehicle systems. The need for these systems has already started a trend toward a new 42-volt (V) standard for vehicle auxiliaries in order to deliver more power. In principle, electric (fuel cell) vehicles and APUs provide an efficient way to meet these power demands.

Fuel cell vehicles are attractive potential replacements for ICE vehicles because they can offer performance similar to that of conventional vehicles, along with several additional advantages. These advantages include better environmental performance; quiet (but not silent) operation; rapid acceleration from a standstill, owing to the torque characteristics of electric motors; and potentially low maintenance requirements. Furthermore, FCVs have the potential to perform functions for which conventional vehicles are poorly suited, such as providing remote electrical power (for construction sites, recreational uses, and so on) and possibly even acting as distributed electricity generators when parked at homes and offices and connected to a supplemental fuel supply. FCVs also provide additional attractions to automakers: by eliminating most mechanical and hydraulic subsystems, they provide greater design flexibility and the potential for using fewer vehicle platforms and therefore more efficient manufacturing approaches.

Market Acceptance and Demand Trajectories

For the FCV to be successful in the marketplace, it must satisfy customer desires and regulatory requirements (see Table 3-1). Fuel cell vehicles will easily meet a few of these desires and requirements. They will excel in fuel economy and emissions reduction. On the negative side, for the foreseeable future they will likely be expensive, have less range, and be more difficult to refuel. Their ability to satisfy other demands and requirements is more ambiguous, depending on perceptions, design decisions, and near-term engineering improvements.

For early fuel cell systems to succeed in the marketplace, they must have special appeal in some market niches, even if these niches are relatively small. One niche might be created by the desire, especially in dense urban areas, to achieve zero tailpipe emissions. The only zero-emission vehicle type other than the direct-hydrogen FCV that is practical at the present time is the battery electric vehicle (EV), which is characterized by short driving ranges, long recharge times, and high costs. To the extent that zero-emission vehicles are encouraged or even mandated in certain areas, direct-hydrogen FCVs may have to compete only with battery EVs and not the entire suite of vehicle technology options. Such a situation could give them a much firmer foothold for breaking in to motor vehicle markets. Another niche might be made up of individuals and businesses that value the large amounts of electrical power carried on board, and that might find a suite of new uses that can only be imagined at this time. And still other niches could include those wanting APUs on trucks or off-road vehicles in areas where noise or pollution is a concern.

One important feature of FCVs that remains crucial for their development is the fact that PEM fuel cells run on either pure hydrogen or a dilute hydrogen gas "reformate" stream (though direct-methanol fuel cells, still in an early stage of development, operate on methanol). This hydrogen can either be stored on board the vehicle in one of several ways, or generated from another fuel with an on-board reformer.

To aid the transition to FCVs without major infrastructure changes, the energy and automotive companies have been working together to develop on-board reformers. On-board reformers convert a liquid (or other gaseous fuel) to hydrogen. Natural gas reforming is more difficult than liquid reforming, and thus the focus has been on liquids for on-board reformers. The most effort has been devoted to methanol and gasoline. DaimlerChrysler was a leader in developing an on-board methanol reformer, and the company unveiled prototype FCVs operating on methanol in the late 1990s. Other companies focused on gasoline reforming. But by 2003, all major automakers had suspended their development of on-board reformers and shifted their FCV efforts to direct hydrogen use. Several oil companies are known to be continuing their development of on-board reformers, which is an appropriate technology to be developed in an industrial R&D laboratory.

On-board reformers are attractive in that they obviate the need to build a hydrogen infrastructure. Methanol is easier to reform than gasoline is, but DaimlerChrysler and others suspended methanol reforming in part because of the challenge of developing a large-scale infrastructure for what was viewed as an interim fuel. More generally, gasoline (and methanol) reforming efforts were suspended by automakers because of several major disadvantages: on-board reformers impose substantial additional cost, add considerable complexity, reduce fuel efficiency, increase emissions, increase "engine" start-up times, and create additional safety concerns. Automakers and others considered these disadvantages to be too large to overcome the advantages of ready gasoline availability, especially when on-board reforming is considered an interim strategy until hydrogen is broadly available.

Most analysts agree that storing hydrogen on board FCVs is the best ultimate solution, but no hydrogen storage system has yet been developed that is simultaneously lightweight, compact, inexpensive, and safe. Further advances in hydrogen storage, so that FCVs can refuel quickly and have driving ranges comparable with those of conventional vehicles, thus constitute a key area for further development. Prototype FCVs have been built that store hydrogen as a cryogenic liquid, as a compressed gas, in metal hydrides, and as sodium borohydrate. (See Chapter 4 for further discussion of hydrogen storage options.)

Market-related aspects of diesels and consumer electronics also deserve mention. Major sales of GHEVs to date have been in Japan and the United States, because Europe has embraced diesel engine technology in recent years as its major fuel-efficiency solution; in the near term (next decade), this trend will likely continue. However, it should be mentioned that European auto manufacturers were also the first to develop major hydrogen programs—for example, BMW with hydrogen ICEs and DaimlerChrysler with fuel cells and hydrogen ICEs—and these programs are continuing.

In many cases, marketplace competition in rapidly changing technologies speeds up the pace of development. A recent example is the development of batteries for consumer electronics applications. During the past decade, most major battery improvements have been driven by the high-volume need for portable power for laptops and cellular phones, among other devices. Some predict that a similar improvement scenario may occur with fuel cells, because today's rechargeable batteries cost approximately $3000 per kilowatt (kW) and have much less energy density than do fuel cells. In fact, a recent article in the *New York Times* noted that methanol-powered fuel cells for laptops might be available within a year (Feder, 2003).

Assuming continuing progress in fuel cell development and the availability of fuel, what are possible scenarios for FCV sales? The recent introduction of GHEVs provides insight. Indeed, the commercialization trajectory of GHEVs provides a "best case" penetration scenario (assuming no major surprises). That is, GHEVs provide a best case because the vehicle attributes were similar to those of a standard, high-volume gasoline ICE vehicle; no fueling infrastructure changes were required; the component technologies were relatively mature; the vehicles were viewed as high-tech and environmentally friendly; and tax benefits aided initial price reductions for the consumers. Table 3-2 shows the actual sales of hybrids through December 2002.

TABLE 3-2 Hybrid Electric Vehicle Sales in North America and Worldwide, 1997 to 2002

Sales Volume	Year 1997	1998	1999	2000	2001	2002
North America[a]	0	0	0	9,600	20,700	35,900
Worldwide	300	17,700	15,500	24,200	42,100	59,300
Total to date[b]	300	18,000	33,500	57,700	99,800	159,100

[a]North American sales—almost all in United States.
[b]Total to date: cumulative worldwide sales.
SOURCE: Personal communication of committee member Daniel Sperling with Toyota and Honda, 2003.

Winter and Kelly (2003) address future possibilities. For example, Toyota has a goal of selling 300,000 GHEVs by 2005, and General Motors Corporation (GM) indicates that it will make hybrid technology "available" on a wide variety of models by 2007. Under the most optimistic scenarios for GHEVs, after a decade in production, the annual volume might approach 2 million vehicles, with a total of 4 million GHEVs on the road. In practice, GHEV sales forecasts are being reduced as of this writing. Toyota and Honda continue to expand sales, but GM and Ford have delayed introduction of their initial offerings. A fundamental concern is cost. Toyota declared in 2003 that it was making a profit on its GHEV, the Prius, selling it at about $3500 more than the cost of a comparable conventional gasoline-fueled vehicle. It is expected that this cost premium will gradually drop over time as sales volume increases and learning takes place. Indeed, the 2004 Prius is more powerful and bigger, with better fuel economy and lower emissions, and sells for the same price as the previous-generation Prius.

The cost premiums for GHEVs are in part a function of the technology used in hybrid vehicles. The Prius is known as a "full" hybrid, in the sense that it relies on a large battery pack and large motor for much of its power and energy. Other models to be introduced by Toyota and other automakers will have smaller batteries and electric motors, implying lower cost and also smaller fuel economy improvements relative to conventional gasoline-fueled ICE vehicles. Full hybrids provide up to a 50 percent fuel economy improvement, while "mild" hybrids, with perhaps only an integrated starter/alternator, will provide only about a 10 percent improvement.

In Chapter 6, it is assumed that GHEVs will represent 1 percent of U.S. sales by 2005 and will increase 1 percentage point per year for the next 10 years and 5 percentage points per year for the following 10 years. The energy efficiency that is used for all hybrids is 45 percent improvement relative to conventional gasoline-fueled ICE vehicles.

The committee estimates that the fuel cell system, including on-board storage of hydrogen, would have to decrease to no more than about $100/kW before a scenario even close to the hybrid scenario postulated here would be realized. The most optimistic estimates project 2010 as the year in which $100/kW can be achieved[2] (Arthur D. Little, 2001) (although this committee has not had the opportunity to evaluate the basis of such estimates—for example, by conducting a part-by-part cost analysis). As the DOE manages its hydrogen program, it is imperative that it understand the components of these cost estimates and, on the basis of these understandings, appropriately evolve its RD&D programs. Because industry is actively pursuing RD&D in fuel cells, particular

[2]The cost includes the fuel cell module, precious metals, the fuel processor, compressed hydrogen storage, balance of plant, and assembly, labor, and depreciation.

DOE attention should be devoted to related fundamental and exploratory research at universities and national laboratories.

Assuming an optimistic scenario for FCVs and numbers of vehicles entering the marketplace similar to those of GHEVs, FCVs could reach 1 percent of U.S. sales by 2015, and then increase by 1 percentage point per year until 2024 and by 5 percentage points per year thereafter until they dominate the market. (It should be noted that the DOE multi-year program plan for hydrogen RD&D [DOE, 2003b] designates 2015 as the year for a "commercialization decision.") Figure 3-1 shows the detailed projection for this scenario. The projection takes into account reasonable transitions for the buildup of GHEV and FCV manufacturing and the associated phaseout of conventional and GHEV manufacturing (see Chapter 6).

Thus, by 2020, the total number of FCVs on the road would be fewer than or equal to 4 million units if the optimistic GHEV penetration scenario was matched. Four million vehicles could not justify a national fuel infrastructure change, although regional infrastructure needs might be high as a result of clustered demand growth; that is, in most locations, marketplace demand would not be the main element in a fuel change by 2020.

The committee's market trajectory for hydrogen fuel cell vehicles reflects what is possible and shows initial market penetration in 2015, growing to 12 percent of new light-duty vehicles sold in 2020 and 40 percent in 2030. Although not directly comparable, there are several other studies that can be compared with the committee's vision of what might happen. For instance, Argonne National Laboratory (Santini et al., 2003) made a market penetration analysis of FCVs that shows 1 percent market share in 2011, growing to 26 percent in 2020, 52 percent in 2025, and reaching 100 percent in 2038. A report of the Pew Center on Global Climate Change (Mintzer et al., 2003) posits similarly high initial market penetration, but slower increases over time—reaching 2.5 percent penetration in 2015 and 5 percent in 2020, and steadily inching upward to 20 percent annual sales in 2035. The 2003 DOE program (DOE, 2003b) assumes initial penetration in 2018, increasing to 27 percent in 2020 and to 78 percent in 2030.

If the committee's FCV projection above is "close to actuality" (or even shifted by some number of years into the future), it indicates that a great deal of thought must be given to the fuel service station scenarios for the decade when FCVs grow from a few thousand to a few hundred thousand—that is, in 2010 to 2020, as shown in Figure 3-1. Today the United States has a dense network of about 180,000 retail fuel stations, serving more than 200 million vehicles. Dense coverage, similar to the number of diesel fuel stations in the United States today, will be required as FCVs grow into the millions. Other parts of this report address technolo-

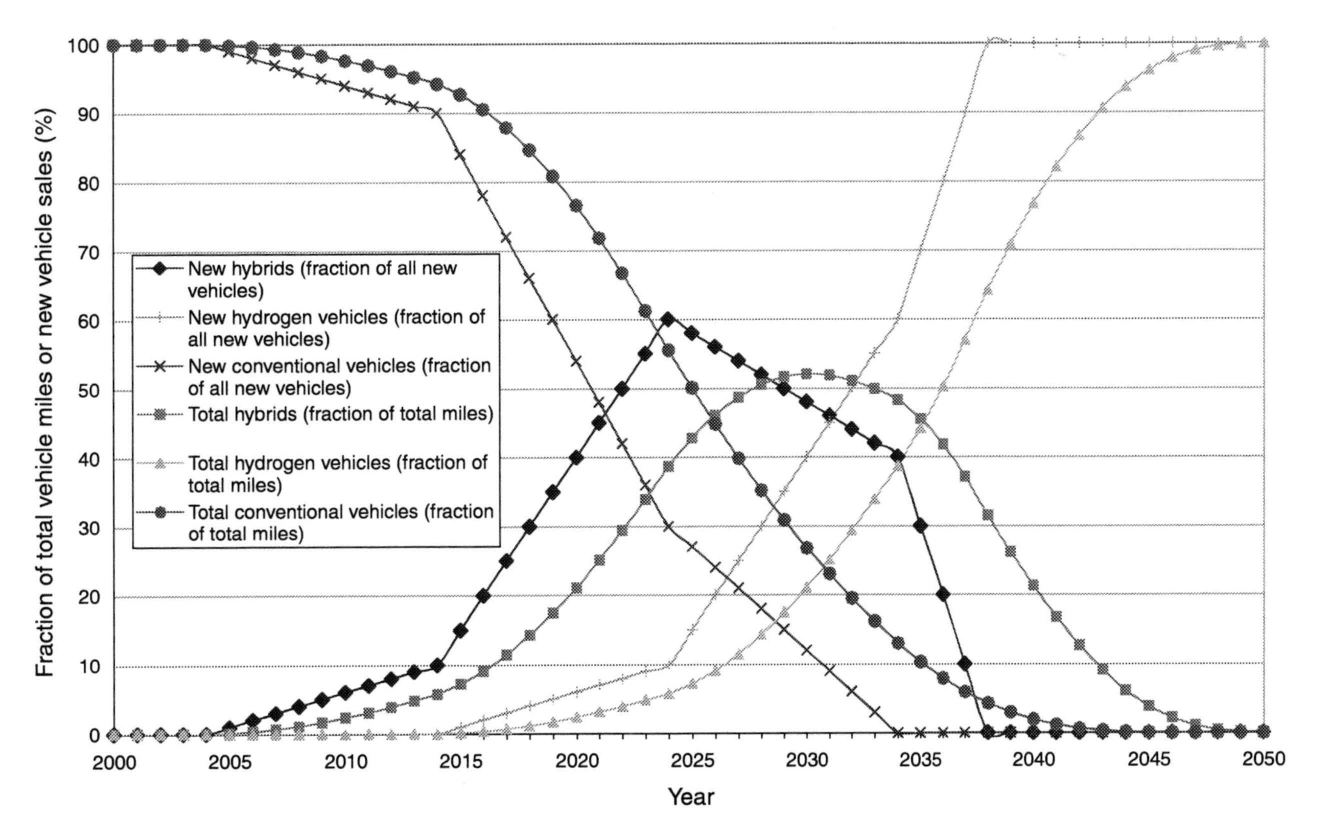

FIGURE 3-1 Possible optimistic market scenario showing assumed fraction of hydrogen fuel cell and hybrid vehicles in the United States, 2000 to 2050. Sales of fuel cell light-duty vehicles and their replacement of other vehicles are shown.

gies that could be employed to bring the relatively large number of retail hydrogen fuel stations online before an extensive network of hydrogen pipelines may be required to be developed.

STATIONARY POWER: UTILITIES AND RESIDENTIAL USES

Introduction

As indicated above, this report and the committee's analysis have been focused on hydrogen production technologies and the demand for hydrogen in the light-duty-vehicle sector. The use of hydrogen in stationary applications may also play an important role in a hydrogen economy, but it was not analyzed in detail by the committee. Distributed generation (DG) of electrical power is projected by some analysts to be a substantial market. Fuel cells and turbines using hydrogen could provide an important future opportunity for hydrogen produced from sources other than natural gas in areas where pipelines are available. This opportunity could help to stimulate hydrogen infrastructure investments, possibly before FCVs reach commercial readiness.

The U.S. electric power system is projected to use large amounts of coal and natural gas for the next 20 years and to produce a significant portion of the nation's CO_2 emissions (EIA, 2003). Advanced fossil-fueled energy plants of the future could produce electricity and/or hydrogen, and achieve high efficiencies using advanced turbines and fuel cells, while also sequestering by-product CO_2 (NRC, 2003b). Hydrogen could be exported from such large plants and used in industrial facilities, either to generate electricity onsite or for process heat; thus, the industrial sector could represent another demand, further stimulating a market for hydrogen production. The committee did not analyze these options and various trade-offs for the use of hydrogen in stationary applications, but the technologies that might be used in stationary applications are addressed in the remainder of this chapter.

In 2001, the U.S. electric power system included about 10,400 generating stations of greater than 60 megawatts (MW), with a total installed capacity of 786 gigawatts (GW). Electricity generation contributes about 40 percent of the CO_2 emissions in the United States (see Figure 2-7 in Chapter 2)—largely as a result of coal's being the source for 50 percent of electricity generation; natural gas is the source for 17 percent. Of the new electricity generation capacity being installed, 80 percent is projected to be with natural gas (EIA, 2003). In comparison with building a comparable fraction of new coal-fueled power plants, this change would reduce carbon emissions, but increase energy demand and imports for natural gas.

Distributed generation is modular generation. DG units are less than 60 MW in size and usually located near the point of use. Technologies available for DG include industrial and aero-derivative gas turbines, reciprocating engines, microturbines, wind turbines, biomass-based generators, solar power and photovoltaic systems, and fuel cells. These technologies offer a greater degree of fuel flexibility than large central power stations do. There are an estimated 10.7 million DG units in place in the United States, of which over 99 percent are small emergency/standby reciprocating engines that are not interconnected with the grid. Currently, 85 percent of the DG units are reciprocating engines; they are fueled primarily by distillate fuel oil or gasoline; combustion turbines make up 5 percent and are fueled by natural gas; and steam turbines constitute 9 percent (Resource Dynamics Corporation, 2003).

DG can be either grid-connected or operated independent of the grid. The aggregate capacity of all DG units in the United States is 169 GW, which is 17 percent of U.S. electricity generation capacity (Resource Dynamics Corporation, 2003). A total of 70 GW of capacity installed prior to 1990 was still operating by the end of 2000. DG units can be used to meet baseload power requirements and needs for peaking power, backup power, remote power, power quality, and heating and cooling. They typically must be able to operate between 40,000 and 50,000 hours without major system overhauls. The market for DG is typically the commercial sector, including hospitals, supermarkets, restaurants, universities, and shopping malls; manufacturing facilities, which need reliable energy; or remote locations where grid power is not available. DG can be customer- or utility-owned.

Direct use of H_2 in stationary systems would provide a new fuel option for DG. It could provide a route for a transition to H_2 that was produced economically but during the time when FCVs were not ready for commercial introduction.[3] The use of hydrogen reformed from natural gas is not likely to displace direct use of natural gas in stationary systems. It is more energy-efficient to use natural gas directly than to convert it to hydrogen in stationary DG applications. (Natural gas is currently the preferred fuel for new DG.)

If economic, small electrolyzers coupled with distributed power-generating devices could replace and supplement batteries in the DG backup power market. So-called regenerative systems in the 1.50 kW scale could convert and store grid power in the form of H_2 that could then be used to regenerate power in fuel cells or in combustion devices. This application may represent a higher-value niche for electrolysis and fuel cells than the transportation market does. It might potentially offer less demanding technical challenges: vehicular fuel cells will be subject to vibration and thermal stresses, whereas stationary backup applications would not, and also would need only short-term reliability.

Fuel cells are currently being developed for distributed generation; most are for applications under 1 MW. Some solid oxide fuel cell/gas turbine hybrid systems are being developed for 5 MW applications. Aside from fuel cells used

[3]Larry Burns, General Motors Corporation, "Fuel Cell Vehicles and the Hydrogen Economy," presentation to the committee, June 11, 2003.

in space applications, stationary fuel cells (phosphoric acid) are the only fuel cells for which there is "production" and market experience. Stationary requirements are usually less stringent than those of transportation with respect to price ($500/kW versus $50/kW) and footprint, but require longer life (40,000 to 50,000 hours versus 3,600 hours).

Central Power

Direct Fired

Gas turbine engines with a conventional natural gas combustion system or water injection combustion system can operate on H_2 or H_2-rich fuels with little or no modifications to the core injectors. Modifications to the fuel delivery system and injectors are required. The volumetric heating value of hydrogen is 10,787 kJ/N-m^3 (274 Btu per standard cubic foot [scf]) as compared with 35,786 kJ/N-m^3 (909 Btu/scf) for methane. In order to supply the required energy input rate to the gas turbine, approximately 3.32 times the volume of hydrogen fuel has to be injected into the primary zone of the combustor to provide the same heating value as that of natural gas fuel.

Large turbines, particularly integrated gasification combined cycle (IGCC) systems, have been run successfully on syngas with volumes up to 62 percent H_2 in process plants in the United States and Europe. For example, General Electric (GE) has 10 IGCC projects running on H_2-rich fuel, with 6 more planned or going into operation. Nine of these projects are associated with refinery operation (Jones and Shilling, 2002; Jon Ebacher, General Electric Power Systems, "SOFCs, Direct Firing, Wind," presentation to the committee, April 23, 2003).

Marketplace Scenarios

Since H_2 can be burned in gas turbines, these turbines could provide an early market for additional H_2 production—assuming that the H_2 is not generated from natural gas. Turbines located at the site of the hydrogen production could generate electricity, which could be transmitted via the usual electrical transmission and distribution (T&D) system to residential, commercial, and industrial users.

For this market there are R&D needs to address issues that include the following: (1) combustion technology to reduce NO_x emissions and achieve higher efficiencies, (2) fuel management and controls for operability and safety requirements, (3) cost-and-efficiency trade-offs, (4) material compatibility of components with H_2 combustion gas, and (5) systems development and optimization.

Fuel Cells for Distributed Generation

Fuel cells offer the potential for very efficient, clean, and quiet distributed power generation. Because the power generation process in fuel cell systems is electrochemical, no emissions from combustion are produced from the power generation itself. These benefits have led to significant federal R&D funding over the past 25 years. Nevertheless, fuel cells are currently more than four times more expensive to install than ICE generators and more than twice as expensive to install as microturbine generators, with which they are frequently compared. The high capital costs of fuel cell systems that have been sold or demonstrated to date have been a major barrier for penetration into the DG market. There are four different fuel cell systems, characterized by their electrolytes, that are potentially suitable for stationary power (Lipman and Sperling, 2003; Shipley and Elliot, 2003). Table 3-3 provides current performance parameters for the various fuel cell types; Table 3-4 presents parameters projected for 2020.

The Energy Information Administration (EIA, 2003) estimates that electricity generation will increase by 2 percent per year to meet increased electrical demands. In 2020, 1.5 trillion kWh of additional electricity-generation capacity will be needed. If 10 percent of the added generation (150 billion kWh) were from hydrogen, it would require 10 million tons of hydrogen, and 20 million tons per year of CO_2 emissions might be avoided, assuming that H_2 is produced from sources other than coal or natural gas or, if other fossil fuels are used, that the CO_2 is sequestered (DOE, 2003a). Of course, existing DG technologies such as microturbines will continue to improve both economically and in terms of achieving higher efficiency; thus, competing technologies are a continual moving target.

The major technical and cost issues for fuel cells regardless of electrolyte or temperature range are (1) stack cost and life, (2) reformer (where needed), and (3) power electronics and overall system integration. Addressing these issues requires basic electrochemistry and material studies. Total funding by the Office of Fossil Energy for its fuel cell activities (phosphoric acid fuel cell [PAFC], MCFC, SOFC) from FY 1978 through FY 2000 was $1167 million (NRC, 2001a), which was cost-shared 20 to 50 percent with industry. The NRC (2001a) study concluded that these funds still did not result in fuel cells' being commercial.

The Office of Energy Efficiency and Renewable Energy has supported PEM stationary fuel cell R&D since FY 2000 and has spent $22 million. The SOFC and MCFC programs are supported by the Office of Fossil Energy and are not part of the DOE hydrogen program, but are considered "associated programs," since they are being developed to operate on natural gas and synfuels. However, these programs could be modified and fueled with H_2 if it were available.

The following subsections treat various types of fuel cells, currently market-deployed or under development, and discuss them in the context of distributed generation, while noting other applications.

TABLE 3-3 Stationary Fuel Cell Systems—Typical Performance Parameters (Current)

Cost and Performance Characteristics	System 1	System 2	System 3	System 4	System 5	System 6
Fuel cell type	PAFC	PEMFC	PEMFC	MCFC	MCFC	SOFC
Nominal electricity capacity (kW)	200	10	200	250	2000	100
Operating temperature (°F)	400	150	150	1200	1200	1750
Internal reforming	No	No	No	Yes	Yes	Yes
Package cost (2003 $/kW)	4500	4700	3120	4350	2830	2850
Total installed cost (2003 $/kW)	5200	5500	3800	5000	3250	3620
Operating and maintenance costs ($/kWh)	0.029	0.033	0.023	0.043	0.033	0.024
Electrical efficiency (%), HHV	36	30	35	43	46	45
Total CHP efficiency (%), HHV	72	69	72	65	70	70
CO_2 (lb/MWh)	1135	1360	1170	950	890	910
Carbon (lb/MWh)	310	370	315	260	240	245
Effective electrical efficiency (%), HHV	65.4	58.6	65.0	59.3	65.6	65.5
Commercial status, 2003	Commercially available	Demonstration	Demonstration	Commercially introduced	Demonstration	Demonstration

NOTE: PAFC = phosphoric acid fuel cell; PEMFC = proton exchange membrane fuel cell; MCFC = molten carbonate fuel cell; SOFC = solid oxide fuel cell; CHP = combined heat and power; HHV = higher heating value.
SOURCE: National Renewable Energy Laboratory (2003).

TABLE 3-4 Stationary Fuel Cell Systems—Projected Typical Performance Parameters (2020)

Cost and Performance Characteristics	System 1	System 2	System 3	System 4	System 5	System 6
Fuel cell type	PAFC	PEMFC	PEMFC	MCFC	MCFC	SOFC
System size (kW)	—	10	200	250	2000	100
Total installed cost ($/kW)	—	2200	1700	1650	1400	1800
Operating and maintenance costs ($/kWh)	—	0.019	0.012	0.020	0.014	0.015
Electrical efficiency (%)	—	35	38	49	50	51
Total CHP efficiency (%)	—	72	75	75	72	72
Effective electrical efficiency (%)	—	65	71	73	69	69
CO_2 emissions (lb/MWh)	—	1170	1140	834	820	801

NOTE: PAFC = phosphoric acid fuel cell; PEMFC = proton exchange membrane fuel cell; MCFC = molten carbonate fuel cell; SOFC = solid oxide fuel cell; CHP = combined heat and power; HHV = higher heating value.
SOURCE: National Renewable Energy Laboratory (2003).

Phosphoric Acid Fuel Cells

The 200 kW phosphoric acid fuel cell (PAFC) was introduced into the market in 1991 by International Fuel Cells/ONSI, now called UTC Fuel Cells. It is the only commercialized fuel cell technology. PAFC units have been installed in various applications—commercial, small industrial, landfill, and military—and some are used for cooling, heating, and power. To date there have been 250 units sold, at roughly $4500/kW. The U.S. Department of Defense (DOD) has cost-shared the purchase of three-quarters of the units sold to date. The units have performed well: they have operated at 95 to 98 percent availability and 99.99 to 99.9999 percent reliability and have served 4 million customers and accumulated 4 million hours of operation. The cost of PAFC units has not decreased and in fact has increased from $3500/kW. These units are not cost-competitive with other DG options, which can provide the same reliability and high-quality power efficiency. Recently, UTC Fuel Cells decided not to manufacture more units and to sell only those in inventory. Current units will continue to be serviced.[4]

What are the lessons learned from the failure of PAFC to become a commercial success and how do these lessons apply to other stationary fuel cell systems in development and demonstration? Was the cause of failure only the high cost relative to the other DG systems? The PAFC systems appeared to perform well. The federal government had spent more than $411 million on PAFC. Should it have continued

[4]John Cassidy, UTC, Inc., "Fuel Cell Commercialization," presentation to the committee, April 24, 2003.

to sponsor R&D to reduce the cost of the commercialized systems? It is often thought that the government can be an early adopter of technology to enable initial volumes to be manufactured and sold. For PAFC, DOD subsidized three-quarters of those produced. Should the government (e.g., the General Services Administration or DOD) make larger purchases of new technologies? Even if costs had been reduced, there are market and regulatory barriers that apply not only to fuel cells, but also to other new DG technologies, such as microturbines.

Proton Exchange Membrane Fuel Cells

The proton exchange membrane fuel cell, which is the fuel cell being considered for vehicle transportation applications, can also be used in DG applications, particularly for small-scale residential and commercial purposes. The PEMFC operating temperature of 150°F is lower than that of the PAFC and much lower than the operating temperatures of the other fuel cell systems in development: the solid oxide fuel cell and the molten carbonate fuel cell. This means that the PEMFC could be used for residential hot water, but not for high-quality steam or combined heat and power (CHP) applications. Many companies (Plug Power, Avista, Ballard, H Power) have been exploring the use of the PEMFC for the 1 to 25 kW market—which would involve residential buildings, including some small multifamily homes. The PEMFC is also being considered in the 50 to 250 kW range. Ballard's first commercial fuel cell product, the 1.2 kW Nexa® power module, was introduced in the market in 2001. Ballard has introduced the Air Gen Unit at 1.2 kW for backup and intermittent-duty applications; this unit has both hydrogen cylinders and cartridges to supply the hydrogen. Ballard's first continuous stationary fuel cell will be introduced in Japan in limited volume by the end of 2004 as a 1 kW CHP unit. PEMFC applications can be considered as a niche market, particularly in the under-25-kW size, because in this size range the PEMFC must compete with existing DG technologies that have heating and cooling system applications and are reliable, durable, and low-cost. If there were a sizable market, DG could provide PEMFC manufacturing experience, enhancing the learning curve for PEMFC and hastening its automotive application, which has much more stringent volume and cost requirements. DG applications require longer life than automotive applications do.

The DOE issued a solicitation in January 2003 for the development of stationary PEMFC for buildings, with the target cost of $1500/kW, design life of 40,000 hours with less than 10 percent degradation, and market entry within the next 3 to 5 years.[5] Recently UTC Fuel Cells announced that it will introduce 150 kW PEMFC units at $1500/kW in early 2004.[6] The company is currently beta testing these units. The Electric Power Research Institute (EPRI) and UTC are currently cofunding a 750 kW PEMFC demonstration that will consist of five 150 kW modules, each with its own processing system, for a projected installed cost of $2600/kW and expected efficiency of 31 percent. The intent is to gain manufacturing experience that would be applicable for PEM automotive fuel cell systems to meet the $50/kW automotive cost target in the 2010 to 2020 time frame. By 2010, UTC expects to have developed an SOFC system, which would be more attractive for DG applications. PEMFCs for stationary applications have similar R&D needs to those for automotive applications, with additional technical challenges related to higher durability (at 40,000 to 50,000 hours), heat utilization (a higher-temperature membrane is needed), power electronics, rapid start-up time for backup power, fuel processing, and development of non-precious-metal catalysts and thermal and water management technologies.

Solid Oxide Fuel Cells

Solid oxide fuel cells have an electrolyte that is solid ceramic and can operate at up to 1000°C. Unlike PEMFC and PAFC systems, there are no noble metals in the anode or cathode. SOFCs can be configured in a tubular or planar configuration and can be operated at high enough temperatures to eliminate a fuel reformer. SOFCs reject high-value waste heat useful for a steam bottoming cycle or available for CHP. These fuel cells can operate on a variety of fuels, including H_2, but current SOFCs are being designed for natural gas as the fuel. There is potentially a broad spectrum of power-generation applications, from small, lightweight, compact devices in the range of watts to kilowatts to larger SOFC/turbine hybrid systems in the megawatt range.

In 2001, the DOE Office of Fossil Energy and industry jointly initiated a Solid State Energy Conversion Alliance (SECA) Program for further SOFC development; the program currently involves six industrial teams. In addition, a parallel core technology program is under way at national laboratories and universities. This effort is to be a $500 million, 10-year program to produce modular, mass-produced fuel cells for stationary, transportation (APUs), and military markets. By 2010, the goal is for the SOFC to have 40 to 50 percent efficiencies and to cost less than $400/kW. The SOFC stack represents 30 percent of projected costs; fuel and air handling are another 30 percent.[7]

In addition to the SOFC as a stand-alone DG or in a CHP system, SOFCs are being developed in an SOFC/gas turbine

[5]U.S. Department of Energy, DOE Solicitation DE-SC02-03CH11137, "R&D for Fuel Cells for Stationary and Automotive Applications," January 24, 2003, p. 2.

[6]John Cassidy, UTC, Inc., "Fuel Cell Commercialization," presentation to the committee, April 24, 2003.

[7]Joseph Strakey, DOE/National Energy Technology Laboratory, "Solid Oxide Fuel Cells," presentation to the committee, April 24, 2003.

hybrid configuration.[8] In the hybrid configuration, the fuel cell converts fuel—for example, either direct hydrogen, syngas from fossil fuels, or biofuels—into electricity and water along with by-product heat. The residual fuel from the fuel cell is then burned by a gas turbine for additional electricity production. The product could be in the 1 to 10 MW range with greater than 65 percent efficiency and with fuel flexibility, and it would be cost-effective when compared with the cost of today's technology.

Applications of the SOFC could include larger commercial sites, industrial manufacturing facilities, and utility substations. As part of its Vision 21 Program, the DOE is sponsoring several hybrid programs, including a 5 MW system (4 MW SOFC, 1 MW gas turbine). GE expects this product to enter the market in 2013. The SOFC hybrid program will utilize the technology advancements from the SECA Program, but there are specific R&D needs related to the hybrid regarding performance, reliability of the life of the stack under a system pressurized operating environment; and optimized system design, controls, and components.

With respect to the marketplace, SOFC and SOFC/gas turbine hybrids are potentially an attractive basis for an efficient, clean, cost-competitive DG system, but they do not depend on having H_2 fuel. However, they could facilitate a transition to a H_2 economy by making use of H_2 for distributed electricity and CHP, while other fuel cells for vehicles are becoming cost-effective, reliable, and efficient. It is important for the DOE to monitor the milestones and goals of the SECA Program and to fully fund it.

Molten Carbonate Fuel Cells

Molten carbonate fuel cells use a mixture of carbonates that are liquid at operating temperature—600°C to 650°C. MCFC, like SOFC, operates at a higher temperature than the PEMFC does; it does not require a fuel reformer; and it can be operated with a hydrogen-rich fuel. The MCFC's liquid electrolyte means more handling issues. It does not have the ability to be pressurized. The MCFC could serve a niche market of data centers and hospitals. FuelCell Energy has recently made a commercial offering of MCFCs. These fuel cells will probably not have the same market penetration potential as SOFCs and thus would likely have little or no impact as a transition strategy for H_2 use.

Direct Use of Hydrogen in Distributed Generation

Small gas turbines, less than 25 MW, can operate on H_2 or H_2-rich fuels with little or no modification, similar to gas turbines for central power generation. There have been some demonstrations of 5 and 10 MW systems with enriched H_2 gas at refineries but not to the same extent as the demonstrations of large GE turbines for processing.[9]

INDUSTRIAL SECTOR

Industry is currently the largest producer and user of H_2 in the United States (3 trillion ft^3 of H_2 annually—which represents less than 3 percent of the energy used by the sector). Steam reforming and water-gas-shift reactions and separations are the primary processes for hydrogen production; they are carried out in refineries and large-scale chemical plants. Natural gas is the primary feedstock for existing hydrogen production. Approximately 50 percent of the H_2 consumed by industry is for ammonia production, 36 percent is for petroleum refining, 8 percent is for the production of methanol, and 6 percent is for other uses (DOE, 2003d).

Combustion offers potential for the industry-wide use of hydrogen. Industrial boilers and process heaters—fueled by the combustion of fossil fuels such as natural gas and coal—use 13.5 quadrillion Btu (quads), more than 75 percent of the total U.S. manufacturing energy. By 2050, the combined industrial energy demand is projected to be more than 26 quads. The use of hydrogen as a combustion fuel source for industrial boilers and process heaters offers the potential for a sizable end-use market for hydrogen—up to 2.6 quads of energy annually by 2050. In addition, there could be improvements in efficiencies—99 percent thermal efficiency versus 80 percent for conventional technology (DOE, 2003d). There is experience in the industrial sector using hydrogen blended with other fuels and diluents; there is little or no experience with H_2-air and H_2-O_2 systems.

Systems studies, as well as conceptual designs and further investigations of component issues related to, for example, combustors, heat exchangers, and flue gas ducting, are needed in order to develop more fully the understanding of the role of H_2 combustion technologies in the industrial sector.

SUMMARY OF RESEARCH, DEVELOPMENT, AND DEMONSTRATION CHALLENGES FOR FUEL CELLS

Despite great improvements in fuel cell technologies over the past decade and demonstration of promising performance, both stationary and automotive fuel cell systems still face large challenges. These primarily involve cost reduction: costs on the order of $500 to $800/kW-peak are required for competitive stationary systems, and costs on the order of $50 to $100/kW-peak are required for competitive FCVs. These cost levels are far below current levels for various fuel cell technologies that are in prototype and low-volume production. Additional challenges include fuel cell

[8]Jon Ebacher, General Electric Power Systems, "SOFCs, Direct Firing, Wind," presentation to the committee, April 23, 2003.

[9]U.R. Brendt, Solar Turbines Incorporated, "Use of Hydrogen Rich Fuels in Gas Turbines, Solar Turbines," private communication to committee member Maxine Savitz, February 2003.

durability: development goals are for 40,000 to 50,000 hours between major overhauls for stationary systems and 4,000 to 5,000 hours for automotive systems; the development of efficient and low-cost fuel reformers (see Chapter 8); and the development of vehicular hydrogen storage systems that are inexpensive, lightweight, compact, safe, and quick to refuel (see Chapter 4).

FINDINGS AND RECOMMENDATIONS

Finding 3-1. The federal government has been active in fuel cell research for roughly 40 years. Proton exchange membrane fuel cells (PEMFCs) applied to hydrogen vehicle systems are a relatively recent development (as of the late 1980s). The Department of Energy has spent more than $1.2 billion since 1978, and there has been considerable private sector investment for all fuel cell types. The DOE has spent $334 million since the 1980s on PEMFCs for transportation applications, most of it at national laboratories. Automakers and suppliers greatly expanded their PEMFC development efforts beginning in the later 1990s. In spite of the large federal and private sector investment, fuel cell prototype costs for light-duty vehicles are still a factor of 10 to 20 times too expensive and these fuel cells are short of required durability. Accordingly, the challenges of developing PEMFCs for automotive applications are large. Furthermore, the DOE's near-term milestones for fuel cell vehicles appear unrealistically aggressive on the basis of the current state of knowledge with respect to fuel cell durability, storage systems, and overall costs. The choice of unrealistic targets can lead to programs that emphasize spending on extensions and expensive demonstrations of current technologies in lieu of breakthroughs that will probably be required if a fuel-cell-based hydrogen economy is to be realized. Industry is expanding its development; thus, the DOE should focus on fundamental research.

Recommendation 3-1a. Given that large improvements are still needed in fuel cell technology and given that industry is investing considerable funding in technology development, increased government funding on research and development should be dedicated to the research on breakthroughs in on-board storage systems, in fuel cell costs, and in materials for durability in order to attack known inhibitors to the high-volume production of fuel cell vehicles.

Recommendation 3-1b. Since a hydrogen transportation economy will probably not emerge without the development of reasonably priced, energy-efficient fuel cells, the transportation portion of the Department of Energy's research, development, and demonstration program should emphasize fuel cells and their associated storage systems at the expense of "transition technologies" such as on-board reformers and hydrogen internal combustion engines. Since transition technologies mainly involve "development," funding for these programs should be provided by industry. Of course, some component breakthrough technologies for reformation might be justified in supply-side programs, and the results might be applicable to on-board reformation.

Finding 3-2. Various fuel cell technologies are attractive for stationary applications. In fact, the major stationary fuel cell research, development, and demonstration programs—in particular, the solid oxide fuel cell and the molten carbonate fuel cell (neither of which requires hydrogen fuel)—are not part of the Department of Energy's integrated direct-hydrogen program. Some private companies have committed to introducing proton exchange membrane stationary fuel cells without DOE funds, and these fuel cells appear to have applicability in a number of niche markets.

Recommendation 3-2. The Department of Energy should discontinue the proton exchange membrane (PEM) applied research and development program for stationary systems. The $7.5 million annual budget (FY 2003 and FY 2004 request) for that program could be applied to PEM fundamental and basic issues (exploratory research) for all applications.

Finding 3-3. During the past 20 years, a number of approaches have been used to encourage the application of alternative fuels and technologies in transportation and stationary systems. Most of these have failed because of the lack of real marketplace pull, shifts in government policies, and the relative disinterest of industry. The role of marketplace pull is especially important, as has been exhibited by the progress in batteries over the past decade to satisfy high-volume consumer electronics demand—for example, the rapid transition from nickel cadmium through nickel metal hydride to today's lithium-ion battery packs.

Recommendation 3-3. As the Department of Energy develops its strategy for the hydrogen economy with respect to the role of public research, development, and demonstration policies, it should sponsor an independent study of lessons learned with respect to the lack of success and widespread market acceptance of previous alternative fuel technologies, as well as other technologies developed for transportation and stationary power systems. The purposes of this study would be as follows: (1) to assess the role of government policy and its stability as it affects industry and consumer behavior, (2) to affect strategies related to the introduction of hydrogen in the end-use sectors, and (3) to avoid repeating the mistakes of prior-technology-introduction programs, such as those for electric and natural gas vehicles and for phosphoric acid fuel cells for distributed generation. In addition, strengths and weaknesses of the Partnership for a New Generation of Vehicles Program and hybrid electric vehicle development should be analyzed, as the FreedomCar Program is structured for the development of fuel cell vehicles.

Finding 3-4. The role and use of hydrogen in stationary applications, such as in large-scale electric power production, in distributed electric generation, or for industrial applications, could be significant—before fuel cell vehicles are commercially viable as well as in the long term. The Committee on Alternatives and Strategies for Future Hydrogen Production and Use did not analyze the opportunities and trade-offs for stationary applications, especially vis-à-vis the transportation sector. Furthemore, as far as the committee can discern, and from reviewing the Department of Energy's hydrogen research, development, and demonstration (RD&D) plan, the DOE has not developed a hydrogen RD&D strategy that systematically incorporates both the stationary and the transportation sectors, nor defined the various trade-offs and opportunities.

Recommendation 3-4. An independent, in-depth study, similar to the present study on the transportation sector, should be initiated to analyze the opportunities for hydrogen in stationary applications and to make recommendations related to a research, development, and demonstration strategy that incorporates considerations of both the transportation and the stationary sectors.

4

Transportation, Distribution, and Storage of Hydrogen

INTRODUCTION

In any future hydrogen-based economy, key economic determinants will be the cost and safety of the fuel distribution system from the site of manufacture of the hydrogen to the end user. This is true of any fuel, but hydrogen presents unique challenges because of its high diffusivity, its extremely low density as a gas and liquid, and its broad flammability range relative to hydrocarbons and low-molecular-weight alcohols. These unique properties present special cost and safety obstacles at every step of distribution, from manufacture to, ultimately, on-board vehicle storage. Also critical is the form of hydrogen being shipped and stored. Hydrogen can be transported as a pressurized gas or a cryogenic liquid; it can be combined in an absorbing metallic alloy matrix or adsorbed on or in a substrate or transported in a chemical precursor form such as lithium, sodium metal, or chemical hydrides. Carbon-bound forms of hydrogen such as today's gasoline, natural gas, methanol, ethanol, and others are not considered in this report, since their properties and use are well understood. However, comparisons with such conventional fuels will be made when necessary to help clarify the issues related to hydrogen.

Any analysis of hydrogen distribution, transportation, and storage must encompass both centralized manufacture at sites remote from the user points (these could include large central station plants or midsize plants for regional markets, cases that are considered in the cost analysis presented in Chapter 5) and distributed manufacture at the vehicle filling facilities. The centralized manufacture of molecular hydrogen requires a means of transportation and distribution as well as intermediate storage capabilities, while distributed manufacture will likely require only storage at the vehicle filling facility. The use of a chemical hydrogen carrier requires centralized manufacture of that material, shipment to the user site, and then disposal or recycling of the waste materials after the hydrogen is released on board the vehicle.

References to storage in the preceding comments relate only to storage in transit from the production site and at the vehicle filling facility. On-board vehicle storage is discussed separately because its requirements are potentially quite different, even though some of the same technologies, modified for vehicle use, may be employed—for example, high-pressure cylinders or liquid hydrogen containers. On-board reforming of fuels such as gasoline, methanol, or ethanol to produce molecular hydrogen is attractive in principle because it allows use of the existing fuel distribution infrastructure and consequently, if practical, could speed the widespread use of fuel cell vehicles without waiting for safe, cost-effective hydrogen storage technologies to be developed. A few companies are pursuing this technology, but significant technical barriers exist, such as size, weight, cost, and long start-up times.[1] (On-board reforming is discussed in Chapter 3.)

The kind of manufacture, transportation, and distribution infrastructure required to support a hydrogen-based fuel cell vehicle will be tied directly to the form of hydrogen used on board the vehicle. For example, on-board storage of molecular hydrogen allows a broad spectrum of raw material precursors to manufacture hydrogen. With a chemical carrier, however, molecular hydrogen may not be needed, and the manufacture, transportation, distribution, and storage systems would be quite different.

In the following sections, various scenarios describe the process of going from the manufacture of hydrogen or its carrier to the on-board storage systems of the vehicle. The major cost and technology barriers to making this process as safe and efficient as possible are presented. Comments are also made on the infrastructure scenarios—that of getting the hydrogen economy started (during the next 10 to 15 years), followed by the intermediate stage as significant numbers of fuel-cell-powered light-duty vehicles are produced (2020 to 2030), and finally, the steady-state scenario

[1]Bill Innes, ExxonMobil Research and Engineering Corporation, "Issues Confronting Future Hydrogen Production and Use for Transportation," presentation to the committee, June 12, 2003.

when such vehicles achieve major market penetration (2050). (See Chapter 6.)

MOLECULAR HYDROGEN AS FUEL

Molecular hydrogen is currently receiving the most attention and financial support as the starting point for fuel cell energy supply. The literature and the many presentations that the committee heard indicate that the manufacture of molecular hydrogen is the consensus approach favored by the majority of leadership within the government, at universities, and in industry. It is favored because it allows the use of a variety of hydrogen sources, ranging from coal and natural gas to biomass, solar, wind, and nuclear energy, as well as a multitude of relatively well understood manufacturing approaches ranging from small to large reformers, water-gas-shift reactors, electrolytic devices, thermal processes, and so on. (See Chapter 8 and Appendix G for a discussion of the various hydrogen production technologies.)

In the early stages of a transformation to a hydrogen economy, molecular hydrogen will probably be obtained from existing sources such as chemical plants and petroleum refineries. Today, about 9 million tons of hydrogen are manufactured annually in the United States[2] and transported for chemical and fuel manufacturing as a low- or high-pressure gas via pipelines and trucks or even as a cryogenic liquid (DOE, 2002a). Much experience worldwide has been achieved over many years to make these transportation modes safe and efficient. However, if the volume of hydrogen use grows, new safety and cost issues will surface, requiring major infrastructure changes. The committee found the analysis presented by Joan Ogden, among others, to be reasonable.[3] These analysts contend that in the very early stage of transition to the hydrogen economy, supplying of hydrogen for use in fuel-cell-powered vehicles would rely predominantly on over-the-road shipment of cryogenic liquid hydrogen or possibly hydrogen in high-pressure cylinders from existing chemical and petroleum refining plants.[4] Because of the high cost of such shipment modes, government subsidies would probably be needed to help fuel-cell-powered vehicles approach cost parity with gasoline-powered cars. It is also possible that pipelines could be used from existing manufacturing facilities, but this would only be possible where location dictated favorable economics as compared with costs for road shipment. The committee believes that as the volume of demand grows, however, this approach will evolve to the use of local distributed hydrogen production based on natural gas reformers and electrolytic units. These alternatives are less capital-intensive than that of building special pipelines coupled to large, dedicated hydrogen manufacturing plants, and are undoubtedly more economic than continued over-the-road shipping.

Whether molecular hydrogen is manufactured centrally or locally, a number of transportation, distribution, and storage requirements pose significant technical, cost, and safety problems. These various requirements could necessitate the use of interim storage facilities at plant sites for inventory or to compensate for demand swings and plant interruptions; the possible use of storage along pipelines and at distribution hubs; storage at the fuel cell vehicle loading stations; and, most critically, storage on board the vehicles themselves. For clarity, on-board vehicle storage is addressed separately from off-board storage, which is associated with distribution from the hydrogen-manufacturing site to the vehicle filling facilities.

The committee notes that resilience to terrorist attack has become a major performance criterion for any infrastructure system. In the case of hydrogen, neither the physical and operating characteristics of future infrastructure systems nor the timing of their construction can be understood in sufficient detail to permit an analysis of their vulnerability. However, the committee does observe that public concerns with terrorism seem likely to influence the choice of any future energy system and that resilience to deliberate attack is best designed in at the beginning.

Centralized Production of Molecular Hydrogen

Table 4-1 underscores key aspects of the costs of moving molecular hydrogen from its place of manufacture to the place where it is used as compared with the same types of costs for today's conventional fuels such as gasoline and natural gas. The table presents a series of cases that the committee developed for purposes of understanding costs and indicating where research or technology development might play a useful role in reducing them. The increased costs for transportation of molecular hydrogen versus those for conventional fuels are the direct result of the fundamental physical and thermodynamic properties of molecular hydrogen compared with today's liquid fuels.

Molecular hydrogen is a uniquely difficult commodity to ship on a wide scale, whether by pipeline, as a cryogenic liquid, or as pressurized gas in cylinders. On a weight basis, hydrogen has nearly three times the energy content of gasoline (120 megajoules per kilogram [MJ/kg] versus 44 MJ/kg), but on a volume basis the situation is reversed (3 megajoules per liter [MJ/L] at 5000 pounds per square inch [psi] or 8 MJ/L as a liquid versus 32 MJ/L for gasoline). Furthermore, the electric energy needed to compress hydrogen to 5000 psi is 4 to 8 percent of its energy content, depending on the starting pressure; to liquefy and store it is of the order of 30 to 40 percent of its energy content.[5] Pipe-

[2]Jim Hansel, Air Products and Chemicals, Inc., personal communication to Martin Offutt, National Research Council, October 3, 2003.

[3]Joan Ogden, Princeton University, "Design and Economics of Hydrogen Energy Systems," presentation to the committee, January 23, 2003.

[4]Joan Ogden, Princeton University, "Design and Economics of Hydrogen Energy Systems," presentation to the committee, January 23, 2003.

[5]Joan Ogden, Princeton University, "Design and Economics of Hydrogen Energy Systems," presentation to the committee, January 23, 2003.

TABLE 4-1 Estimated Cost of Elements for Transportation, Distribution, and Off-Board Storage of Hydrogen for Fuel Cell Vehicles—Present and Future

Case	Production Costs ($/kg)	Distribution Costs ($/kg)	Dispensing Costs ($/kg)	Total Dispensing and Distribution Costs ($/kg)	Total Costs ($/kg)	Total Energy Efficiency (%)
Centralized Production, Pipeline Distribution						
Natural gas reformer						
Today	1.03	0.42	0.54	0.96	1.99	72
Future	0.92	0.31	0.39	0.70	1.62	78
Natural gas + CO_2 capture						
Today	1.22	0.42	0.54	0.96	2.17	61
Future	1.02	0.31	0.39	0.70	1.72	68
Coal						
Today	0.96	0.42	0.54	0.96	1.91	57
Future	0.71	0.31	0.39	0.70	1.41	66
Coal + CO_2 capture						
Today	1.03	0.42	0.54	0.96	1.99	54
Future	0.77	0.31	0.39	0.70	1.45	61
Distributed Onsite Production						
Natural gas reformer						
Today					3.51	56
Future					2.33	65
Electrolysis						
Today					6.58	30
Future					3.93	35
Liquid H_2 Shipment						
Today		1.80	0.62	2.42		
Future		1.10	0.30	1.40		
Gasoline (for reference)	$0.93/gal refined			$0.19/gal	$1.12/gal	Well to tank: 79.5%

NOTES: The energy content of 1 kilogram of hydrogen (H_2) approximately equals the energy content of 1 gallon of gasoline. Details of the analysis of the committee's estimates in this table are presented in Chapter 5 and Appendix E of this report; see the discussion in this chapter.

line transmission of hydrogen is expected to be more capital-intensive than pipeline transmission of natural gas because of the need for pipes at least 50 percent greater in diameter to achieve the equivalent energy transmission rate, and because of the likelihood that more costly steel and valve metal seal connections will be required for pipelines for hydrogen in order to avoid long-term embrittlement and possibilities of leakage. As the shipments of hydrogen grow from today's low levels to the amounts required to support full-fledged fuel cell vehicle use, major transportation safety code revisions will undoubtedly be required (see Chapter 9).

Table 4-1 presents selected data from the committee's estimates for the costs to deliver hydrogen to fuel cell vehicles (see Chapter 5 and Appendix E). The table summarizes the committee's assessment of today's technology costs and possible future costs based on improvements through development and research for the following cases:

- *Centralized production*, followed by pipeline distribution and dispensing of gaseous molecular hydrogen. Natural gas and coal are the raw materials, and costs are given with and without CO_2 by-product capture and storage.[6]
- *Distributed onsite production* by natural gas reforming or electrolysis of water.
- *Over-the-road shipment costs* of cryogenic liquid hydrogen. This mode is expected to be used in the early stages of hydrogen supply to filling depots and stations.
- *Gasoline distribution and dispensing* via today's infrastructure is shown for reference.

[6]The cost of capturing CO_2 in a natural-gas-to-hydrogen plant is roughly three times that of a coal-gasification-to-hydrogen plant owing to greater added capital costs related to CO_2 capture in the natural gas plant (monoethanolamine [MEA] scrubber plus CO_2 compressor) versus that of the coal plant (compressor only). In addition, the natural gas reformer plant pays a greater efficiency penalty than does the coal plant (relative to the case in which CO_2 is vented), so its increase in variable costs (feed and fuel) is greater.

Obviously the future costs given in Table 4-1 are speculative and were based on the committee's consensus views of what might be possible. They are to some extent optimistic. Table 4-1 also includes a column on overall efficiency from raw material to final product at the pump, which is interesting for showing how difficult it is to approach today's gasoline refining and delivery efficiencies.

The complete cost data sets with assumptions for the cases in Table 4-1 are given in Appendix E. These cost estimates also include estimates of future improvements through technology refinements and basic research; these results are not listed in Table 4-1 because they do not change the overall conclusions with respect to where the critical areas for cost improvement lie for the distribution and dispensing of hydrogen in a future fuel cell economy.

According to the committee's analysis, the most efficient means of producing hydrogen in the long run is via large-scale, centralized plants that use pipeline distribution networks. Strikingly, while hydrogen can be produced today at costs ranging from $1.22 to $1.03/kg H_2 from natural gas, and from coal at $1.03 to $0.96/kg H_2 with and without carbon sequestration, respectively, pipeline shipment and dispensing adds an estimated cost of $0.96/kg H_2, which is essentially equal to the cost of production. Even with possible future improvements in shipping and distribution, this cost is much more than today's gasoline dispensing and distribution costs, at $0.19/gal. This analysis demonstrates the realities of shipping H_2 gas versus the much more efficient shipment of a liquid.

If and when extensive new hydrogen transmission pipelines are needed in the decades ahead, research in such areas as lower-cost pipeline materials, technology for dual-use natural gas-and-hydrogen pipeline connection techniques, layout optimization, and even pipeline emplacement technologies may be of significant value. However, the committee sees this as a priority research area only to the extent that such efforts directly benefit distributed production techniques, which are expected to dominate over the next 20 to 30 years.

The energy needed to pressurize hydrogen for pipeline transmission and for local storage at filling facilities where it is stepped up to vehicle on-board storage needs will be significant in terms of capital and electricity; this area may benefit from the development of new technologies. Those used today are mature and have not been improved significantly for many years. Here, too, the committee believes that this is not a near-term priority research area unless it is related to distributed hydrogen production systems, as mentioned above.

In the initial phases of hydrogen infrastructure development, the transportation of cryogenic liquid hydrogen via trucking or rail could play a significant role. Table 4-1 shows that over-the-road shipment of liquid hydrogen and dispensing at a vehicle filling site is estimated to add anywhere from $2.42 to $1.40/kg H_2 to the production costs. The process of liquefying molecular hydrogen consumes up to 40 percent of the energy content of the weight shipped and may represent an opportunity for technology development. If that could be reduced to a 20 percent loss through some sort of breakthrough, there could be an incremental decrease in cost relative to today's liquefaction costs, somewhere in the range of $0.20/kg.

Research to reduce the liquefaction costs for hydrogen could potentially benefit its cost of shipment by truck, ship, or rail, but could also be advantageous for storage at plant sites to guard against unplanned shutdowns. The committee views this research as more appropriate for nearer-term investment, since this mode of shipment could dominate in the early stages of fuel cell vehicle introduction.

In addition to the shipping considerations already discussed, the centralized manufacture of molecular hydrogen will require a series of storage facilities as it makes its way to the consumer. A large-volume, centrally placed manufacturing plant site will require storage for 1 to 5 days' supply of production to accommodate demand fluctuations and short-term outages. If hydrogen were stored as a pressurized gas, the most economical method at the manufacturing site would probably be underground caverns. A few such caverns have been used in Europe, although they depend for their utility on appropriate underground formations, such as depleted petroleum reserves or wet salt caverns (Ogden, 1999). Clearly, widespread use of such storage would engender much government regulation and careful permitting procedures that in the long run might render them uneconomic as compared with the more-capital-intensive insulated tanks that use liquefied hydrogen as the plant buffer.

Whether the hydrogen was stored as pressurized gas or liquid hydrogen, there would also be a need for local storage at the filling facilities and possibly secondary regional distribution sites. For local storage of liquid hydrogen, there would be the need for insulated tanks with tall evaporation dispersement stacks or other means to capture and reliquefy the vaporized hydrogen. For gaseous hydrogen, arrays of high-pressure cylinders probably would be needed. Shipment of compressed hydrogen gas also requires local step-up compressors to bring the pipeline-delivered pressures (100 to 200 psi) or the mobile truck cylinder pressure (2,500 psi) to the needed on-board vehicle pressures of 5,000 to 10,000 psi. The capital and energy-loss costs of all these steps present formidable obstacles to justifying hydrogen as an energy carrier when compared with today's liquid fuels.

Safety issues related to the placement of filling facilities near population centers are also of major concern. Measures to address safety should be a major part of near-term R&D expenditures (see Chapter 9).

At a briefing to the committee from representatives of DOE's Office of Energy Efficiency and Renewable Energy (EERE) on June 10, 2003, cost ranges were given for pipeline and liquid shipment of hydrogen that were somewhat higher than the results shown in Table 4-1. Comparison of the assumptions used for EERE's and the committee's calculations reveals that the difference lies principally in the length of the transmission pipes, their diameters, and their cost compared with natural gas pipelines. Additionally, EERE's calculations lumped costs for dispensing with those for transmission and did not include costs of buffer storage at the centralized production facility. Both groups' assumptions are reasonable, and both lead to the same conclusion for future research targets.

Distributed Manufacture of Molecular Hydrogen

In the intermediate stages of expansion of fuel cell vehicle use (in the 2010–2020 time frame), local distributed generation with small-scale natural gas reformers or by electrolysis of water will probably make the most economic sense before large, central, dedicated plants with pipeline distribution can be justified economically. The delay of large capital investments for centralized H_2 production through distributed manufacture is a significant advantage when fuel cell vehicle density is low, but there are drawbacks in terms of the higher costs associated with current distributed H_2 generation technology as well as in the inability to capture CO_2 emissions in the case of local reformers. There will undoubtedly also be many new safety and code issues related to the manufacture of hydrogen adjacent to or in urban areas.

In the case of local manufacture, however, there appears to be opportunity for important technological improvements in costs and efficiencies for distributed reformers and electrolytic hydrogen generators. Over the next 5 years, improved small reformers with lower operating costs, higher energy efficiency, and lower investment deserve priority (see Chapter 8). If economic means of capturing CO_2 on a small scale could also be found, this capability might be a strong incentive for local manufacture in the long run. The committee believes that reformer research aimed at the distributed market should be emphasized now in order to provide hydrogen manufacturing options in the 2010–2030 period. Exploratory research to improve electrolyzer efficiency should also be supported. If it were possible to develop electrolyzers that could lower the cost of local ancillary equipment, such as compressors, or reduce the need for components of storage facilities and improve safety, such advances could significantly benefit the intermediate stages of a hydrogen economy. The committee believes that distributed manufacturing technologies deserve significantly increased research investment over the next 10 to 15 years (see Chapter 9).

Solid-State Transport of Hydrogen and Off-Board Hydrogen Storage

Means other than pressurized gas or cryogenic liquid theoretically exist for useful transportation and storage of molecular hydrogen. They principally include pressurized absorption in metallic alloys and on or in carbon or other substrates. There are many possibilities, perhaps hundreds (see Thomas [2003] and DOE [2003e] for excellent assessments of the many possibilities under study or suggested as areas for future work). None of these technologies are serious contenders for shipment from centralized manufacturing sites because they are inefficient on a weight and/or volume basis in comparison with cryogenic liquid hydrogen and pipeline-transmitted hydrogen. However, they are still in contention for possible local storage or on-board vehicle storage. Some of the technologies in this category have been used in demonstration projects, but none have come close to being practical for light-duty vehicles. Problem areas include the overall weight of the storage alloys, the limited capacity of the alloys and carbon materials, the difficulties in liberating hydrogen from the carriers, and the high overall system costs. Nevertheless, the committee believes that absorption, adsorption, and related dense-phase hydrogen carrier technologies are a fruitful area for sustained exploratory research primarily because of their promise of safety for off-board and on-board vehicle applications.

Almost as important as the need to study this area is the need to narrow the field of technology options as quickly as possible rather than spreading a limited development budget too thinly. The committee makes this point based on the observation of the great number of proposed concepts vying for support. The committee is pleased that the requested DOE budgets in these areas have been increased substantially over the next several years (DOE, 2003a), but it is concerned that continuing existing programs on pressurized tanks and liquid hydrogen approaches may limit more exploratory areas (described above and in the next subsection).

On-Board Storage of Molecular Hydrogen

Viable options to provide acceptable and adequate on-board vehicle storage of molecular hydrogen for at least a 300-mile driving range follow directly from the preceding discussion. These options include, for example, containment in high-pressure cylinders, in cryogenic dewars with controlled bleed-off and the ability to accommodate significant pressure buildup to slow losses, and in metal alloy matrices or some type of solid absorbent or adsorbent.

In the case of 5,000 to 10,000 psi cylinders, the principal issues are concern for public acceptance of their safety, the cost to manufacture such containers (which today are made as multishelled structures that use fiber-wound composite technologies), the time and complexity of the filling operations, and the space that such tanks with the needed capacity would occupy on board the vehicle (see Table 4-2). For example, for more than a 200-mile driving range, today's natural gas vehicles usually require two tanks, which use up much of the trunk. A hydrogen-fueled vehicle with 5,000 psi tanks would probably require two tanks, or, if the tank was 10,000 psi, a small vehicle might need one tank. Several companies are trying to develop these tanks, but none has achieved the required performance. Table 4-2 summarizes the minimum performance needs for hydrogen on-board storage as expressed by representatives of the automotive industry (DOE, 2002b). Table 4-2 also includes the targets established by the DOE with the FreedomCAR Hydrogen Storage Technical Team (DOE, 2003b).

Cryogenic pressurized storage technologies are less developed than high-pressure gas storage cylinders are, but have been used in some demonstration vehicles. The use of liquid hydrogen as fuel on board a light-duty passenger vehicle seems unlikely to meet the capacity and size requirements acceptable to the automotive industry. In addition, further obstacles to this approach include the high energy requirements for liquefying molecular hydrogen, safety concerns related to continuous hydrogen boil-off, and the escalating number of delivery trucks that would be on the road to meet demand in the middle years of scale-up.

If molecular hydrogen is to be used on board small personal vehicles, it seems most likely that some sort of reversible solid system must be developed. Currently, many concepts are under study for this type of system. These include a wide variety of metal alloys that form reversible hydrides, hydrogen adsorbers based on various forms of carbon and other high-surface-area materials, high-energy chemical compounds such as sodium borohydride that react with water or even alcohols, and a whole series of early concept ideas that aim to store and then liberate hydrogen when it is heated or reacted (Thomas, 2003). None of the concepts under study has achieved the minimum objectives set by industry (see Table 4-2). Even if the capacity and percent-by-weight goals can be demonstrated, there are major issues around costs of the carrier materials, filling times, and heat management during filling and hydrogen liberation to meet the fluctuations in electrical demand associated with normal driving.

Heat management during hydrogen uptake (fueling) and hydrogen desorption during vehicle operation need further study. If the heat of desorption per mole of molecular hydrogen is large, two important implications follow. First, a large surface area for the heat exchangers is required, and it will add weight and volume; if waste heat is not available at the needed temperature and rate, a significant fraction of the fuel energy will be wasted. This also means that the fuel cell must operate at a higher temperature than the desorption temperature for hydrogen. Current proton exchange membrane fuel cells (PEMFCs) operate at approximately 80°C; consequently, the desorption temperature must be substantially lower. This relationship suggests that important research is needed either to raise the fuel cell operation temperature or to lower the H_2 desorption temperature. New materials concepts have an important role to play in finding a solution for the hydrogen release problem. Heat management during uptake and release is a critical area requiring attention. Device designs that can load vehicles in an acceptable time with fail-safe safety controls and then release hydrogen at the rates demanded are vital to the success of this approach. The committee views these areas, although still in their infancy, as very important.

In summary, the committee questions the use of high-pressure tanks aboard mass-marketed private passenger vehicles from cost, safety, and convenience perspectives. The committee is also concerned about the complexity and capital intensity of the filling station equipment. The committee has a similar view of the use of liquid hydrogen. Exploratory budgets for the development of dense-phase materials as hydrogen carriers are being expanded, as mentioned above, but goals for this research need to be sharpened toward the objective of focusing on a few options that have real promise, and then on accelerated early-stage development. Without such a commitment to show encouraging progress in this critical area, private sector enthusiasm toward the development of fuel-cell-powered light-duty vehicles could wane substantially.

Alternatives to Molecular Hydrogen Transportation, Distribution, and Storage

The preceding discussion is based on the assumption that the cost and safety problems associated with transportation, distribution, and on- and off-vehicle storage can be satisfactorily solved with molecular hydrogen at every stage of its scale of use, and that there is no better approach available. However, the committee was presented with several intriguing "game-changing" possibilities (JoAnn Milliken, Department of Energy, "Hydrogen Storage," presentation to the

TABLE 4-2 Goals for Hydrogen On-Board Storage to Achieve Minimum Practical Vehicle Driving Ranges

Energy Density	General Motors Minimum Goals	Compressed/Liquid Hydrogen (Currently)	DOE Goal
Megajoules per kilogram	6	4/10	10.8
Megajoules per liter	6	3/4	9.72

NOTES: Energy densities are based on total storage system volume or mass. Energy densities for compressed hydrogen are at pressures of 10,000 psi.
SOURCES: DOE (2002b, 2003b).

committee, December 2, 2002; Thomas, 2003) that it believes should be vigorously examined for their potential. Here again, narrowing the field as quickly as possible to focus on those few prospects with the most potential is a vital component of any research investment strategy.

All alternatives to molecular hydrogen relate to the manufacture of energetic metals or their hydrides, which, when reacted with water, emit hydrogen (Thomas, 2003). These materials would be shipped from centralized manufacturing sites by conventional truck, rail, or ship and distributed to consumer fuel cell vehicle filling facilities. Vehicles would be equipped with devices for reacting the compounds with water in order to generate fuel-cell-quality hydrogen and for storing the waste reactants. Waste would then need to be recycled or disposed of in an environmentally acceptable manner.

The principal game-changing features of these materials are the elimination of most safety and cost issues that high-pressure or cryogenically liquefied molecular hydrogen has, and the possibility of a major safety and range enhancement for on-board storage of hydrogen. Several small-vehicle demonstrations of the efficacy of this approach and its ability to provide acceptable driving range, hydrogen purity, and delivery rate and vehicle space efficiency have been successfully made (Bak, 2003). The use of 20 to 30 percent by weight of alkali-stabilized aqueous solution of sodium borohydride as fuel, which is pumped over a catalyst to generate hydrogen instantaneously, was demonstrated recently by Daimler-Chrysler in its Chrysler Town and Country Natrium fuel cell minivan vehicle.[7] This approach demonstrated the potential for meeting vehicle mileage, weight, and volume goals.[8]

The principal current shortcomings of these chemical methods for generating hydrogen are the high cost of manufacture of the chemicals and the not-yet-demonstrated technology for recycling or disposing of waste products effectively. Secondary issues include catalyst longevity over the vehicle life, fuel stability on board the vehicle, and the ability to meet automotive range and reliability requirements. However, all of these shortcomings, with the exception of the cost of recycling and initial manufacture, have had encouraging real-world demonstrations in full-sized passenger vehicles, as for example with the Natrium fuel cell vehicle.

The committee believes that this is an important area for further research and that it should be pursued vigorously to find the best chemicals for this use and to improve the economics of their manufacture and regeneration. The DOE should also continue to encourage other game-changing concepts because of the pivotal importance of this need to the future of fuel-cell-powered vehicles.

[7]The spent fuel cartridges would be regenerated at a central location.

[8]Additional information is available online at www.h2cars.biz/artman/publish/article_144.shtml. Accessed December 4, 2003.

THE DEPARTMENT OF ENERGY'S HYDROGEN RESEARCH, DEVELOPMENT, AND DEMONSTRATION PLAN

The committee was pleased to be given an early draft of the DOE Office of Energy Efficiency and Renewable Energy's "Hydrogen, Fuel Cells and Infrastructure Technologies Program: Multi-Year Research, Development and Demonstration Plan" (dated June 3, 2003) (DOE, 2003b). The following are the committee's comments on this document regarding the areas of off-board storage, transportation, and distribution of hydrogen (see DOE [2003b], pp. 3-30 through 3-55).

Fundamentally, the committee agrees with the DOE's assessment of the research needs in these important areas, especially those relative to pipeline costs and the need to improve the energetics of hydrogen compression and liquefaction. The committee differs with the DOE on near-term priorities. The committee believes that the requested increased funding in these areas should be prioritized to strongly favor solid or dense-phase storage of hydrogen, especially for on-board vehicle use, since on-board storage appears to be one of the primary obstacles to fuel cell vehicle practicality, along with the needed fuel cell cost reduction and reliability improvements.

FINDINGS AND RECOMMENDATIONS

The following findings and recommendations are based on the idea that some research and technology investments are at present more important than others in criticality and in time. This prioritization reflects the need to invest in overcoming the technology gaps that might be major stumbling blocks to immediate progress and to delay or reduce investment in those activities that, while very important, can wait for several years because they are not critical to near-term progress.

Finding 4-1. It seems likely that in the relatively near term (the next 10 to 30 years), distributed rather than centralized production of hydrogen will be a driver for the continued expansion of fuel-cell-powered private vehicles. Needs in the very early period are expected to be covered by shipment of pressurized or liquefied molecular hydrogen, but as volume requirements grow, such an approach may be deemed too expensive and/or too hazardous for continued widespread use. Distributed manufacture of molecular hydrogen seems most likely to be best done with small-scale natural gas reformers or by electrolysis of water. At present both technologies are capital-intensive and relatively energy-inefficient. Without such distributed manufacture, it seems likely that the very large centralized production and pipeline distribution investments will be difficult to justify and could slow conversion to hydrogen markedly. It seems possible that, in comparison with today's state-of-the-art technology, the new technology for distributed manufacture may reduce production costs through efficiency improvements and possibly by enabling reduced capital requirements for ancillary storage and filling equipment.

Recommendation 4-1. Increased research and development investment in support of breakthrough approaches should be made in small-scale reformer and electrolyzer development with the aim of increasing efficiency and reducing capital costs. A related goal should be to increase the safety and reduce the capital intensity of local hydrogen storage and delivery systems, perhaps by incorporating part or all of these capabilities in the hydrogen-generating technologies.

Finding 4-2. It is clear that the vast majority of current private and governmental investments in the manufacture of hydrogen for fuel cell vehicles are aimed at the direct use of molecular hydrogen. Because of the inherent difficulties in the transportation, distribution, and storage of molecular hydrogen, it is apparent that other approaches for hydrogen generation may have advantages for transportation and for on- and off-board storage. The latter include compounds that, on reaction with water or some other reactant, generate hydrogen, and solid-state carriers that contain high concentrations of adsorbed or absorbed hydrogen that liberate the stored hydrogen through the application of heat. Many possibilities exist in these categories, but few have received significant research support. Solid-state hydrogen carriers will probably not be useful for the transportation and distribution of hydrogen, but may be valuable for local and/or on-board vehicle storage. The committee strongly supports the requested Department of Energy budget increases in the vital area of hydrogen storage. The committee believes, however, that major shifts in emphasis should be made immediately in order to make sure that the many new ideas currently available are properly examined—because without relatively near-term confidence by industry and government leaders, interest in continuing the pursuit of fuel cell vehicle transportation uses is likely to wane over time.

Recommendation 4-2. The Department of Energy should halt efforts on high-pressure tanks and cryogenic liquid storage for use on board the vehicle. These technologies are in a pre-commercial development phase, and in the committee's view they have little promise of long-term practicality for light-duty vehicles. The DOE should apply most if not all of its budgets to the new areas described in Finding 4-2 with the objective of identifying as quickly as possible a relatively few, promising technologies. Where relevant, efficient waste-recycling studies for the chemically bound approaches should be part of these studies. Even during this winnowing process the DOE should continue to elicit new concepts and ideas, because success in overcoming the major stumbling block of on-board storage is critical for the future of transportation use of fuel cells.

Finding 4-3. The evolution of the transportation and delivery and storage systems for hydrogen will transition several times as hydrogen demand increases over many decades. This would of necessity mean continuous and overlapping shifts from small-scale delivery and storage, to distributed manufacture and storage, to centralized production with extensive pipeline, distribution, and storage networks. Such a complex evolution would likely benefit from systems analysis to help guide the optimum research and technology investment strategies for any given stage of the evolution and thus enable the most effective progress toward the long-term end states.

Recommendation 4-3. Systems modeling for the hydrogen supply evolution should be started immediately, with the objective of helping guide research investments and priorities for the transportation, distribution, and storage of molecular hydrogen. In addition, parallel analysis of the many alternatives for other means of supplying hydrogen to fuel-cell-powered facilities and vehicles should be performed; such analysis is needed to prevent wasteful expenditures and to help focus attention on viable technology that would potentially compete with the direct supply and delivery of molecular hydrogen and that might be useful for all or portions of the future hydrogen economy.

Finding 4-4. Hydrogen is particularly difficult to ship from a manufacturing site to filling facilities for vehicle servicing. In fact, the cost to ship and store can easily equal the costs of production. These costs are directly related to molecular hydrogen's thermodynamic properties, low molecular weight, and consequently high diffusion capabilities, and to its great flammability and ability to form explosive mixtures over a wide range of concentrations. Particular concerns relate to the energy losses during compression and liquefaction and to the tendency of hydrogen to embrittle some current pipeline materials.

Recommendation 4-4. Research and technology development should be carried out in support of novel concepts that promise major improvements in the cost and efficiency of compressors for molecular hydrogen and reductions in the cost of pipeline materials, valves, and other leak-prone components of its distribution system. Initial research should focus on those components that are directly related to distributed hydrogen production. In later years, research should shift to components for large, centralized production plants with extensive pipeline and storage facilities. The committee believes that current Department of Energy plans call for research that relates primarily to centralized molecular hydrogen manufacture—a need that is many decades in the future—and consequently may shortchange other, more immediate needs.

5

Supply Chains for Hydrogen and Estimated Costs of Hydrogen Supply

The supply chain for hydrogen comprises the processes necessary to produce, distribute, and dispense the hydrogen. Currently, most hydrogen is produced from natural gas close to where it is needed for industrial purposes. A variety of potential hydrogen supply chain pathways are considered in this chapter. The major factors that will affect the cost of delivered hydrogen are these:

- The feedstock and/or the major energy source from which the hydrogen is produced,
- The size of the facility at which the hydrogen is produced and the transportation requirements to deliver it to the customer,
- The state of the technology used—whether current or to be improved by future developments, and
- Whether or not the carbon dioxide (CO_2) by-product is sequestered when hydrogen is produced from fossil fuel.

This chapter presents estimates of the costs of hydrogen, measured in terms of dollars per kilogram of hydrogen, for the most likely supply chain pathways.

HYDROGEN PRODUCTION PATHWAYS

Feedstocks and Energy Sources

Hydrogen must be chemically separated from some other material. Currently, natural gas is the most common feedstock, but coal is also used. Biomass could be used in the future. The full costs of the production, processing, and purification of these hydrocarbon feedstocks are included in this analysis. When these materials are used to produce hydrogen, the required energy is embedded in the feedstock. (See Chapter 8 and Appendix E for more details.)

Hydrogen also can be separated from water via electrolysis or high-temperature chemical reactions. Electricity can be taken from the grid (from a variety of sources) or generated by wind turbines or photovoltaics that feed the hydrogen production facility directly.[1] Nuclear energy might be used in high-temperature chemical reactions.[2]

Scale of Production

The estimates presented here are developed at three different scales of hydrogen generation, referred to as central station (CS), midsize (MS), and distributed (Dist). Central station plants are assumed to have a production capacity of 1,200,000 kilograms per day (kg/d) and to operate with a 90 percent or higher capacity factor, therefore producing on average 1,080,000 kg/d H_2 supporting nearly 2 million cars. Midsize plants are assumed to have a production capacity[3] of 24,000 kg/d; operating with a 90 percent capacity factor, they produce on average 21,600 kg/d H_2 and support about 40,000 cars. The distributed plants have different production capacities corresponding to the differing capacity factors. Those that operate with a 90 percent capacity factor are assumed to have a production capacity of 480 kg/d H_2, producing on average 432 kg/d. Those operating with lower capacity factors are assumed to have the larger production capacities, so that each distributed unit produces on average 432 kg/d H_2, supporting about 800 cars.

For each feedstock (or energy source), the committee selected the scales of generation that could be appropriate, given its analysis of the nature of the technology and its cost estimations. Table 5-1 shows the combinations examined in

[1]The committee did not consider hydroelectric power explicitly except as part of the electricity grid mix. The remaining renewable energy resources—except wind, solar, and biomass—were not considered owing to their current small fraction of total primary energy supply or small projected growth.

[2]Nuclear fission energy was considered by the committee, but not nuclear fusion, since the DOE projects commercialization of fusion in about 2050 (DOE, 2003g), which is beyond the time frame considered in this analysis.

[3]These production capacities correspond to 497,400,000 standard cubic feet per day (scf/d) for the central station plants and 9,948,000 scf/d for the midsize plants.

TABLE 5-1 Combinations of Feedstock or Energy Source and Scale of Hydrogen Production Examined in the Committee's Analysis

	Feedstock or Primary Energy Source						
Scale of Production	Natural Gas	Coal	Nuclear Energy	Biomass	Photo-voltaics	Wind	Grid-Based Electric Energy (from any source)
Central station plant	Steam reforming	Gasifier	Thermal splitting of water				
Midsize plant	Steam reforming			Gasifier or direct conversion			
Distributed	Steam reforming				Electrolysis	Electrolysis	Electrolysis

developing this analysis.[4] Appendix E contains the data for each technology case analyzed.[5]

The costs and energy requirements for distributing the hydrogen to the "filling" station and then dispensing it into the vehicle can be a significant fraction of the total. For central station plants, it is assumed that the distribution system uses pipelines. For midsize plants, it is assumed that distribution would be by cryogenic truck, because the low volumes of hydrogen involved would not justify a pipeline system. Distributed technologies generate hydrogen at the filling station itself and do not require a distribution system.

State of Technology Development

Almost all[6] of the cost estimates are developed for two different states of technology development. One state, referred to as *current,* is based on technologies that could in principle be implemented in the near future. No fundamental technological breakthroughs would be needed to achieve the performance or cost estimates, although normal processes of design, engineering, construction, and system optimization might be needed to achieve costs as low as those estimated in this analysis.

The second state, referred to as *possible future,* is based on technological improvements that may be achieved if the appropriate research and development (R&D) are successful. These improvements are not predicted to occur; rather, they may result from successful R&D programs. Some may require significant technological breakthroughs. The nature of the improvements in each particular technology is discussed in Chapter 8; additional detail is provided in Appendix G. Generally these future technologies are assumed to be available at a significantly lower cost than that of the current technologies using the same feedstock.

Carbon Dioxide Sequestration

Some of the technologies in the analysis are further differentiated by whether carbon dioxide resulting from hydrogen generation is separated and sequestered. In particular, the midsize and the central station production facilities at which production is based on natural gas, coal, or biomass are examined both with and without the sequestration of carbon dioxide.

Summary of Technologies Considered

The hydrogen supply chain pathways that are considered in this chapter are identified in Table 5-2. They do not include all combinations of the factors listed above (e.g., coal as a feedstock in a distributed plant, or sequestration in a photovoltaic-driven electrolyzer plant). Intermittent technologies (wind, photovoltaics) can be used independently or in combination with the electric grid in order to allow hydrogen production when the renewable technology is not producing power. The results presented here are for the latter case, representing the average output of these intermittent technologies, as discussed later in this chapter. The cases for 100 percent renewables are presented in Appendix E. An all-grid-based system is included here.

CONSIDERATION OF HYDROGEN PROGRAM GOALS

Although the unit cost of producing and delivering hydrogen from the various technologies is critically important

[4]In the graphs in this chapter (Figures 5-1 through 5-13), all of the combinations listed in Table 5-1 are included except midsize generation of hydrogen based on natural gas. The analysis suggests that this alternative would be dominated by either distributed or central station use of natural gas, and thus those estimates are not reported.

[5]Solar-photovoltaic (PV) and wind technologies were examined by the committee only at distributed scale. These technologies do not benefit from scale economies to the same extent as do single-train processes, such as gasification (of biomass or coal) and steam methane reforming. For example, in the case of solar-PV, twice the structural supports will be required for a solar field of twice the generating capacity (watts)—a linear scaling. Wind farms require multiple turbines to reach capacities above a few megawatts.

[6]Evaluation of a current nuclear thermal reforming of water is not included because no such technology exists at the present time.

TABLE 5-2 Hydrogen Supply Chain Pathways Examined

	Primary Energy Source						
Scale	Natural Gas	Coal	Nuclear	Biomass	Grid-Based Electricity (from any source)	Wind	Photovoltaics
Central station	*CN*: CS NG-C *FN*: CS NG-F *CY*: CS NG-C-Seq *FY*: CS NG-F-Seq	*CN*: CS Coal-C *FN*: CS Coal-F *CY*: CS Coal-C-Seq *FY*: CS Coal-F-Seq	*F*: CS Nu-F				
Midsize				*CN*: MS Bio-C *FN*: MS Bio-F *CY*: MS Bio-C-Seq *FY*: MS Bio-F-Seq			
Distributed	*C*: Dist NG-C *F*: Dist NG-F				*C*: Dist Elec-C *F*: Dist Elec-F	*C*: Dist WT-Gr Elec-C *F*: Dist WT-Gr Elec-F	*C*: Dist PV-Gr Elec-C *F*: Dist PV-Gr-Elec-F

NOTES: C = current technology; Y = sequestration (hydrocarbon feedstock in central station and midsize plants only); F = future technology; N = no sequestration. The abbreviations for the hydrogen supply chain pathways (e.g., CS NG-C) are from the detailed spreadsheets in Appendix E of this report.

in determining their likely competitive success, other characteristics are important as well. One of the important goals of the nation's hydrogen program is to reduce emissions of CO_2 into the atmosphere. Therefore, it is important to estimate whether shifts from gasoline-fueled automobiles to hydrogen-fueled vehicles or other substitutions from direct use of fossil fuels to hydrogen would reduce CO_2 emissions and, if so, by how much. For each of the technological pathways considered, estimates were developed regarding the amount of CO_2 that would be released into the atmosphere per kilogram of hydrogen produced. As a point of comparison, estimates were made of the CO_2 that would be released into the atmosphere per gallon of gasoline use.

Since a goal of the committee's analysis is to compare costs and CO_2 release from gasoline with those from hydrogen, it was important to adjust the gasoline costs and CO_2 releases to account for engine efficiency differences between gasoline-powered and fuel cell vehicles (FCVs). For gasoline-powered vehicles, the committee chose a gasoline hybrid electric vehicle (GHEV).

A second important goal of the hydrogen program is to improve energy security by substituting secure domestic resources for imported energy resources, particularly those that may be traded in unstable international markets. In motor vehicles, the use of hydrogen reduces the use of gasoline and therefore could reduce the imports of crude oil or petroleum products. However, if natural gas is the feedstock used to produce hydrogen, this substitution will increase the importation of natural gas, a commodity that may be subject to international market instability just as in the petroleum markets. On the other hand, if coal, biomass, wind, or solar energy are used to produce hydrogen, energy security could be improved. The committee developed estimates of the amount of natural gas that would be needed for technologies using natural gas to produce hydrogen; those data are presented in Chapter 6.

COST ESTIMATION METHODS

For each hydrogen production pathway and for both states of technology development (current and possible future), the committee developed engineering–economic models to estimate the primary inputs of feedstocks, of electricity or other energy, and of capital equipment for each standard-sized plant and to estimate the resulting outputs of H_2 and CO_2. Within the models, a distinction is made between pathways in which the CO_2 is sequestered and those in which it is released back into the atmosphere. Additional costs of CO_2 separation, capture, compression, transport, and sequestration are included for processes in which most of the CO_2 is sequestered.

Prices of feedstocks and electricity, costs of major pieces of capital equipment, operation and maintenance (O&M) costs, and rates of return on investment are used to translate physical measures of inputs to total costs of operating the plant annually. The total annual cost and the total annual average hydrogen output together give the cost per kilogram of hydrogen produced.

The original engineering–economic models were developed for the committee by SFA Pacific (an engineering and economic consulting firm located in Mountain View, California), working closely with a member of this committee. Committee members extensively reviewed all of the original models and subsequently modified or replaced many of them. Most of the models of current technologies using fossil fuels still correspond closely to the original models, although the committee made some changes in these models. The models of possible future technologies were modified greatly to correspond with the best judgments of the committee members about technological possibilities and the economic parameters. The final models used to analyze renewable technologies for hydrogen production were based almost entirely on analysis by committee members. Thus, the final versions of the models and the resulting cost estimations reflect the overall judgment of the committee.

Committee judgments, and thus the final parameters in the models, are based on a combination of information derived from many presentations by experts and industry representatives, SFA Pacific data, the expertise and experience of committee members, and committee follow-up on specific issues with outside experts. Many components of the cost estimates rely heavily on technical and economic judgments by members of the committee and on the information gathered during the course of the study. Thus, ultimately, the quantification represents collective judgments of the committee members. *As such, the estimates, although they may look precise, are simply estimates.*

There remains significant uncertainty about what the actual costs of the technologies would be under current conditions. Costs are site-specific, particularly for wind and solar-based technologies; only single representative costs are reported. And the uncertainty about possible future technologies is substantially greater. In addition, because these cost estimates are so heavily dependent on the judgment of committee members, other people may well make very different technical and economic judgments, particularly about the possible future technologies. Therefore, costs could be either higher or lower than the committee's estimates.

The committee's analysis generally is based on the assumption that critical technology development programs will be successful. The committee needed estimates of what might possibly be achieved with concerted research and development in order to determine the impact on petroleum consumption and CO_2 emissions of an optimistic but plausible future. The committee is not predicting that the requisite research and development will be pursued, or that all of these technical advances necessarily will be achieved, even with a concerted R&D program. The committee simply needed a framework for its further analysis. If the research goals are not met, there will be less (or even no) hydrogen in the nation's energy system.

The committee chose not to provide sensitivity tests for the various parameters. A complete range of sensitivity tests would increase the report to unmanageable proportions and would still depend on the technological judgments of the committee members. However, the committee is making the spreadsheets containing the underlying data (see Appendix E) publicly available. These spreadsheets can be used by interested parties to conduct complete sensitivity tests based on their own technical and economic judgments.

In addition, the committee qualitatively estimated the sensitivity of supply chain costs to various parameters of the model. Table 5-3 includes these estimates, labeling the sensitivity as "low," "medium," or "high." A blank cell in a column means that there is very low or no sensitivity to the particular parameter. More details about the cost estimations appear in Appendix E. The technologies are described in Chapter 8 and Appendix G.

In the following section, graphical estimates are presented to show the costs per kilogram of hydrogen production for many of the technologies examined in the study. Comparable information covering all of the technologies in the analysis appears in Appendix E.

UNIT COST ESTIMATES: CURRENT AND POSSIBLE FUTURE TECHNOLOGIES

Current Technologies

Figure 5-1 presents the unit cost estimates—the cost per kilogram of hydrogen—for the current technologies state of development for 10 of the current hydrogen supply pathways included in Table 5-2. (The possible future case is discussed later in this chapter.) For each pathway, the cost includes production, distribution (for CS and MS plants), and dispensing costs. State and federal fuel taxes—commonly called gasoline taxes—are not included. For central station production (4 pathways) and midsize production (2 pathways), the cost is separated into five components: (1) production cost (cost including production and storage onsite), (2) distribution cost (cost of transporting hydrogen by pipeline or cryogenic truck to the filling station), (3) dispensing cost (cost of compressing and storing hydrogen at the filling station and cost of dispensing hydrogen into vehicles), (4) CO_2 disposal cost (cost of transporting and sequestering CO_2 for technologies involving CO_2 sequestration), and (5) an imputed cost[7] for CO_2 released into the atmosphere[8] (with the imputed cost of $50 per metric ton [tonne] of carbon). For distributed production (4 pathways), there are no pipeline or cryogenic truck costs; compression and storage cannot be separated between production and dispensing. Thus, for distributed production, the first three cost components are combined into the total distributed cost; the imputed cost for CO_2 released into the atmosphere is shown separately. There is no CO_2 disposal cost included for distributed technologies; it is assumed that all of the CO_2 is vented to the atmosphere.

In order to facilitate the comparison of total supply chain hydrogen costs with costs of gasoline, the gasoline efficiency adjusted (GEA) cost for a GHEV is introduced as a separate bar in Figure 5-1. The GEA allows head-to-head comparison of the total supply chain cost of amounts of gasoline and hydrogen that provide equal vehicle miles traveled (VMT) when consumed in a GHEV or an FCV, respectively. Thus, included within the GEA calculation is an adjustment for the efficiency of the respective vehicle. The estimate of GEA cost in Figure 5-1 is based on an assumed crude oil price of $30 per barrel (bbl) and a 66 percent efficiency gain of the FCV over the GHEV (see Chapter 6, especially Figure 6-2, for a detailed explanation). The gasoline cost is estimated as $1.27 per gallon (gal) of gasoline. Thus, the GEA cost of gasoline is $2.12 per kilogram of hydrogen (calculated as $1.27 × 1.66).[9]

Figure 5-1 shows that the cost per kilogram of hydrogen for the four central station technologies is similar to the GEA cost of gasoline in hybrid electric vehicles, once these plants are operating at full capacity. This suggests that with current technologies, if the cost and functionality of a hydrogen-fueled vehicle could be made similar to the cost and functionality of a hybrid electric vehicle, then hydrogen generated at central stations, using natural gas or coal as feedstocks, could be roughly comparable in overall cost to gasoline used in hybrid electric vehicles, once plants were

[7]In Figure 5-1, there is a negative imputed cost of carbon for the generation of hydrogen from biomass with sequestration. That occurs because growing the feedstock takes CO_2 from the atmosphere, CO_2 that is ultimately sequestered. In the graph, that negative imputed cost appears as the part of the bar below the y = $0 line. The total cost would be reduced by this amount.

[8]Often this imputed cost is referred to as a carbon tax. However, the committee chose to use other terminology because it does not make a prediction as to whether the United States will legislate a carbon tax, issue tradable permits for CO_2 releases, or not implement any such carbon controls. However, the cost to the environment per ton of carbon released is not dependent on whether such instruments are adopted. Thus the somewhat clumsy phrasing "imputed cost is used for carbon dioxide released into the atmosphere," or the shortened version, "imputed cost of carbon dioxide," is neutral on the particular instruments that might be adopted. The committee uses a $50 per metric ton cost of carbon. If the United States does not impose carbon restrictions, that cost will not be incorporated into the prices facing the producers of hydrogen. Likewise, if global climate changes turn out to be more severe than posited in some analyses, that cost may be an underestimate.

[9]In calculating the GEA cost, the cost of hydrogen included production, distribution, dispensing costs, and the imputed cost of carbon released into the atmosphere. The estimate of gasoline price excludes state and federal gasoline taxes. Similarly, the various components of gasoline cost—production, distribution, dispensing, and imputed carbon cost—are scaled in the same manner to calculate hydrogen-equivalent costs.

TABLE 5-3 Sensitivity of Results of Cost Analysis for Hydrogen Production Pathways to Various Parameter Values

Parameter	CS NG	CS NG Seq	CS Coal	CS Coal Seq	CS Nu	MS Bio	MS Bio Seq	Dist NG	Dist Elec	Dist WT-Gr Ele	Dist PV-Gr Ele	WT Ele	PV Ele
Capital investment	Low	Low	Low	Low	High	High	High	High	High	High	High	High	High
Electrolyzer cost										High	High	High	High
Electricity price	Low	Low	Low	Low		Low	Low	Low	High	High	High	Low	Low
Natural gas price	Medium	Medium						Medium					
Costs of photovoltaics/wind										Medium	Medium	High	High
Nonfuel O&M costs	Low	Low	Low	Low	Low	Low	Low	Low	Low	Low	Low	Low	Low
Distribution costs	Medium	Medium	Medium	Medium	Medium	High	High						
Dispensing costs	Medium	Medium	Medium	Medium	Medium	Medium	Medium	Medium	Medium	Medium	Medium	Medium	Medium
Sequestration unit cost	Low	Low	Low	Low		Low	Low	Low	Low	Low	Low	Low	Low
CO_2 imputed cost	Low		Low			Low		Low	Low	Low	Low		

NOTES: O&M = operation and maintenance. A blank cell means that there is very low or no sensitivity to a parameter.

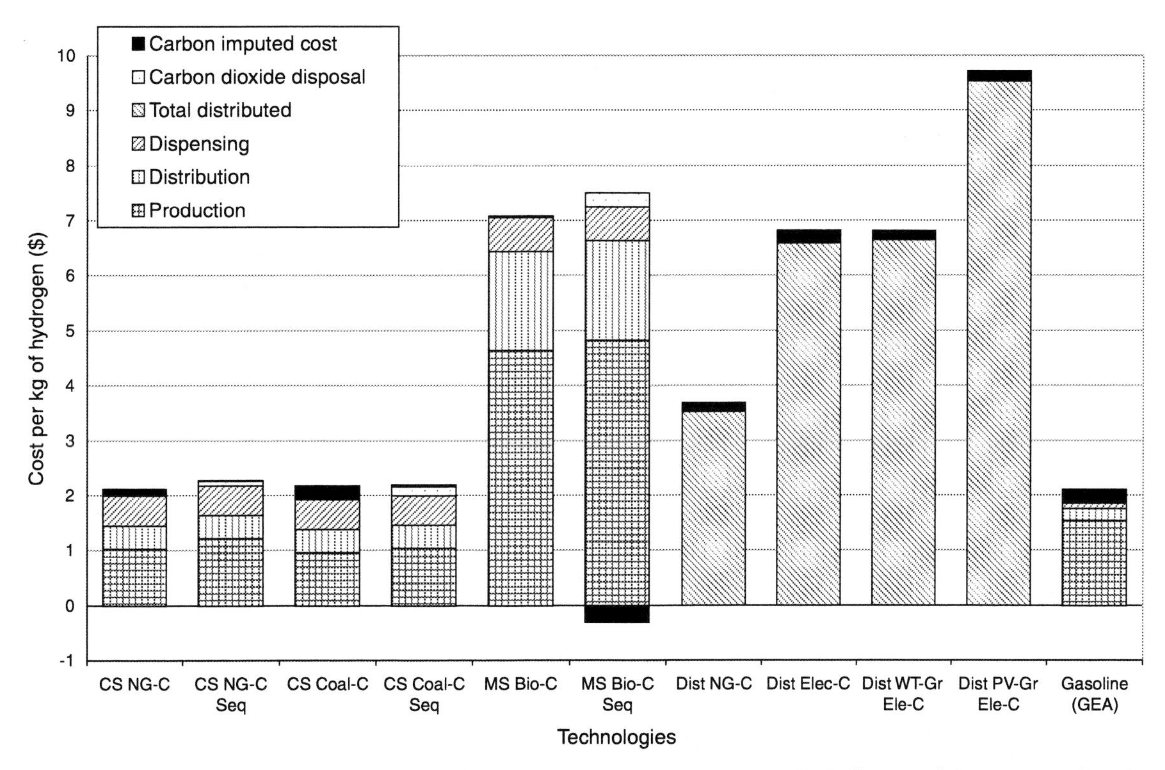

FIGURE 5-1 Unit cost estimates (cost per kilogram of hydrogen) for the "current technologies" state of development for 10 hydrogen supply technologies. Evaluation of a current technology case for nuclear thermal reforming of water is not included because no such technology exists at the present time. See Table 5-2 and discussion in text. NOTE: GEA = gasoline efficiency adjusted.

operating at full capacity. Each central station plant could provide enough hydrogen to fuel about 2 million vehicles. Thus, until there were several million vehicles operated within the service territory of a central station plant, these plants would operate at less than full capacity, and the average costs would exceed those estimated here.

Figure 5-1 also shows that with current technologies, the costs of generating hydrogen with any of the distributed technologies or the midsize biomass technologies would greatly exceed the gasoline costs.

The cost of hydrogen distribution and dispensing is important in assessing the overall economics of hydrogen production. Figure 5-1 shows that for the central station natural gas and coal technologies, the production cost is likely to be only one-half of the total cost of hydrogen; the cost of distribution plus dispensing is roughly as large as the production cost. Therefore, any estimation of the costs of supplying hydrogen must include the costs of distribution and dispensing or else risk sharply underestimating total supply costs.

Figure 5-1 also shows that CO_2 disposal costs of $10 per tonne of CO_2, and the carbon imputed cost of $50 per tonne of carbon (C), have very little impact on the comparative cost across technology options.

Figure 5-2 provides detail underlying the cost estimates. It includes each of the same technologies but disaggregates the production cost for central station and midsize technologies into five components: (1) capital charges, (2) feedstocks, (3) electricity, (4) nonfuel operation and maintenance, and (5) fixed costs. The costs of dispensing, distribution, CO_2 disposal, and the imputed cost of carbon are not further disaggregated here, but their disaggregation is shown in Appendix E. For distributed technologies, the total cost is disaggregated to the same five components listed above.

Figure 5-2 shows that for the central station plants, feedstock costs play major roles in natural gas technologies, while capital costs are a very significant percentage in coal technologies. For biomass technologies, both feedstock and capital costs are high, resulting in hydrogen costs greater than $7.00/kg. Figure 5-2 shows that for the midsize and the distributed technologies, with the exception of distributed natural gas technologies, the capital costs alone exceed $2.00/kg. To calculate this capital cost in this analysis, the committee used a levelized annual capital cost equal to 15.9 percent of the capital investment cost for central station and midsize plants and equal to 14.0 percent of the capital investment cost for distributed generation.[10] Central station and midsize plants were assumed to have a 2.5-year construction time, while distributed plants were assumed to have

[10]These capital cost factors were based on an assumption that each technology faces an 11 percent nominal interest rate, with 2 percent inflation in the economy, a marginal tax rate of 33 percent, a 10-year tax life, and a 20-year project life.

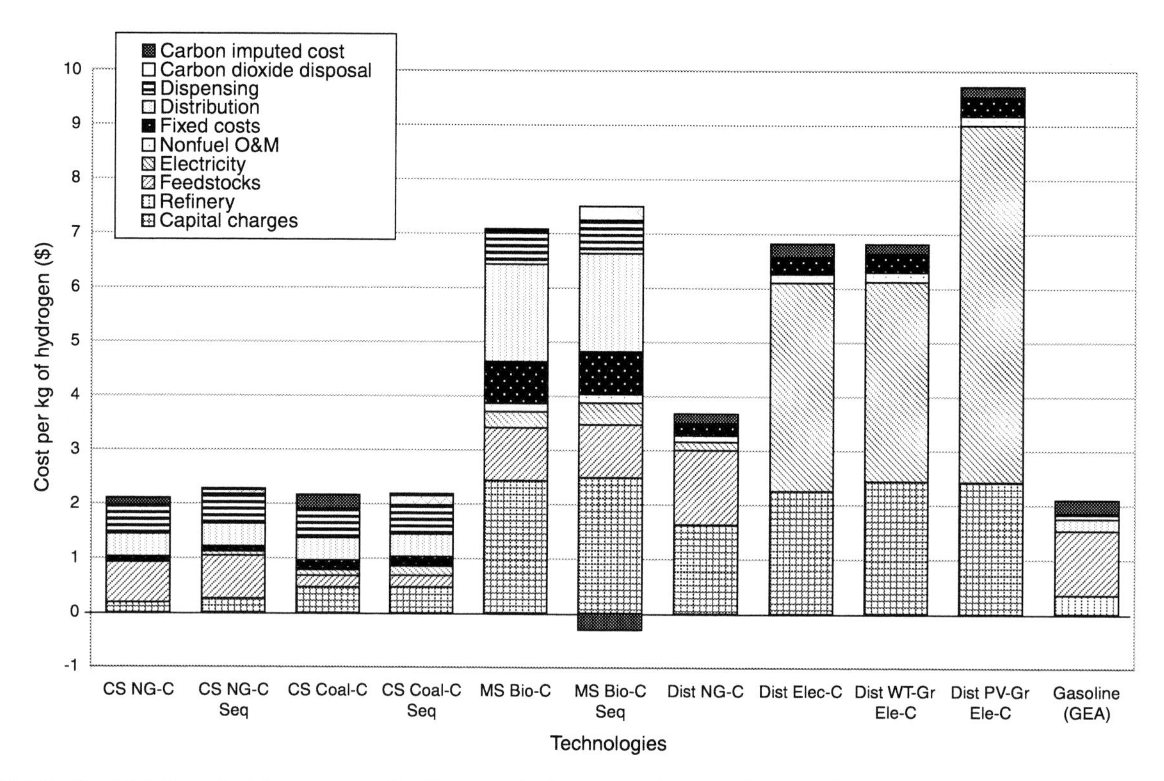

FIGURE 5-2 Cost details underlying estimates for 10 current hydrogen supply technologies in Figure 5-1. Evaluation of a current technology case for nuclear thermal reforming of water is not included because no such technology exists at the present time. See Table 5-2 and discussion in text. NOTE: O&M = operation and maintenance; GEA = gasoline efficiency adjusted.

a 1-year construction time.[11] The differences in construction times result in the 1.9 percent differential in the annual capital cost factors.

The estimated costs for the three electrolysis-based distributed technologies are dominated by the electrolyzer capital costs and electricity costs, either grid-delivered electricity or electricity generated by wind turbines or photovoltaics. Therefore, the per-kilowatt-hour (kWh) cost of purchasing or generating electricity is an important determinant of the overall cost of supplying hydrogen using distributed electrolysis. This analysis assumes that grid-delivered electricity is available all of the time at a delivered price of $0.07/kWh, photovoltaic-derived electricity is available 20 percent of the time at an average cost of $0.32/kWh, and wind-turbine-generated electricity is available 30 percent of the time at an average cost of $0.06/kWh.

Future Technology Cases

The costs of supplying hydrogen might be significantly reduced if research and development directed toward reducing these supply costs were successful. Figure 5-3 provides cost estimates for the possible future technologies, based on judgments by committee members about possible technological progress. This figure presents cost estimates for each hydrogen production process shown in Figures 5-1 and 5-2, plus hydrogen generated by dedicated nuclear plants, using a thermal process to decompose water (CS Nu-F)—in all, 11 technologies. The distributed electrolysis based on wind turbines (Dist WT Ele-F) now is assumed to use *only* electricity generated by wind turbines, in contrast to the current technologies analysis, in which it was assumed that most of the electricity was purchased from the grid. The wind machines and the electrolyzer are assumed to be made large enough that sufficient hydrogen can be generated during the 40 percent of the time that the wind turbines are assumed to provide electricity.[12] The vertical scale is the same as the scale in the two previous graphs.

Figure 5-3 shows the committee's estimation that, with this assumed technical state, hydrogen generated from natu-

[11]In some cases, the time needed for procurement and installation of "off the shelf" or built-to-order distributed production units may be less than 1 year, though during a period of expansion the increased demand for such units could incur delays due to permitting, connecting to electricity or natural gas (for methane conversion units), and so on.

[12]The assumed reductions in the cost of the electrolyzer and the cost of wind-turbine-generated electricity make this option less costly than using a smaller electrolyzer and purchasing grid-supplied electricity when the wind turbine is not generating electricity. However, with current technologies, hydrogen generation is estimated to be less costly if the facility purchases grid-supplied electricity when the wind turbine is not generating enough electricity. In both cases the lower cost option is used.

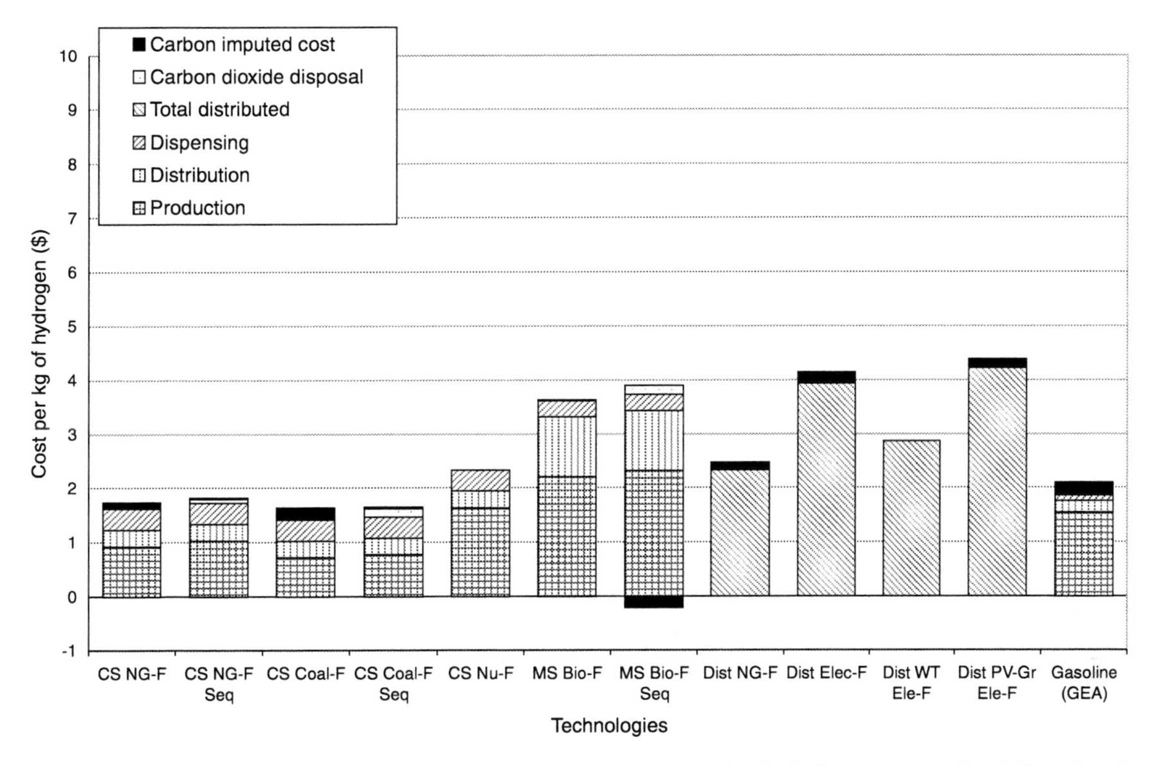

FIGURE 5-3 Unit cost estimates for 11 possible future hydrogen supply technologies, including generation by dedicated nuclear plants. See Table 5-2 and discussion in text. NOTE: GEA = gasoline efficiency adjusted.

ral gas or coal in central stations would be approximately the same cost to a cost lower than that for gasoline used in GHEVs. The gasoline cost assumes no increases in refining efficiency, and crude oil stays at $30/bbl.[13] The committee estimates that hydrogen generated by central station nuclear energy, distributed natural gas steam reforming, and distributed electrolysis using wind-turbine-generated electricity would have costs within about $1.00/kg of the equivalent cost of gasoline used in GHEVs. Figure 5-3 shows that hydrogen generated using grid-delivered electricity or photovoltaic-derived electricity or using biomoss as a feedstock would be substantially more costly. This figure suggests that, if technology does advance as much as assumed possible, then several different technologies, using several different domestically available feedstocks, might become economically competitive with gasoline.

Figure 5-4 shows the detailed cost components for the possible future technologies. For fossil and nuclear technologies, distribution and dispensing costs are still a significant part of the costs. And feedstock costs are high for natural gas conversion. This figure, compared with Figure 5-2, shows that reduced capital costs and reduced electricity costs are the most important differences. The reduced electricity costs result from reduced costs of generating electricity using wind turbines or photovoltaics and estimated increases in the efficiency of electrolyzers.

This figure also suggests that because the electricity cost remains such an important component of overall cost, the price of electricity purchased from the grid and the costs of generating electricity using photovoltaics or wind turbines will be extremely important factors in determining the economic competitiveness of distributed electrolysis. For these possible future technologies, the estimates of the cost of delivered electricity generated using wind turbines[14] decreases to $0.04/kWh (from $0.06/kWh), and using photovoltaics to $0.098/kWh (from $0.32/kWh). The price of grid-delivered electricity is kept at $0.07/kWh, the default estimate, under the assumption that advances in hydrogen-production tech-

[13]Reductions in oil imports can be expected to put downward pressure on the world oil price. However, over the time horizon of this study, the committee expects that the excess production capacity in the world oil market will disappear and that oil prices will be determined by costs of new oil resources. Thus, although the committee does not expect there to be a very large impact due to hydrogen on world oil prices, the committee does not attempt to examine the magnitude of this feedback.

[14]These delivered costs include a 10 percent transmission cost from the wind farms to the distributed hydrogen facility. This transmission cost is consistent with the wind farms' being located in the geographical vicinity of the hydrogen facility, but not at the facility.

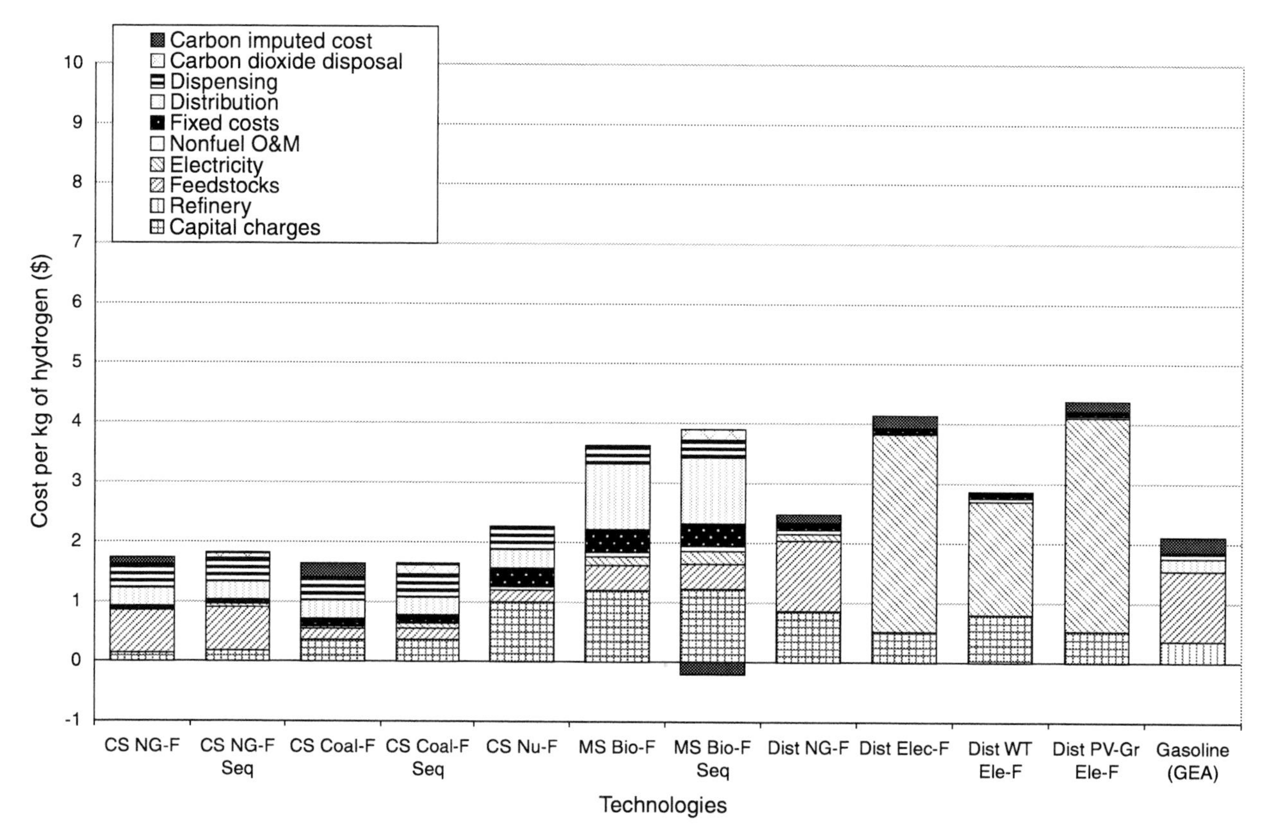

FIGURE 5-4 Cost details underlying estimates in Figure 5-3 for 11 future hydrogen supply technologies, including generation by dedicated nuclear plants. See Table 5-2 and discussion in text. NOTE: O&M = operation and maintenance; GEA = gasoline efficiency adjusted.

nologies and in wind turbines and photovoltaics will have small impact on the price of grid-delivered electricity.

COMPARISONS OF CURRENT AND FUTURE TECHNOLOGY COSTS

In order to facilitate comparisons between costs of current technologies and those of possible future technologies, both sets of costs can be displayed in a single graph. Figures 5-5 through 5-8 provide such graphs, with technologies grouped by primary feedstock from which the hydrogen is generated.

Distributed Electrolysis

Figure 5-5 shows the various distributed electrolysis technologies. This graph shows that the committee conceives of large reductions in hydrogen costs with technology advances. Most of the reduction comes from reduced electrolysis capital costs. The reduced capital cost is primarily the result of the assumption that the costs of proton exchange membrane (PEM) electrolyzers should decline by almost 90 percent, with successful research and development that parallels the advances in PEM fuel cells. The cost of solar photovoltaic electricity also decreases by 50 percent, owing to significant efficiency and manufacturing cost enhancements.[15] Wind electricity also decreases, but by a smaller amount owing to its advanced state of current development.

For wind-turbine-derived electricity, both production using grid-delivered electricity when wind turbines are not providing electricity (Dist WT-Gr Ele-C and Dist WT-Gr Ele-F) and production relying exclusively on wind-turbine-generated electricity (Dist WT Ele-C and Dist WT Ele-F) are included. Capital cost decreases by a larger percentage for electrolysis using wind turbines exclusively. This particularly large capital cost decrease occurs because, for this technology, the capacity of the electrolyzer is inversely proportional to the capacity factor of the wind turbines that supply the electricity. It is assumed that current wind turbines supply electricity 30 percent of the time and that the possible future wind turbines supply electricity 40 percent of the time owing to better technology for utilizing a wider variation in wind speeds. In practice, these figures would be very site-specific, with some sites having higher capacity and others

[15]Photovoltaic costs, in the committee's analysis, are for installed panels inclusive of structures to mount the solar panels themselves. A modular approach is expected to reduce the cost of such structures, although their contribution to the total system cost will continue to be significant owing to the size of the solar field that is required.

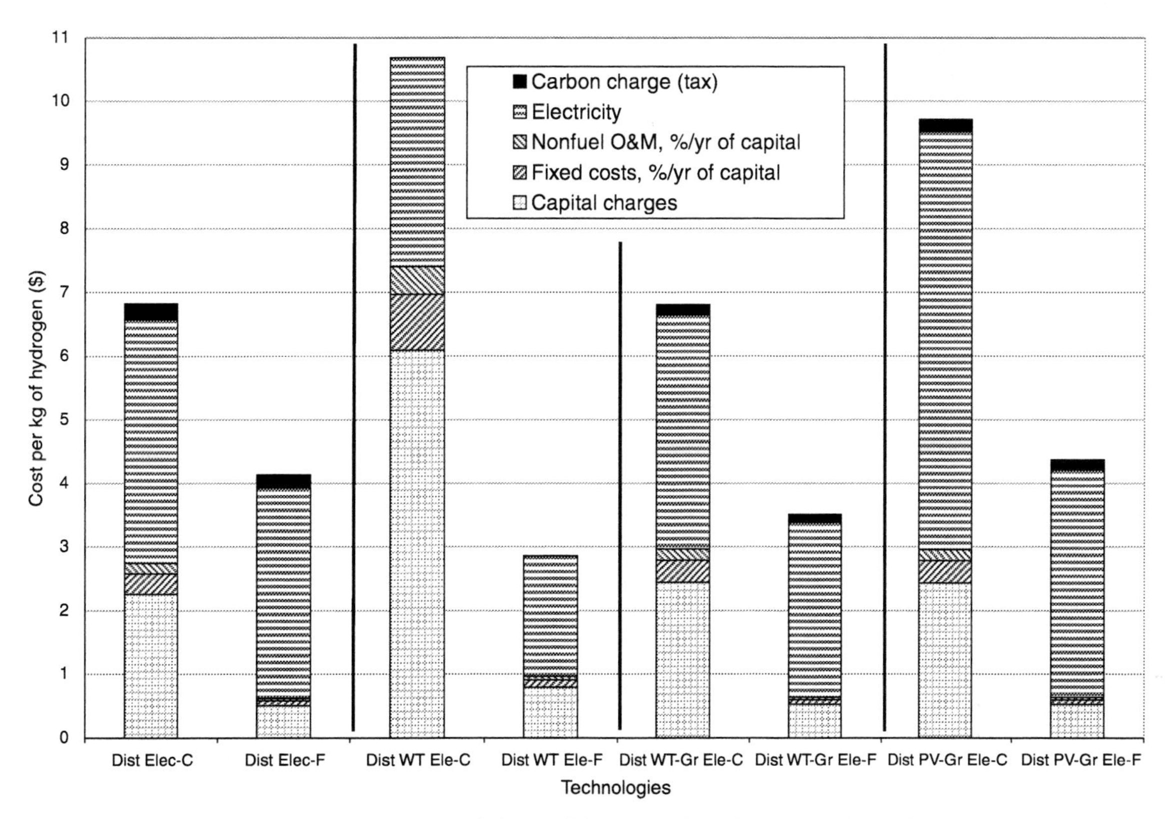

FIGURE 5-5 Unit cost estimates for four current and four possible future electrolysis technologies for the generation of hydrogen. See Table 5-2 and discussion in text. NOTE: O&M = operation and maintenance.

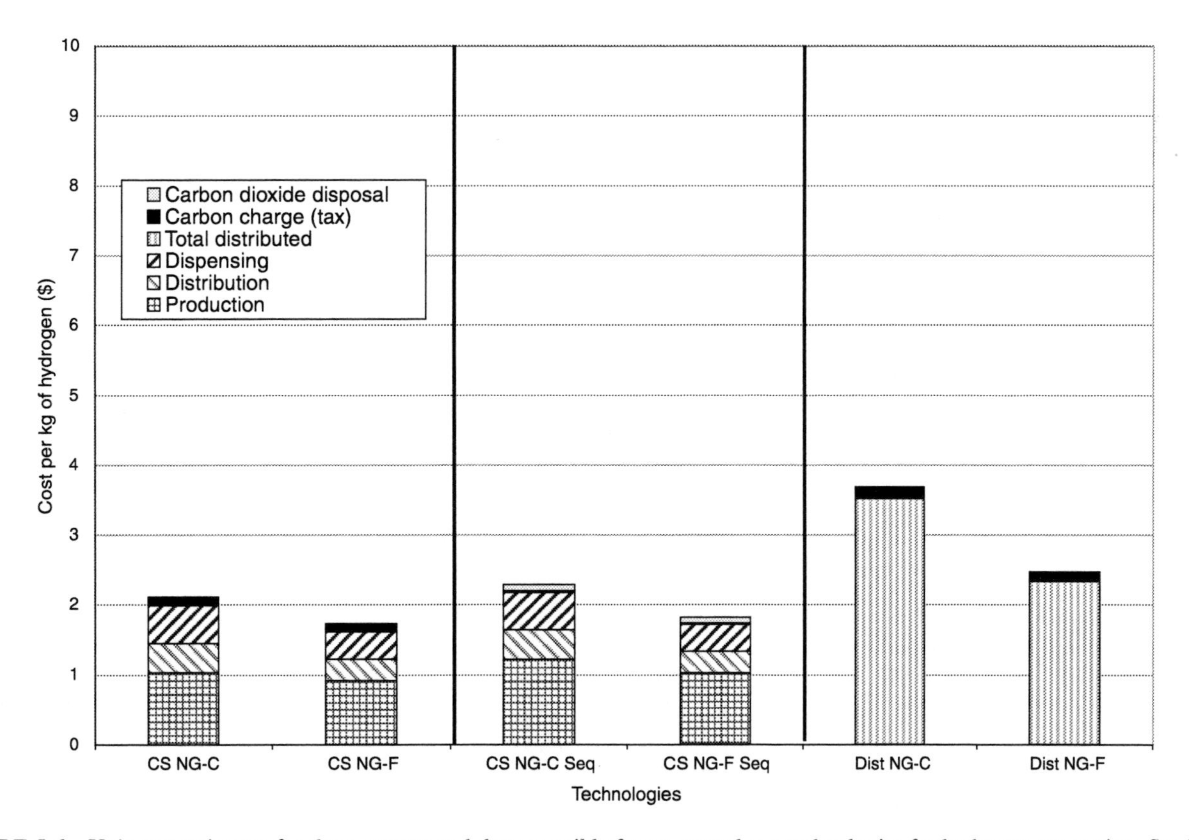

FIGURE 5-6 Unit cost estimates for three current and three possible future natural gas technologies for hydrogen generation. See Table 5-2 and discussion in text.

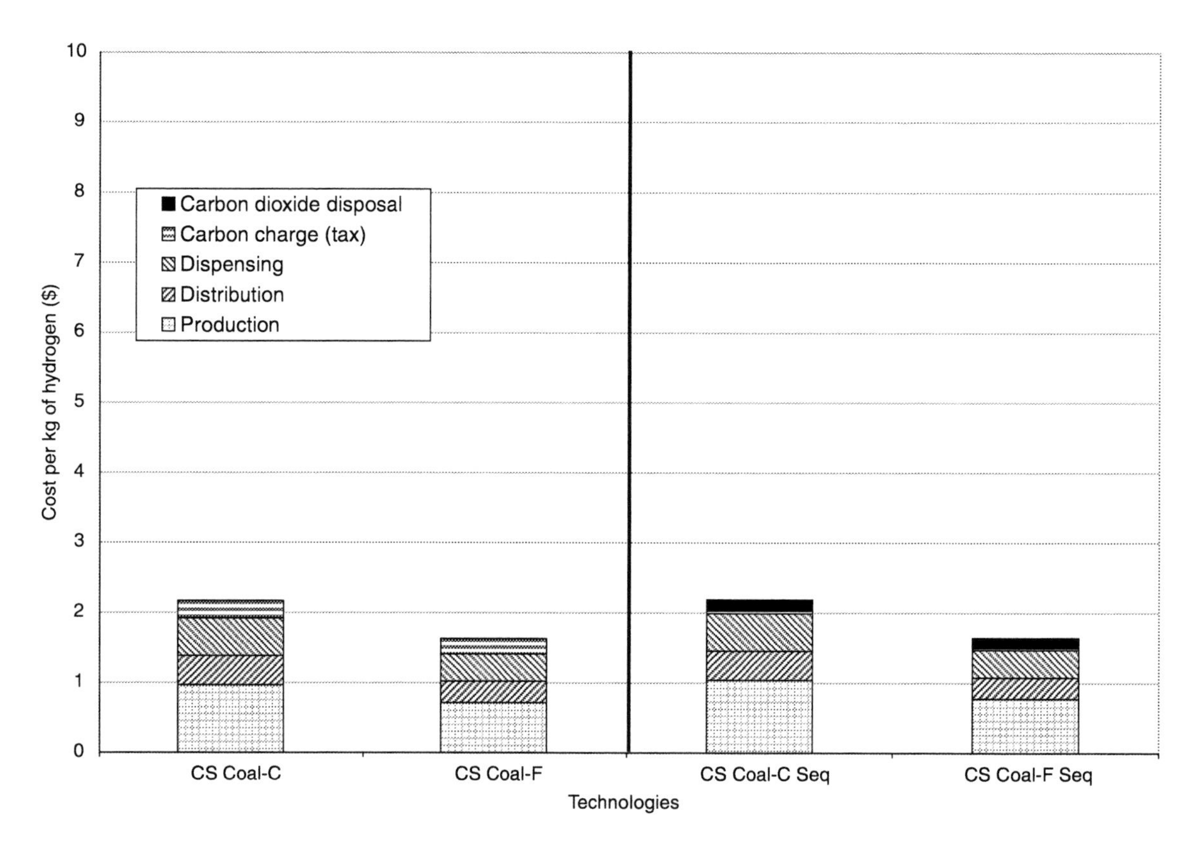

FIGURE 5-7 Unit cost estimates for two current and two future possible coal technologies for hydrogen generation. See Table 5-2 and discussion in text.

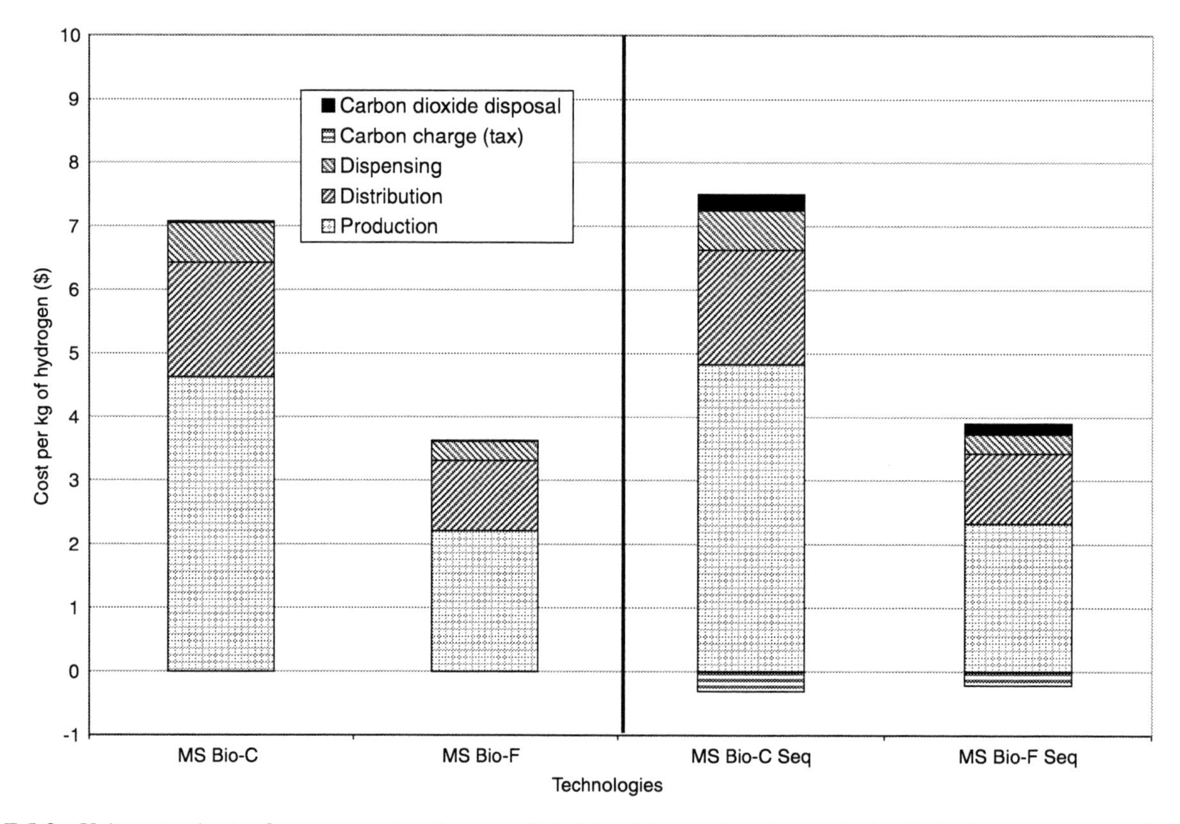

FIGURE 5-8 Unit cost estimates for two current and two possible future biomass-based technologies for hydrogen generation. See Table 5-2 and discussion in text.

lower capacity than is assumed here. Thus, the possible future technology electrolyzers need only be 75 percent as large as the current technology electrolyzers. The combination of the decreased cost of electrolyzers of a given size and the decreased size of required electrolyzers leads to the large reduction in estimated capital costs. The electricity cost also decreases more for electrolysis exclusively using wind turbines than for the technologies that rely on the grid to supply a large amount of the electricity, because a constant $0.07/kWh price of grid-supplied electricity is assumed.[16]

Steam Reforming of Natural Gas

Figure 5-6 shows the various natural-gas-based hydrogen technologies, including both central station and distributed units based on steam reforming of natural gas. As can be seen, technology advances in central station plant hydrogen from natural gas will have a relatively small impact (15 to 20 percent) on hydrogen costs, while advances will have a larger impact on distributed electrolysis hydrogen costs (see Figure 5-5). The cost difference between distributed reforming and central station technologies comes about primarily for three reasons: (1) Capital costs per kilogram of hydrogen are considerably larger for the small steam reformer that would be used in a distributed operation. Central station reformers are assumed to be 2500 times as large as the distributed reformers, but cost only 333 times as much in total. Thus, the capital cost per kilogram of hydrogen is almost 8 times as large for the distributed unit. (2) Delivered natural gas prices for small-volume distributed units would probably differ from delivered prices for large-volume central station units. The committee assumes that the central station units would be able to purchase natural gas at a liquid natural gas parity price of $4.50 per million Btu (EIA, 2003), but that the distributed units would need to pay $6.50 per million Btu because of smaller volumes. (3) The cost advantage of the distributed unit, that no distribution costs would be required to transport the hydrogen from the point of production, is small compared with these two cost disadvantages of the distributed unit.

Coal

Figure 5-7 shows a graph, similar to Figure 5-6, for the central station generation of hydrogen using coal as the feedstock. Technology advances could improve the costs of hydrogen from coal by 25 percent. Under the assumptions of the costs of CO_2 sequestration and the assumption of a $50 per tonne imputed cost of carbon released into the atmosphere, the total costs of coal-based hydrogen production would be almost identical with and without sequestration of CO_2. This occurs because the analysis of the additional costs of CO_2 separation and sequestration suggests that these costs would be very similar to the imputed cost of CO_2 released into the atmosphere. If an imputed cost of carbon of more than $50 per tonne of carbon is used, sequestration of the carbon would be the less costly overall option, whereas if a smaller imputed cost of carbon is used, venting the CO_2 into the atmosphere would be the less costly option.

Biomass

Finally, Figure 5-8 shows the cost comparisons for the hydrogen technologies using biomass as a feedstock. These technologies all assume the following: crops, such as switchgrass, would be grown and used as the feedstock, the biomass would be gasified, and the resultant syngas would be processed to separate the hydrogen. The cost differences between the possible future and the current technologies primarily stem from two factors: (1) The gasifiers are assumed to be reduced in cost and become more efficient—from 50 percent to 70 percent, with the appropriate successful research and development. (2) In addition, the growing of the biomass is assumed to become more productive with the genetic engineering of crops and other productivity advances, so the possible future technologies cases assume that 50 percent more crop could be grown per acre of land.

General Observations

In Figures 5-6, 5-7, and 5-8, the cost of distribution and dispensing from central station and midsize plants is a large part of the overall costs. In this analysis, it is assumed that some reductions in these costs will occur with future technologies, owing to the complex logistical issues in transporting hydrogen and delivering it into the end-use devices, the vehicles. As mentioned in Chapter 4, radically different methods of distribution and dispensing need to be developed to overcome these hurdles. The committee chose not to assume how much these breakthroughs would reduce costs.

As mentioned, this analysis assumes an imputed cost of $50 per tonne C released into the atmosphere and a $10 per tonne CO_2 disposal cost. The committee concludes that technology choices for supplying hydrogen would not be significantly influenced by these costs, as they are small components of the overall costs.

As noted in Chapter 4, in the committee's vision of a possible hydrogen future, the demand for hydrogen will likely be met using distributed production during the first couple of decades of transition. The total cost of hydrogen from the various distributed methods can be compared using Figure 5-5 and the last two bars on the right of Figure 5-6. These data show that with current technology, distributed electrolysis (Figure 5-5) produces hydrogen at a total cost much greater than that for hydrogen produced by distributed natural gas reforming. If competitive electrolysis is not avail-

[16]The committee follows the Energy Information Administration's estimation from *Annual Energy Outlook 2003* (AEO) that electricity is likely to stay roughly constant over the AEO time horizon (to 2025) (EIA, 2003).

able during the transition, use of distributed natural gas may be necessary during the transition period, until centralized facilities and the required distribution system are built.

UNIT ATMOSPHERIC CARBON RELEASES: CURRENT AND POSSIBLE FUTURE TECHNOLOGIES

Characteristics other than the unit cost of producing hydrogen from the various technologies are important as well. Regarding the important hydrogen program goal of reducing emissions of CO_2 into the atmosphere, this analysis incorporates one measure of the goal by including the imputed cost of $50 per tonne of carbon for releasing CO_2 into the atmosphere. But a general consensus has not been reached about the appropriate magnitude of this imputed cost of carbon, or equivalently, about the value of reducing carbon emissions. For that reason, the committee provides here its primary estimates of the unit impacts of introducing various hydrogen technologies into the energy system. In particular, estimates are developed of the amount of CO_2 that would be released into the atmosphere per kilogram of hydrogen produced for each of the technological pathways considered. And, for comparison, similar estimates are also included, on a hydrogen-equivalent basis, of the amount of CO_2 released from the combustion of gasoline in light-duty GHEVs (passenger cars and light-duty trucks). This information is used in Chapter 6 to provide estimates of the amount by which shifts from gasoline-fueled automobiles to hydrogen-fueled vehicles might change CO_2 emissions.

Figures 5-9 and 5-10 provide estimates of the amount of carbon, in the form of CO_2, that would be released into the atmosphere per kilogram of hydrogen produced. Figure 5-9 provides estimates for the current state of technology and Figure 5-10 for the possible future technologies.

The bars in Figures 5-9 and 5-10 are divided into two segments, as applicable, to indicate contributions from direct and indirect releases of CO_2. One segment represents the direct release of CO_2 from the generation of hydrogen. But many of the hydrogen-generation processes use significant amounts of electricity, and generation of that electricity itself releases carbon dioxide into the atmosphere. Estimates of these indirect releases are shown in Figures 5-9 and 5-10 in a second segment of each bar. For the indirect releases, it is assumed that the new electric generation facilities will release much less CO_2 than the current grid does. For these estimates, it is assumed that electricity generation from new facilities releases 0.32 kg CO_2 (0.087 kg C) per kilowatt of electricity, in contrast to the current system, which releases on average about 0.75 kg CO_2 (0.205 kg C) per kilowatt of electricity.[17] The estimates for the new facilities are used in the calculations, since it is expected that new facilities will represent the marginal impacts on the system. The two segments of a given bar together show the total release.

In order to compare carbon releases for hydrogen production with carbon releases from the use of gasoline, a gasoline estimate is included in Figures 5-9 and 5-10, in the same way that gasoline cost comparison is shown in Figures 5-1 through 5-4. This carbon emissions estimate for gasoline can be interpreted as the carbon emission for a GHEV on a "gasoline efficiency adjusted" basis. It is estimated that a gallon of gasoline, when used in an internal combustion engine, would release 2.42 kg C (or 8.87 kg CO_2). The supply chain (reservoir to pump) for gasoline is about 79.5 percent efficient. Therefore, about 3.0 kg C is released into the atmosphere per gallon of gasoline consumed (3.0 is calculated as the ratio of 2.42 to 0.795). Thus, the carbon emission of gasoline is 5.0 kg C per kilogram of hydrogen (calculated as 3.0×1.66).

Figures 5-9 and 5-10 show that for all of the technologies except those involving electrolysis, the direct release of CO_2 is far greater than the indirect release. However, for those involving electrolysis, there is no direct release of CO_2; all releases are indirect, through electricity generation.

These figures show that whether or not the production of hydrogen would reduce CO_2 emissions in comparison with emissions from gasoline-fueled vehicles depends on the particular hydrogen supply chain and on the characteristics of the gasoline-fueled vehicles.

Figure 5-9 shows that two current technologies—the central station coal facility without CO_2 sequestration (CS Coal-C) and the distributed electrolysis system (Dist Elec-C)—would release about as much CO_2 into the atmosphere as would the GHEV. This results from the higher energy efficiency of the FCV over the GHEV, offsetting the higher carbon content of the coal vented to the atmosphere during electricity generation from the coal.

Figure 5-9 also shows that using natural gas as a feedstock would reduce CO_2 emissions by 30 percent (Dist NG-C) or 50 percent (CS NG-C) versus emissions from a GHEV, even though the CO_2 from distributed natural gas reforming is assumed not to be sequestered. The use of wind turbines (Dist WT-Gr Ele-C) and photovoltaics (Dist PV-Gr Ele-C) for electrolysis would reduce the CO_2 emissions to the extent that these renewables were the source of electricity rather than grid-supplied electricity. (In the committee's analysis, these technologies rely on the power grid as backup.) However, because in these systems either 70 or 80 percent of the electricity is grid-supplied, these systems would reduce CO_2 emissions by only 30 or 20 percent, as Figure 5-9 shows. Only with CO_2 sequestration or with biomass as a feedstock would the current technology emissions be driven to near zero. And biomass with CO_2 sequestration (MS Bio-C Seq) could lead to substantial negative net emissions of carbon dioxide: the CO_2 taken from the atmosphere while growing the biomass would greatly exceed the residual amount released back into the atmosphere at the time of hydrogen production.

[17]It is assumed that high-efficiency, natural gas combined-cycle units would be installed to replace retired power generation.

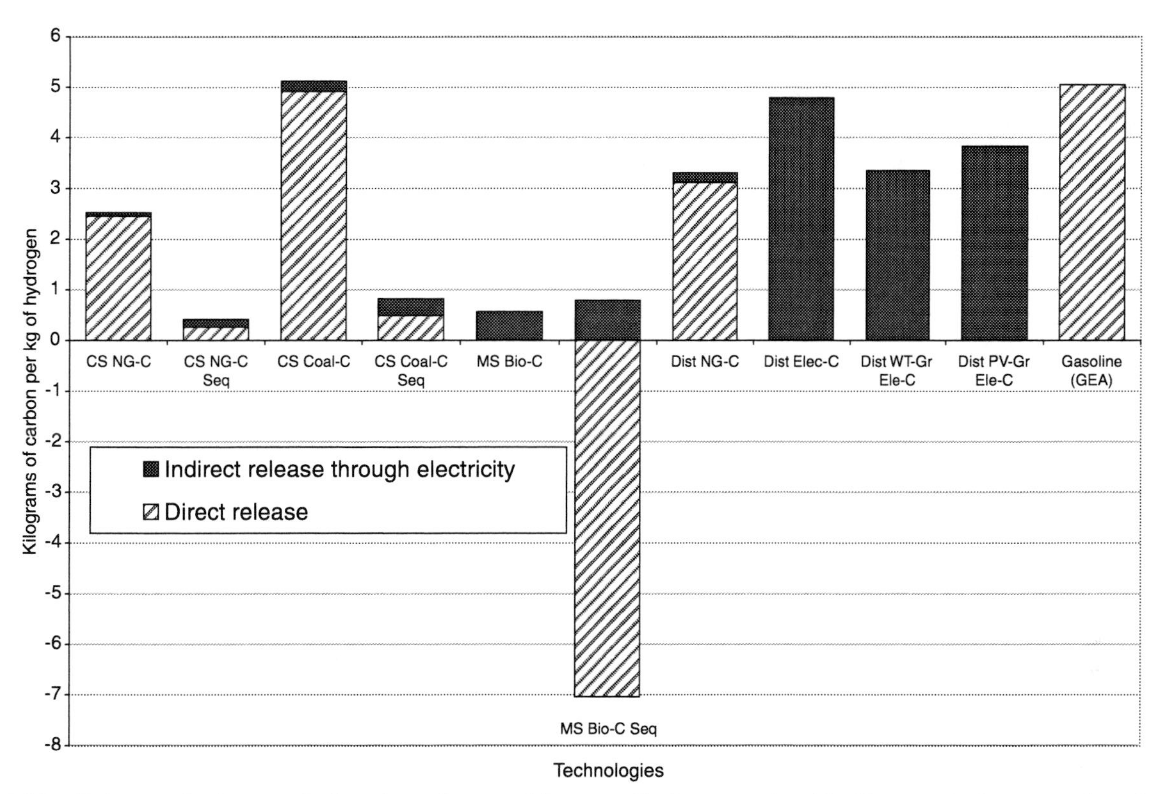

FIGURE 5-9 Estimates of unit atmospheric carbon release per kilogram of hydrogen produced by 10 current hydrogen supply technologies. See Table 5-2 and discussion in text. NOTE: GEA = gasoline efficiency adjusted.

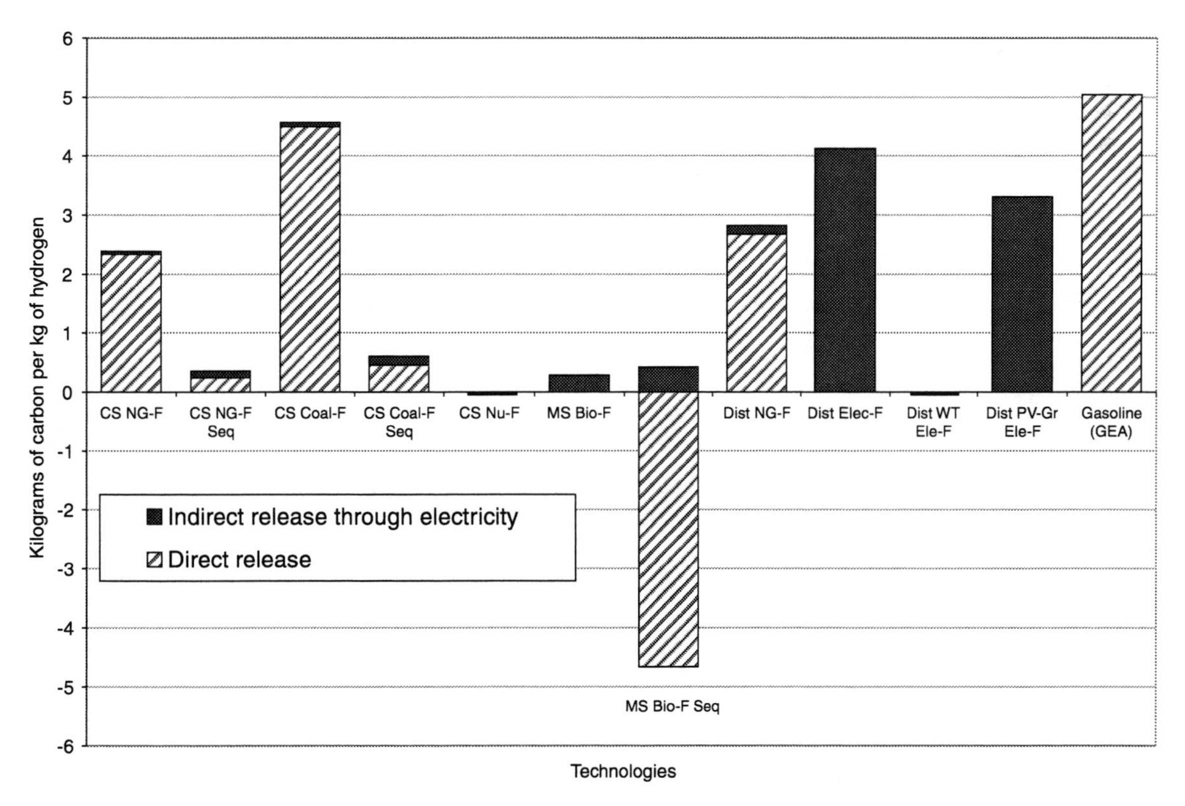

FIGURE 5-10 Estimates of unit atmospheric carbon release per kilogram of hydrogen produced by 11 future possible hydrogen supply technologies, including generation by dedicated nuclear plants. See Table 5-2 and discussion in text. NOTE: GEA = gasoline efficiency adjusted.

Figure 5-10 shows that the implications are similar with the possible future technologies. There are important differences. The first and most significant is the impact of the assumed large reductions in the capital cost of the electrolyzers. It would be less costly to purchase much larger electrolyzers, generate all electricity from wind turbines while they were generating electricity (rather than purchasing most of the electricity from the grid), and leave the electrolyzer idle the rest of the time. The generation of all of the electricity from the wind turbines implies that no CO_2 would be released into the atmosphere.

Second, the carbon from sequestered biomass would be reduced in magnitude, becoming less negative. This reduction would be the result of the increased efficiency of hydrogen generation with the new technologies. A more efficient process implies that less biomass is needed per kilogram of hydrogen and thus less CO_2 is removed from the atmosphere.

Finally, for all other technologies there are only small differences in the CO_2 generation between the current and future cases, thus indicating that, in terms of CO_2 releases, the choice of technology is more important than the technology advances that have been assumed.

Figure 5-11 plots unit carbon emissions (kilograms of carbon per kilogram of hydrogen produced) versus unit costs (dollars per kilogram of hydrogen) for each of the hydrogen production technologies depicted. Two key drivers—low cost and low net carbon emissions—can thus be compared in one graph. In the figure, the current technology is plotted as a square and the possible future technology as a triangle for each hydrogen production method.

WELL-TO-WHEELS ENERGY-USE ESTIMATES

One measure of the performance of a supply chain is its energy efficiency. For vehicles, such a measure is the well-to-wheels calculation of the amount of energy used[18] per mile driven. Figures 5-12 and 5-13 provide these estimates for current technologies and possible future technologies, respectively, with PEM fuel cell vehicles. For the distributed wind-turbine-based electrolysis and photovoltaic-based hydrolysis, the committee assigned zero energy use for electricity from the wind turbine and photovoltaic arrays. Electricity from the grid, where applicable, is assumed to be 50 percent efficient.

The energy used per mile driven[19] depends on the weight, aerodynamic resistance, and other physical characteristics of vehicles. Therefore, any measure of the energy used per mile driven must be standardized to these characteristics. The measurement in this analysis is based on a 27 miles-per-gallon conventional gasoline-fueled vehicle (CFV).

Figure 5-12 shows that for current technologies, some technologies, such as the biomass-based or 100 percent grid-electric-based electrolysis technologies,[20] would use more energy per mile driven than would the conventional vehicle and considerably more energy than would a GHEV. However, biomass uses renewable solar energy, and if enough land is available, the lower efficiency may not be particularly important. Other technologies—the electrolysis processes that use a combination of renewable wind power or photovoltaic electricity plus grid-based electricity—would use less energy per mile driven than would the conventional vehicle, but more than a GHEV would use. Still others—such as natural-gas-based or coal-based units—would use significantly less energy per mile driven than a conventional gasoline vehicle would, but would use only slightly less energy per mile driven than a GHEV would. Thus, with current technologies, hydrogen vehicles would not significantly increase the overall energy efficiency beyond the increase available with hybrid electric vehicles.

Figure 5-13 shows that energy efficiency would be increased with the possible future technologies, so that all of the hydrogen technologies would use less energy per mile driven than would the conventional gasoline-fueled passenger car. Natural gas, coal, or nuclear-based technologies would be more energy-efficient than GHEVs, but even these technologies would not substantially reduce energy use per mile driven. Only the system that uses 100 percent of its electricity from wind turbines would sharply reduce well-to-wheels energy use, in this case down to near zero.

FINDINGS

Several findings emerge from the analysis in this chapter:

Finding 5-1. Hydrogen from central station plant natural gas or coal, used in fuel cell vehicles, can be roughly cost-equivalent to gasoline in a hybrid electric vehicle, on a "gasoline-efficiency-adjusted" (GEA) basis. For natural gas and coal, the differences between current and possible future technologies are relatively small, in comparison to the committee's estimation accuracy.

[18]Energy is not used up, but is transformed into kinetic energy and thermal energy, and ultimately to thermal energy released into the environment. However, energy is used in the present context to mean the amount of useful energy in the supply chain that is so transformed.

[19]For the hydrogen technologies, these measurements are not strictly well-to-wheels. The energy used is from the point of feedstock delivery to the conversion facility and ignores energy used to produce the feedstock or to transport the feedstock from the point of extraction ("well") to the conversion facility. Because this use of energy is small compared with the total energy delivered to the point of use, the committee's calculations only underestimate the energy use of the hydrogen technologies by a small percentage for all cases except those that rely on liquefied natural gas (LNG). The energy loss associated with LNG would be about 10 percent (8 percent to 12 percent). Thus, those natural gas technologies that use gas from LNG would have well-to-wheels energy use about 10 percent larger than that shown in these graphs.

[20]The committee did not make these calculations for electrolysis based on photovoltaics or wind turbines, since the appropriate measurement of energy used has not been generally accepted.

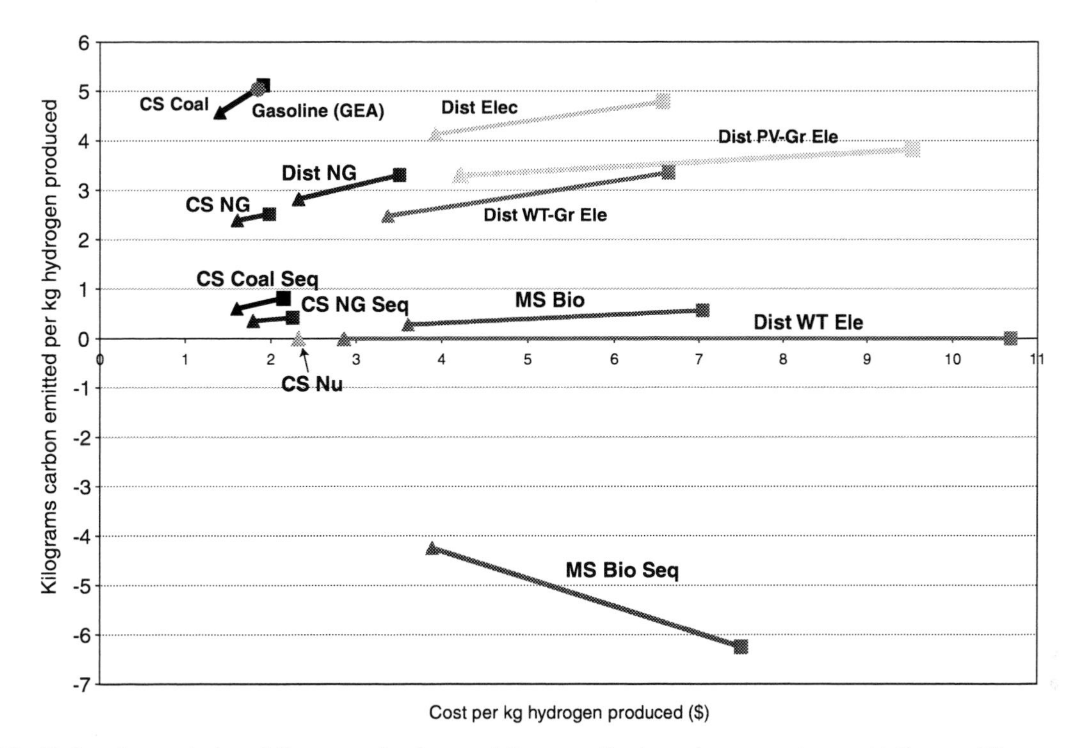

FIGURE 5-11 Unit carbon emissions (kilograms of carbon per kilogram of hydrogen) versus unit costs (dollars per kilogram of hydrogen) for various hydrogen supply technologies, in both current (■) and possible future (▲) states. See Table 5-2 and discussion in text. NOTE: GEA = gasoline efficiency adjusted.

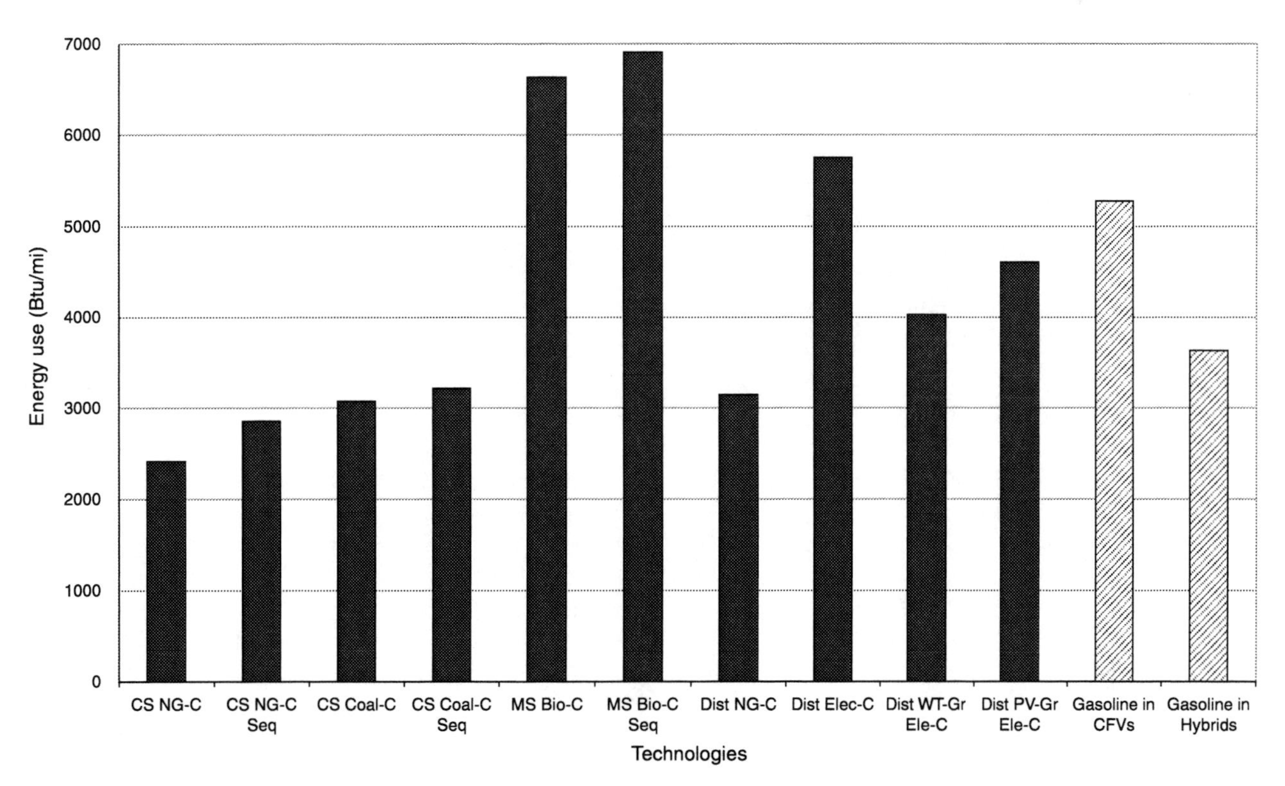

FIGURE 5-12 Estimates of well-to-wheels energy use (for 27 miles-per-gallon conventional gasoline-fueled vehicles [CFVs]) with 10 current hydrogen supply technologies. See Table 5-2 and discussion in text.

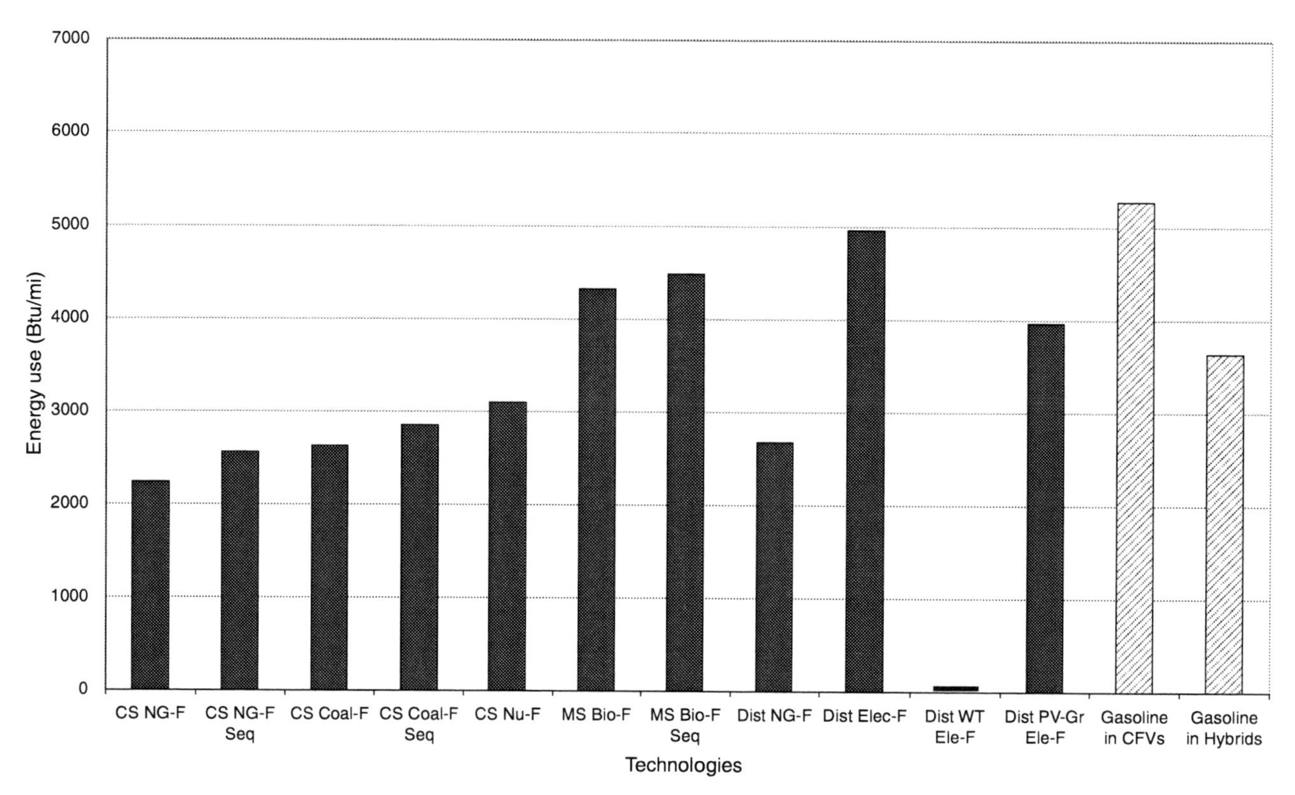

FIGURE 5-13 Estimates of well-to-wheels energy use (for 27 miles-per-gallon conventional gasoline-fueled vehicles [CFVs]) with 11 possible future hydrogen supply technologies, including generation by dedicated nuclear plants. Well-to-wheels energy use for wind-turbine-based electrolysis (Dist WT Ele-F) is near zero (narrow bar), as wind turbines have been assigned zero energy use. See Table 5-2 and discussion in text.

Finding 5-2. With the possible future technology advances, hydrogen generated by central station nuclear energy, distributed natural gas steam reforming, and distributed electrolysis using wind-turbine-generated electricity could have costs within about $1.00 per kilogram of gasoline costs on a gasoline-efficiency-adjusted basis.

Finding 5-3. Even with the possible technology advances, hydrogen from distributed electrolysis using photovoltaics or grid-supplied electricity, or hydrogen using gasification of biomass would have gasoline-efficiency-adjusted costs significantly higher than the gasoline cost. Thus, technological breakthroughs, even beyond the optimistic assumptions of the committee, would be needed to make these technologies competitive.

Finding 5-4. Distribution and dispensing costs will continue to be a significant component of total hydrogen supply chain costs for all production pathways except those based on distributed generation. Ignoring these costs would significantly underestimate total supply chain costs for hydrogen.

Finding 5-5. Using estimated carbon dioxide disposal costs of $10 per tonne of carbon dioxide, and the carbon imputed cost of $50 per tonne of carbon released into the atmosphere, these two costs of carbon management would have only a small impact on the relative costs of the various technologies.

Finding 5-6. Whether distributed electrolysis becomes economically viable will depend critically on the cost of the electricity used in the electrolysis. Therefore, the price of electricity purchased from the grid and the costs of generating electricity using photovoltaics or wind turbines will be extremely important factors in determining the economic competitiveness of distributed electrolysis.

Finding 5-7. Hydrogen can be produced by electrolysis using wind turbines as the source of the electricity. Whether this technology would be competitive on a gasoline-efficiency-adjusted basis with gasoline depends critically on whether the capital cost of the proton exchange membrane electrolyzers declines by the 90 percent assumed by the committee. With very low cost of electrolyzers, installation of very large electrolyzer units could fully compensate for the intermittent nature of wind-produced electricity. Costs of wind-produced electricity include the full capital costs of wind turbines, even though the wind turbine would produce electricity only some of the time.

Finding 5-8. Solar-based hydrogen does not appear viable even with currently envisioned cost decreases in photovoltaic cells and in electrolyzers.

Finding 5-9. Most of the hydrogen supply chain pathways would release significantly less carbon dioxide into the atmosphere than would gasoline used in hybrid electric vehicles. Only coal-based nonsequestered production and grid-based electrolysis are comparable to gasoline in this respect. The higher efficiency of fuel cell vehicles compensates for the high carbon dioxide content of the fossil fuels.

Finding 5-10. The technology advances envisioned by the committee would not significantly reduce the carbon dioxide emissions from fossil fuels, absent sequestration.

Finding 5-11. Carbon dioxide emissions could be brought down to near zero with biomass, with electrolysis depending exclusively on wind turbines or photovoltaics, with nuclear energy, or with the successful sequestration of carbon dioxide from the production of hydrogen from fossil fuels. Carbon dioxide emissions could be made negative if the hydrogen was produced from biomass and the carbon dioxide from production was separated and sequestered.

Finding 5-12. With current technologies, hydrogen vehicles would not significantly increase the "well-to-wheels" energy efficiency significantly beyond the increase available with gasoline hybrid electric vehicles. Well-to-wheels energy efficiency would be increased with the possible future technologies, and so all of the hydrogen technologies would use less energy per mile driven than would the conventional gasoline-fueled passenger vehicle. Fuel cell vehicles that derive their hydrogen from natural gas, coal, or nuclear-based technologies would be more energy-efficient than hybrid electric vehicles would, but even these technologies would not substantially reduce energy use per mile driven. Only the system that uses 100 percent of its electricity from wind turbines and solar power would sharply reduce well-to-wheels energy use, in this case down to near zero.

6

Implications of a Transition to Hydrogen in Vehicles for the U.S. Energy System

In this chapter, estimates are provided of the possible impacts of a successful transition to hydrogen in vehicles, focusing on the potential economic and environmental impacts and on those related to energy security and domestic resource use. The analysis is structured around a vision of transition to the use of hydrogen in light-duty vehicles (passenger cars and light-duty trucks). Although there are other proposed uses of hydrogen—for example, in heavy-duty trucks and buses, electricity generation, and stationary home applications—the focus here is on one use in order to gain a sense of the potential quantitative significance of a transition to hydrogen.

In this analysis, it is assumed that many problems of hydrogen use in vehicles are solved: low-cost and durable fuel cells are available; high density of energy storage on vehicles allows reasonable range and quick refilling of the vehicles; vehicles have the same functionality, reliability, and cost[1] associated with their gasoline-fueled competitors; hydrogen-fueled vehicles are as safe as gasoline-fueled vehicles. (These problems are discussed more fully in Chapter 3.)

This vision is not a prediction of the diffusion of hydrogen technologies into the fleet of vehicles, depending as it does on such a large number of factors that are inherently uncertain. However, it is offered to allow some specificity in the analysis of the possible implications for the U.S. energy system of a transition to hydrogen in vehicles.

Starting with this optimistic vision, estimates are made of the consumption of gasoline and of hydrogen for the first half of this century. This estimation depends on assumptions of the growth in vehicle miles; the average fuel efficiency over time of conventional gasoline-fueled vehicles, gasoline hybrid electric vehicles (GHEVs), and hydrogen vehicles; the sales of new vehicles; and the operational life of vehicles once purchased.

The analysis of Chapter 5 is combined with this estimation of hydrogen consumption over time. For each particular hydrogen-producing technology, an examination is made of the economic, environmental, energy security, and domestic resource use implications, under the pure case assumption that *all* of the hydrogen is generated through that individual technology. This analysis is conducted for both the "current" state of technology development and the "possible future" state of technology development.

Although the analysis is conducted on the basis of the pure cases of 100 percent of the hydrogen's being generated from a particular technology, the committee does not believe that the system would evolve that way. If there is a successful transition to hydrogen, the committee expects hydrogen to be produced using multiple technologies. The committee has chosen in this study not to create a single scenario in which the proportions of production using the various technologies are postulated. But the interested reader can examine the implications of such scenarios by taking weighted averages of the impacts estimated from the pure strategies.

In developing the analyses, the committee made quantitative estimates of some of the impacts believed to be most important, but it was not able to examine all of the possible impacts. The environmental impacts examined are associated with potential global climate change caused by carbon dioxide (CO_2) emissions from light-duty vehicles under the various technology pathways (see Table 5-2 in Chapter 5). The committee does not attempt to estimate any impacts on

[1]With respect to vehicle cost for the three vehicle types considered in the analysis—hydrogen vehicles, conventional gasoline-fueled vehicles, and gasoline hybrid electric vehicles (GHEVs)—the committee has assumed that vehicles having equivalent performance will have equal cost. This cost equivalence is a goal for the auto industry. In making this assumption, however, the committee has not conducted its own analysis or projection of whether this goal will be achieved. The advantage of assuming equivalence among the three vehicle types is that it permits comparisons strictly of fuel supply systems without judgments as to the success or failure of vehicle developments underway. However, the total cost of a hydrogen economy compared with a hybrid or conventional vehicle economy is left undetermined.

global climate change of hydrogen leakage or of changes in the quantities of other greenhouse gases released into the atmosphere, nor does it examine the impacts on emissions of criteria pollutants[2] from vehicles.

The economic impacts examined are the costs to the United States as a whole from fueling the fleet of light-duty vehicles. Under the committee's maintained assumption that the costs of the vehicles themselves are equivalent to the costs of the vehicles for which they substitute, differences in the costs of fueling the fleet will translate into differences in the total costs of driving the fleet of light-duty vehicles. Costs of the infrastructure to fuel the vehicles are included in the supply costs for hydrogen. Therefore, although the committee does not explicitly separate the infrastructure costs from the fuel costs, the infrastructure costs are part of the total. However, because the development of infrastructure may involve large investments concentrated over a small number of years, calculations should not be interpreted as capturing the time dimension of the physical investments themselves. And the committee does not examine any of the redistributional consequences of a shift to hydrogen. In particular, such a massive transition will lead to economic opportunities for some established companies, many new companies, and many individuals, while reducing the economic opportunities for some established companies and individuals. The committee does not examine these potentially important consequences.

The energy security implications examined are related to the imports of energy, in particular, petroleum and natural gas. The committee examined the impacts on the use of gasoline, impacts that can be expected to translate directly to impacts on the imports of crude oil or petroleum products. Impacts on the use of natural gas were examined. An increase in demand would cause an increase in price, which in turn could increase domestic supply. Thus, it is not clear what fraction of this increase in natural gas use would translate into increases in natural gas imports. However, it is assumed that most of this increase in natural gas use would translate directly into increases in natural gas imports, consistent with projections in *Annual Energy Outlook 2003* (EIA, 2003). The committee did not try to quantify other impacts on energy security associated with changes in the vulnerability of the energy infrastructures to human error, mechanical breakdown, or terrorism. However, the committee does recognize that choices of distributed production versus central station production, choices of particular hydrogen transportation options, and choices of precise locations of new plants can have significant impacts on energy security.

The committee analyzed several implications relative to domestic resource use. For biomass production, it examined the amount of land that would be required to grow the crops used as feedstocks. For coal-based hydrogen production, it examined the amount of coal that would be used over time. For technologies involving sequestration, it examined the amount of CO_2 that would be sequestered on a year-by-year basis and the cumulative quantity sequestered. The committee did not try to quantify several other resource use impacts: it did not examine the amount of land that would be required for wind farms, production facilities, or distribution infrastructure; it did not examine the impacts on water use for steam reforming processes or for biomass production; it did not attempt to examine any labor force issues; nor did it examine the needs for metals or other materials for fuel cells, electrolyzers, or production facilities, or the number of pipelines, or other infrastructure.

HYDROGEN FOR LIGHT-DUTY PASSENGER CARS AND TRUCKS: A VISION OF THE PENETRATION OF HYDROGEN TECHNOLOGIES

Starting with the assumption that the many problems related to the use of hydrogen in vehicles are solved, a plausible but optimistic vision of the penetration of hydrogen technologies into the fleet of vehicles was created. In this vision, as described in Chapter 3, the committee assumes that GHEVs initially begin capturing market share from conventional vehicles, reaching 1 percent in 2005 and growing by 1 percentage point per year until hybrids reach 10 percent market share in 2014. With the introduction of hydrogen vehicles in 2015, initially the market share of GHEVs grows by 5 percentage points per year, while that of hydrogen vehicles grows by 1 percentage point annually. During this period, the market share of conventional vehicles declines by 6 percentage points annually. As hydrogen vehicles continue to grow in popularity, with their market share increasing, the market share of GHEVs peaks in 2024 at 60 percent and then begins declining by 2 percentage points annually. After reaching a 10 percent market share in 2024, hydrogen vehicles begin increasing their market share by 5 percentage points per year until capturing a 60 percent market share in 2034. In that year, hybrids capture 40 percent of the market, and conventional vehicles are no longer purchased. From that point on, hydrogen vehicles increase their market share by 10 percentage points per year, until hydrogen vehicles ultimately capture 100 percent of the market for new vehicles in 2038. The committee considers this vision to repre-

[2]Criteria pollutants are air pollutants emitted from numerous or diverse stationary or mobile sources for which National Ambient Air Quality Standards have been set to protect human health and public welfare. The original list of criteria pollutants, adopted in 1971, consisted of carbon monoxide, total suspended particulate matter, sulfur dioxide, photochemical oxidants, hydrocarbons, and nitrogen oxides. Lead was added to the list in 1976, ozone replaced photochemical oxidants in 1979, and hydrocarbons were dropped in 1983. Total suspended particulate matter was revised in 1987 to include only particles with an equivalent aerodynamic particle diameter of less than or equal to 10 micrometers (PM_{10}). A separate standard for particles with an equivalent aerodynamic particle diameter of less than or equal to 2.5 micrometers ($PM_{2.5}$) was adopted in 1997.

sent an optimistically fast rate of penetration of hydrogen vehicles into the marketplace.

In order to examine the impacts of the hydrogen introduction, the committee examined a case in which no hydrogen vehicles are introduced, but hybrids capture the entire market share that would have been captured by hydrogen vehicles. In this case, the time path of conventional vehicles remains the same as in the committee's plausible but optimistic vision. For every additional hydrogen vehicle in this analysis, there is one fewer hybrid electric vehicle.

The market shares of new vehicle sales of the three classes of vehicles in the vision are shown in Figure 6-1.

Once new automobiles are sold, they are driven for many years.[3] Thus, the fraction of miles driven by each class of vehicles lags well behind the market share of new vehicle sales. Figure 6-1 shows the fractions of all miles assumed to driven by each class in the committee's vision, in addition to the fractions of new vehicles sold by each class. The fractions of all miles are calculated as the fractions of all vehicles on the road, adjusted by the assumption that new vehicles are driven more than old vehicles are.

During the years in which it is driven, each type of vehicle must use the fuel for which it is was designed. And, in the committee's analysis, it also assumed that the fuel economy of each vehicle is determined at the year the vehicle is sold, and that the fuel economy remains constant during the lifetime of the vehicle.

Figure 6-2 shows the fuel economy assumed for the three classes of vehicles over time. New and existing conventional vehicles are assumed to achieve, on average, 21 miles per gallon (mpg) of gasoline in 2002. However, this average fuel efficiency is assumed to increase by 1 percentage point per year during the entire time horizon. No assumptions are made about whether this increase is determined by regulations such as changing corporate average fuel economy standards, improved technologies, market forces, or some combination of factors. The committee notes that historic trends in light-duty-vehicle fuel economy, on a fleetwide basis, reached a plateau in the mid-1980s (EPA, 2003).

New GHEVs are estimated to have a 45 percent higher fuel economy than that of conventional vehicles in any year (see Chapter 3 for a discussion of efficiency differences); new hydrogen vehicles are estimated to have 2.4 times the fuel economy of conventional vehicles (or a 66 percent higher fuel economy than that of GHEVs). (For both types of vehicles, the average fuel efficiency is assumed to increase by 1 percentage point per year during the entire time horizon.[4]) Thus, the ratio of miles per kilogram for new hybrid vehicles to miles per gallon for gasoline-fueled vehicles remains constant over time, with all fuel economies growing steadily. This assumption about the relative efficiencies is designed to provide an optimistic view of the fuel efficiency of hydrogen vehicles.

In the committee's analysis, both the number of new cars sold and the total vehicle miles traveled increase at 2.3 percent per year, consistent with the Energy Information Administration's (EIA's) Reference Case forecast of growth in vehicle miles traveled for light-duty vehicles.[5] (This forecast rate of increase is consistent with recent historical trends, but the committee recognizes that it could be subject to alteration by many factors.) The total vehicle miles traveled for each type of car is proportional to the number of each type on the road, adjusted so that new cars are assumed to be driven more than older cars are.

Taken together, the assumptions about new-car sales, new-car fuel economy, proportions of the different types of vehicles, and vehicle miles traveled allow the committee to estimate the amount of hydrogen and of gasoline that would be used for light-duty vehicles if those assumptions in the optimistic vision came to pass. Figure 6-3 shows the consumption of hydrogen by light-duty vehicles estimated for this vision. By the year 2050, light-duty vehicles would be consuming 101 billion kilograms, or 111 million tons, of hydrogen per year. The consumption can be compared with the current U.S. industrial production of hydrogen of about 8 billion kilograms annually (see Chapter 2). In the committee's vision of the possible penetration of hydrogen vehicles into the marketplace, light-duty vehicles could be consuming 8 billion kilograms of hydrogen annually by the year 2027.

In contrast, gasoline consumption would continue to rise only until the year 2015, after which it would begin declining until it reached zero in 2050. This trajectory of gasoline consumption is shown in Figure 6-4. Note that this figure includes two scales, measuring gasoline use in millions of barrels per day (right scale) and in quadrillion British thermal units (Btu) per year (left scale).[6]

Figure 6-4 also displays two other trajectories of gasoline consumption. The first shows an estimate of gasoline consumption in the absence of either hybrid electric vehicles or hydrogen vehicles. It shows that gasoline consumption would continue increasing at rates consistent with historical

[3]In the committee's analysis, automobiles are driven for 14 years, with annual vehicle miles (per car from the given vintage) declining as the vehicles get older. New vehicles are assumed to be driven 15,000 miles annually; 5-year-old vehicles, 14,490 miles; 10-year-old vehicles, 7,758 miles; 14-year-old vehicles, 603 miles. This decline reflects both the scrapping of vehicles over time and the reduced mileage of older vehicles.

[4]Note that the increase in vehicle fuel efficiency (for all three types of vehicles) is assumed in all of the analyses and is independent of the choice of supply technology, that is, "current" or "possible future."

[5]New car sales have grown less rapidly. But the committee's estimates are most sensitive to vehicle miles. Therefore, the model was calibrated to vehicle miles data from the EIA. Estimates were made for year 2000 vehicle miles traveled to be 2523 billion miles for light-duty vehicles, using the estimate from *Annual Energy Outlook 2003* (EIA, 2003).

[6]Quadrillion Btu = 10^{15} Btu.

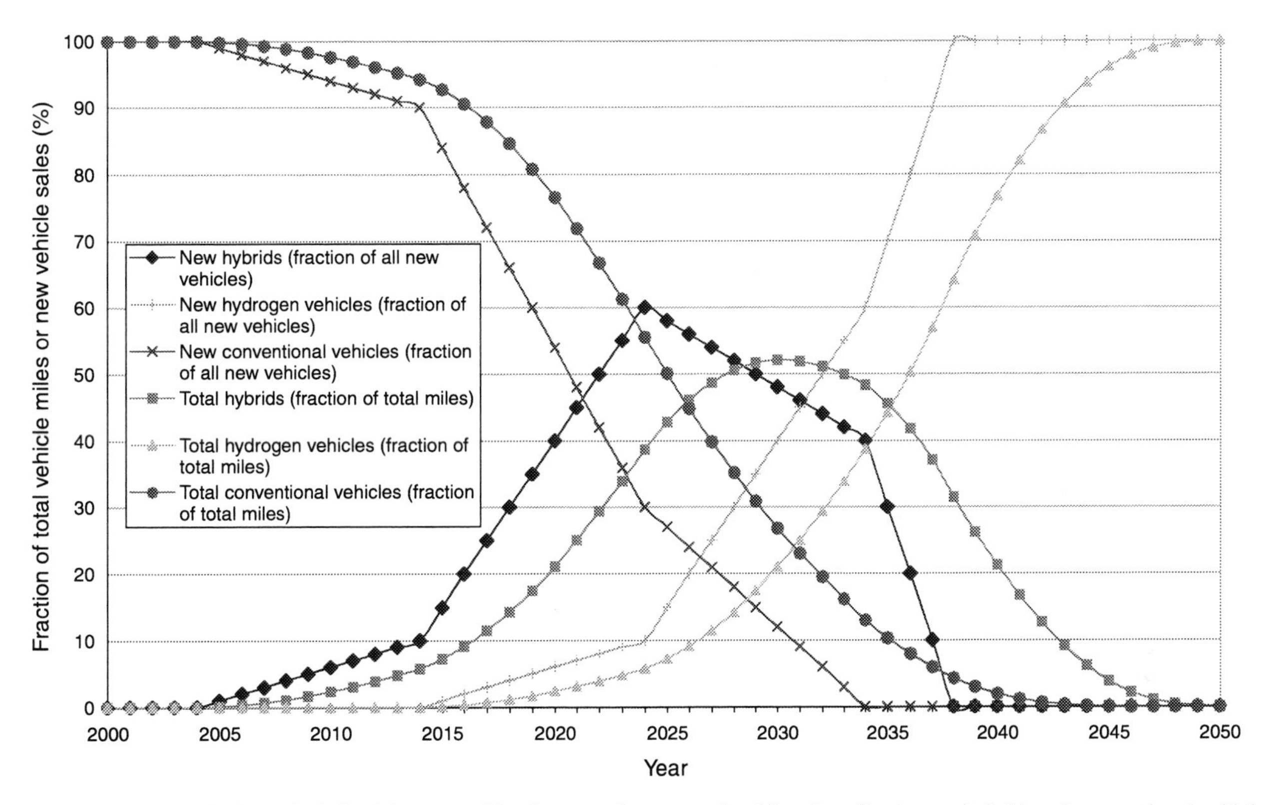

FIGURE 6-1 Demand in the optimistic vision created by the committee: postulated fraction of hydrogen, hybrid, and conventional vehicles, 2000–2050.

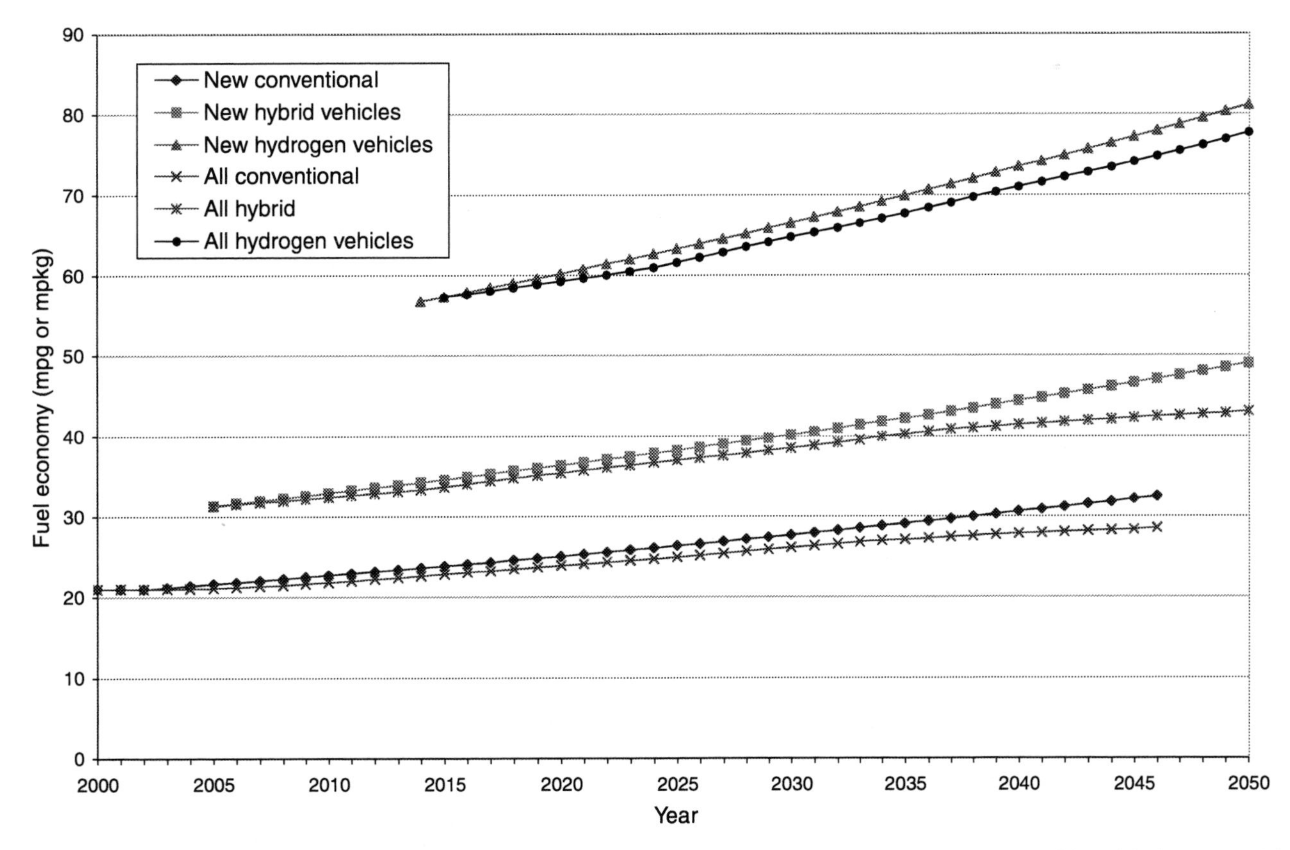

FIGURE 6-2 Postulated fuel economy based on the optimistic vision of the committee for conventional, hybrid, and hydrogen vehicles (passenger cars and light-duty trucks), 2000–2050.

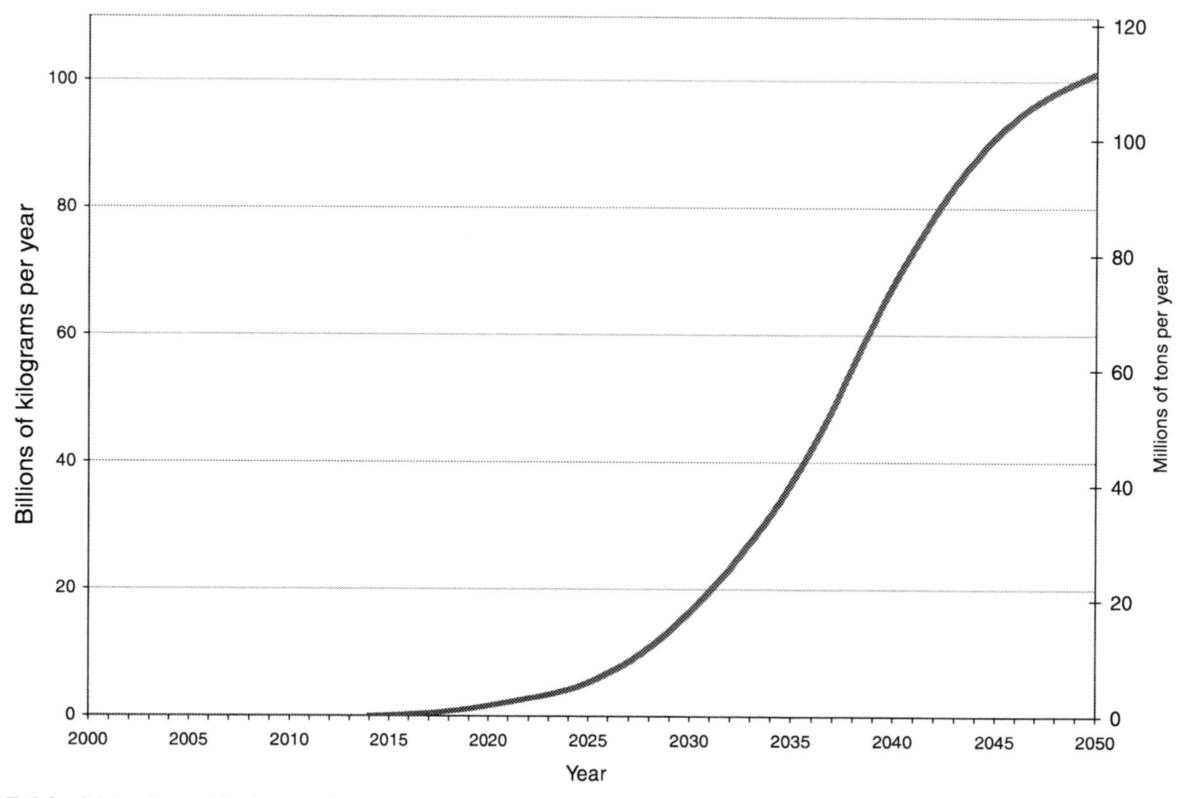

FIGURE 6-3 Light-duty vehicular use of hydrogen, 2000–2050, based on the optimistic vision of the committee.

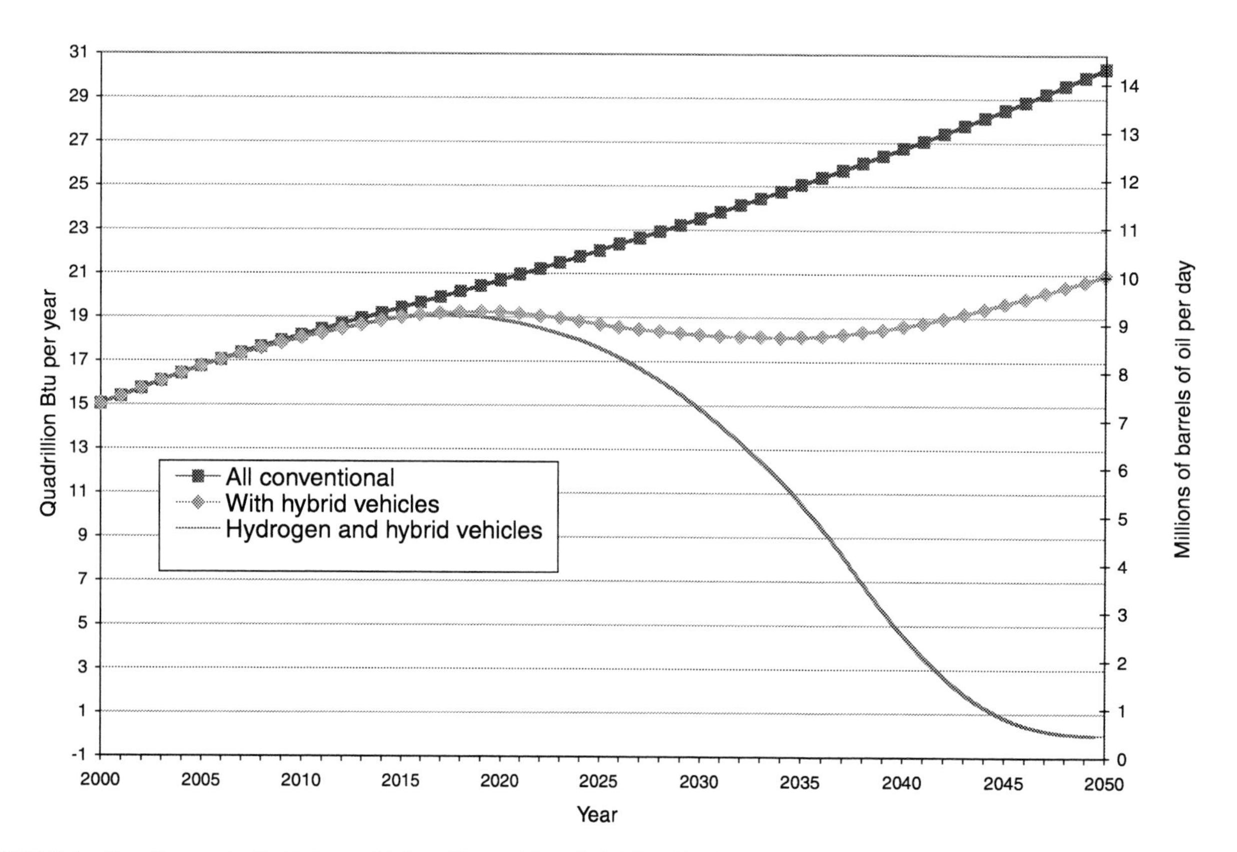

FIGURE 6-4 Gasoline use by light-duty vehicles with or without hybrid and hydrogen vehicles, 2000–2050, based on the optimistic vision of the committee.

experience, taking into account the increased mileage per gallon of conventional vehicles. The second trajectory shows the estimated consumption of gasoline if hydrogen vehicles were never adopted and hybrids captured the entire market share that would have been captured by hydrogen vehicles. In the discussions that follow, the committee considers the impact on gasoline consumption of a transition to hybrid vehicles. That impact can be seen as the difference in Figure 6-4 between the gasoline consumption and the consumption with hydrogen and hybrid vehicles.

In order to put the figures showing gasoline use in context, the committee can plot the projections from the Energy Information Administration (EIA, 2003) of U.S. oil consumption, production, and imports along with the committee's estimates of gasoline consumption in the three cases. This superposition of the gasoline consumption estimates with the EIA projections of oil supply, demand, and imports appears in Figure 6-5. This figure shows that automotive consumption of gasoline is a large fraction of total oil consumption but is less than 50 percent of total U.S. use of crude oil and petroleum products. Thus, a transition to hydrogen in light-duty vehicles could lead to a large reduction in oil imports, although the United States would continue to import crude oil or petroleum products to be used in large trucks, airplanes, and other industrial uses.

It should be noted that none of the estimates in Figures 6-1 through 6-5 depends on which technologies are used to produce hydrogen, but rather on whether hydrogen vehicles are introduced into the marketplace and on the rate at which they are adopted. However, the environmental, energy security, economic, and domestic resource use implications depend significantly on which technologies are used to generate the hydrogen. These issues are examined in subsequent sections of this chapter.

CARBON DIOXIDE EMISSIONS AS ESTIMATED IN THE COMMITTEE'S VISION

As noted in Chapter 5, one of the important goals of the hydrogen program is to reduce the emissions of carbon dioxide into the atmosphere, given the impacts of possible global climate change associated with releases of greenhouse gases. Therefore, it is important to estimate the amount by which shifts from gasoline in automobiles to hydrogen for fueling vehicles would change CO_2 emissions. In order to put the committee's estimates in context, Figure 6-6 shows EIA projections of U.S. carbon emissions in the form of CO_2, broken down by energy-consuming sectors and by fossil fuels. The EIA projects that by the year 2025, the United States will be emitting more than 2200 million metric tons of carbon, over

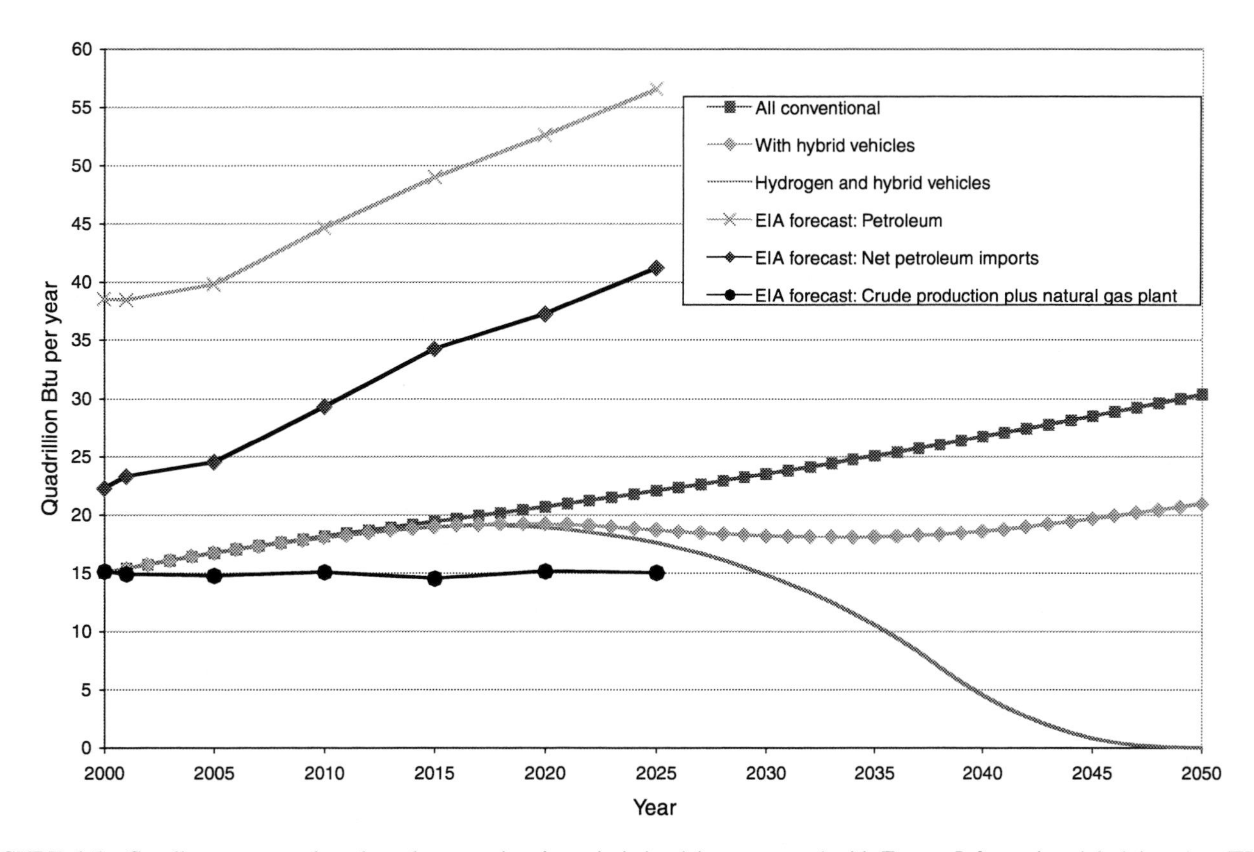

FIGURE 6-5 Gasoline use cases based on the committee's optimistic vision compared with Energy Information Administration (EIA) projections of oil supply, demand, and imports, 2000–2050. SOURCE: EIA (2003) for EIA projections.

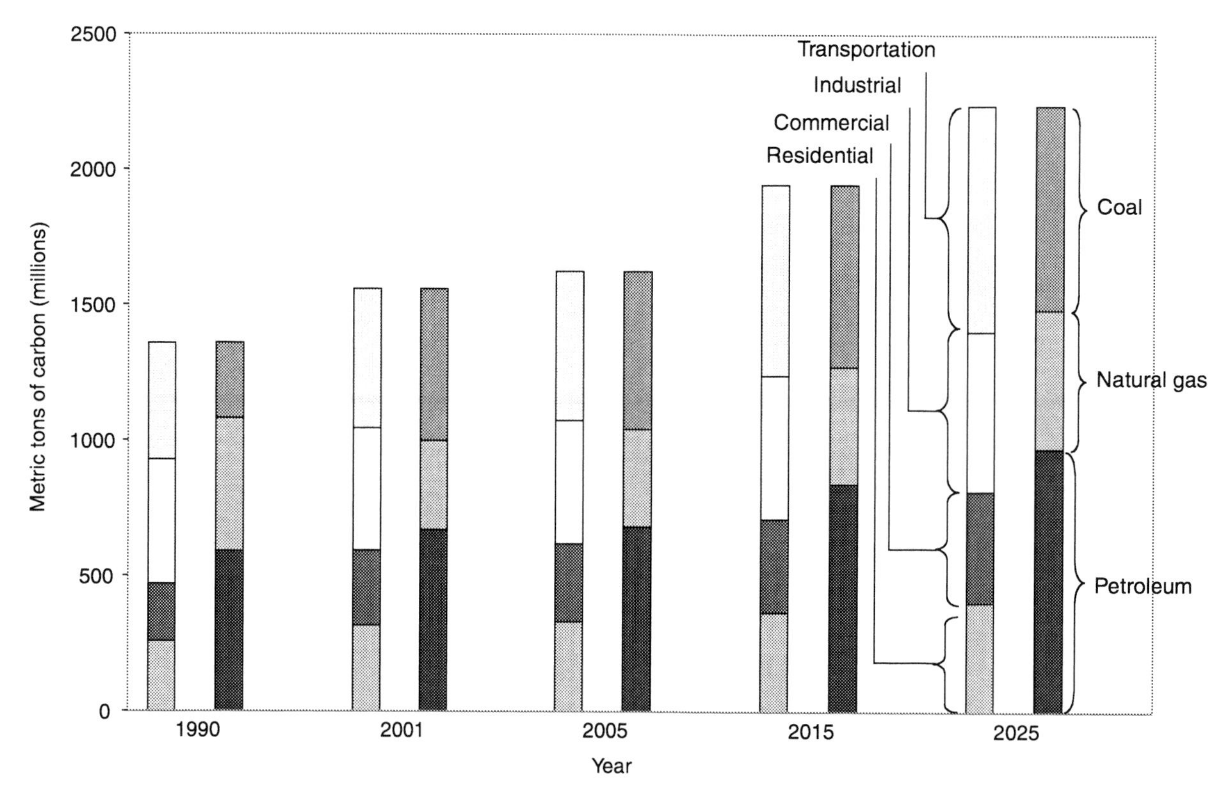

FIGURE 6-6 Projections by the Energy Information Administration (EIA) of the volume of carbon releases, by sector and by fuel, in selected years from 1990 to 2025. SOURCE: EIA (2003).

one-third of which is projected to be from petroleum use (EIA, 2003). The projections show that the entire transportation sector, not simply the light-duty vehicles, will account for 37 percent of these emissions. Thus, gasoline use in light-duty vehicles is an important component of the release of CO_2 into the atmosphere, comprising roughly two-thirds of the carbon emissions from the transportation sector (EIA, 2002), but it is not the dominant component.

In Chapter 5, the committee presented estimates of the amount of CO_2 that would be released into the atmosphere per kilogram of hydrogen produced for each of the technological pathways considered; it also gave estimates of the amount of CO_2 that would be released into the atmosphere per gallon of gasoline used. These estimates can be applied to the committee's estimates of gasoline consumption and hydrogen consumption over time in order to estimate the impacts of a transition to hydrogen on the carbon releases into the atmosphere. These estimates appear in Figures 6-7 and 6-9 for current hydrogen production technologies and in Figures 6-8 and 6-10 for possible future technologies.

Figures 6-7 through 6-10 show that a transition from conventional fueled vehicles to hybrids alone, without the introduction of hydrogen-fueled vehicles, would reduce carbon emissions by 200 million metric tons annually by 2050. A further transition from GHEVs to hydrogen vehicles would have sharply different impacts, depending on which technology was utilized. At one extreme, the use of coal without sequestration or of distributed electrolysis using grid-supplied electricity would lead to little or no further reductions in CO_2 releases than would occur through a transition to GHEVs.

Distributed generation of hydrogen by electrolysis using photovoltaics or wind turbines when they were available, and using grid-supplied electricity when the wind turbines or photovoltaics were not supplying electricity, could further reduce CO_2 emissions by a moderate amount (on the order of 100 million to 150 million metric tons per year by 2045). The reductions in CO_2 emissions from the possible future technologies could be somewhat greater than those obtainable using the current technologies, but the differences between the two are not great. However, distributed electrolysis using electricity exclusively from wind turbines could bring CO_2 emissions down to zero by 2050 if it were possible to generate all of the hydrogen by this means. The committee shows this particular technology for the possible future state of technology development and shows wind turbines combined with grid-supplied electricity for the current state of development.[7]

[7]The committee shows the particular technologies in this way because for the current state of technology development it will be less costly to have the grid-based electricity used with wind-based electricity, and for the possible future technologies it would be less costly to have an entirely wind-based system without the use of electricity from the grid.

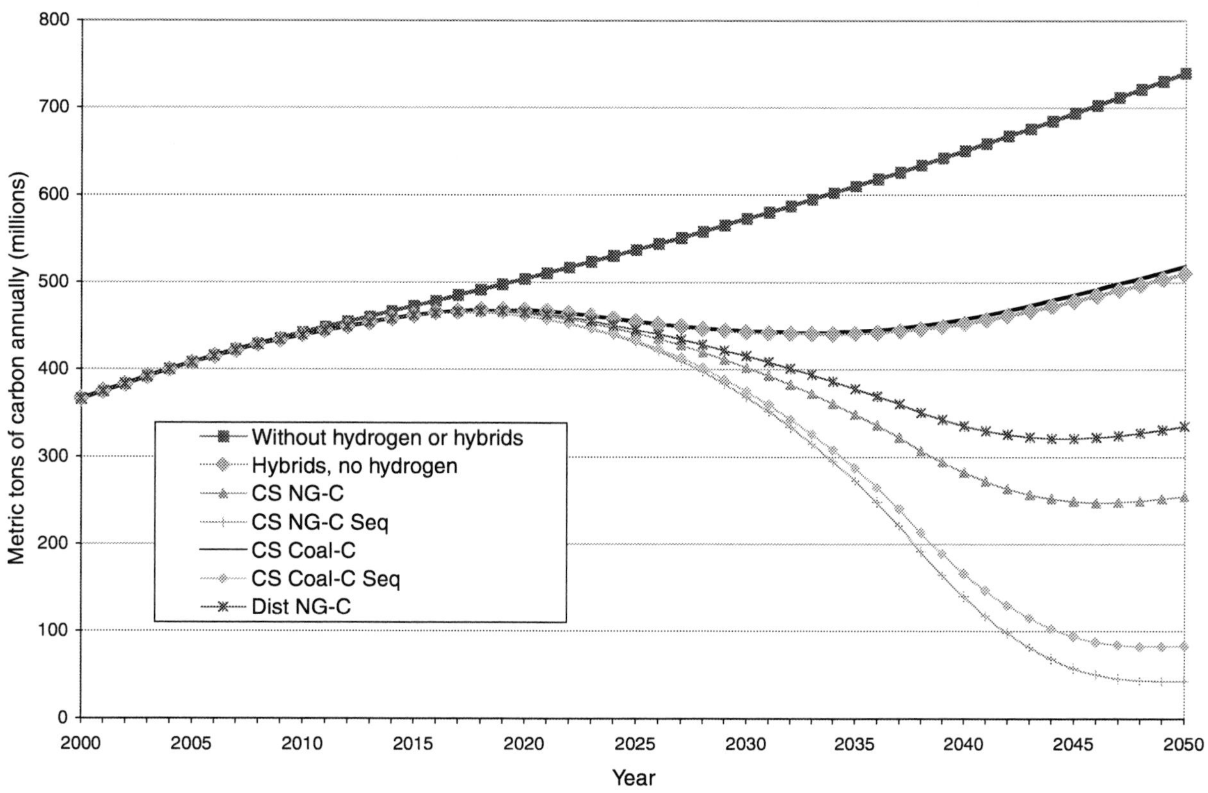

FIGURE 6-7 Estimated volume of carbon releases from passenger cars and light-duty trucks: current hydrogen production technologies (fossil fuels), 2000–2050. See Table 5-2 in Chapter 5 and discussion in text.

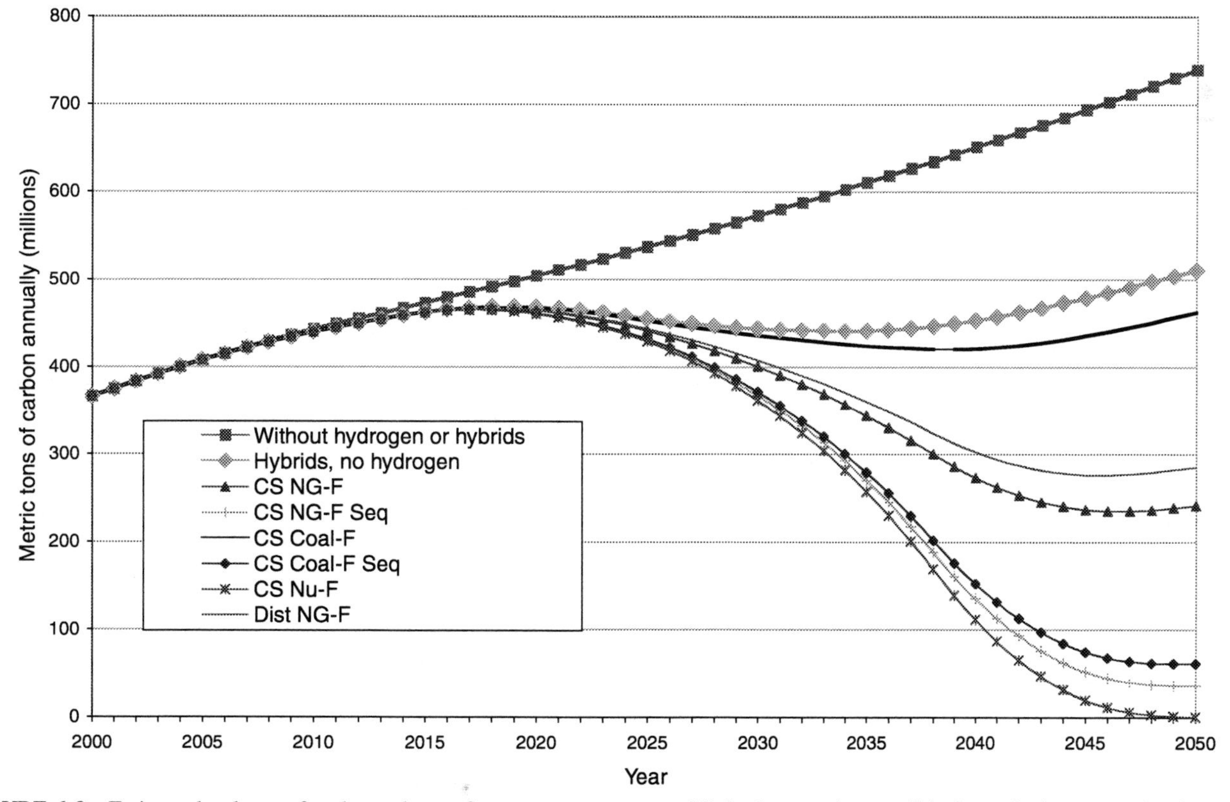

FIGURE 6-8 Estimated volume of carbon releases from passenger cars and light-duty trucks: possible future hydrogen production technologies (fossil fuels and nuclear energy), 2000–2050. See Table 5-2 in Chapter 5 and discussion in text.

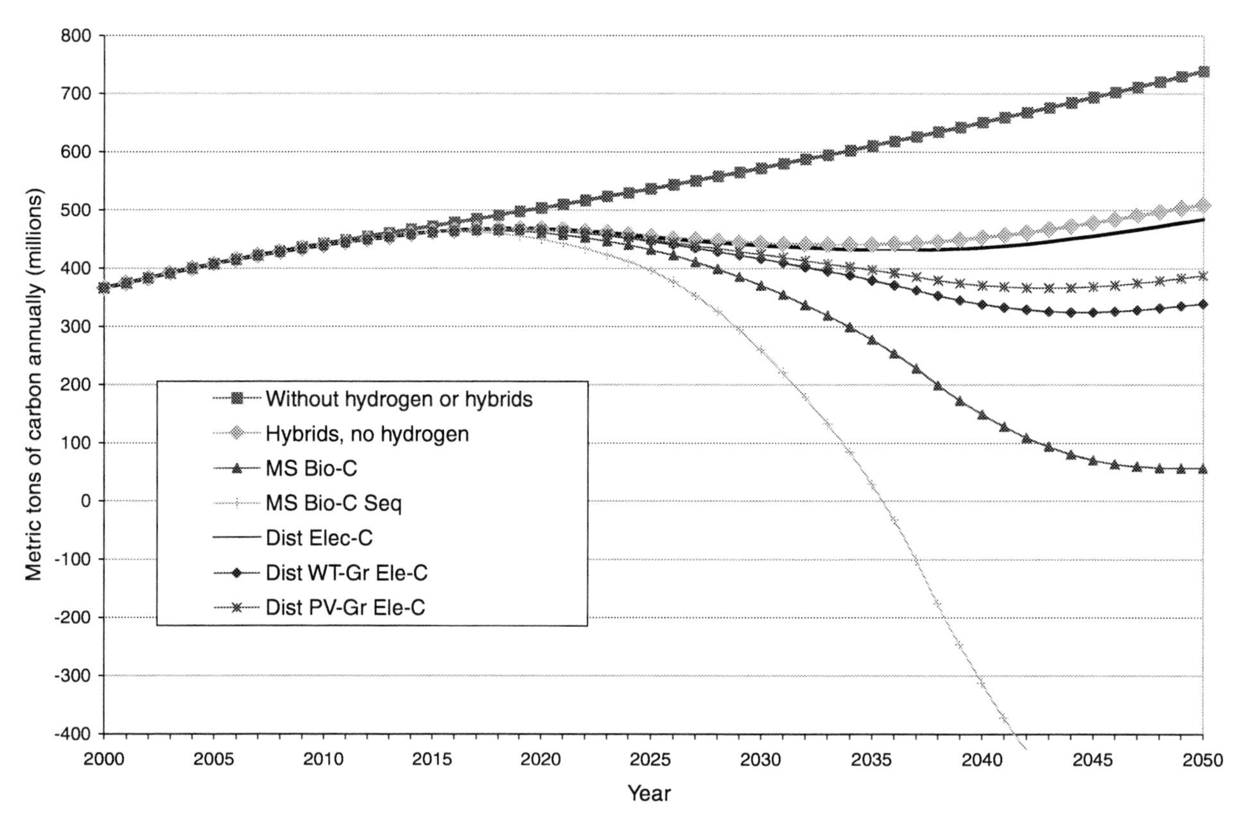

FIGURE 6-9 Estimated volume of carbon releases from passenger cars and light-duty trucks: current hydrogen production technologies (electrolysis and renewables), 2000–2050. See Table 5-2 in Chapter 5 and discussion in text.

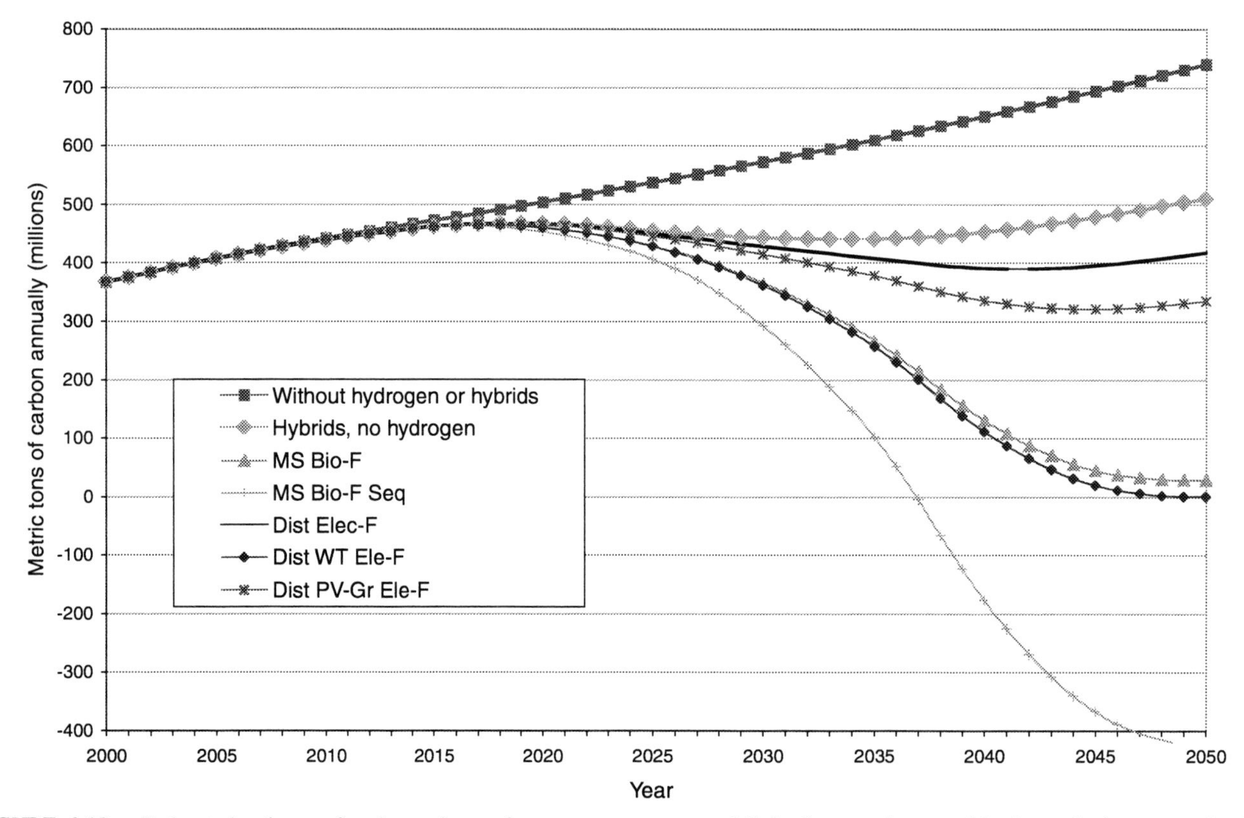

FIGURE 6-10 Estimated volume of carbon releases from passenger cars and light-duty trucks; possible future hydrogen production technologies (electrolysis and renewables), 2000–2050. See Table 5-2 in Chapter 5 and discussion in text.

Steam reforming using natural gas in a central station or distributed facility could reduce CO_2 emissions on the order of 200 million metric tons per year by 2050, in either state of technology development. Also, sharp reductions in CO_2 emissions would occur if all of the hydrogen was generated using biomass as a feedstock, or nuclear power as a heat source, or if the CO_2 from a coal-based or a natural-gas-based technology was separated and sequestered.

At the other extreme, if all of the hydrogen could be generated using biomass as a feedstock and all of the CO_2 could be separated at the point of hydrogen production and sequestered, there would be negative net emissions of CO_2 into the atmosphere after 2036. That is, on net, the process would take significant amounts of CO_2 out of the atmosphere.[8]

SOME ENERGY SECURITY IMPACTS OF THE COMMITTEE'S VISION

As noted, a second important goal of the hydrogen program is to improve energy security by substituting secure domestic resources for imported energy resources, particularly those that may be traded in unstable international markets. Figure 6-4 shows that a transition to hydrogen in light-duty vehicles could sharply reduce the use of gasoline and thus could reduce the importation of oil. Some of the technologies would use domestic resources without increasing the importation of other energy from potentially unstable parts of the world. Technologies based on coal, biomass, nuclear power, or entirely on renewables, such as wind turbines and photovoltaics, would not lead to significant energy imports. A transition to hydrogen could improve energy security if the hydrogen were generated from such domestic feedstocks.

Other technologies, however, would use natural gas, a commodity which, although produced domestically, is also imported in significant quantities and would be subject to some of the same international market instability that occurs in the petroleum markets. Additional uses of natural gas would lead to additional imports. In this case, whether energy security is improved or harmed depends on whether the security benefits from reduced oil imports are greater than the security costs of increased natural gas imports.

In order to examine this issue, estimates were developed of the amount of natural gas that would be used if all of the hydrogen were generated using one of the natural-gas-based technologies. These estimates appear in Figure 6-11, which includes estimates for both current and possible future technologies. This figure also includes the EIA projections of natural gas supply, demand, and imports in order to put the estimates from the committee's vision in context.

Figure 6-11 shows that if all of the hydrogen were generated using one or more of the natural-gas-based technologies, the increase in natural gas consumption would be a significant fraction of the projected domestic production. It also shows that, according to EIA projections, the United States will be importing a significant fraction of this natural gas in the years 2010 through 2025. Given the magnitude of the use of natural gas for hydrogen production, it can be reasonably expected that most of the additional consumption will result in additional imports of natural gas once the United States gets beyond a transition period. However, during the transition period (through 2030), natural gas imports would not increase significantly.

The additional use of natural gas can be compared with the reduced use of gasoline. Figure 6-12 provides this comparison for the current technologies, and Figure 6-13 provides the comparison for possible future technologies. Both of these graphs plot, on the same scale, the gasoline reductions associated with the penetration of hydrogen vehicles in place of hybrid electric vehicles, and the natural gas use increases for the central station natural-gas-based technologies, with and without sequestration, and the distributed reforming of natural gas.

Figures 6-12 and 6-13 show that the increases in natural gas use, measured in quads, are of similar magnitude to the decreases in gasoline use, although the natural gas increases with the possible future technologies will be somewhat smaller than the decreases in gasoline use will be. These figures suggest that it is unlikely that a transition to hydrogen based on natural gas would significantly increase energy security.

It must be stressed, however, that the issue raised here would not be relevant for the other domestically produced resources or if large new sources of domestic natural gas are found. Technologies based on coal, biomass, nuclear power, or the two renewables—wind turbines and photovoltaics—would not result in such compensating increases of energy imports. A transition to hydrogen using these feedstocks could thus improve energy security.

A sharp reduction in gasoline use would require important adjustments in U.S. petroleum refining. These adjustments themselves could have energy security implications. Existing refineries swing between summer and winter differences in demand for gasoline and distillate fuels. However, if gasoline use is reduced to a very small portion of refined products, new refining processes may be needed. Alternatively, U.S. refiners might continue importing crude oil, making gasoline for exportation. The implications of such a scenario, or of alternative responses, are worthy of examination.

---

[8]Less carbon is sequestered in the possible future biomass technology case than in the current technology case (i.e., carbon emissions become less negative). This reduction would be the result of the increased efficiency of hydrogen generation with the new technologies. A more efficient process implies that less biomass is needed per kilogram of hydrogen and thus less CO_2 is removed from the atmosphere and fixed as organic carbon in the biomass.

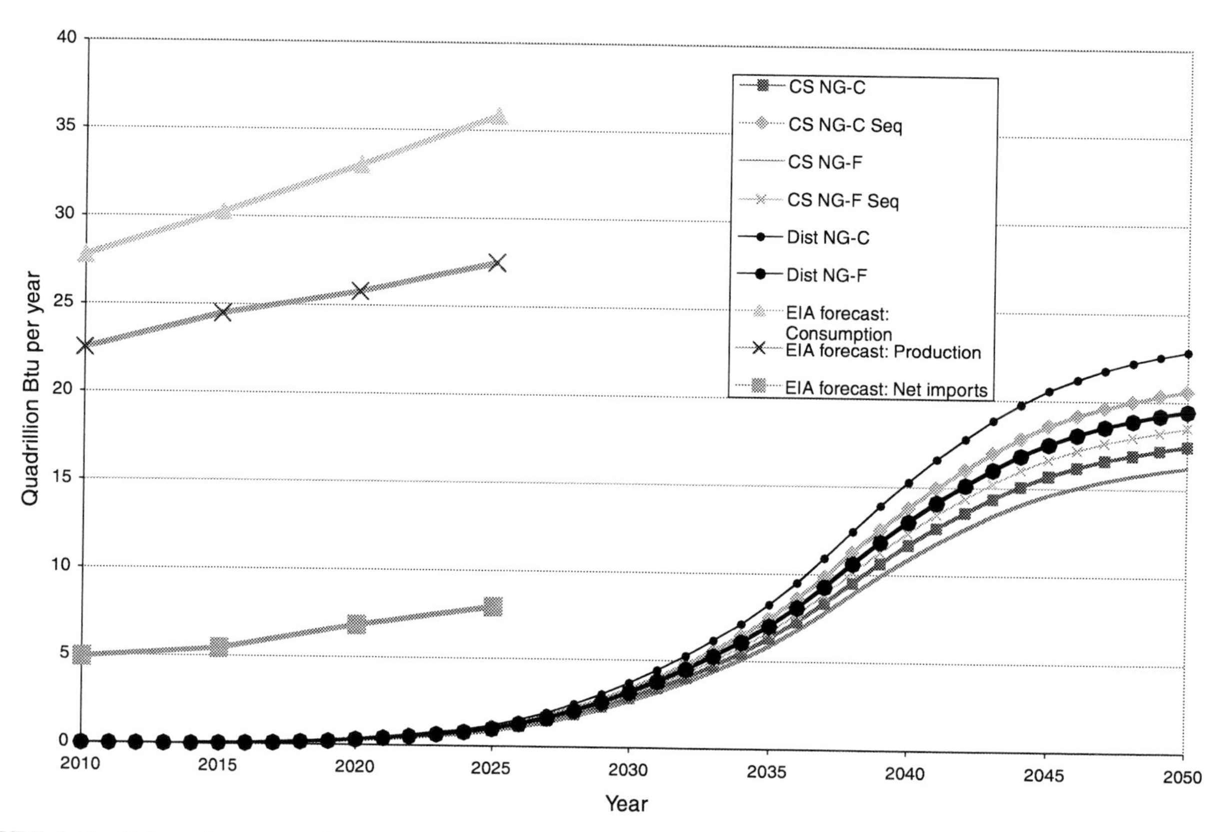

FIGURE 6-11 Estimated amounts of natural gas to generate hydrogen (current and possible future hydrogen production technologies) compared with projections by the Energy Information Administration (EIA) of natural gas supply, demand, and imports, 2010–2050. See Table 5-2 in Chapter 5 and discussion in text. SOURCE: EIA (2003) for EIA projections.

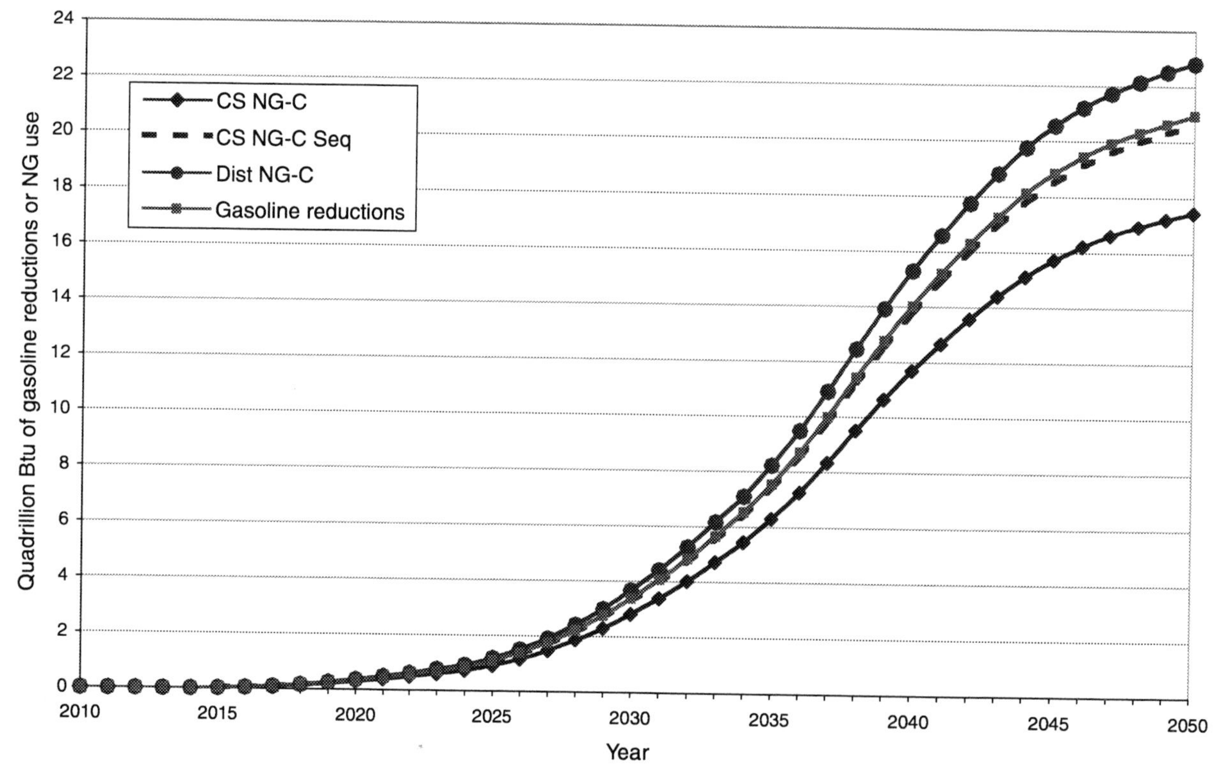

FIGURE 6-12 Estimated gasoline use reductions compared with natural gas (NG) use increases: current hydrogen production technologies, 2010–2050. See Table 5-2 in Chapter 5.

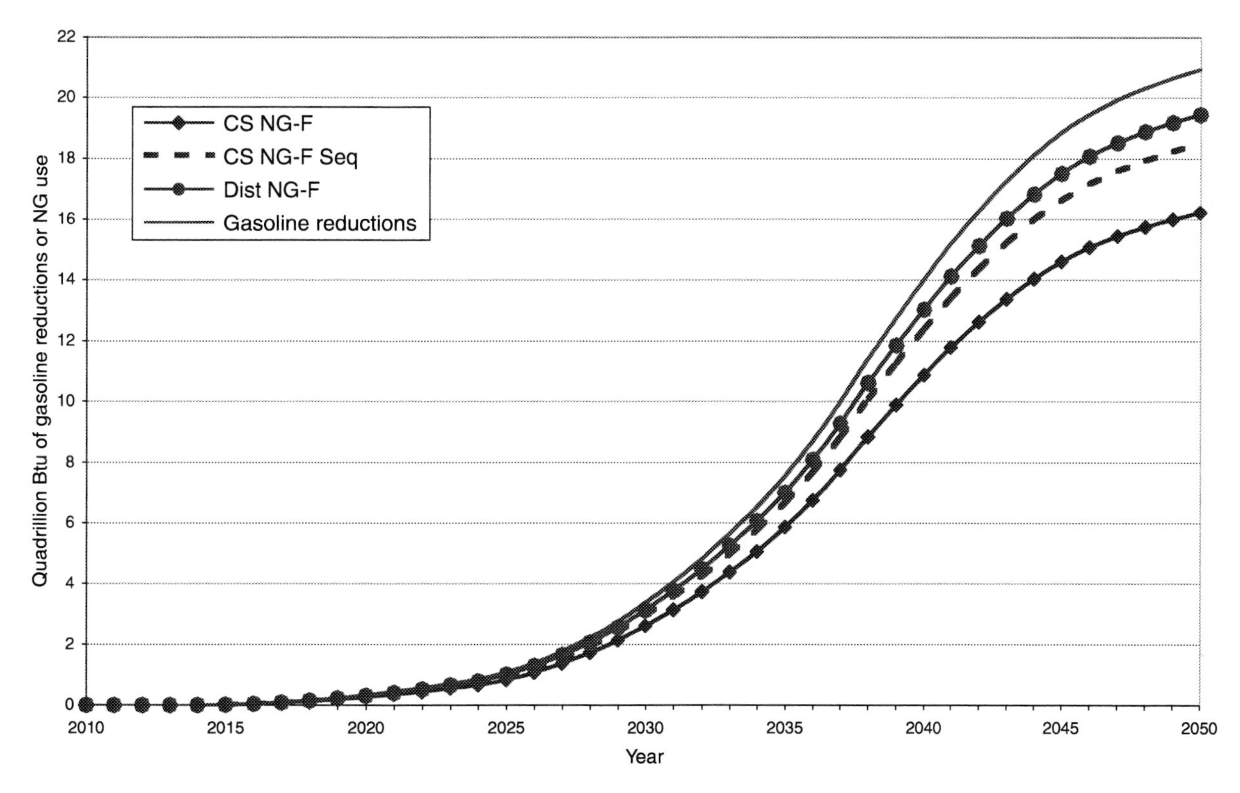

FIGURE 6-13 Estimated gasoline use reductions compared with natural gas (NG) use increases: possible future hydrogen production technologies, 2010–2050. See Table 5-2 in Chapter 5.

OTHER DOMESTIC RESOURCE IMPACTS BASED ON THE COMMITTEE'S VISION

In addition to impacts on natural gas, the committee has estimated impacts on several other domestic resources. Coal-based hydrogen generation would require increased U.S. production of coal. Biomass-based hydrogen production would require the use of land. The sequestration of CO_2 would require infrastructure for sequestration as well as domestic resources into which the sequestered CO_2 could be prominently placed. The committee summarizes here some of the most important of these impacts on such domestic resources. It continues to maintain the discussion about pure options in which all of the hydrogen is produced from a given feedstock. The reader should be reminded that, more realistically, if the challenges of hydrogen are mastered, the transition will not be to such a pure system but rather to a system in which many different supply chains are used to provide the hydrogen.

Hydrogen generation using only coal as a feedstock could be expected to significantly increase the use of coal in the United States. Figure 6-14 provides those estimates for both current and possible future technologies that use coal as a feedstock, either with or without CO_2 sequestration. The figure puts these estimates in perspective by including the EIA forecast of U.S. consumption and production of coal.[9]

Figure 6-14 shows that, by 2050, hydrogen production could use between 13 quadrillion and 15 quadrillion Btu per year of coal, with slightly smaller quantities for possible future technologies and slightly larger quantities for technologies involving CO_2 sequestration. The figure shows that, at least through 2035, the use of coal for hydrogen production can be expected to be a relatively small fraction of total coal production. However, by 2050, if hydrogen were generated exclusively using coal-based technologies, its use for hydrogen production would be a substantial portion of the industry.

Technologies that use biomass as a feedstock require substantial acreage in order to grow the biomass. In the models developed for the study, it is assumed that under current technology conditions, 4.0 tons of bone-dry biomass can be grown per year for each acre of land and that each ton of biomass has an energy content of 16 million Btu. Under possible future technology conditions, it is assumed that the growing of a biomass becomes more productive, so that 6.0

[9]Figure 6-14 shows the EIA projection that domestic production and consumption of coal will remain equal to one another, so there will be no net imports of coal.

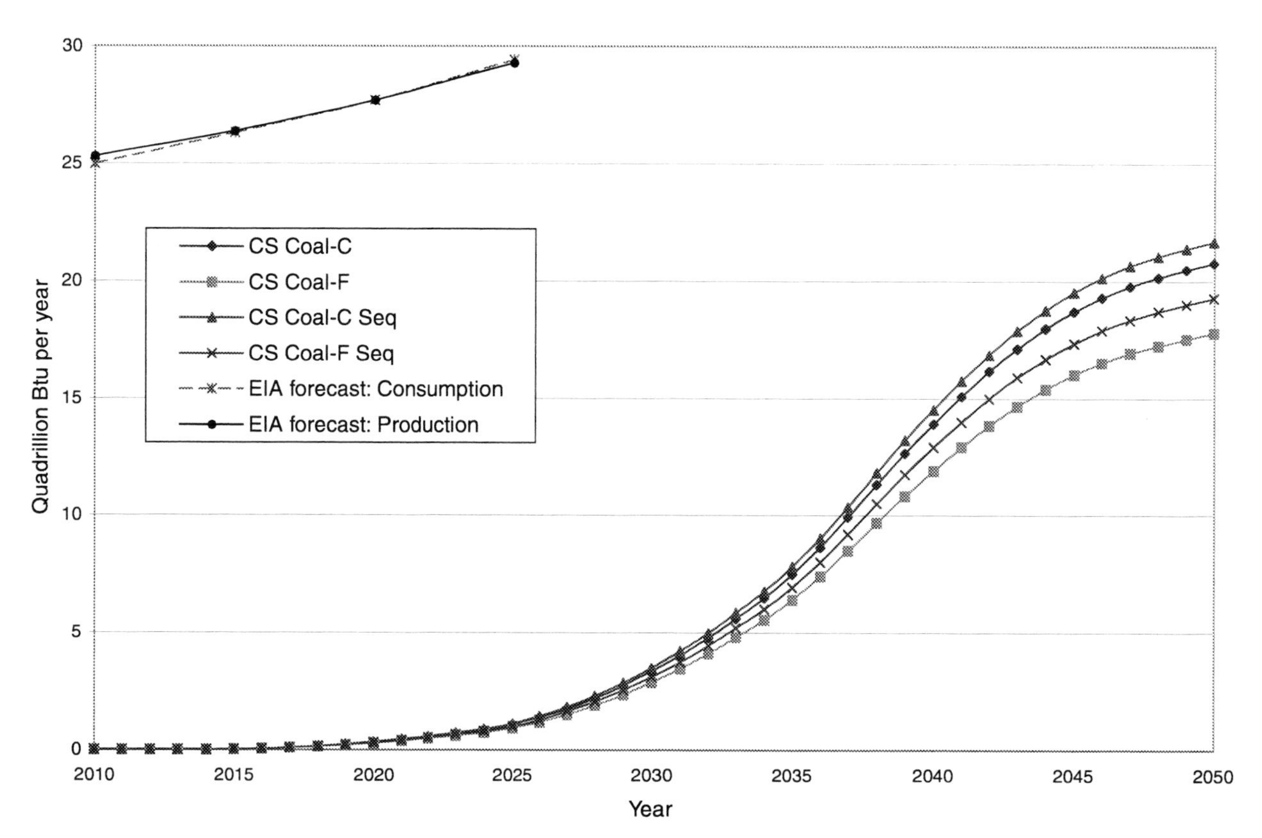

FIGURE 6-14 Estimated amounts of coal used to generate hydrogen (current and possible future hydrogen production technologies) compared with Energy Information Administration (EIA) projections of coal production and use, 2010–2050. See Table 5-2 in Chapter 5 and discussion in text. SOURCE: EIA (2003) for EIA projections.

tons of bone-dry biomass can be grown per year for each acre of land. These assumptions allow the committee to develop estimates of the amount of land that would be required if biomass were the feedstock for 100 percent of the hydrogen production. Figure 6-15 provides those estimates for both the current and possible future technologies, both with and without CO_2 sequestration.

Figure 6-15 shows that under current technology conditions, if all of the hydrogen were generated from biomass, in 2050 the United States would be using about 650,000 mi^2 of land to grow the biomass needed to fuel the light-duty fleet of vehicles. However, with the possible future technologies, the nation would need a substantially smaller amount, about 280,000 mi^2. The difference between the two estimates of land use results from differences in the assumed productivity of land and differences in the efficiency of the gasifier under the two states of technology development.

For comparison purposes, the United States is estimated to have roughly 700,000 mi^2 of cropland and 900,000 mi^2 of rangeland or pastureland (Vesterby and Krupa, 1997). If the biomass can be grown on land that currently serves as rangeland or pastureland, which the committee believes is unlikely because of water-use restrictions, then under possible future technology conditions, by 2050 biomass production would account for about 16 percent of this land, even if all of the hydrogen were made using biomass as a feedstock. However, if the biomass requires land that currently serves as cropland, then by 2050 under possible future technology conditions, biomass production could use about 33 percent of all current cropland.

For those technologies that rely on CO_2 sequestration, the committee examined the amount of CO_2 that would be sequestered annually and the cumulative sequestration. The models assume that 90 percent of the CO_2 for a given plant can be separated and sequestered and that 10 percent of the CO_2 will escape into the atmosphere. Figures 6-16 and 6-17 respectively provide estimates of the annual and cumulative amounts of CO_2 that would be sequestered with current technologies, for central station natural gas and coal plants and midsize biomass plants. Figures 6-18 and 6-19 respectively provide annual and cumulative sequestration estimates for possible future technologies.

Figures 6-16 through 6-19 show the massive amount of CO_2 sequestration that would be required, both annually and cumulatively, in order to use fossil fuels as hydrogen feedstocks while sharply reducing the amount of CO_2 released into the atmosphere. By 2050 the United States would need

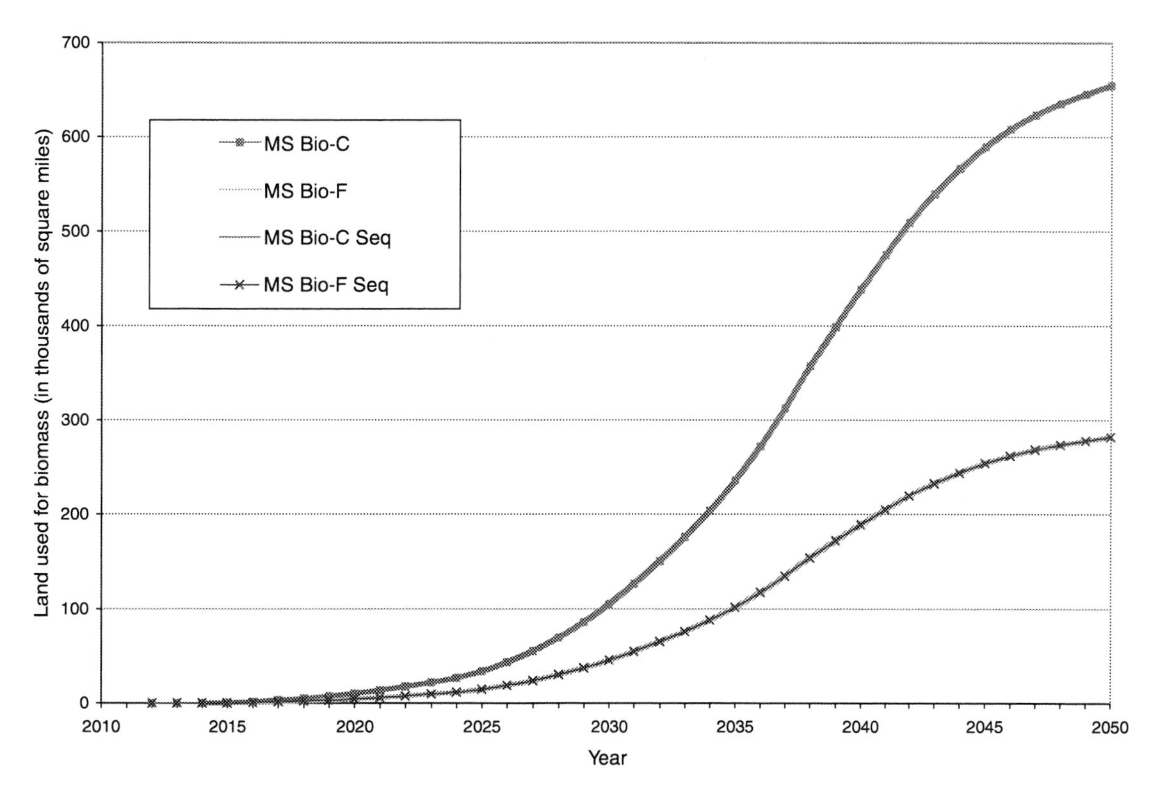

FIGURE 6-15 Estimated land area used to grow biomass for hydrogen: current and possible future hydrogen production technologies, 2010–2050. See Table 5-2 in Chapter 5 and discussion in text. NOTE: The curves for current midsize biomass with sequestration (MS Bio-C Seq) and without sequestration (MS Bio-C) are identical, as are the curves for possible future midsize biomass with sequestration (MS Bio-F Seq) and without sequestration (MS Bio-F).

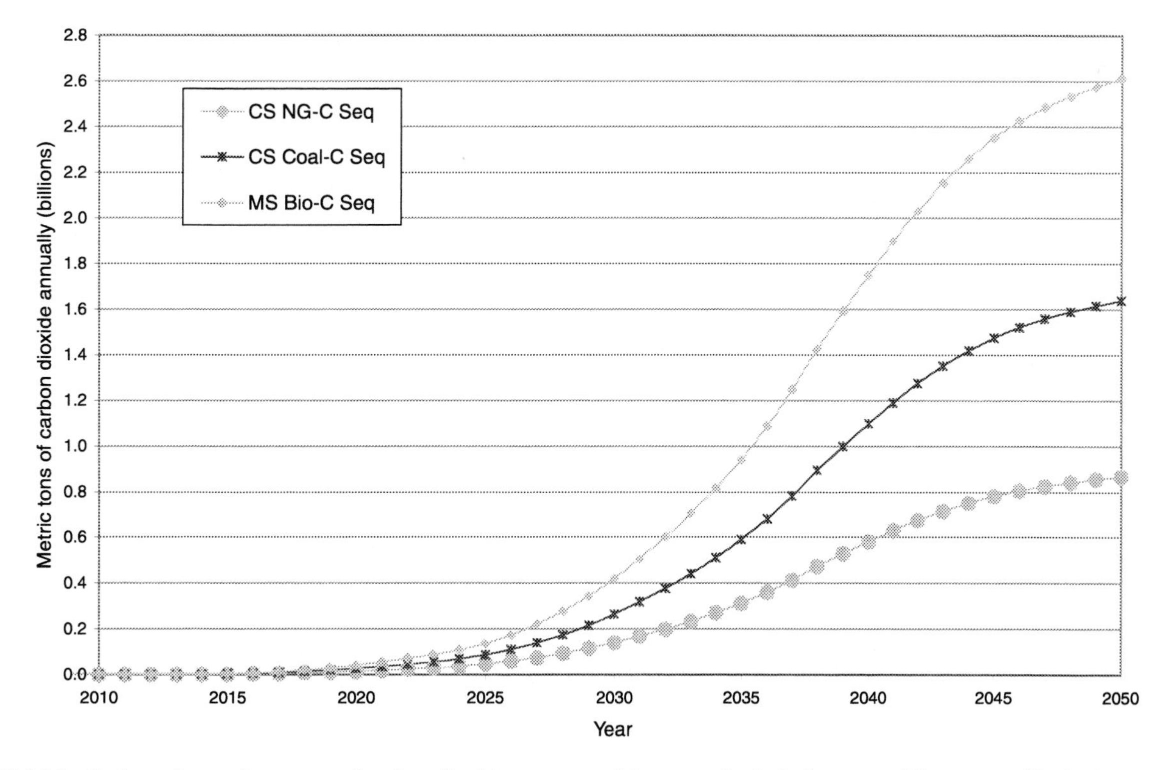

FIGURE 6-16 Estimated annual amounts of carbon dioxide sequestered from supply chain for automobiles powered by hydrogen: current hydrogen production technologies, 2010–2050. See Table 5-2 in Chapter 5 and discussion in text.

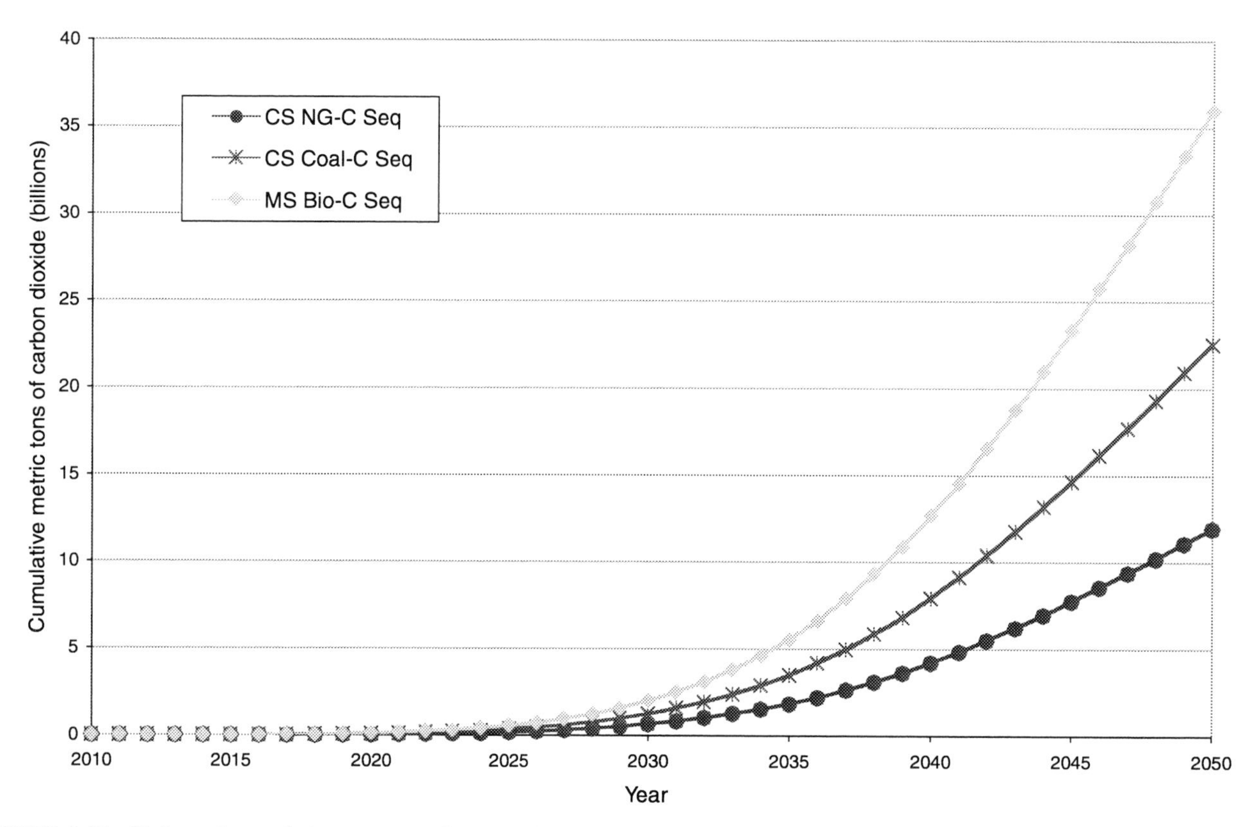

FIGURE 6-17 Estimated cumulative amounts of carbon dioxide sequestered from supply chain for automobiles powered by hydrogen: current hydrogen production technologies, 2010–2050. See Table 5-2 in Chapter 5 and discussion in text.

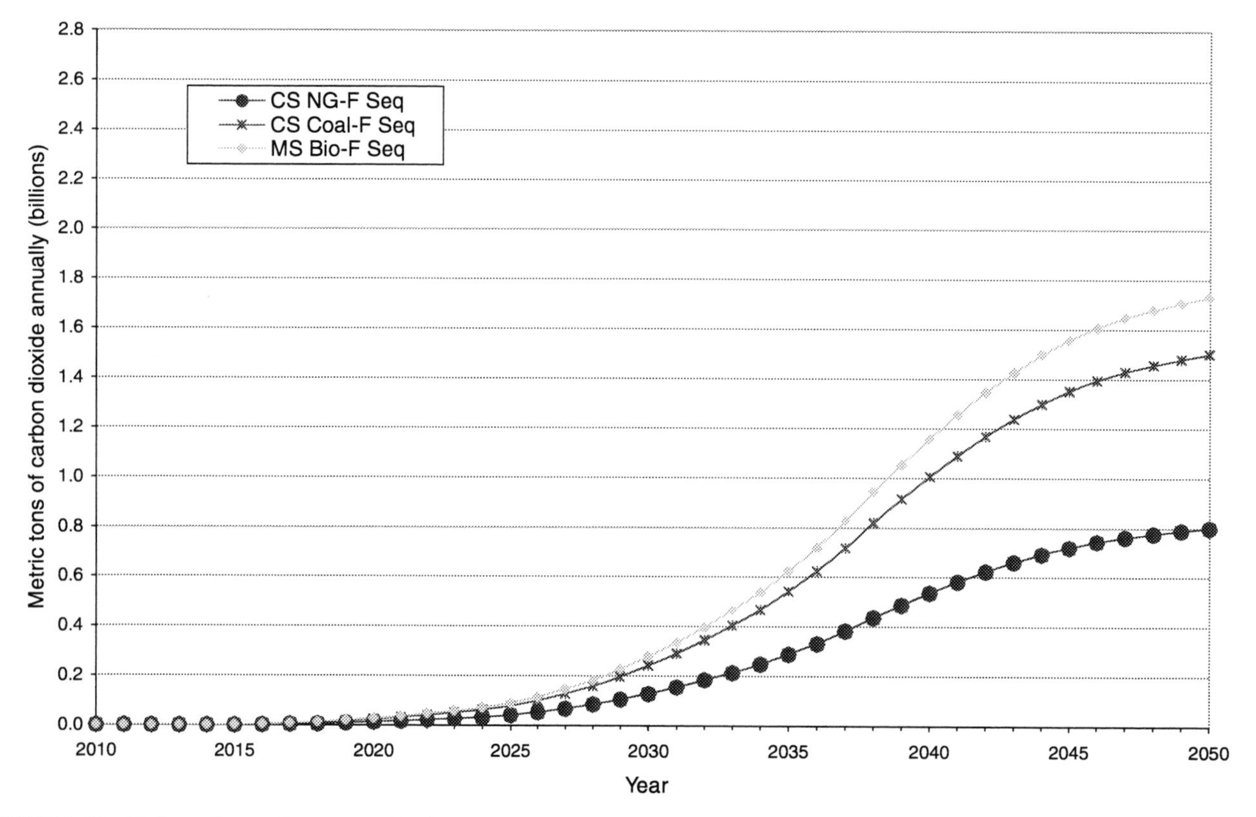

FIGURE 6-18 Estimated annual amounts of carbon dioxide sequestered from supply chain for automobiles powered by hydrogen: possible future hydrogen production technologies, 2010–2050. See Table 5-2 in Chapter 5 and discussion in text.

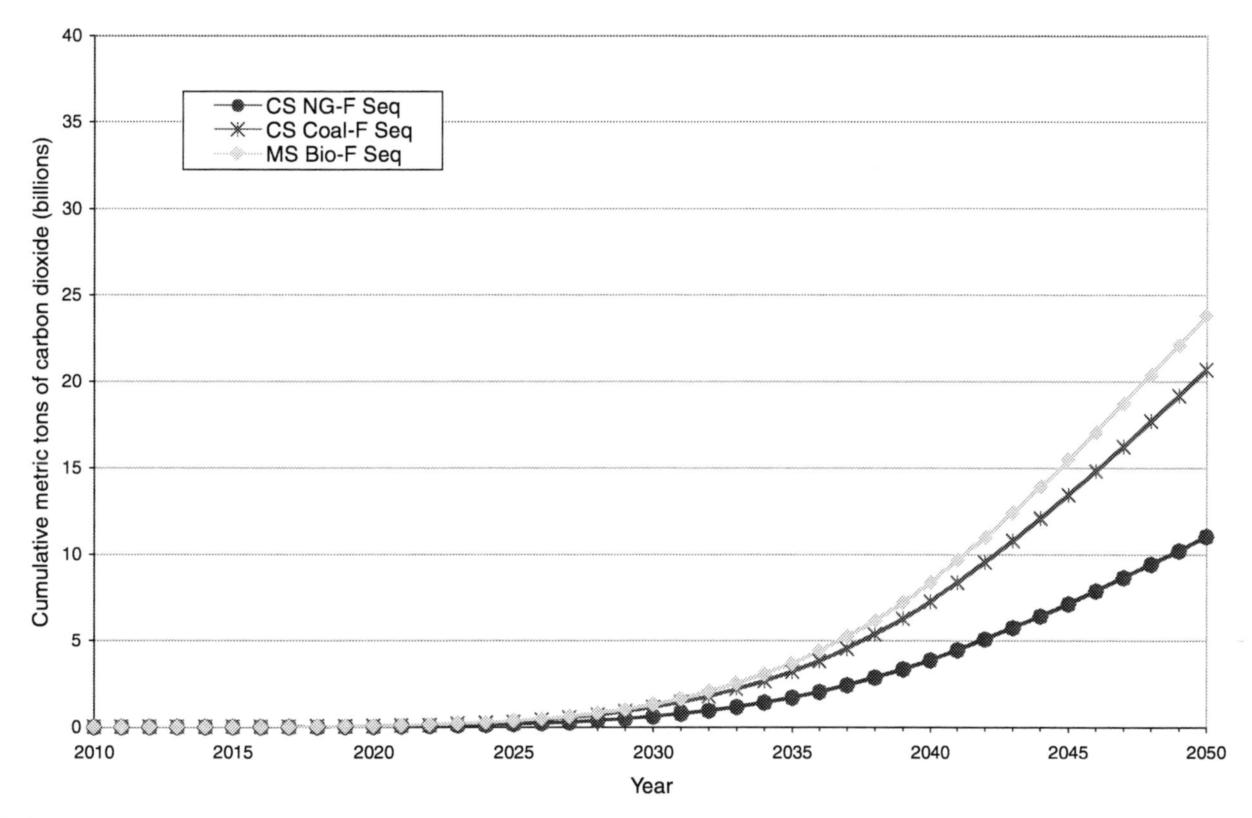

FIGURE 6-19 Estimated cumulative amounts of carbon dioxide sequestered from supply chain for automobiles powered by hydrogen: possible future hydrogen production technologies, 2010–2050. See Table 5-2 in Chapter 5 and discussion in text.

to sequester about 10 billion metric tons of CO_2, cumulatively, if hydrogen was generated using natural gas as a feedstock, and about twice as much with coal as a feedstock. These figures also suggest that, except for biomass-based hydrogen, there is relatively little difference in the amount of sequestration needed between current and possible future technologies. Both the rate of sequestration and the cumulative amount of sequestration needed can be expected to pose very great challenges.

These estimates can be compared with the available estimates of the geological sequestration capacities of potential locations. North American storage capacity is estimated at between 5 and 500 gigatons (GT) CO_2 in Holloway (2001), and the same review article notes that the capacity of a single aquifer has been estimated at 9 to 43 GT CO_2.

To put the annual volumes of sequestration in context, one can compare them with the movement of natural gas. The EIA (2003) projections for the year 2025 of natural gas consumption at 36 quadrillion Btu per year translates to roughly 0.7 billion metric tons of natural gas moved per year. Thus, sequestration of CO_2 from coal-based or biomass-based hydrogen production in 2050 (see Figures 6-16 and 6-18) would require the movement of a mass of CO_2 twice the amount that the EIA projects to be the mass of natural gas moved in 2025.

IMPACTS OF THE COMMITTEE'S VISION FOR TOTAL FUEL COSTS FOR LIGHT-DUTY VEHICLES

Finally, the committee considered the economic impacts of the alternative hydrogen production technologies. Taking into account the estimate of the consumption of gasoline over time, the consumption of hydrogen over time, the cost of gasoline,[10] and the cost of hydrogen from the various technologies, estimates were made of the total cost per year of fueling the fleet of automobiles. Under the assumption that hydrogen-fueled vehicles have the same production and maintenance costs as those for gasoline-fueled vehicles, differences in the total cost per year of fueling the fleet of automobiles translates directly into differences in the total economic costs of the transition to hydrogen.

Figures 6-20 and 6-21 provide these total annual costs for the current technologies, for fossil fuels, and for renewables and distributed electrolysis, respectively. Figures 6-22 and 6-23 provide similar data for future technologies for fossil fuels and nuclear thermal energy and for renewables and distributed electrolysis, respectively. In each of Figures 6-20 through 6-23, there is a curve displaying an estimation of the annual fuel cost with only conventional vehicles, with no

[10]The gasoline cost is estimated as $1.27 per gallon.

GHEVs and no hydrogen vehicles. A second line provides an estimate of total annual fuel costs if GHEVs ultimately capture 100 percent of the market share and hydrogen-fueled vehicles are never introduced. The other lines assume that hydrogen-fueled vehicles capture the market shares over time (at the rates shown in Figure 6-1) and that all of the hydrogen is produced using the particular technology denoted; GHEVs are being phased in and then out of the market using the estimates in Figure 6-1.

Figures 6-20 and 6-21 show the large impact of the penetration of GHEVs into the marketplace. These figures suggest that by 2050, the movement from conventional vehicles to GHEVs alone could reduce the fuel cost by about $75 billion per year, without the introduction of hydrogen-fueled vehicles.

Figures 6-20 and 6-21 show that most of the current technologies would lead to total costs that are higher than the amount drivers would face if GHEVs ultimately dominated the fleet. However, central station coal-based or natural-gas-based hydrogen production could keep total costs almost identical to the costs with GHEVs. Hydrogen based on distributed natural gas would be somewhat more costly. But Figure 6-21 shows that if the system were to be based on distributed electrolysis, biomass, or distributed photovoltaics, the total cost would be substantially greater than would be possible with even hybrid vehicles or conventional vehicles. For example, in 2050 the cost of using these technologies would exceed the cost of using gasoline in GHEVs by more than $400 billion annually.

Figures 6-22 and 6-23 show the great importance of possible future technologies on the total cost of the system. They show that if the possible future technologies are successfully developed and have costs consistent with the committee's estimates, all but the biomass and the grid-electric or photovoltaic-based electrolysis technologies could be operated at costs less than those that would characterize a system of gasoline-fueled conventional vehicles. The central station coal-based and natural-gas-based technologies would be lower in cost than that of operating a system of gasoline-fueled hybrid electric vehicles. But the technologies based on distributed electrolysis operating either entirely on grid-supplied electricity or partially on photovoltaic-supplied

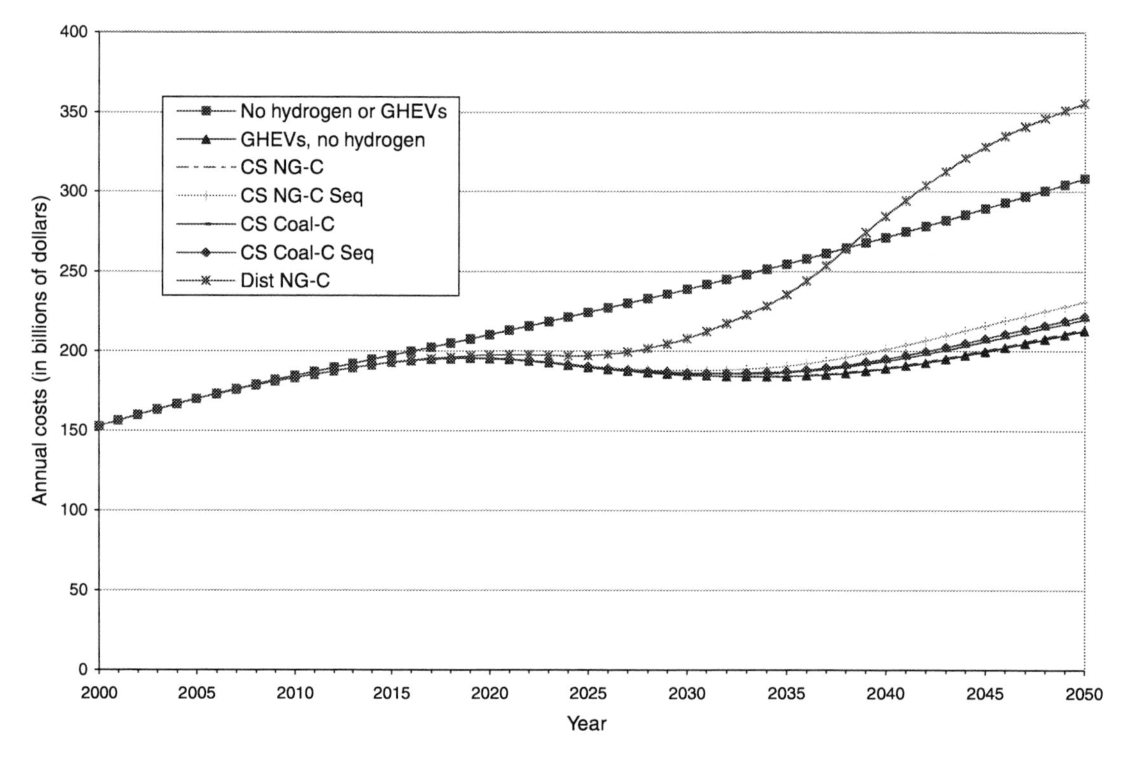

FIGURE 6-20 Estimated total annual fuel costs for automobiles: current hydrogen production technologies (fossil fuels), 2000–2050. Each line for the various hydrogen production technologies assumes that hydrogen-fueled vehicles capture the market shares over time (at the rates shown in Figure 6-1) and that all of the hydrogen is produced using the particular technology denoted (e.g., CS NG-C, CS Coal-C, and so on); gasoline hybrid electric vehicles (GHEVs) are being phased in and then out of the market using the estimates in Figure 6-1. Two other cost curves are provided, one displaying an estimation of the annual fuel cost with only conventional vehicles (no hydrogen or GHEVs). A second line provides an estimate of total annual fuel costs if GHEVs ultimately capture 100 percent of the market share and hydrogen-fueled vehicles are never introduced (GHEVs, no hydrogen). See Table 5-2 in Chapter 5 and discussion in text. NOTE: The cost curve for central station natural gas (CS NG-C) is obscured by the cost curve for GHEVs (GHEVs, no hydrogen), and the cost curve for central station coal with sequestration (CS Coal-C Seq) is partly obscured by the cost curve for coal without sequestration (CS Coal-C).

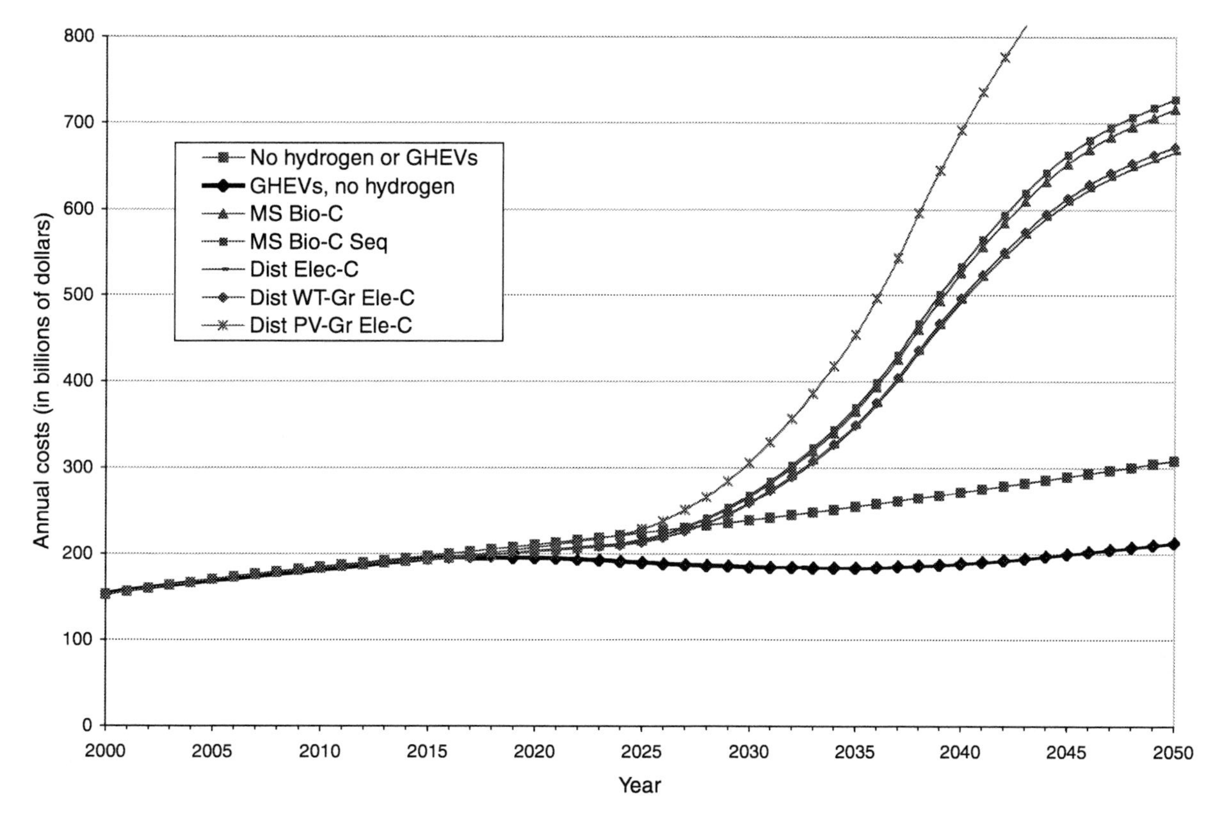

FIGURE 6-21 Estimated total annual fuel costs for light-duty vehicles: current hydrogen production technologies (electrolysis and renewables), 2000–2050. Each line for the various hydrogen production technologies assumes that hydrogen-fueled vehicles capture the market shares over time (at the rates shown in Figure 6-1) and that all of the hydrogen is produced using the particular technology denoted (e.g., MS Bio-C, Dist Elec-C, and so on); gasoline hybrid electric vehicles (GHEVs) are being phased in and then out of the market using the estimates in Figure 6-1. Two other cost curves are provided, one displaying an estimation of the annual fuel cost with only conventional vehicles ("no hydrogen or GHEVs"). A second line provides an estimate of total annual fuel costs if GHEVs ultimately capture 100 percent of the market share and hydrogen-fueled vehicles are never introduced ("GHEVs, no hydrogen"). See Table 5-2 in Chapter 5 and discussion in text. NOTE: The cost curve for distributed electrolysis (Dist Elec-C) is obscured by the cost curve for distributed wind turbine/grid hybrid electrolysis (Dist WT-Gr Ele-C), since these two cost estimates are virtually identical.

electricity have substantially greater costs than those for a system with no hybrids or conventional vehicles. Likewise, the biomass technologies are substantially more costly. Therefore, if research and development are successful and the possible new technologies are developed consistent with the committee's estimations, and if the challenges associated with fuel cell vehicles themselves are solved, almost all of these technologies might be able to compete successfully with conventional gasoline-fueled vehicles, and most would lead to total costs that are roughly comparable with the costs of operating GHEVs.

The committee expects that, absent hydrogen, GHEVs, not conventional vehicles, will come to dominate the fleet of automobiles; nonetheless, the cost of conventional vehicles provides a useful benchmark. This cost is consistent with current cost per mile driven and growth in vehicle miles influenced by population and income growth. Thus, the conventional vehicle cost is consistent with what Americans have been willing to pay for the fuel costs of driving (itself only a fraction of the total costs of driving). Thus, Figures 6-22 and 6-23 show that if the possible future technologies are successfully developed and have costs consistent with the committee's estimates, all but the biomass and the grid-electric or photovoltaic-based electrolysis technologies could be operated at costs less than what Americans have been willing to pay for the fuel costs of driving.

SUMMARY

In this chapter, the committee examined its vision of how the energy system might operate if hydrogen-fueled vehicles were broadly adopted in place of gasoline-fueled vehicles. The implications of broad adoption of hydrogen for other purposes, such as electricity generation, are not examined in depth. However, this examination of the transition to hydrogen for light-duty vehicles suggests that the implications for the energy system could be profound, depending on which technologies were adopted. Some technologies could lead to sharp reductions in the amount of CO_2 released into the atmosphere, but not all could lead to such environmental

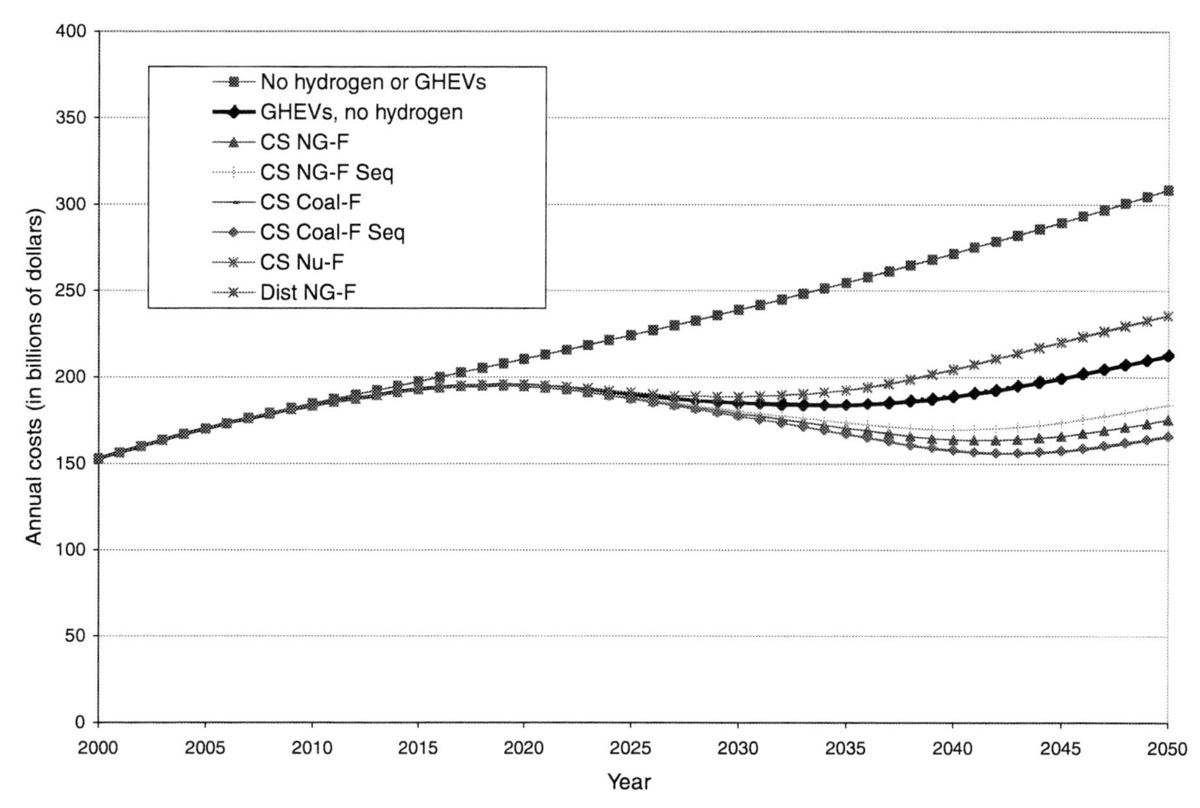

FIGURE 6-22 Estimated total annual fuel costs for light-duty vehicles: possible future hydrogen production technologies (fossil fuels and nuclear energy), 2000–2050. See Table 5-2 in Chapter 5 and discussion in text. NOTE: The cost curve for nuclear thermal energy (CS Nu-F) is obscured by the cost curve for distributed generation from natural gas (Dist NG-F), since these two cost estimates are virtually identical.

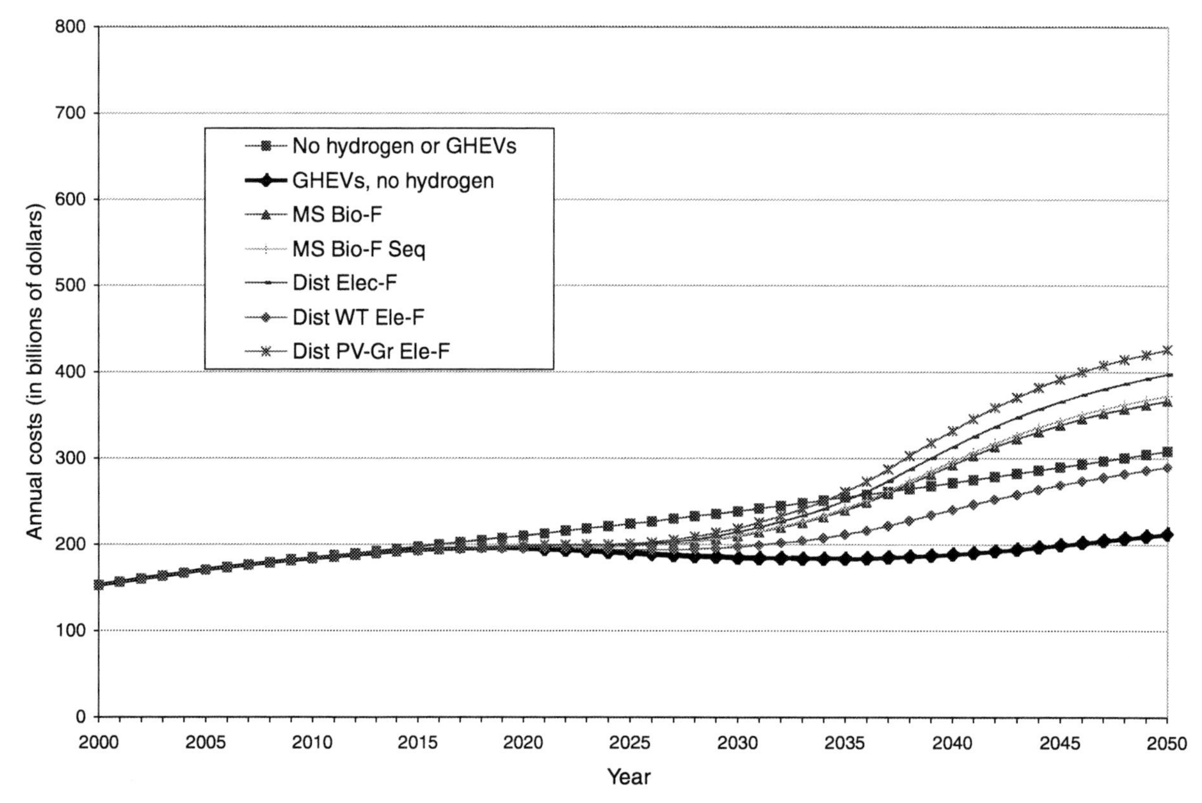

FIGURE 6-23 Estimated total annual fuel costs for light-duty vehicles: possible future hydrogen production technologies (electrolysis and renewables), 2000–2050. See Table 5-2 in Chapter 5 and discussion in text.

improvements. Some technologies could lead to sharp reductions in the amount of energy imported from unstable parts of the world, but not all could lead to such energy security improvements. And some technologies could reduce the cost of driving, but not all could lead to such economic benefits. Thus, the ultimate configuration of a hydrogen supply system will depend crucially not only on the technological successes but also on the trade-offs that individuals and societies are willing to make.

Chapter 8 contains a detailed examination of the various production technologies and of the possible technical advances that might give society and individuals the opportunity to make those choices. The discussion of production technologies in Chapter 8 makes use of the analyses presented in this chapter, together with those of Chapter 5, to make recommendations on the DOE hydrogen program.

FINDINGS

The following findings are drawn from the analysis in this chapter:

Finding 6-1. In the committee's optimistic vision of a transition to hydrogen for light-duty vehicles, 25 years from now the demand for hydrogen for these vehicles would be about equal to the current production of 9 million tons per year. This is only a small fraction of the 110 million tons required for full replacement of gasoline for light-duty vehicles.

Finding 6-2. A transition to a light-duty fleet of vehicles fueled entirely by hydrogen would dramatically reduce U.S. oil consumption and imports.

Finding 6-3. If coal, renewable energy, or nuclear energy is used to produce hydrogen, a transition to a light-duty fleet of vehicles fueled entirely by hydrogen would reduce total energy imports by the amount of oil consumption displaced. However, if natural gas is used to produce hydrogen, and if, on the margin, natural gas is imported, there would be little if any reduction in total energy imports, because natural gas for hydrogen would displace petroleum for gasoline.

Finding 6-4. The exclusive use of coal to produce hydrogen could significantly increase the scale of this domestic industry. This increase in coal production is unlikely to have any significant impact on coal imports, since all of the coal can reasonably be produced domestically. Oil imports could decrease greatly over time, resulting in a significant net decrease in energy imports.

Finding 6-5. Using current technologies, the land required for biomass to produce all of the hydrogen for fueling light-duty vehicles would be as large as the total amount of cropland in the United States. However, significant improvements in the yield of lands and the efficiency of biomass gasifiers could significantly reduce that figure. But because biomass does not appear to be economically viable even with technology advances, the committee does not expect much if any hydrogen to be produced by biomass gasification.

Finding 6-6. Using only natural gas for the production of the hydrogen for fueling light-duty vehicles would have a great impact on natural gas consumption and imports by the end of the 50-year time horizon in the committee's analysis, but it would have only an insignificant impact during the next 25 years.

Finding 6-7. Carbon dioxide emissions from the total supply chain of hydrogen-fueled vehicles can be cut significantly only if the hydrogen that is produced is based entirely on renewables, or nuclear energy, or with sequestration of CO_2 from fossil fuels. However, emissions of CO_2 from all modes of transportation are projected to account for only 37 percent of total U.S. CO_2 emissions; emissions from light-duty vehicles are only two-thirds of this 37 percent. Thus, sharply reducing overall CO_2 releases would require carbon reductions in other parts of the economy, particularly in electricity production. Technology advances, other than those that make the use of nuclear energy or sequestration economic, would have very little additional impact on carbon releases.

Finding 6-8. Sequestration would involve a massive movement of carbon dioxide. While the United States probably has sufficient geological capacity to sequester these amounts, the sequestration of all the carbon dioxide possible could involve moving twice as much carbon dioxide as the amount of natural gas that the nation anticipates moving.

Finding 6-9. The penetration of hybrids into the marketplace, even absent hydrogen-fueled vehicles, could reduce fuel costs by tens of billions of dollars per year in the United States. Most current hydrogen production technologies would lead to a total driving cost higher than the total cost if hybrid electric vehicles ultimately dominated the fleet, but central station coal-based or natural-gas-based hydrogen production could keep total costs similar to those with hybrid electric vehicles.

Finding 6-10. With the possible future technology advances, all but biomass and grid-electric or photovoltaic-based electrolysis technologies could be operated at costs less than those that Americans have been willing to pay in fuel costs for driving gasoline-fueled conventional vehicles.

Finding 6-11. Although a transition to hydrogen could greatly transform the U.S. energy system in the long run, the impacts on oil imports and CO_2 emissions are likely to be minor during the next 25 years. Thus, hydrogen—although it could transform the energy system in the long run—does not represent a short-run solution to any of the nation's energy problems.

7

Carbon Capture and Storage

THE RATIONALE OF CARBON CAPTURE AND STORAGE FROM HYDROGEN PRODUCTION

Hydrogen is used extensively for its chemical properties, and on a global basis most hydrogen is produced from fossil fuels. As shown in Chapter 5, the committee believes that a transition to a hydrogen energy system would likely rely on hydrogen (H_2) from the reforming of natural gas and from electrolysis powered by a grid mix inclusive of coal-fired power plants. The fossil resources combusted or used as feedstock during such a transition will produce by-product carbon dioxide (CO_2). Thus, converting to a hydrogen economy based on fossil fuels would have no advantage in reducing CO_2 emissions unless the CO_2 can be isolated indefinitely.

Such a strategy is called carbon capture and storage (CCS).[1] Figures 6-7 through 6-10 in Chapter 6 compare emissions of carbon, with and without carbon capture and storage, that would be associated with the fleet of U.S. light-duty vehicles as fuel cell vehicles (FCVs) penetrate the market over the period 2020 to 2050, replacing gasoline-fueled vehicles by 2050. Comparison of the curves in the year 2050 permits head-to-head comparison of projected carbon emissions that would be associated with fleets composed entirely of one power plant (that is, the conventional gasoline-fueled vehicle, gasoline hybrid electric vehicle [GHEV], or fuel cell vehicle) and, in the case of hydrogen, one of several possible production technologies. Even though the hydrogen fuel cell vehicle is assumed in the committee's analysis to be 66 percent more efficient than GHEVs are, emissions in 2050 are almost the same when coal is the hydrogen source and CCS is not used versus GHEVs.

CCS may turn out to be one of the critical strategies for reducing net carbon emissions,[2] or it may turn out to have shortcomings that limit its role. Work is under way worldwide to develop various separation and storage technologies, reduce projected costs, and understand risks.

Scale of Carbon Capture and Storage Associated with Hydrogen Production in a Hydrogen Economy

Table 7-1 presents the ratios of emitted carbon (C) to produced hydrogen used in this report for natural gas reforming and coal gasification, for both current and possible future technology cases (see Chapters 5 and 6 for details). Thus, for example, it is estimated that 5.1 kg C will be emitted to the atmosphere for each kilogram of hydrogen produced from coal gasification using current technology.

Hydrogen production from coal gasification results in about twice the carbon emissions of hydrogen production from the reforming of natural gas. CCS technology achieves a reduction of about 85 percent in atmospheric carbon emissions from either feedstock.

One of the main rationales for moving toward a hydrogen economy is to address the goal of improved environmental quality, including reduction in CO_2 emissions. Carbon can be captured at central station production plants and stored, consistent with this goal. However, the quantity of carbon to be handled is enormous. (Capture is virtually impossible in vehicles burning gasoline or diesel fuel.) As discussed in Chapter 6, if hydrogen FCVs dominate the market in 2050, they will require the production of 100 million metric tons

[1]CCS is also called carbon sequestration. The two kinds of sequestration are sometimes distinguished, one from the other, by calling the removal of carbon already in the atmosphere *indirect sequestration* and the prevention of carbon from reaching the atmosphere *direct sequestration*. This discussion uses the term CCS.

[2]The terms *carbon emissions* and *CO_2 emissions* are used interchangeably, because virtually all carbon emitted in the hydrogen production cases considered by the committee is in the form of CO_2. It should be understood that CO_2 weighs 3.7 times as much as carbon.

TABLE 7-1 Estimated Carbon Emissions as Carbon Dioxide Associated with Central Station Hydrogen Production from Natural Gas and Coal

State of Technology Development	Technology	Ratio[a] Without Carbon Capture and Storage	Ratio[a] With Carbon Capture and Storage
Current	Natural gas reforming	2.51	0.42
	Coal gasification	5.12	0.82
Possible future	Natural gas reforming	2.39	0.35
	Coal gasification	4.56	0.60

[a]Mass of carbon emitted divided by mass of hydrogen produced.

(Mt) of hydrogen per year.[3] Assuming possible future technologies, without CCS, the associated annual carbon emissions from natural gas production would be 255 Mt C, and 518 Mt C from coal plants. CCS technology reduces emissions to 42 Mt C and 83 Mt C for natural gas and coal, respectively (less than 100 percent of the CO_2 will be removed by any competitive technology). Thus, the scale of the task of capturing and storing most of the CO_2 associated with an all-hydrogen U.S. fleet of light-duty vehicles in 2050 could be approximately 200 to 400 Mt C/yr (assuming insignificant hydrogen production from nuclear or renewable energy sources).

Today's annual U.S. CO_2 emissions from light-duty vehicles, about 320 Mt C, are about 20 percent of U.S. CO_2 emissions and 5 percent of global emissions from all sources. It was beyond the scope of this report to generalize beyond transportation to the role of CCS in hydrogen production for use in other sectors of the U.S. economy (hydrogen use in industry and in buildings, for example).

Capturing and storing 200 to 400 Mt C annually by 2050 is a huge task. On the order of a thousand projects the size of the first two CO_2 geological storage demonstration projects (discussed below, in the subsection "CCS Demonstration Projects") would be required. These projects each capture and store about 300,000 metric tons C/yr. This is also the scale of CCS planned for the proposed FutureGen Project (see below).

Carbon Emissions Associated with Current Hydrogen Production

At the present time, global crude hydrogen production relies almost exclusively on processes that extract hydrogen from fossil fuel feedstock (see Figure 7-1). It is not current practice to capture and store the by-product CO_2 that results

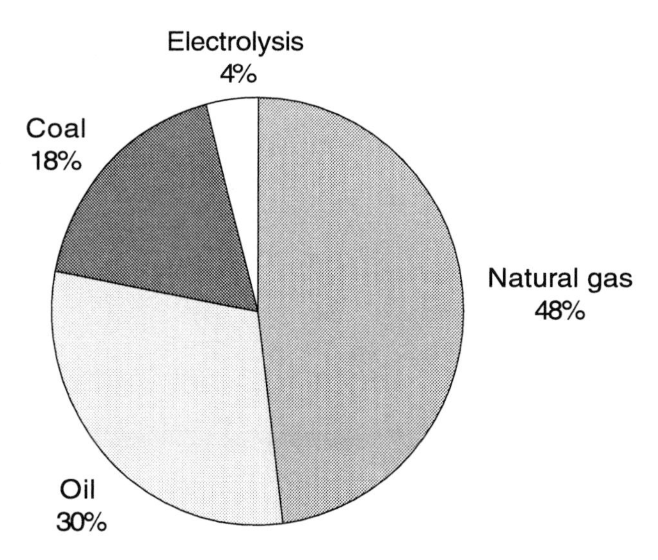

FIGURE 7-1 Feedstocks used in the current global production of hydrogen. SOURCE: Courtesy of Air Products and Chemicals (2003).

from the production of hydrogen from these feedstocks. Consequently, more than 100 Mt C/yr are vented to the atmosphere as part of the global production of roughly 38 Mt of hydrogen per year.[4]

Some but not all of the CO_2 is emitted in concentrated streams at elevated pressure. At an intermediate step in the production of hydrogen from fossil fuels, there is a mixture of CO_2 and H_2 under pressure, together with impurities. At this stage, there are two options:

1. Extract CO_2 at high purity and elevated pressure, which is well matched to CCS technology. Left behind is a less-than-pure hydrogen, but it is of high enough concentration to be commercially useful. Such hydrogen is sometimes called *refinery hydrogen*. It is suitable for applications such as hydrogen-fueled internal combustion engines, but some CO_2 and other impurities will be emitted.
2. Extract hydrogen at high purity; such hydrogen is sometimes called *fuel-cell-grade hydrogen*. The residue still contains some hydrogen, which can be burned to produce steam or electricity. If the residue is burned in air, the exhaust stream contains the CO_2 from the conversion, but also a large fraction of nitrogen from the atmosphere. Then, the CO_2 is dilute and more difficult to capture than in option 1.

There is a recent trend in refineries away from the first option and toward the second, because of the falling cost of hydrogen separation from gas mixtures via pressure-swing absorption. If the CO_2 from refinery production of hydrogen

[3]Corresponding to 7.9 trillion vehicle miles of annual travel (three times the current value) at an average fuel economy equivalent to 80 mpg. There is similar energy content in 1 kg H_2 and 1 gallon of gasoline.

[4]Based on estimated global annual production of hydrogen of 15.9 trillion cubic feet (ORNL, 2003).

is to be captured and stored, this trend will complicate the process. Similarly, fuel cells require high-purity hydrogen. Plants built to produce this hydrogen *and* to capture and sequester CO_2 will require equipment and processing not used in current plants. The joint goals of high-purity hydrogen and low-cost CCS may demand new capture technology.

CCS Demonstration Projects

Two CCS demonstration projects are currently under way, one in Norway and one in Canada:

- At the Sleipner gas field in the North Sea off Norway, natural gas is produced containing about 10 percent CO_2. The CO_2 concentration exceeds the concentration allowed in the European natural gas grid by about a factor of 4. Normally in such cases, CO_2 is stripped from the gas onshore and vented to the atmosphere. Since 1996, motivated by a carbon tax of about $140/t C (in U.S. dollars), the field operators are stripping CO_2 offshore and injecting it into the nearby Utsira formation, a very large saline aquifer that does not contain hydrocarbons.
- At the Weyburn oil fields in Saskatchewan, Canada, CO_2 captured at a coal-to-methane plant in North Dakota and piped across the border is being injected, with the joint objective of enhanced oil recovery (EOR) and CO_2 storage.

Both projects store about 0.3 Mt C/yr.

Several additional CCS demonstration projects are being planned at roughly the same scale. Notable among them is the FutureGen Project, proposed in the spring of 2003 by the Bush administration. In this project, a 275 megawatt (MW) coal-fired power plant would produce both electricity and hydrogen and would capture CO_2 for offsite storage. Announced at the same time and coupled with FutureGen is the U.S. Carbon Sequestration Leadership Forum, aimed at developing international partnerships for the commercialization of CCS technology.

In addition to demonstration projects, experience bearing on CCS technologies is being gained through EOR projects that inject CO_2 into partially depleted oil fields. The first EOR project was in West Texas in 1972, and most of the EOR in the world today is still concentrated there.

Only about one-quarter of the CO_2 used in EOR is derived from industrial sources (Hill, 2003, p. 25). Most is extracted from natural CO_2 formations. The value of CO_2 for EOR has been sufficient to warrant the construction of several multi-hundred-kilometer (km) CO_2 pipelines, including one of 800 km from the McElmo Dome in southwest Colorado to the Permian basin oil fields in West Texas.

Until recently, EOR was not thought of as a carbon storage strategy. Once the CO_2 has done its work and production is concluded, EOR project managers have not considered whether the CO_2 would remain belowground for decades, centuries, or millennia. Joint optimization of EOR and long-term CO_2 storage could lead to revisions in EOR practices. Thus, EOR to date provides only partial precedents for CCS.

Capture

Hydrogen production involves the transfer of most of the energy from the feedstock chemical compound to the product molecular hydrogen. Adding the objective of CO_2 capture complicates the design of the equipment and increases the costs of production. However, most of the plant components required to capture CO_2 are already required to produce hydrogen, so the fractional increment in the cost of hydrogen owing to CO_2 capture is not large: in the committee's estimates (see below), between 10 and 20 percent for natural gas and less than 10 percent for coal. The fractional increment is substantially smaller than would be incurred if CO_2 capture were added to electricity production.

An intriguing concept is that of the co-capture and co-storage of impurities with the CO_2, saving the costs currently incurred to prevent these impurities from becoming air or water pollutants. For example, sulfur in coal can be co-captured with the CO_2 (as either hydrogen sulfide [H_2S] or sulfur dioxide [SO_2], depending on the plant configuration), piped with the CO_2 to a storage location, and co-stored as a single fluid.

Table 7-2 shows estimated hydrogen production costs for large-scale plants.[5] The "with carbon capture and storage" cases also show the estimated costs of storage. Both sets show an imputed cost of CO_2 emissions on a $50 per metric ton (tonne) C basis.[6]

The estimated cost of carbon associated with either type of plant—natural gas or coal—is calculated as the sum of storage costs plus capture costs: the storage costs are shown in Table 7-2, and capture costs are defined as the difference between the plant production costs with and without CCS. It is interesting to note that the assumed cost of carbon emissions is insufficient to justify CCS for natural gas plants, whereas it approximately balances the additional costs for CCS in a coal plant. The capture costs for coal are a smaller percentage of the plant costs for the technologies assumed in this report (see Chapter 8); for example, the CO_2 in this study's coal gasification plant is available for capture at a higher partial pressure, which reduces the cost of capture. The capture costs, as a percentage of the cost of production at the plant without CCS, are 18 percent and 11 percent for

[5]The scale of this study's large plants is 1200 t H_2/day, or 2.0 GW H_2 (higher heating value), or 1.7 GW H_2 (lower heating value).

[6]The fuel costs assumed (see Chapter 5 and Appendix E) are as follows: $4.50 per million Btu (higher heating value) for natural gas; $1.22 per million Btu (higher heating value) for coal; and 4.5 cents/kWh for electricity. The cost of storage is highly uncertain at this time and has not been a focus of this committee's analysis. The committee assumed $37/t C, which is consistent with the range of current estimates. The imputed cost of CO_2 emissions is even more uncertain. The $50/t C cost arbitrarily chosen in this report is a point of departure for many analyses (see Chapter 5).

TABLE 7-2 Estimated Plant Production Costs and Associated Outside-Plant Carbon Costs (in dollars per kilogram of hydrogen) for Central Station Hydrogen Production from Natural Gas and Coal

		Without Carbon Capture and Storage ($/kg H_2)		With Carbon Capture and Storage ($/kg H_2)		
State of Technology Development	Energy Source	Plant	Carbon Emissions at $50/t C	Plant	Storage Costs at $37/t C	Carbon Emissions at $50/t C
Current	Natural gas	1.03	0.13	1.22	0.09	0.02
	Coal	0.96	0.26	1.03	0.16	0.04
Possible future	Natural gas	0.92	0.12	1.02	0.08	0.02
	Coal	0.71	0.23	0.77	0.15	0.03

natural gas, current and possible future cases, respectively, and they are 8 percent for coal, both current and possible future cases. The storage costs, on the other hand, are roughly twice as large for coal as for natural gas, because roughly twice as much CO_2 must be stored per unit of hydrogen produced. As a result, the capture costs are comparable to the storage costs for natural gas, but they are much less than the storage costs for coal.

It is interesting to express carbon costs in various units. The cost of storage, for example, at $10 per metric ton of carbon dioxide, or $37 per metric ton of carbon, is approximately as follows (assumptions are in footnotes):

- 9 cents per gallon of gasoline,[7]
- $4 per barrel of oil,[8]
- $24 per U.S. ton of coal,[9]
- 54 cents per 1000 standard cubic feet (scf) of natural gas,[10]
- 0.8 cent per kWh of electricity from coal,[11]
- 0.4 cent per kWh of electricity from natural gas,[12]
- 18 cents/kg H_2 from coal,[13]
- 9 cents/kg H_2 from natural gas,[14] and
- $220 billion per year implicit flow through the global energy system.[15]

Storage

After CO_2 leaves the plant gate, it can be directed to several kinds of destinations where it can be stored in several chemical forms. The destinations that have received most attention are geological strata deep belowground and the deep ocean. Both destinations may have storage capacity for a century of aggressive global carbon management. Among the other concepts are storage as stable carbonates (produced by the exothermic reaction of CO_2 with certain magnesium and calcium minerals) and storage as elemental carbon (charcoal)[16]—in both cases, presumably, piled aboveground.

It is often suggested that CO_2 should not be stored but somehow recycled. Indeed, there are commercial uses for CO_2 in industrial process cleaning, in dry ice, in carbonated beverages, and elsewhere in the economy, but these typically delay the release of CO_2 to the atmosphere by days or weeks, so recycling is not an alternative to CO_2 storage. It is also possible to convert CO_2 to hydrocarbons or other carbon-containing compounds, but significant energy would be required (producing more CO_2), and few uses of these compounds will keep the carbon from the atmosphere for as long as a year. Once carbon is brought out of the ground in fossil fuels, the industrial economy provides very few incidental routes to long-term storage; exceptions include asphalt roadbeds and shingles. Only deliberate storage can prevent fossil fuel carbon from becoming atmospheric CO_2.

As seen in Table 7-2, if the total CO_2 storage cost is as little as $37/tonne C, it is a small fraction of the cost of H_2 produced from fossil fuels at large scale. This estimate was derived from the cost of pipelines and wells. It is possible, however, that other costs will dominate, such as those of permitting and monitoring (see below).

Infrastructure

The first step in CO_2 storage is transporting the CO_2 from the production site to the storage site. An entirely new infrastructure will be required for this purpose, presumably a

[7]1 m^3 = 264.2 U.S. gal; 630 kg C/m^3 gasoline.

[8]1 bbl = 42 U.S. gal; 730 kg C/m^3 crude oil.

[9]1 U.S. ton = 907 kg; 0.71 kg C/kg coal.

[10]1 N-m^3 = 37.24 scf; 0.549 kg C/N-m^3 natural gas.

[11]29.3 gigajoules (GJ)/metric ton coal (12,600 Btu/lb); 40 percent conversion efficiency.

[12]55.6 GJ/metric ton natural gas; 0.75 kg C/kg natural gas; 50 percent conversion efficiency.

[13]5 t C/t H_2.

[14]2.5 t C/t H_2.

[15]6 Mt C/yr anthropogenic carbon emissions from the energy sector.

[16]By this method, only part of the available energy in the hydrocarbons is extracted, but the resulting solid form of carbon may present less of a challenge for capture and storage.

pipeline system with trunk lines and branch lines. There is already considerable experience with CO_2 pipelines for EOR, but institutional issues, including financing, construction, operation, and monitoring of the infrastructure, may be quite different when the goal is storage.

In a recent analysis of the cost of a large CO_2 pipeline leading to a disposal well 100 km from a large coal-to-hydrogen plant, the pipeline itself is found to contribute $13/t C (Ogden, 2003). This is one-third of this study's assumed total storage cost. At 5 t C/t H_2, a mass ratio appropriate for coal (see Table 7-1), this cost is equivalent to 6.5 cents/kg H_2.

Because of economies of scale in pipelines, the costs per unit of CO_2 stored or per unit of hydrogen generated increase as the scale of the hydrogen production system decreases. Unit costs vary approximately as the inverse square root of the flow rate. The pipeline discussed here has a capacity of 10,000 t CO_2/day; roughly two of these large pipelines are required to handle the CO_2 from this study's central station coal plant. Midsize production of hydrogen from natural gas also may turn out to be compatible with CCS. However, it is very likely that if hydrogen is generated in the distributed natural gas systems modeled in this report, the costs of storage (and probably capture as well) will be prohibitive. The dividing line is not clear: further work is required to understand the size of the smallest hydrogen production plant at which CCS is economically feasible. Unless unexpected breakthroughs in technology are realized, however, small-scale production of hydrogen from natural gas will be consistent with the objective of climate change only if it is considered a stepping-stone to production strategies that limit emissions of CO_2.

Geological Storage

Storage belowground can be in sedimentary formations that are either hydrocarbon bearing or largely free of hydrocarbons ("brine formations"). Hydrocarbon-bearing formations include oil and gas reservoirs and coal seams. Thanks to experience with EOR, much has been learned about migration of supercritical CO_2 as a distinct phase, dissolution into hydrocarbons and brine, and chemical interaction with host rock. EOR probably provides the lowest-cost early opportunities for CO_2 storage in many areas of the world. Storage can also be in reservoirs that were once sources of commercial hydrocarbons but that are no longer productive; CO_2 injection would not offer any economic benefits, but, overall, these destinations may offer more total storage capacity than EOR sites do, with comparable permanence of storage.

Brine formations appear to contain much larger storage capacity than do hydrocarbon reservoirs. The Sleipner project referred to above is an example of such storage. Hydrocarbon reservoirs have held buoyant fluids for geological epochs; brine formations have no such track record. Accordingly, the durability of storage in brine formations is under intense investigation at present.

In geological storage, the CO_2 is injected as a supercritical fluid, with roughly half the density of water.[17] To achieve the necessary supercritical pressures, CO_2 is injected at high pressure at least 800 m below the surface.

Deep-Ocean Storage

The storage of CO_2 in the deep ocean is an area of active modeling and some experiments. At issue is the retention time for various sites as well as the biological impacts. Advocates of deep injection point out that even in the absence of deliberate injection, the oceans already receive a portion of the carbon released from fossil fuels because of the continuous exchange of CO_2 between the atmosphere and the oceans. Add CO_2 to the atmosphere and some of it will move naturally to the ocean, as equilibrium is sought at the ocean surface.[18] Also under study are the biological impacts of additional CO_2 in the near-surface ocean—for example, the impacts on coral reefs.

Storage in Biological Carbon

Biological strategies to remove CO_2 from the atmosphere, although not the focus of this report, deserve mention. These strategies remove carbon from the atmosphere by photosynthesis. On land, storage usually takes place at the same site as that of capture—for example, in a tree. Increases in atmospheric CO_2 enhance plant growth up to a point that is yet to be fully understood. In the ocean, capture is via various organisms at the surface that then sink into the deep ocean.

The initial costs of biological carbon sequestration on land are small, and joint gains from land improvement plus carbon storage may be available. However, the earth's capacity for biological carbon sequestration is limited, particularly for long-term storage (many decades or centuries), which could conflict with food production and other uses of the land. The mass of carbon aboveground in terrestrial vegetation is roughly equal to the mass of carbon in the atmosphere, about 700 billion tons of carbon (IPCC, 2001a). Imagine the future production of 700 billion tons of carbon from fossil fuel resources belowground, and imagine that the CO_2 is to be captured in forests and grasslands instead of building up in the atmosphere. The result would be to double the biological carbon stock. Such a huge change is likely to be inconsistent with the retention of ecosystem quality. Local environmental impacts of biological carbon storage also need to be considered.

---

[17]The critical pressure, temperature, and density of CO_2 are 73.9 bar, 304 K, and 467 kg/m^3, respectively.

[18]A major removal process for CO_2 depends on the transfer of carbon content of near-surface waters to the deep ocean, which has a century timescale (NRC, 2001b).

The Carbon Storage Market

In a carbon management regime designed to mitigate climate change, a value will be put on not emitting carbon into the atmosphere. Those capable of storing carbon will compete for the opportunity to do so, creating a CO_2 storage market. Low-cost storage will be available at only a few facilities in only a few locations; storage at higher cost will be available at many more locations. Owners of low-cost storage capacity will be at a commercial advantage. As the demand for storage grows, the clearing price for storage may climb as a result of greater demand or fall as providers accumulate knowledge that translates into lower costs.

The lowest-cost storage is likely to be at EOR sites. Today, the oil industry pays about \$10 to \$15/t CO_2 (roughly, \$40 to \$60/t C) for CO_2 delivered to the EOR site (Kuuskraa and Pekot, 2003). In the United States in 2001, about 10 Mt C/yr were injected for EOR, enhancing domestic oil production by 180,000 barrels per day (bbl/d) (Hill, 2003), or about 2 barrels of oil produced for each metric ton of CO_2 injected.

The next-lowest-cost storage is likely to be at depleted oil and gas fields, where reservoir geology is already known, wells suitable for injection of CO_2 may have already been drilled, relevant permits may already exist, and subsurface rights may be well defined.

Those purchasing CO_2 storage will take into account the proximity and capacity of the storage site. Buyers and sellers will allocate costs for site characterization, leakage monitoring, and liability insurance.

Storage sites will differ in many respects. The value of a storage site may increase when the site can be demonstrated with high probability to be effective for a long time, when the loss of storage integrity can be easily detected, and when damage from the loss of storage integrity is small. One can imagine storage sites rated much as bonds are rated, with lower-quality storage being valued less.

The supplier of carbon storage may be able to gain further revenue by providing additional services—for example, offering co-storage of carbon and sulfur, relieving the purchaser of storage of aboveground sulfur management costs.[19]

Permitting

A CO_2 storage regime can emerge only if public acceptance of the concept is widespread. Among the critical issues are these (Socolow, 2003):

- *Trust.* Public trust is critical. To what extent will openness, lack of bias, fairness, and vigilance be achieved?
- *Goals.* What constitutes success? Will society be relaxed about the loss of 1 percent of the stored CO_2 each year through slow leaks? What about the loss of 1 percent per year from 10 percent of the sites?
- *Permissiveness.* The level of leakage allowed during the first few decades of storage can probably be greater than that allowed in later decades, not only because experience will permit improvements, but also because the total quantities stored will increase over time. How can the permitting process be made sufficiently permissive?
- *Reversibility.* Should the storage system be one that future generations can undo?
- *Storage integrity at individual sites.* Concentrations of more than a few percent of CO_2 in air are dangerous, so bulk releases of CO_2 must be guarded against. Under some conditions, safety may be an issue, as evidenced by the lakes in Africa that erupted with CO_2, asphyxiating many local residents.[20] Upward migration of injected CO_2 could contaminate hydrocarbon reservoirs or surface drinking water supplies, so certain slow releases may also be of concern. How should such risks be addressed?
- *Property rights to storage space.* Are ownership rights belowground clear? What about below the ocean floor and in the ocean? The market in carbon storage will generate requirements for well-defined property rights, attribution of ownership, liability rules, insurance, monitoring, and, in some cases, active intervention to limit damage.
- *Net carbon.* It will be important to quantify, for various technologies and energy supply systems, the additional carbon that will be brought out of the ground to provide the energy necessary to capture and store the CO_2.
- *Monitoring.* Can infrastructure and storage be designed in ways that facilitate attribution and monitoring (e.g., by adding tracers to the injected gas)? What techniques are available to respond constructively to evidence that stored materials are not behaving as expected? How can long-term monitoring be institutionalized?
- *Precedents.* There are two obvious precedents in the United States for the storage of CO_2: underground injection of hazardous waste and nuclear waste storage. Both offer lessons to the designers of CO_2 storage. The underground injection of hazardous wastes is governed by an Environmental Protection Agency (EPA)-regulated permitting process based on detailed modeling intended to prove that nothing serious will happen belowground after injection, followed by little, if any, post-injection monitoring and verification of what is actually happening belowground. The program to store nuclear waste began with promises of leakproof, very

[19]Today, in Alberta, Canada, H_2S and CO_2 are routinely removed from natural gas between wellhead and transmission pipeline and then co-stored belowground.

[20]For example, in 1986, Lake Nyos in Cameroon erupted in a massive outgassing of CO_2, killing 1800 people nearby. The cold bottom water of the lake, saturated with CO_2 seeping up from the earth below, became unstable. The CO_2-laden air, heavier than ordinary air, filled nearby valleys. Further information is available online at http://helios.physics.uoguelph.ca/summer/scor/articles/scor158.htm (accessed December 11, 2003) and http://perso.wanadoo.fr/mhalb/nyos/index.htm (accessed December 11, 2003).

long term storage. Skepticism that such promises could be kept has severely hampered progress.

FINDINGS AND RECOMMENDATIONS

Finding 7-1. It is highly likely that fossil fuels will be the principal sources of hydrogen for several decades. It follows that an expanded role for hydrogen will have a much larger positive impact on the mitigation of climate change if carbon capture and storage technologies are successfully integrated into the fossil fuel production of hydrogen.

The majority of the early carbon capture and storage projects might not involve hydrogen, but could instead involve the capture of the CO_2 impurities in natural gas, the capture of CO_2 produced at electric plants, or the capture of CO_2 at ammonia and synfuels plants. All of these routes to capture, however, share carbon storage as a common element, and it is carbon storage that raises the most difficult institutional issues and issues of public acceptance.

As of 2002, for the first time, the Department of Energy's programs related to carbon sequestration are listed as an associated program of its hydrogen program.

Recommendation 7-1. The U.S. Department of Energy's hydrogen program needs to be well integrated with the carbon capture and storage program, to assure that any expanded role for hydrogen produced from fossil fuel has a positive impact on the mitigation of climate change. Such integration will enable the hydrogen program to identify critical technologies and research areas that can enable hydrogen production from fossil fuels with CO_2 capture and storage.

Tightening the coupling of the two programs should facilitate setting priorities in those portions of the hydrogen program addressing hydrogen production from fossil fuels and biomass. It should also promote the exploration of overlapping issues—for example, the co-capture and co-storage with carbon dioxide of pollutants such as sulfur during hydrogen production.

Because of the hydrogen program's large stake in the successful launching of carbon capture and storage activity, the hydrogen program should participate in all of the early carbon capture and storage projects, even those that do not directly involve carbon capture during hydrogen production. These projects will address the most difficult institutional issues and the challenges related to issues of public acceptance, which have the potential of delaying the introduction of hydrogen in the marketplace.

Finding 7-2. The Department of Energy's recently announced FutureGen Project is intended to demonstrate the production of hydrogen and electricity from coal at a large scale, while capturing CO_2. FutureGen may become the world's earliest carbon capture and storage project integrated with hydrogen production. This project should provide an opportunity to integrate the development of advanced technologies for the production of hydrogen with CO_2 capture and storage.

Recommendation 7-2. The FutureGen Project should be managed to encourage the development of technologies that integrate hydrogen production with carbon dioxide capture. FutureGen should have strong research and development components.

Finding 7-3. The successful integration of carbon capture and storage will depend at least as much on institutional factors and public acceptance as on engineering prowess and geological opportunities. Institutional factors include property rights at storage sites, the management of infrastructure, insurance and liability, and the funding of monitoring and verification, including those efforts over the very long term. Public acceptance will depend on achieving and maintaining trust, which will require processes that are regarded as open and fair. The public discussion of local environmental and health risks in this realm has hardly begun, nor has public discussion regarding criteria for long-term storage, such as criteria for durability and verifiability. It may be possible and desirable to achieve broad consensus that early criteria are more permissive and later ones are tougher.

Recommendation 7-3. The Department of Energy should foster public discussion of institutional factors affecting carbon capture and storage, including property rights at storage sites, the management of infrastructure, insurance and liability, and the funding of monitoring and verification, including those efforts over the very long term.

The DOE should foster the identification of the issues likely to have the greatest impact on public acceptance of carbon capture and storage. It should encourage public discussion of local environmental and health risks. It should encourage public discussion of what constitutes "adequate" storage, from the standpoint of durability and verifiability. It should explore the merits of broad agreement that early criteria be more permissive and later ones tougher.

8

Hydrogen Production Technologies

This chapter discusses in more detail the various technologies that can be used to produce hydrogen. These technologies have already been identified in previous chapters and the cost analyses presented in Chapter 5 enumerate them (see Table 5-2). In this chapter, the committee addresses the following technologies: (1) reforming of natural gas to hydrogen, (2) conversion of coal to hydrogen, (3) use of nuclear energy to produce hydrogen, (4) electrolysis, (5) use of wind energy to produce hydrogen, (6) production of hydrogen from biomass, and (7) production of hydrogen from solar energy. The following sections—one for each technology—include a brief description of the current technology; its major technical challenges; possible improvements for future technology; references to Chapter 5 and Appendix E (spreadsheet data from the committee's cost analyses), where applicable; the potential advantages and disadvantages of using the technology for hydrogen production; comments on the Department of Energy's hydrogen research, development, and demonstration (RD&D) plan; and recommendations.

The committee emphasizes that it made recommendations about research and development (R&D) and priorities in the context of hydrogen production and the possible future "hydrogen economy." The committee understands that the DOE programs outside the Office of Hydrogen, Fuel Cells, and Infrastructure Technologies have other objectives and priorities besides those related to hydrogen, and the committee did not review those other programs vis-à-vis that of producing hydrogen. For example, the committee identified R&D needs for producing hydrogen from wind and solar-based technologies, but did not review the wind program or the solar technologies program, which also (as does the Office of Hydrogen, Fuel Cells and Infrastructure Technologies) reside within the Office of Energy Efficiency and Renewable Energy (EERE), or consider the objectives and priorities within those offices. Appendix G contains more extensive discussion of each of the technology areas covered in this chapter.

In general, in developing estimates about possible future technologies, the committee systematically adopted an optimistic posture. The state of development referred to as *possible future* is based on technological improvements that may be achieved if the appropriate R&D is successful. The committee is not predicting that these technical advances will be achieved; however, they may be the result of successful R&D programs. And they may require significant technological breakthroughs. Generally, these possible future technologies are available at significantly lower cost than that of *current technologies* using the same feedstocks.

HYDROGEN FROM NATURAL GAS

Compared with other fossil fuels, natural gas is a cost-effective feed for making hydrogen, in part because it is widely available, is easy to handle, and has a high hydrogen-to-carbon ratio, which minimizes the formation of by-product carbon dioxide (CO_2). However, as pointed out elsewhere in this report, natural gas is imported into the United States today, and imports are projected to grow. Thus, increased use of natural gas for a hydrogen economy would only increase imports further, and as a result the committee considers natural gas to be a transitional fuel for distributed generation units, not a long-range fuel for centralized plants for the hydrogen economy.

The primary ways in which natural gas, mostly methane, is converted to hydrogen involve reaction with either steam (steam reforming), oxygen (partial oxidation), or both in sequence (autothermal reforming). In practice, gas mixtures containing carbon monoxide, as well as carbon dioxide and unconverted methane, are produced and require further processing. Reaction of carbon monoxide with steam (water-gas shift) over a catalyst produces additional hydrogen and carbon dioxide, and after purification, high-purity hydrogen is recovered. In most cases, carbon dioxide is vented to the atmosphere today, but there are options for capturing it in centralized plants for subsequent sequestration. For distributed generation, the cost of sequestration appears prohibitive (DiPietro, 1997). Release of carbon dioxide from distributed generation plants during the period of a transition to a hydrogen economy may be a necessary consequence unless an alternative such as hydrolysis with electricity from renewable resources becomes sufficiently attractive or R&D significantly improves distributed natural gas production systems. Further information on the technology and the economics of conversion is given in Appendix G.

Distributed generation from natural gas could be the lowest-cost option for hydrogen production during the transition. However, it has never before been achieved in a manner that meets all of the special requirements of this application. The principal challenge is to develop a hydrogen appliance with demonstrated capability to be mass-produced and operated in service stations reliably and safely with only periodic surveillance by relatively unskilled personnel (station attendants and consumers). The capability for mass production is needed in order to meet the demand during the transition, when thousands of these units would be needed, and in order to minimize manufacturing costs. These units need to be designed to maximize operating efficiency and to include the controls, "turndown" capability, and hydrogen storage required to meet the variable demand for hydrogen during a 24-hour period. They must also be designed to meet the hydrogen purity requirements of fuel cells. Steam reforming process technology is available for this application, and companies have already provided one-of-a-kind units in the size range of interest.[1] Whether it will be possible to utilize partial oxidation or autothermal reforming for the distributed generation of hydrogen appears to depend on developing new ways of recovering oxygen from air or separating product hydrogen from nitrogen. This is needed because conventional, cryogenic separation of air becomes increasingly expensive as unit size is scaled down. Membrane separations, in contrast, appear amenable to this application and may provide the means for producing small, efficient hydrogen units.

Currently, there is little if any market for mass-produced hydrogen appliances such as those described, and it is clear to the committee that the DOE should stimulate development of these devices. The primary challenges involve the development and demonstration of the following:

- A mass-produced hydrogen appliance suitable for distributed generation in fueling stations, and
- A complete hydrogen system for fueling stations, capable of meeting variable demand for hydrogen on a 24-hour basis.

Each of these challenges is discussed below.

The committee estimates that, with further research and development, the unit capital cost of a typical distributed hydrogen plant producing 480 kilograms per day (kg/d) of hydrogen could be reduced from $3,847/kg/d to $2,000/kg/d, and the unit cost of hydrogen reduced from $3.51/kg to $2.33/kg. These hydrogen unit costs are based on a natural gas price of $6.50 per million British thermal units (Btu); a change in natural gas price of plus or minus $2.00 per million Btu would change hydrogen cost by about 12 percent with current technology. Improved plants could reduce CO_2 emissions from an estimated 12.1 to 10.3 kg per kilogram of hydrogen, and overall thermal efficiencies could improve from 55.5 to 65.2 percent, in each case without sequestration. Additional information on these estimates as well as estimates for central station (i.e., large, centralized) hydrogen generators using natural gas is included in Appendixes E and G.

The DOE program publications indicate that the program on distributed generation will include demonstration of a "low-cost, small-footprint plant" (DOE, 2003a, b). However, it is not clear whether the program gives priority to distributed generation or includes an effort to demonstrate the benefits of and specific designs for mass production in the specified time frame of the program. The latter would involve concomitant engineering, including design for manufacturing engineering to guide research and prepare for mass production of the appliance. It would also include development of a system design for a typical fueling facility, including the generation appliance, compression, high-pressure storage incorporating the latest storage technology, and dispensers. With today's technology, the ancillary systems cost about 30 percent as much as the reformer. The committee believes that these costs can be reduced by over 50 percent and that efficiency can be improved through system integration and incorporation of the latest technology. Compression and high-pressure storage are examples of areas in which significant improvements are expected.

The DOE program is positioned to stimulate the development of newer concepts such as membrane separation coupled with chemical conversion, and this seems appropriate to the committee. However, most of the effort appears to be directed toward partial oxidation or autothermal reforming. The committee believes that steam reforming could be the preferred process for this application and that it should also be pursued in parallel with the effort on partial oxidation.

Finally, the committee notes that the DOE program places significant emphasis on centralized hydrogen plants using natural gas and believes that this effort should be limited, given the increasing importation of natural gas, to those developments that would be applicable to distributed generation.

Recommendation 8-1. The Department of Energy should focus its natural gas conversion program on the develop-

[1] Dennis Norton, Hydro-Chem, "Hydro-Chem," presentation to the committee, June 11, 2003; Marvin A. Crews and Howe Baker, "Small Hydrogen Plants for the Hydrogen Economy," presentation to the committee, June 11, 2003.

ment of a hydrogen generation appliance that can be mass-produced and operated reliably and safely in a typical fueling station with only periodic attention, with the goal of having prototype designs in 5 to 7 years. Two prototype designs, one incorporating partial oxidation (or autothermal reforming) and the other steam methane reforming, should be pursued. Funding should be adjusted to ensure that this goal is achieved. In addition, the DOE should downsize its efforts on centralized generation, pursuing only those developments that would be applicable to distributed generation.

Recommendation 8-2. The committee recommends that the Department of Energy give appropriate attention in its program to the development of an integrated fueling facility, including the generation appliance and its ancillary subsystems, to minimize cost and to improve efficiency, safety, and reliability.

HYDROGEN FROM COAL

This section presents the basics of making hydrogen from coal in large centralized plants. Appendix G presents a detailed discussion of making hydrogen from coal. Many of the issues and technologies associated with making hydrogen from coal are similar to those associated with making electric power from coal. These subjects are closely linked to one another and should be considered in concert. This is particularly the case for gasification, a clean coal technology, which will be required for making hydrogen and which also offers the best opportunity for making low-cost, high-efficiency, and low-emission power production through the integrated gasification combined cycle (IGCC) process. The lowest-cost hydrogen coal plants are likely to be ones that coproduce power and hydrogen.[2]

Coal is a viable option for making hydrogen in very large, centralized plants when the demand for hydrogen becomes large enough to support an associated very large distribution system. The United States has enough coal to make all of the hydrogen that the economy will need for more than 200 years, a substantial coal infrastructure already exists, commercial technologies for converting coal to hydrogen are available from several licensors, the cost of hydrogen from coal is among the lowest available, and technology improvements are identified to reach the future DOE cost targets. The major consideration is that the CO_2 emissions from making hydrogen from coal are larger than those from any other way of making hydrogen. This puts an added emphasis on the need to develop carbon sequestration techniques that can handle very large amounts of CO_2 before the widespread use of coal to make hydrogen is implemented.

[2]David Gray and Glen Tomlinson, Mitretek Systems, "Hydrogen from Coal," presentation to the committee, April 24, 2003.

Gasification Technology

The key to the efficient and clean manufacture of hydrogen from coal is to use gasification technology, which is a clean coal technology, as opposed to the combustion process used in conventional coal-fired power plants. Gasification systems typically involve partial oxidation of the coal with oxygen and steam in a high-temperature and elevated-pressure process. This creates a synthesis gas, a mix of predominantly carbon monoxide (CO) and H_2 with some steam and CO_2. This synthesis gas (syngas) can be further reacted with water to increase H_2 yield. The gas can be cleaned in conventional ways to recover hydrogen and a high-concentration CO_2 stream that is easily isolated and sent for disposal. Syngas produced from current gasification plants can be used in a variety of applications, often with multiple applications from a single facility. These applications include use as a feedstock for chemicals and fertilizers, use for making hydrogen for hydro-processing in refineries, or use for generating electricity by burning the syngas in a gas turbine.

Research and Development Needs

In terms of its stage of development, coal gasification is a less mature commercial process than other coal processes and other hydrogen generation processes using other fossil fuels, especially with respect to capturing CO_2 and providing flexibility in both H_2 and electricity production. In the committee's analysis, the current production cost of making hydrogen from coal in central station (i.e., large, centralized) plants is estimated to be $1.03/kg. The potential for improvement through technology development is significant, as indicated below:

- R&D for current technology should be directed at the following: capital cost reduction; standardization of plant design and execution concept; and improvements in reliability, gas cooler designs, process integration, oxygen plant optimization, and acid gas removal technology. With success in these areas, the production cost of hydrogen from coal is estimated to drop to $0.90/kg.
- The potential also exists for new technologies to make larger improvements in the efficiency and cost of making hydrogen from coal. For new gasification technologies, the best opportunities for R&D appear to be for new reactor designs (entrained bed gasification) and improved gas separation (hot gas separation) and purification techniques (membrane purification).

These new technologies and the concept of integrating them with one another into a complete operating plant are in very early development phases and will require longer-term development to verify the true potential and to reach commercial readiness. With success, the estimated hydrogen production cost can be reduced to $0.77/kg.

Department of Energy Programs for Coal-to-Hydrogen Production

The DOE programs for making hydrogen from coal reside in the Office of Fossil Energy (FE) and are related to programs for making electricity from coal. The overall goal of the Hydrogen from Coal Program is to have an operational, zero-emissions coal-fueled facility in 2015 that coproduces hydrogen and electricity with 60 percent overall efficiency (DOE, 2003c). Major milestones to reach this goal include the following:

- *2006*—Identification of advanced hydrogen separation technology including membranes tolerant of trace contaminants;
- *2011*—Demonstration of hydrogen modules for a coal gasification combined-cycle co-production facility; and
- *2015*—Demonstration of a zero-emission, coal-based plant producing hydrogen and electric power (with sequestration) that reduces the cost of hydrogen by 25 percent compared with the cost of current coal-based plants.

To reach these milestones, R&D activities within the Hydrogen from Coal Program are focused on the development of novel processes that include the following:

- Advanced water-gas-shift reactors using sulfur-tolerant catalysts,
- Novel membranes for hydrogen separation from carbon dioxide,
- Technology concepts that combine hydrogen separation and water-gas shift, and
- Reduction of steps needed to separate impurities from hydrogen.

Beyond the DOE's Hydrogen from Coal Program, two other significant DOE coal R&D programs are ongoing and important to the hydrogen program: Vision 21 and FutureGen. Several years ago the DOE initiated an R&D program called Vision 21, which is up and running and was reviewed by the National Research Council most recently in early 2003 (NRC, 2003b). Major aspects of the Vision 21 program include the following areas that will be applicable to making hydrogen from coal and will lead to more efficient and lower-cost hydrogen:

- Advanced ion transport membrane technology for oxygen separation from air,
- Advanced cleaning of raw synthesis gas,
- Improvements in gasifier design, and
- Carbon dioxide capture and sequestration technology.

Making hydrogen from coal produces a large amount of CO_2 as a by-product. A part of the DOE program is aimed at developing safe and economic methods of sequestering CO_2 in a variety of underground geologic formations. Indeed, a sequestration R&D program was initiated in the Office of Fossil Energy a number of years ago and is now supported at a significant level. The new coal-based power systems being developed under the Vision 21 program are aimed at coupling power plant with sequestration systems.

Beyond the Vision 21 program, the DOE recently announced its intention to proceed with FutureGen, a large coal-to-electricity-and-hydrogen verification plant with coupled sequestration. This plant is now in the early stages of detailed planning. In addition to demonstrating coproduction of electricity and hydrogen with sequestration, the system is also intended to act as a large-scale testbed for innovative new technologies aimed at reducing systems costs.

Recommendations

Recommendation 8-3. Coal is a viable option for making hydrogen in large, centralized plants when the demand for hydrogen becomes large enough to support an associated distribution system. Thus, coal should be a significant component of any domestic research and development program aimed at producing large quantities of hydrogen for a possible U.S. hydrogen economy.

Recommendation 8-4. Because there are a number of similarities between the integrated gasification combined cycle process and the coal-to-hydrogen process, the committee endorses the continuation of both programs in tandem at budget levels that are determined to be adequate to meet the programs goals.

Recommendation 8-5. The committee commends the Department of Energy on its initiative in undertaking the FutureGen Project and recommends that the DOE move ahead with the project because of its promise of demonstrating coal-to-hydrogen production coupled with sequestration at a significant scale and its use as a large-scale testbed for related process improvements. As costs can be very high for this type of demonstration, the overall project size and complexity should be closely monitored by the DOE.

HYDROGEN FROM NUCLEAR ENERGY

Why Nuclear?

Nuclear energy is a long-term energy resource that can serve the United States and the world for centuries. With major uranium supplies in the United States, Canada, and Australia, increased reliance on nuclear fuel supplies adds to U.S. energy security. Nuclear power reactors do not involve any CO_2 emissions to the atmosphere, nor do they emit any toxic air pollutants such as are emitted by fossil-fueled power plants.[3] The development of more efficient nuclear power stations requires technologies with high-temperature coolants—developments that are also required for efficient application of nuclear technology to hydrogen generation. The United States is making progress toward establishing a geologic repository for the spent fuel used in a once-through nuclear fuel cycle, while other fuel cycles are being investigated to optimize resource utilization and reduce the waste burden. Nuclear fuel cycles involving separation of fissile materials leave open the possibility of improper access to those materials (e.g., plutonium) through theft or diversion, but this risk can be mitigated through international cooperation (PCAST, 1999).

Status of Nuclear Power Technology

The United States derived about 20 percent of its electricity from nuclear energy in 2002 (EIA, *Electric Power Monthly,* 2003). The 103 power reactors operating today have a total capacity of nearly 100 gigawatts electric (GWe) and constitute about 13 percent of the installed U.S. electric generation capacity. The current U.S. plants use water as the coolant and neutron moderator (hence called light-water reactors, or LWRs) and rely on the steam Rankine cycle as the thermal-to-electrical power conversion cycle. Other countries use other technologies—notably CO_2-cooled reactors in the United Kingdom and heavy-water-cooled reactors (HWRs) in Canada and India.

In the past 20 years, several advanced versions of the LWR, collectively called ALWRs, have been designed, but only one type has been built: the advanced boiling water reactor (ABWR), which was built in Japan. New versions of light-water reactors are now under review for safety certification by the U.S. Nuclear Regulatory Commission (USNRC). It is expected that a high-temperature helium-cooled reactor, if built in South Africa, would become of interest to U.S. utilities and would also be reviewed by the USNRC for certification.

In 2002, several reactor concepts were selected by an international team representing 10 countries, including the United States, as promising "Generation IV (GEN IV) technologies" that should be further explored for availability beyond 2025. The goals for the proposed advanced reactor systems are to improve the economics, safety, waste characteristics, and security of the reactors and the fuel cycle. The emphasis in the development was given to six options (see Appendix G), to be later narrowed to fewer options. The helium-cooled very high temperature reactor (VHTR) is an extension of the helium-cooled reactors built in the United States and in other countries so as to reach higher temperatures and to use gas turbines for their power generation.

[3]Nuclear power reactors do emit trace amounts of radionuclides, chiefly noble gases (USNRC, 1996), which cause small doses of radiation to persons offsite, the total annual risk of which is less than one part in 1 million.

Hydrogen Production Using Nuclear Energy

Hydrogen can be produced using reactors for water splitting by electrolysis or by thermochemical processes without any CO_2 emissions. Potentially more efficient hydrogen production may be attained by significantly raising the water temperature before splitting its molecules using either thermochemistry or electrolysis. Such approaches require temperatures in the range of 700°C to 1000°C. Current LWRs and near-term, water-cooled ALWRs produce temperatures under 350°C and cannot be used for such purposes. However, other coolants of several Generation IV reactor concepts are proposed to reach such high temperatures (above 700°C) and may be coupled to thermochemical plants (Brown et al., 2003; Doctor et al., 2002; and Forsberg, 2003). A recent report by the Electric Power Research Institute (EPRI) pointed out that the use of nuclear reactors to supply the heat needed in the steam methane reforming (SMR) process is potentially more economic than their use for water splitting (Sandell, 2003). Nuclear-assisted SMR would reduce the use of natural gas in the process as well as the CO_2 emissions. The various options for nuclear hydrogen production are compared in Table 8-1.

High-Temperature Electrolysis of Steam

The increased demand for thermal energy is offset by a decrease in the electrical energy demand, which improves the overall thermal-to-hydrogen heat conversion efficiency. Higher temperatures also help lower the cathodic and anodic overvoltages. Thus, the high-temperature electrolysis of steam (HTES) is advantageous from both thermodynamic and kinetic standpoints. The HTES overall efficiency appears less sensitive to temperature than the thermochemical processes appear to be. However, much about the technology needs to be investigated. The durability of the electrode and electrolyte materials is not known and needs to be investigated. Also, the effect of high pressure is to increase the overvoltage needed and reduce the size of the chemical units and transmission lines. The scale-up of the size of the electrolysis cell should be sought.

Thermochemical Reactions

A recent screening of several hundred possible reactions (Besenbruch et al., 2001) has identified two candidate thermochemical cycles for hydrogen production from water (i.e., cycles that enable chemical reactions to take place at high temperatures) with high potential for efficiency and practical applicability to nuclear heat sources. These are the sulfur-iodine (S-I) and calcium-bromine-iron (Ca-Br) cycles. Also, Argonne National Laboratory (ANL) has identified the copper-chlorine (Cu-Cl) thermochemical cycle for this purpose (Doctor et al., 2002). A hybrid sulfur-based process that does not require iodine but has a single electrochemical

TABLE 8-1 An Overview of Nuclear Hydrogen Production Options

	Approach			
	Electrolysis		Thermochemistry	
Feature	Water	High-Temperature Steam	Methane Reforming	Water Splitting
Required temperature (°C)	>0	>300 for LWR >600 for Cu-Cl cycle	>700	>850 for S-I cycle >600 for S-AGR
Efficiency (%) of chemical process	75–80	85–90	70–80	>45, depending on temperature
Efficiency (%) coupled to LWR	27	30	Not feasible	Not feasible
Efficiency (%) coupled to HTGR, AHTR, or S-AGR	Below 40	40–60, depending on temperature	>70	40–60, depending on cycle and temperature
Advantages	Proven technology with LWRs Eliminates CO_2 emissions	Can be coupled to reactors operating at intermediate temperatures Eliminates CO_2 emissions	Proven chemistry 40% reduction in CO_2 emissions	Eliminates CO_2 emissions
Disadvantages	Low efficiency	Requires high-temperature reactors Also requires development of durable HTES units	CO_2 emissions are not eliminated Depends on methane prices	Aggressive chemistry Requires development

NOTE: LWR = light-water reactor; S-AGR = supercritical CO_2 advanced gas reactor; S-I = sulfur-iodine; Cu-Cl = copper-chlorine; HTGR = high-temperature gas-cooled reactor; AHTR = advanced high-temperature reactor; HTES = high-temperature electrolysis of steam.

low-temperature step to produce H_2 and reforms sulfuric acid has also been proposed by researchers at Westinghouse.[4] The low-voltage electrolysis step (low power compared with electrolysis of water) may allow much larger scale-up of the electrochemical cells. The temperatures required for these reactions are generally higher than those provided by LWRs and ALWRs. However, several of the GEN IV reactors would be able to provide the needed heat and high temperatures. The high temperature (above 700°C) would also enable the use of nuclear heat in connection with the SMR process. Reaching higher temperatures would increase the reaction efficiencies. (See Appendix G for details.)

Advanced Reactor Technologies

Several aspects of nuclear energy technology development for electricity production are also useful for hydrogen production and are not detailed here. Among these is the development of high-temperature reactors that can provide coolants at temperatures higher than 800°C. This seems most readily achievable using the helium-cooled gas reactor technology of high-temperature gas-cooled reactors (HTGRs). Irradiation effects at higher temperatures need to be examined. Operation and control of the helium power cycle at very high temperatures are yet to be demonstrated. Development of a supercritical CO_2 power cycle should also be given a high priority. It could potentially allow achieving high power cycle efficiencies at lower temperature than that of the helium gas turbines. This will help the high-temperature electrolysis approach.

Research and Development Priorities for Hydrogen Production

Research and development priorities for nuclear hydrogen production include the following:

1. The efficiency of thermochemical schemes to accomplish water splitting without any CO_2 emissions should be examined at a laboratory scale for the promising cycles such

[4]Charles Forsberg, Oak Ridge National Laboratory, "Production of Hydrogen Using Nuclear Energy," presentation to the committee, January 22, 2003.

as the S-I cycles. The R&D program should include the following areas:

- Materials compatibility issues at high temperature,
- Catalysts to enhance the reaction at lower temperatures, and
- Determination of the efficiency of the integrated processes, and how to optimize it through careful thermal management of the heat and mass flows.

2. Development of the high-temperature steam electrolysis process should be pursued. The following issues should be investigated:

- Materials durability for electrodes and electrolytes,
- Reduction of overvoltages,
- Effects of the operating pressure, and
- Separation of gas products in an efficient and safe manner.

3. The safety issues of coupling the nuclear island to the hydrogen-producing chemical island need to be examined in order to establish the guidelines necessary for avoiding accident propagation from one island to the other. Such guidelines would be needed even if the first application of nuclear hydrogen production was based on the nuclear-assisted SMR approach.

Summary

Hydrogen can be produced from current nuclear reactors using electrolysis of water. More efficient hydrogen production may be attained by thermochemical splitting of water or electrolysis of high-temperature steam. Another possibility is the use of nuclear energy as the source of heat for steam methane reforming (SMR). The water-splitting approach releases no carbon dioxide. Efficient water-splitting processes and nuclear-SMR all require temperatures well above 700°C. Current water-cooled reactors produce temperatures under 350°C and cannot be used for efficient hydrogen production. Advanced reactors, such as gas-cooled reactors, can achieve the required high temperatures. The committee supports a nuclear-energy-to-hydrogen research program as a small incremental effort to the nuclear-to-power program. The nuclear-to-power program is justified on its own merits.

The research budget for the nuclear-to-hydrogen program, which was requested at the level of $4 million for 2004 (DOE, 2003b), appears to be modest. The examination of several options for promising cycles, including the process kinetics, the ability of materials to withstand the aggressive chemistry and temperatures, the separation of fluids, and overall efficiency of the systems requires a higher level of funding for a few years, to determine if a feasible H_2 technology concept can be identified. The research program should allow innovative exploration of other processes, such as direct photolysis generation of hydrogen using intense radiation sources.

The research portfolio should also include safety aspects of integrating the nuclear reactor with the chemical plant for hydrogen production. This aspect of the program is an important ingredient in establishing guidelines for the designs to avoid potential accident propagation. The involvement of industry in assessing the practicality and cost of the technology is recommended.

Recommendations

Recommendation 8-6. The Department of Energy's nuclear hydrogen program should focus on the options to accomplish water splitting without any CO_2 emissions. At this early stage of laboratory-scale investigations, the program should involve several options for promising cycles covering catalysts to enhance the reactions at lower temperatures and materials-compatibility issues. Development of the high-temperature steam electrolysis process should be pursued in balance with the thermochemical cycles. The issues of materials durability, reduction of overvoltages, operating pressure effects, and the separation of gas products in an efficient and safe manner should be investigated. If research is successful, one or two processes should be selected for demonstration of the integrated process in a few years.

Recommendation 8-7. A portfolio of research that advances the near-term technologies while examining the innovative approaches needs to be maintained. The total budget covering the thermochemical, electrochemical, and other alternatives should be increased for a few years in order to allow for selecting the most promising approaches for demonstration. The Department of Energy should promote industry involvement in assessing the economic potential of the various options.

Recommendation 8-8. The Department of Energy's research and development program should involve safety elements of the nuclear-chemical integrated system and aim to establish guidelines to arrest accident propagation from one part of the system to another.

HYDROGEN FROM ELECTROLYSIS

Electrolysis to dissociate water into its separate hydrogen and oxygen constituents has been in use for decades, primarily to meet industrial chemical needs. While more expensive than steam reforming of natural gas, electrolysis may play an important role in the transition to a hydrogen economy because small facilities can be built at existing service stations. In addition, electrolysis is well matched to intermittent renewable technologies. Finally, electrolyzers can allow distributed power systems to manage power during peak-demand hours by using stored hydrogen to generate additional power; this hydrogen can be generated during off-peak hours.

Technology Options

Current electrolysis technologies fall into two basic categories: (1) solid polymer using a proton exchange membrane (PEM) and (2) liquid electrolyte, most commonly potassium hydroxide (KOH). In both technologies, water is introduced into the reaction environment and subjected to an electrical current that causes dissociation, after which the resulting hydrogen and oxygen atoms are put through an ionic transfer mechanism that causes the hydrogen and oxygen to accumulate in separate physical streams.

A PEM electrolyzer is literally a PEM fuel cell operating in reverse mode. When water is introduced to the PEM electrolyzer cell, hydrogen ions (protons) are drawn into and through the membrane, where they recombine with electrons to form hydrogen molecules. Oxygen gas remains behind in the water. As this water is recirculated, oxygen accumulates in a separation tank and can then be removed from the system. Hydrogen gas is separately channeled from the cell stack and captured.

Liquid electrolyte systems typically use a caustic solution to perform functions analogous to those of a PEM electrolyzer. In such systems, oxygen ions migrate through the electrolytic material, leaving hydrogen gas dissolved in the water stream. This hydrogen is readily extracted from the water when directed into a separating chamber.

The all-inclusive costs of hydrogen from PEM and KOH systems today are roughly comparable. Reaction efficiency tends to be higher for KOH systems because the ionic resistance of the liquid electrolyte is lower then the resistance of current PEM membranes. But the reaction efficiency advantage of KOH systems over PEM systems is offset by higher purification and compression requirements, especially at small scale (1 to 5 kilograms per hour). Further details are provided in Appendix G.

Electrolysis may be particularly well suited to meeting the early-stage fueling needs of a fuel cell vehicle market. Electrolyzers scale down reasonably well; the efficiency of the electrolysis reaction is independent of the size of the cell or cell stacks involved. The compact size of electrolyzers makes them suitable to being placed at or near existing fueling stations, and they can use existing water and electricity infrastructures, minimizing the need for new infrastructure.

Future Electrolysis Technology Enhancements

The DOE goal for electrolysis is a capital cost of $300/kW for a 250 kg/d plant (at 5000 pounds per square inch [psi] with 73 percent system efficiency, lower heating value basis [DOE, 2003b, p. 3-15]). Such a plant could be integrated with a renewable energy source to produce hydrogen at $2.50/kg by 2010. A large, central station plant could then produce hydrogen at $2.00/kg (DOE, 2003b, p. 3-16). The DOE research program focuses on ways to reduce costs, improve efficiency, and integrate electrolysis plants with renewable electricity sources. The DOE is also continuing development of reversible solid oxide electrolyzer materials, which can operate at higher temperatures than PEMs can, and at potentially very high efficiencies. The DOE reported that its FY 2004 budget request included approximately $3.2 million for electrolysis-to-hydrogen research.[5,6]

The committee finds it plausible that PEM electrolyzer capital costs can fall by a factor of eight—from $1000/kW in the near term to $125/kW over the next 15 to 20 years, contingent on similar cost reductions occurring in PEM fuel cells. Should capital costs decrease to this level, the committee estimates that hydrogen could be produced for about $4/kg using grid electricity and electrolysis, making it attractive during the transition period 2010–2030, until centralized facilities and the required distribution system are built. The DOE's multi-year research, development, and demonstration plan (DOE, 2003b) includes a technical plan on fuel cells, which addresses technology and cost barriers—barriers that, if overcome, will benefit electrolyzers as well. Elements of the fuel cell plan include, for example: development of high-temperature membranes for PEM fuel cells, development of lower-cost polymer membranes having higher ionic conductivity, and development of alternative catalyst formulations and structures.

In addition, electrolyzer system efficiencies may rise from the current 63.5 percent to 75 percent (lower heating value) in the future. Among research priorities that can improve the efficiency and/or reduce the cost of future electrolysis fueling devices and could become part of the DOE's electrolysis program are the following:

1. *Reducing other (parasitic) system energy losses.* A variety of parasitic loads, such as power conditioning, can be reduced through system redesign and optimization.
2. *Reducing current density.* Conversion efficiencies are a function of electric current density, so the substitution of more electrolyte or more cell surface area has the impact of reducing overall power requirements per unit of hydrogen produced.
3. *Development of electrolysis/oxidation hybrids.* The hybrid concept uses the oxidation of natural gas as a means of intensifying the migration of oxygen ions through the electrolyte and thereby reducing the effective amount of electric energy required to transport the oxygen ion. The concept

---

[5]Pete Devlin, Department of Energy, "DOE's Hydrogen RD&D Plan," presentation to the committee, June 11, 2003.

[6]The committee understands that of the $78 million subsequently appropriated for hydrogen technology for 2004 in the Energy and Water appropriations bill (Public Law 108-137), $37 million is earmarked for activities that will not particularly advance the hydrogen initiative.

appears to offer the potential for significantly improved net electrochemical efficiency, but will require several technical breakthroughs in harnessing solid oxide technology.

Recommendations

Recommendation 8-9. The Department of Energy's electrolysis technology program should continue to target cost reduction, enhanced system efficiency, and improved durability for distributed-scale hydrogen production from electricity and water. These technology objectives can be advanced through research into (1) lower-cost membranes, catalysts, and other cell and system components; (2) membranes and systems that can operate at higher temperatures and pressures; and (3) improved system design and integration with an eye toward low-cost manufacturing. Specifically, the DOE should increase emphasis on electrolyzer development with a target of $125 per kilowatt and a significant increase in efficiency toward a goal of over 70 percent (lower heating value basis).

Recommendation 8-10. The Department of Energy should emphasize component development and systems integration to enable electrolyzers to operate from inherently intermittent and variable-quality power derived from wind and solar sources.

HYDROGEN PRODUCED FROM WIND ENERGY

The production of hydrogen from renewable energy sources is often stated as the long-term goal of a mature hydrogen-based economy (Turner, 1999). Of all renewable energy sources, using wind-turbine-generated electricity to electrolyze water, particularly in the near to medium term, has arguably the greatest potential for producing pollution-free hydrogen. The issues for its successful development and deployment are threefold: (1) further reducing the cost of wind turbine technology and the cost of the electricity generated by wind, (2) reducing the cost of electrolyzers, and (3) optimizing the wind turbine-electrolyzer with a hydrogen storage system. The current study considered only distributed-scale wind-to-hydrogen production systems. For a more in-depth discussion, see Appendix G.

Wind energy is one of the most cost-competitive renewable energy technologies available today, and in some places it is beginning to compete with new fossil fuel electricity generation. A principal parameter determining the economic success of wind turbines is the annual energy output, which is most sensitive to wind speed and the on-stream capacity factor. The current cost of generating electricity from wind at good wind sites falls in the range of 4 to 7 cents per kilowatt-hour (kWh) (without financial incentives), with capacity factors of about 30 percent. Analysts generally forecast that these costs will continue to drop and capacity factors increase as the technology improves further and the market grows.

There are obvious advantages to hydrogen produced from wind energy. It is essentially emission free, producing no CO_2 or criteria pollutants, such as oxides of nitrogen (NO_x) and sulfur dioxide (SO_2), and it is a domestic source of energy. Thus, it addresses both of the main concerns motivating the current drive toward a hydrogen economy. But wind energy is not free of problems. There are environmental, siting, and technical issues that must be dealt with. Wind energy's most serious drawback continues to be its intermittence and mismatch with demand, an issue both for electricity generation and hydrogen production.

Hydrogen Production by Electrolysis from Wind Power

The committee's analysis considered wind-energy-to-hydrogen systems deployed on a distributed scale, which thus bypasses the extra costs and requirements of hydrogen distribution. For distributed wind-electrolysis-hydrogen production systems, it is estimated that using today's technologies, hydrogen can be produced at good wind sites (class 4 and above, without financial incentives) for approximately $6.64/kg H_2. The committee's analysis considers a system that uses grid electricity as backup for when the wind isn't blowing to alleviate the capital underutilization of the electrolyzer. This hybrid wind-to-hydrogen production system has pros and cons. It reduces the cost of producing the hydrogen, which without grid backup would be $10.69/kg H_2, but it also incurs CO_2 emissions from what would otherwise be an emission-free hydrogen production system.

In the future, the wind-electrolysis-hydrogen system could be substantially optimized. The wind turbine technology will improve, with a resulting decrease in the cost of electricity generated and an increase in the turbine's capacity factor, and the electrolyzer's efficiency will increase and its capital costs decrease (see the section above, entitled "Hydrogen from Electrolysis"). With the assumptions used in this study, the committee finds that the wind energy system and the electrolyzer can be designed to be large enough that sufficient low-cost hydrogen can be generated and stored when the wind is blowing, without grid backup. This is a lower-cost option than using a smaller electrolyzer and purchasing grid-supplied electricity when the wind turbine is not generating electricity. With future estimated improvements in the technology, hydrogen produced from wind without grid backup is estimated to cost $2.86/kg H_2, while for a system with grid backup it is $3.38/kg H_2 (all without financial incentives). Furthermore, this stand-alone system has the added advantage of a hydrogen production system that is CO_2-emission-free. The results of the committee's analysis are summarized in Table 8-2.

Wind-electrolysis-hydrogen production systems are currently far from optimized. For example, better integration of the wind turbine and electrolyzer power control system

TABLE 8-2 Results from Analysis Calculating Cost and Emissions of Hydrogen Production from Wind Energy

	Current Technology		Future Technology	
	With Grid Backup	No Grid Backup	With Grid Backup	No Grid Backup
Average cost of electricity (cents/kWh)	6	6	4	4
Wind turbine capacity factor (%)	30	30	40	40
Hydrogen ($/kg)	6.64	10.69	3.38	2.86
Carbon emissions (kg C/kg H_2)	3.35	0	2.48	0

is needed, as is hydrogen storage tailored to the wind turbine design. Furthermore, there is the potential to optimize coproduction of electricity and hydrogen, which under the right circumstances could be more cost-effective and provide broader system utility.

Conclusions

Wind energy has some very clear advantages as a source of hydrogen. It fulfills the two main motivations that are propelling the current push toward a hydrogen economy, namely, reducing CO_2 emissions and reducing the need for hydrocarbon imports. In addition, it is the most affordable renewable technology deployed today, with expectations that costs will continue to decline. Of all renewable energy technologies, wind is the closest to practical widespread utilization with the technical potential to produce a significant amount of hydrogen in the future. Yet, it still faces many barriers to deployment and therefore deserves continued as well as more focused attention in the DOE's hydrogen program. In particular, there is a need to partner with industry to develop optimized wind-to-hydrogen systems and to help identify the R&D needed to advance such systems to the next level.

Energy security and environmental quality, including the reduction of carbon dioxide emissions, are strong factors motivating a hydrogen economy. These two goals can both be addressed by wind-energy-to-hydrogen systems. Thus, wind has the potential to play an important role in a future hydrogen economy, particularly during the transition and potentially in the long term. Wind technology is likely to continue to improve. Such improvements would include enhanced performance at variable wind speeds, thereby permitting capture of the maximum amount of wind according to local wind conditions, and better grid compatibility, matching supply with demand. These advancements can occur through better turbine design and optimization of rotor blades, more efficient power electronic controls and drive trains, and better materials.

Wind-electrolysis-hydrogen systems have yet to be fully optimized. There are integration opportunities and issues with respect to wind energy systems, electrolyzers, and hydrogen storage that need to be creatively explored. For example, coproduction of electricity and hydrogen can potentially reduce costs and increase the function of the wind-energy-to-hydrogen system. This could facilitate the development of wind energy systems that are more cost-effective and have broader utility.

Department of Energy's Research, Development, and Demonstration Plan

There is little mention of hydrogen production from wind throughout the entire June 2003 draft entitled "Hydrogen, Fuel Cells and Infrastructure Technologies Program: Multi-Year Research, Development and Demonstration Plan" (DOE, 2003b), or in the July 2003 *Hydrogen Posture Plan: An Integrated Research, Development, and Demonstration Plan* (DOE, 2003a). An RD&D program for hydrogen production from wind power needs to be developed and integrated into the overall hydrogen strategic RD&D plan. Clear and strong crossover is needed between the research program on renewable-based electricity generation, such as wind energy, and the hydrogen production R&D program.

Recommendation 8-11. Wind energy for hydrogen production does not appear at present in the Department of Energy's multi-year research, development, and demonstration plan. Wind-energy-to-hydrogen systems need to be an important element in the DOE's hydrogen program and need to be integrated into the hydrogen production strategy. The plan should address the technical issues related to costs and capacity factor, particularly wind sites in class 3 and below.

Recommendation 8-12. The Department of Energy's multi-year research, development, and demonstration plan should address how best to partner with industry to create robust, efficient, and cost-effective wind-electrolysis-hydrogen systems that will be ready for deployment as the distributed hydrogen infrastructure begins to develop. This is particularly important as there needs to be a research and development emphasis on optimizing wind-electrolysis-hydrogen systems. The role that the coproduction of hydrogen and electricity from wind can play needs to be further analyzed and integrated into future hydrogen production strategies so that potential synergies can be better understood and utilized.

HYDROGEN PRODUCTION FROM BIOMASS AND BY PHOTOBIOLOGICAL PROCESSES

Renewable solar energy is the primary energy source for hydrogen production from biomass or by direct photobiological processes (Turner, 1999). Two types of biomass feedstock are available to be converted into hydrogen: (1) dedicated bioenergy crops and (2) less expensive residues, such as organic waste from regular agricultural farming and wood processing (biomass residues). In direct photobiological hydrogen production, water is directly cleaved by photosynthetic (micro)organisms, without biomass formation as intermediate. Hydrogen production by these biological means is attractive because solar energy is a renewable energy source.

Biomass Costs and Availability

Hydrogen production from biomass is a thermodynamically inefficient and expensive process, in which approximately 0.2 percent to 0.4 percent of the total solar energy is converted to hydrogen at a price of currently about \$7.05/kg H_2 by gasification in a midsize plant (see Figure 5-2 in Chapter 5 and Appendix E). This price reflects higher feedstock, distribution, and fixed and capital costs relative to other production methods such as natural gas reforming (Figure 5-2). In its economic analysis, the committee did not consider fertilizer costs and the environmental impact associated with the production, harvest, and transport of biomass. In addition, the analysis did not quantify any potential degradation in land quality associated with intensive bioenergy crop farming.

All renewable technologies and feedstocks for hydrogen production compete for land area with other societal needs, such as agricultural goods and services, recreation, and land conservation. To minimize this competition, biological processes for H_2 production need to be thermodynamically efficient and kinetically fast to reduce land use. In the committee's possible future technology case, crop yield is projected to improve *by* 50 percent, and efficiency is projected to improve *to* 40 percent, up from 26 percent in the current technology case (inclusive of conversion, distribution, and dispensing energy efficiencies). In a future, all-fuel-cell-vehicle economy, the amount of biomass required to satisfy 100 percent of the hydrogen demand would require roughly 282,000 square miles (mi^2) for bioenergy crop farming.[7] (See also Figure 6-15 in Chapter 6.) This comprises an area of land that is approximately 40 percent of cropland currently used for crops in the United States.[8] Farming of bioenergy crops on such a scale is likely to have significant environmental impacts on soil, water sources, biodiversity, and eutrophication, and might also affect the price for agricultural goods. The committee estimates the possible future technology price for hydrogen from gasification of biomass to be \$3.60/kg H_2, which is noncompetitive relative to other hydrogen production technologies (see Figure 5-4).

Biomass Conversion

Current technologies for converting biomass into molecular hydrogen include gasification or pyrolysis of biomass coupled to subsequent steam reformation.[9] The main conversion processes are (1) indirectly heated gasification, (2) oxygen-blown gasification, (3) pyrolysis, and (4) biological gasification (anaerobic fermentation). Plants dedicated for biomass gasification are designed to operate at low pressure and are limited to midsize-scale operations,[10] due to the heterogeneity of biomass, the localized production of biomass, and the relatively high costs of gathering and transporting biomass feedstock. Therefore, in addition to the relatively high feedstock costs, dedicated biomass gasification plants are associated with capital costs that, in the current technology case, the committee estimates to be \$2.44/kg H_2, compared with about \$0.46/kg H_2 for large, central station coal gasification (see Figure 5-2). The committee considered possible improvements in biomass gasification technology in its possible future technology case and projected capital costs of about \$1.20/kg H_2, which would be still more than twice that for current coal gasification (see Figure 5-4). In addition, the energy efficiency of the biomass gasification plants for converting feedstock into hydrogen will be lower than that of coal plants.

After capital costs (which include the gasifier), feedstock costs and distribution costs are the two largest cost components for the production of hydrogen from biomass by gasification, in the committee's analysis (see Figures 5-2 and 5-4). The size of these two cost components is an inherent consequence of the low density of biomass as collected from

[7]Calculated using the biomass "possible future technology case" (and its higher efficiency and crop yield).

[8]The U.S. Department of Agriculture's National Agricultural Statistics Service (NASS, 2003) estimated that 349 million acres (545,000 mi^2) of cropland were used for crops (i.e., cropland harvested, crop failure, and cultivated summer fallow) in 1997, the most recent year for which data were available.

[9]Roxanne Danz, Department of Energy, "Hydrogen from Biomass," presentation to the committee, December 2, 2002.

[10]Biomass gasification has been demonstrated at 100 to 400 tons of biomass per day (Margaret Mann and Ralph Overend, National Renewable Energy Laboratory, "Hydrogen from Biomass: Prospective Resources, Technologies, and Economics," presentation to the committee, January 22, 2003). The committee's midsize biomass-to-hydrogen gasification technology cases fall within this range (see Appendix E).

bioenergy crop farming, the distributed nature of biomass residue collection, and the cost of transportation to the biomass gasification plant. Because of the costs associated with harvesting low-density bioenergy crops, biomass gasification plants will be limited to a midsize scale. Consequently, such gasification plants will not make use of the economy of scale, which will keep the costs for distribution high and will inherently limit the plant's energy efficiency.

Coproduction (biorefinery) of, for example, phenolic adhesives, polymers, waxes, and other products with hydrogen production from biomass is being discussed in the context of biomass gasification plant designs to improve the overall economics of biomass-to-hydrogen conversion.[11] The technical and economic viability of such coproduction plants is unproven and was not considered in this analysis.

Biomass Gasification as a Means for Net Reduction of Atmospheric Carbon Dioxide

Biomass gasification could play a significant role in meeting the DOE's goal of greenhouse gas mitigation. It is likely that both in the transition phase to a hydrogen economy and in the steady state, a significant fraction of hydrogen might be derived from domestically abundant coal. In co-firing applications with coal, biomass can provide up to 15 percent of the total energy input of the fuel mixture. The DOE could address greenhouse gas mitigation by co-firing biomass with coal to offset the losses of carbon dioxide to the atmosphere that are inherent in coal combustion processes (even with the best-engineered capture and storage of carbon). Since growth of biomass fixes atmospheric carbon, its combustion leads to no net addition of atmospheric CO_2 even if vented. Thus, co-firing of biomass with coal in an efficient coal gasification process, affording the opportunity for capture and storage of CO_2, could lead to a net reduction of atmospheric CO_2. The co-firing fuel mixture, being dilute in biomass, places lower demands on biomass feedstock. Thus, cheaper, though less plentiful, biomass residue could supplant bioenergy crops as feedstock. Using residue biomass would also have a much less significant impact on the environment than would farming of bioenergy crops.

Advanced Direct Photobiological Hydrogen Production

Hydrogen production by direct oxidative cleavage of water, mediated by photosynthetic (micro)organisms, without biomass as intermediate, is an emerging technology at the early exploratory research stage.[12] By circumventing biomass formation and subsequent gasification, the yield of solar energy conversion to hydrogen by direct photobiological processes is theoretically more efficient than is biomass gasification by 1 to 2 orders of magnitude. The direct photobiological hydrogen release could be on the order of 10 percent (see Appendix G), compared with efficiencies of 0.5 to 1 percent for biomass-to-hydrogen conversion. It is conceivable that bioengineering efforts on the light harvesting complex and reaction center chemistry (see Appendix G) could improve this efficiency severalfold over the coming decades, and thereby bring the overall efficiency (solar-to-hydrogen) of direct photobiological hydrogen production into the range of 20 to 30 percent. However, substantial, fundamental research needs to be undertaken before photobiological methods for large-scale hydrogen production are considered.

Department of Energy Research and Development Program

According to the draft (June 3, 2003) "Hydrogen, Fuel Cells and Infrastructure Technologies Program: Multi-Year Research, Development and Demonstration Plan" (DOE, 2003b), the DOE's Office of Energy Efficiency and Renewable Energy is focused on biomass gasification/pyrolysis and has set technical targets for the years 2005, 2010, and 2015 to reduce costs for biomass gasification/pyrolysis and subsequent steam reforming. Specific goals include reduction of costs for (1) biomass feedstock, (2) gasification operation (including efficiency), (3) steam reforming, and (4) hydrogen gas purification. The DOE reported that its FY 2004 budget request included approximately $5.4 million for biomass-to-hydrogen research. While the breakdown of this amount was not further specified, the DOE did list its six priority items for hydrogen production, included among which was the following biomass-to-hydrogen priority item: "Low cost and high efficiency gasifier/pyrolyser and reforming systems."[13] The technical targets that the DOE has set for biomass gasification and biomass pyrolysis are ambitious; they include cost reduction from $3.60 to $2.00/kg H_2 by 2015 for gasification and from $3.80 to $2.40/kg H_2 by 2015 for pyrolysis (DOE, 2003b). The EERE program seems to support photobiological hydrogen research, but the DOE does not specify the amount beyond noting a total FY 2004 request for photolytic systems of $3.2 million—a total that includes *photoelectrochemical* hydrogen production methods in addition to photobiological methods.[14] The DOE's R&D targets for increasing the utilization efficiency of absorbed light and hydrogen production are very ambitious (a factor-of-four improvement to 20 percent by 2010) (DOE, 2003b).

---

[11]Margaret Mann and Ralph Overend, National Renewable Energy Laboratory, "Hydrogen from Biomass: Prospective Resources, Technologies, and Economics," presentation to the committee, January 22, 2003.

[12]Catherine E. Grégoire Padró, National Renewable Energy Laboratory, "Hydrogen from Other Renewable Resources," presentation to the committee, December 2, 2002.

[13]Pete Devlin, Department of Energy, "DOE's Hydrogen RD&D Plan," presentation to the committee, June 11, 2003.

[14]Pete Devlin, Department of Energy, "DOE's Hydrogen RD&D Plan," presentation to the committee, June 11, 2003.

Challenges

The most significant challenges for hydrogen production by biological technologies are these:

- The low thermodynamic efficiency of biomass-to-hydrogen conversion, the high costs of bioenergy crop production and biomass gasification, and the significant demand for and impact on land use and natural resources for bioenergy crop farming; and
- The engineering of (micro)organisms and of processes for direct photobiological hydrogen production without biomass as an intermediate at high themodynamic efficiency and high kinetic rates.

Recommendation 8-13. The committee recommends that the Department of Energy deemphasize the current biomass gasification program and refocus its bio-based program on more fundamental research on photosynthetic microbial systems to produce hydrogen from water at high rate and efficiency. The DOE should encourage innovative approaches and should make use of important breakthroughs in molecular, genomic, and bioengineering research. Research and development for co-firing of biomass, for example, with coal, coupled to subsequent carbon sequestration should continue. The DOE should resist pressure for premature demonstration projects of developing technologies.

HYDROGEN FROM SOLAR ENERGY

It has been estimated that solar energy has the potential to meet global energy demand well into the future (Turner, 1999). Hydrogen from solar energy can be produced through two methods. In one method, solar energy is converted into electricity using a photovoltaic (PV) cell and then hydrogen is generated through the electrolysis of water. In the alternate method, photoelectrochemical cells are used for the direct production of hydrogen. The photoelectrochemical methods are still in the early stages of development.

Approximately 85 percent of the current commercial PV modules are based on single-crystal or polycrystalline silicon. A second type of PV technology is based on deposition of thin films of amorphous as well as microcrystalline silicon, and on compounds based on group II-VI and group I-III-VI elements of the periodic table. Thin-film technology appears to hold greater promise for cost reduction. However, in spite of its promise, the thin-film technology has been unable to reduce the cost of solar modules, owing to low deposition rates. This problem translates to low solar cell production rates. The current low film deposition rates lead to a decreased rate of solar cell production and directly translate into a significant cost because of the lower productivity of the expensive deposition machines. As yields and throughputs are low, the plants need better inline controls and easier and faster deposition techniques enhancing reproducibility. Also, there is a substantial drop in efficiency of the solar cell from the laboratory scale to the module scale.

The current cost of solar modules is in the range of \$3 to \$6 per peak watt (W_p). For solar cells to be competitive with the conventional technologies for electricity production alone, the module cost has to come down below \$1/W_p.

The committee estimated the cost to produce hydrogen using electricity from solar PV devices to power electrolyzers. In the current technology case, with a favorable installed cost of about \$3.28/W_p, the electricity cost is estimated to be about \$0.32/kWh and the hydrogen cost to be \$28.19/kg (scenario Dist PV-C and Dist PV Ele-C of Chapter 5 and Appendix E). In the possible future technology case, the installed capital cost of \$1.011/W_p provides an electricity cost of \$0.098/kWh and a hydrogen cost of \$6.18/kg (scenario Dist PV-F and Dist PV Ele-F). The \$6.18 possible future cost of hydrogen, in the committee's analysis, is the sum of \$4.64/kg for PV-generated electricity and \$1.54/kg, mostly for capital charges associated with producing (via electrolysis), storing, and dispensing hydrogen. The total supply chain cost is thus about a factor of four higher than that of the central station coal plant in its possible future case (CS Coal-F), which the committee estimates to be \$1.63/kg H_2, inclusive of delivery and dispensing. For the PV-electrolyzer combination to be competitive in the future, either the cost of PV modules has to be reduced by an order of magnitude from current costs, or the electrolyzers' cost has to come down substantially from the low cost of \$125/kW already assumed in the committee's future technology case. A factor contributing to this need for low electrolyzer cost is the low utilization of the electrolyzer capital (solar energy is taken to be available 20 percent of the time). *Therefore, while electricity at \$0.098/kWh from a PV module can be quite attractive for distributed applications in which electricity is used directly at the site, hydrogen costs via PV-electricity and electrolysis will not be competitive.* Energy is consumed in the manufacture of solar modules. It has been estimated by the National Renewable Energy Laboratory (NREL) that for a crystalline silicon module, the payback period of energy is about 4 years. For amorphous silicon modules this period is currently around 2 years, with the expectation that it will eventually be less than 1 year.

Various developments are likely to improve the economic competitiveness of solar PV technology. The current research on thin-film deposition techniques is leading to higher deposition rates and efficiencies. Better barrier materials to eliminate moisture ingress in the thin-film modules will prolong the module life span. Robust deposition techniques will increase the yield from a given type of equipment. Inline detection and control methods will help to reduce the cost. Some of this advancement will require creative tools and methods. The anticipated improvements from this thin-film technology research were included in the possible future technology case calculations.

Alternate concepts to thin films, such as dye-sensitized solar cells, also known as Grätzel cells, are being explored

(O'Regan and Grätzel, 1991). In these cells, a dye is incorporated in a porous inorganic matrix such as TiO_2, and a liquid electrolyte is used for positive charge transport. This type of cell has a potential to be low-cost. However, the efficiencies at present are quite low, and the stability of the cell in sunlight is unacceptable. Research is needed to improve performance in both respects.

Another area of intense research is that on the integration of organic and inorganic materials at the nanometer scale into hybrid solar cells. The current advancement in conductive polymers and the use of such polymers in electronic devices and displays provides the impetus for optimism. The nano-sized particles or rods of the suitable inorganic materials are embedded in the conductive organic polymer matrix. Once again, the research is in the early phase and the current efficiencies are quite low. However, the production of solar cells based either solely on conductive polymers or hybrids with inorganic materials has much potential to provide low-cost solar cells. It is hoped that one would be able to cast thin-film solar cells of such materials at a high speed, resulting in low cost.

Research is being done to create aqueous photoelectrochemical cells for direct conversion of solar energy to hydrogen (Grätzel, 2001).[15] In this method, light is converted to electrical and chemical energy. A solid inorganic oxide electrode is used to absorb photons and provide oxygen and electrons. The electrons flow through an external circuit to a metal electrode, and hydrogen is liberated at this electrode. The candidate inorganic oxides are $SrTiO_3$, $KTaO_3$, TiO_2, SnO_2, and Fe_2O_3. If successful, such a method holds the promise of directly providing low-cost hydrogen from solar energy.

It seems that a photoelectrochemical device in which all of the functions of photon absorption and water splitting are combined in the same equipment may have better potential for hydrogen production at reasonable costs. However, a quick "back of the envelope" analysis shows that in order to compete with the hydrogen produced from fossil fuels, photoelectrochemical devices should recover hydrogen at an energy equivalent of \$0.4 to \$0.5/W_p. This cost challenge is similar to that for electricity production from solar cells.

Challenges and Research and Development Needs

Large-scale use of solar energy for a hydrogen economy will require research and development efforts on multiple fronts:

- In the short term, there is a need to reduce the cost of thin-film solar cells. To do this will require the development of deposition techniques of thin films such as microcrystalline silicon and other materials that are robust and provide high throughput rates without sacrificing film efficiencies. In the short run, thin-film deposition methods can potentially gain from a fresh look at the overall process from the laboratory scale to the manufacturing scale. The research in this area is expensive. For such research, some additional centers in academia with industrial alliances could be beneficial. It will be necessary to collect interdisciplinary teams from different science and engineering disciplines for such studies.
- In the midterm to long term, organic polymer-based solar cells hold promise for mass production at low cost. They have an appeal for being cast as thin films at very high speeds using known polymer film casting techniques. Currently, the efficiency of such a system is quite low (in the neighborhood of 3 to 4 percent or lower), and stability in sunlight is poor. However, due to the tremendous development in conducting polymers and other electronics-related applications, it is anticipated that research in such an area has a high potential for success. Similarly, the search for a stable dye material and better electrolyte material in dye-sensitized cells (Grätzel cells) has a potential to lead to lower-cost solar cells. There is a need to increase the stable efficiency of such cells.
- In the long run, the success of directly splitting water molecules by using photons is quite attractive. Research in this area can be very fruitful.

Department of Energy Programs for Solar Energy to Hydrogen

The current DOE target for photoelectrochemical hydrogen production in 2015 is \$5/kg H_2 at the plant gate.[16] Even if this target is met, solar-energy-to-hydrogen is unlikely to be competitive. Therefore, a much more aggressive cost target for hydrogen production by photoelectrochemical methods is needed.

Since photoelectrochemical hydrogen production is in an embryonic stage, a parallel effort to reduce the cost of electricity production from PV modules must be made. A substantial reduction in PV module cost (lower than \$0.5/W_p) coupled with similar reductions in electrolyzer costs (about \$125/kW at reasonable high efficiency of about 70 percent on a lower heating value basis) can provide hydrogen at reasonable cost. The potential research opportunities listed in the preceding subsection for PV solar cells along with electrolyzers must be actively explored.

Summary

All of the current methods and the projected technologies for producing hydrogen from solar energy are much more expensive (greater than a factor of three) when compared with hydrogen production from coal or natural gas plants. This is due partly to the lower annual utilization factor of about 20 percent (as compared with, say, wind of 30 to 40 percent). This creates enormous pressure to reduce the cost

[15] Nathan Lewis, California Institute of Technology, "Hydrogen Production from Solar Energy," presentation to the committee, April 25, 2003.

[16] Pete Devlin, Department of Energy, "DOE's Hydrogen RD&D Plan," presentation to the committee, June 11, 2003.

of a solar energy recovery device. While an expected future installed module cost of about $1/$W_p$ is very attractive for electricity generation and deserves a strong research effort in its own right, this cost fails to provide hydrogen at a competitive value. It is apparent that there is no one method of harnessing solar energy that is a clear winner. However, it appears possible that new concepts may emerge that would be competitive.

In the future, if the cost of the fuel cell system approaches $50/kW, the cost of the electrolyzer is also expected to approach a low number (about $125/kW). Such low capital costs for electrolyzer units, together with levelized electricity costs in the neighborhood of $0.02 to $0.03/kWh, would result in a competitive hydrogen cost. It is also estimated that for a photoelectrochemical method to compete, its cost needs to approach $0.04 to $0.05/kWh. The order-of-magnitude reductions in cost for both hydrogen processes are similar.

Recommendations

Recommendation 8-14. Because of the large volume of hydrogen potentially available from solar energy and its carbon dioxide-free hydrogen, multiple paths of development should be pursued until a clear winning technology emerges for hydrogen production. There is a need for more basic research to provide a low-cost option. More specifically, alternate new technologies for harnessing solar energy should be developed, as well as new and novel methods to substantially reduce the manufacturing cost of some of the promising known technologies.

Recommendation 8-15. While basic research in photoelectrochemical as well as other methods that directly convert solar energy to hydrogen should be actively pursued, the route of solar electricity generation coupled with use of an electrolyzer for hydrogen production should also be pursued in a balanced Department of Energy solar program. A more aggressive target for photovoltaic solar of about $0.02 per kilowatt-hour (roughly $200 to $300 per kilowatt for the solar module) requires research on new and novel approaches. This will be especially important if improvements in battery storage density and cost are not achieved and hydrogen usage becomes dominant.

9

Crosscutting Issues

This chapter addresses several issues that have implications for Department of Energy (DOE) hydrogen program activities. These issues are not specific to any one hydrogen production or end-use technology, but instead concern the integration of the DOE's various hydrogen program elements with societal goals and with the capabilities of various sectors—industrial and academic—that are stakeholders in a transition to a hydrogen economy.

PROGRAM MANAGEMENT AND SYSTEMS ANALYSIS

Program Management

The director of the DOE Office of Hydrogen, Fuel Cells and Infrastructure Technologies—charged with coordinating hydrogen programs across the DOE—provided the committee with an organization chart that shows how the department set up various coordinating committees and lines of authority to help manage the overall DOE hydrogen program. The challenge of managing across office boundaries is very significant, as cooperation between various offices in the DOE has often been less than harmonious, owing to conflicts over budgets and technology.

The situation is further complicated because a number of technologies are being pursued within the DOE for the future production of electricity as well as for their possible use in hydrogen generation. Examples include coal gasification, which can be used for the production of electricity, synthesis gas, synthetic liquid fuels, and/or hydrogen. The various parts of the Office of Fossil Energy's (FE's) coal gasification program are funded at a very significant level, but only a relatively small portion of the total is included in the DOE hydrogen program budget. Similarly, wind systems and photovoltaics are being developed for electric power production, but in principle both also have potential application for the production of hydrogen via electrolysis. Accordingly, challenges involving coordination and cooperation are significant and require careful monitoring by DOE senior management to ensure that the needed cooperation and balance are maintained.

Throughout this report the committee has emphasized the challenges and complexity of developing a viable hydrogen energy system. As the lead government agency in this investigation, the DOE faces a larger management problem than it ever faced in the development of other domestic energy technologies. Recognizing that the tools for managing such a complex effort have been developed and utilized in both the National Aeronautics and Space Administration (NASA) and the Department of Defense (DOD), the DOE has initiated a new-to-the-department systems integration activity to assist in the management of the remarkable complexities associated with hydrogen system development. In a draft charter on systems integration, the DOE defines this "Systems Integration" function as "a disciplined approach applied to the design, development, and commissioning of complex systems that ensures that requirements are identified, validated, and met while minimizing the impact on cost and schedule of unanticipated events and interactions."[1] In a manner similar to the assignment of responsibilities in complex system development at NASA and DOD, the DOE states that its systems integrator will carry out the following:

- Define and validate program requirements.
- Identify and manage interfaces.
- Identify risks and propose management options for mitigation.
- Support informed decision-making.
- Verify that products meet systems requirements.[2]

The DOE differentiates the functions of systems integration from those of systems analysis. While mutually supportive, they are separate functions.

[1]"DOE Hydrogen Program Systems Integrations Charter (draft)," presentation to committee (partial), June 15, 2003.

[2]"DOE Hydrogen Program Systems Integrations Charter (draft)," presentation to committee (partial), June 15, 2003.

Additionally, it is important for the DOE to separate the programs that it includes under systems integration management from its exploratory research programs. Exploratory research requires a dramatically different management approach. As noted in the committee's interim report (see its Letter Report, reprinted in Appendix B) and throughout this report, exploratory research is absolutely essential if the DOE is to identify and develop the dramatically new technologies necessary to the eventual viability of a hydrogen economy. Indeed, without breakthroughs that can only come from exploratory research, the likelihood of a hydrogen economy's coming into being would be greatly diminished.

Systems Analysis

On April 4, 2003, the committee provided the DOE with its interim report, which includes four management-related recommendations focused on the areas of systems analysis, exploratory research, safety, and organization (see Appendix B). The committee is pleased to acknowledge that the director of the DOE Office of Hydrogen, Fuel Cells, and Infrastructure Technologies and others in the DOE immediately began action to respond to those recommendations. In this chapter, the committee presents some additional perspectives related to systems analysis and management.

The Office of Energy Efficiency and Renewable Energy (EERE) has assigned responsibility for the establishment of an independent hydrogen systems analysis program to the National Renewable Energy Laboratory (NREL). The director of the DOE Office of Hydrogen, Fuel Cells, and Infrastructure Technologies informed the committee that there will be a "firewall" between normal NREL activities and the systems analysis function in order to minimize the possibility of undue influence favoring renewable technology interests. The director of the DOE Office of Hydrogen, Fuel Cells, and Infrastructure Technologies and the director of NREL agreed to seek an experienced, objective, systems analysis professional from outside NREL, and a national search was initiated.

In parallel, the Office of Fossil Energy has established a new, independent systems analysis function at the National Energy Technology Laboratory. The primary focus of this effort will be on hydrogen production from coal, and the intent here is also to perform comparative analyses with other options for hydrogen production, which is appropriate in order to understand the capabilities and economics of alternative technologies.

With the participation of staff from the National Research Council, members of the committee held separate, informal meetings with personnel from FE and EERE so that some of the committee members with related expertise could share their individual thoughts on the important elements of effective systems analysis. No committee positions were provided beyond those presented in the committee's interim report. Among the items discussed, which the committee endorses, are the following:

- The most important ingredient in systems analysis is the people who do the work. There are relatively few with the training, talent, and background to be able to properly identify, evaluate, trade off, and deal with the myriad of technical and economic parameters characteristic of complex energy technologies. A core of specially selected people is therefore essential.
- A viable systems analysis function must be managed independently of the various DOE line research and development (R&D) programs in order to minimize both the existence and even the appearance of technology bias.
- There are many envelopes for systems analysis. At one end of the spectrum are unit operations, such as the analysis of all elements of a production concept. At the opposite end is a fully integrated national system, including fuel acquisition for a production plant, production operations, transportation and storage of product, distribution, end use, and other related considerations. Detailed analysis within envelopes and across the national energy system must all be part of an effective systems analysis program, particularly in the case of a potentially vast, future energy system based on hydrogen.
- A viable systems analysis program for a wide-ranging effort such as the hydrogen program is a very significant activity, requiring substantial funding—on the order of $10 million per year on a continuing basis. Without such a comprehensive effort, research priorities may be less well justified, and the full meaning of research results will be less well understood.
- A few of the topics to be studied in systems analysis are the following: (1) systems and subsystems currently under development; (2) the character of competitive approaches to providing energy services—electricity, for example—and how such systems are likely to change over time; (3) an examination of different future energy scenarios and forcing functions that may impact the nation; and (4) the development of an understanding of how proposed technologies might fit into the national system.
- The benefits that can accrue from a properly managed systems analysis program include (1) an independent, consistent, unbiased description of technologies as they are and might be; (2) a fact-based prescription to guide the selection and evaluation of research projects; and (3) a sound basis for estimating the potential benefits of research programs.
- A number of potential pitfalls must be avoided in doing effective systems analysis. Poor-quality or biased results have the potential to severely damage institutional credibility, in addition to providing faulty direction to programs. Perhaps most significant is the need to guard against outside influences, because an independent systems analysis function will almost certainly be inundated by people wanting to protect their own preferences or projects.

For various reasons, the establishment of a viable hydrogen systems analysis program will not be an easy task. For one thing, the DOE has had relatively little experience with effective systems analysis in the past. Additionally, such analysis programs are expensive. Also, the results of sound systems analysis can upset vested interests.

Finding 9-1. The pathway to achieving a hydrogen economy will be neither simple nor straightforward. Indeed, significant risks lie between the present vision for a hydrogen economy and the actual achievement of that vision. The chief technical challenges include these: safe, durable, and economic hydrogen storage; cost-effective, durable fuel cell technology; economic and publicly acceptable carbon capture and sequestration; breakthroughs in hydrogen distribution systems; and cost-effective, energy-efficient renewable and distributed hydrogen generation systems. These challenges can only be addressed by research and development, and there is no guarantee that such efforts will be successful. However, even more-challenging issues will arise from a larger set of economic and social concerns, especially those of enabling investment in hydrogen distribution and logistic systems and of public acceptance, which will likely be heavily influenced by hydrogen safety concerns. Thus, it is appropriate for the Department of Energy to advance hydrogen and fuel cell options at this time. In addition, given the important role of other, nonhydrogen energy technologies and strategies, including conservation, during the coming decades and given the possibility that some, such as battery storage, might more rapidly advance many of the goals set out for the hydrogen economy, the DOE's hydrogen research programs must be measured and managed against progress in these nonhydrogen fields so that an appropriate balancing of overall energy policy can be achieved.

Recommendation 9-1. The Department of Energy should continue to develop its hydrogen initiative as a potential long-term contributor to improving U.S. energy security and environmental protection. The program plan should be reviewed and updated regularly to reflect progress, potential synergisms within the program, and interactions with other energy programs and partnerships (e.g., the California Fuel Cell Partnership). In order to achieve this objective, the committee recommends that the DOE develop and employ a systems analysis approach to understanding full costs, defining options, evaluating research results, and helping balance its hydrogen program for the short, medium, and long term. Such an approach should be implemented for all U.S. energy options, not only for hydrogen.

As part of its systems analysis, the DOE should map out and evaluate a transition plan consistent with developing the infrastructure and hydrogen resources necessary to support the committee's hydrogen vehicle penetration scenario or another similar demand scenario. The DOE should estimate what levels of investment over time are required—and in which program and project areas—in order to achieve a significant reduction in carbon dioxide emissions from passenger vehicles by midcentury.

Finding 9-2. The effective management of the Department of Energy hydrogen program will be far more challenging than any activity previously undertaken on the civilian energy side of the DOE. That being the case, the use of management tools employed elsewhere in the government has the potential for a very high payoff in terms of the effective use of taxpayer funds and the development of the most efficient pathways to hydrogen systems success. In that regard, the adoption of systems integration techniques used elsewhere in the government has the potential for significant value. However, the DOE's hydrogen exploratory research program must be managed in a very different manner—independent of the projects covered by systems integration management.

Recommendation 9-2. The Department of Energy should identify potentially useful management tools and capabilities developed elsewhere in the government for managing complex programs and should evaluate their potential for use in the hydrogen program. While such techniques are known to exist, it may well be that they will need to be modified to account for the overriding importance of economics in energy system development.

Finding 9-3. An independent, well-funded, professionally staffed and managed systems analysis function, separated by a "firewall" from technology development functions, is essential to the success of the Department of Energy's hydrogen program.

Recommendation 9-3. An independent systems analysis group should be established by the Department of Energy to identify the impacts of various hydrogen technology pathways, to assess associated cost elements and drivers, to identify key cost and technological gaps, to evaluate the significance of actual research results, and to assist in the prioritization of research and development directions.

HYDROGEN SAFETY

Safety-Related Issues

High market penetration of devices that use hydrogen to deliver energy services can expose the general public to unaccustomed hazards. These hazards pose three challenges that will become manifest from the earliest days of any transition to a hydrogen economy:

1. The requirement to protect human life and property;
2. The need to develop codes and standards for hydrogen devices, production technologies, and logistic systems that

allow the economic siting of facilities and that enable commercial innovation; and

3. The need to develop hydrogen safety competence among local emergency-response and zoning officials.

Timely attention to these safety-related issues cannot, by itself, draw hydrogen into the marketplace. Left unaddressed, however, these issues will raise formidable barriers to the wide-scale use of hydrogen in the consumer economy.

As noted in previous chapters, hydrogen is widely used in industry at the present time. About 41 million short tons of hydrogen are produced each year in industrial facilities around the world (ORNL, 2003). From this industrial experience, the committee concludes that the safety record of professionally managed hydrogen compares favorably with that of similar industrial processes, and that hydrogen can be manufactured and used by trained professionals under controlled conditions with acceptable safety. Thus, safety issues are most likely to arise when hydrogen is used by consumers with neither special training nor the discipline of industrial procedures. In particular, three areas of consumer use merit special attention: (1) the fueling process for consumer-owned, hydrogen-powered vehicles; (2) the garage storage of hydrogen-powered vehicles, either in homes or in public facilities; and (3) the use of hydrogen-powered vehicles in tunnels and similar structures.

Safety Implications of the Properties of Hydrogen

The risk to the public during consumer end use of hydrogen derives from the possibility of accidental fire and explosion, a direct consequence of the physical and chemical properties of hydrogen. These properties help to define the kinds of safety issues that must be addressed, the fundamental design goals for hydrogen systems, and the operational limitations of these systems. Table 9-1 summarizes the properties of hydrogen in contrast with those of other commonly used fuels.

With regard to flammability, hydrogen has an unusually wide range of flammable concentrations in air—between a lower limit of 4 percent and an upper limit of 75 percent by volume. As a result, the release of any volume of hydrogen presents a larger probability of ignition than would be the case for a similar volume of other gaseous fuels commonly in use. On the other hand, hydrogen's high buoyancy and high diffusion rate in air lead any flammable hydrogen-air mixture to disperse more rapidly than other gases would. Thus, its ignition potential tends to decrease faster than that of other gases. Designs that minimize leaks and allow for the dispersal of the gas that does escape are more important for hydrogen than for other fuels.

With regard to ignitability, a flammable hydrogen-air mixture can be ignited either by a spark or by heating the mixture to its autoignition temperature. The minimum spark energy required for the ignition of hydrogen in air is low enough that a static electricity spark produced by the human body (in dry conditions and within the ideal fuel/air range) would be sufficient to ignite hydrogen. Among other considerations, this means that proper grounding of the vehicle and its operator will be essential for safe fueling. The risk is the presence of static electricity coincident with possible hydrogen from a leak at the fueling connection to the vehicle. By contrast, gasoline vapors are always present during gasoline fueling.

TABLE 9-1 Selected Properties of Hydrogen and Other Fuel Gases

Property	Hydrogen	Natural Gas	Propane	Gasoline Vapor
Density relative to air	0.07	0.55	1.52	4.0
Molecular weight	2	16	44	107
Density (kg/m^3)	0.084	0.651	1.87	4.4
Diffusion coefficient (cm^2/s)	0.61	0.16	0.12	0.05
Explosive energy (MJ/m^3)	9	32	93	407
Flammability range (% by volume)	4 to 75	5 to 15	2 to 10	1 to 8
Detonation range (% by volume)	18 to 59	6 to 14	3 to 7	1 to 3
Minimum ignition energy (mJ)	0.02	0.29	0.26	0.24
Flame speed (cm/s)	346	43	47	42

SOURCE: Nyborg et al. (2003).

With regard to explosion, all gaseous fuels can support detonation. However, hydrogen's higher flame speed and wider flammable concentration range make detonation more likely, all else being equal. The geometry of the confining space strongly influences the likelihood of detonation and presents special concerns in places such as enclosed, poorly ventilated garages or tunnels.

The extremely low density of hydrogen, 0.084 kilogram per cubic meter, challenges the design of any transfer operation, especially vehicle refueling. To accomplish the transfer of gaseous hydrogen in a timely manner, current technologies require high pressure—perhaps 5,000 to 10,000 pounds per square inch. This pressure also creates a heat load that limits refueling rates and must be dispersed.

Finally, hydrogen is difficult for humans to detect by sense of sight or smell. Unlike natural gas, familiar to consumers, hydrogen is unodorized. It burns with a pale blue, nearly invisible flame and becomes easily visible only as it ignites dust or other materials in the air. The gas itself is odorless, and the small size of the hydrogen molecules does not accommodate well the presence of chemical odorants. Further, many common odorants would poison the catalyst in hydrogen fuel cells.

These special characteristics do not by themselves preclude a transition of hydrogen into the consumer economy. But they do imply that the safety practices and skills that have evolved through many years' handling of other fuels cannot be applied uncritically. Developing the unique safety codes, standards, and practices to support any emerging hydrogen economy will thus become an essential public task.

The DOE is well aware of the safety issues surrounding hydrogen and has initiated relevant programs. The committee recognizes and supports the emphasis on participatory development of codes and standards found in the "Hydrogen, Fuel Cells and Infrastructure Technologies Program: Multi-Year Research, Development and Demonstration Plan" (DOE, 2003b).

Finding 9-4. Safety will be a major issue from the standpoint of commercialization of hydrogen-powered vehicles. Much evidence suggests that hydrogen can be manufactured and used in professionally managed systems with acceptable safety, but experts differ markedly in their views of the safety of hydrogen in a consumer-centered transportation system. A particularly salient and underexplored issue is leakage in enclosed structures, such as home and commercial garages. Hydrogen safety, from both a technological and a societal perspective, will be one of the major hurdles that must be overcome to achieve the hydrogen economy. Greater concerns, however, arise from its widespread use in the consumer economy. Safety issues become manifest in two forms: (1) concern with loss of human life and property, and (2) zoning codes that restrict the location of hydrogen facilities and vehicles. The current program is focusing on codes and standards and does not yet include a strong component of exploratory research and development.

Recommendation 9-4. The committee believes that the Department of Energy's program on safety is well planned and should be a priority. However, the committee emphasizes the following:

- *Safety policy goals.* Safety policy goals should be proposed and discussed by the Department of Energy with stakeholder groups early in the hydrogen technology development process. The safety issue should not be framed as an absolute standard but rather in comparison with the chief consumer alternatives—gasoline for vehicular use or natural gas for home use. The standard for consumer acceptability should be a level of safety equal to or greater than that of these alternative fuels.
- *Distributed production.* The DOE should continue its work with standards development organizations to ensure that timely codes and standards are available to enable a transition strategy that emphasizes the distributed production of hydrogen.
- *Inclusion of safety principles.* In weighing the merits of alternative hydrogen systems, the independent systems analysis group that the DOE is establishing in response to an earlier committee recommendation should specifically include safety among the essential components. In consultation with independent safety experts, the DOE should develop systems preferences to encourage inherent risk avoidance rather than relying solely on the layering of multiple protections.
- *Local capability building.* The training of local fire and rescue officials in the special procedures required for dealing with any emergencies involving hydrogen should proceed in step with the development and deployment of the technology. Model safety training programs should be prepared on the national level but in consultation with local officials.
- *Physical testing.* The DOE understanding of hydrogen safety should be reinforced by a rigorous testing program that is informed in part by reasoned but vigorous skeptics of hydrogen safety. The goal of the physical testing program should be to identify, quantify, and resolve safety issues in advance of the commercial use of the technology rather than to convince the public or local officials that hydrogen is safe.
- *Leak detection.* Low-cost, reliable sensors should be emphasized. Non-instrument detection methods, such as odorants, available to most consumers should also be explored.
- *Public education.* The DOE's public education program should continue to focus on hydrogen safety, particularly the safe use of hydrogen in distributed production and in consumer environments.

EXPLORATORY RESEARCH

Areas Needing Increased Emphasis

In its April 2003 interim report to the DOE, the committee recommended that fundamental and exploratory research should receive additional budgetary emphasis (see Appendix B). In May 2003, a hydrogen workshop was sponsored jointly by the Office of Science (SC) and the Office of Energy Efficiency and Renewable Energy, and a workshop report was published (DOE, 2003e). However, in none of the research and development documents reviewed by the committee and in no discussions regarding the FY 2005 budget was there any indication that the SC will be establishing a meaningful program of basic and exploratory research related to hydrogen.

In order to execute the hydrogen program, there need to be far more basic and exploratory research centers than there are now involved in the DOE program. The hydrogen initiative needs "a thousand points of innovation" in order to reach its full potential. The committee believes that the recently expanded hydrogen storage program has many characteristics of a strong program with the right balance of basic research.

Finding 9-5. The committee is impressed by the breadth and thoroughness of the Department of Energy's hydrogen research, development, and demonstration (RD&D) plans. However, the committee has a number of related concerns. First, the transition to a hydrogen economy involves challenges that cannot be overcome by research and development and demonstrations alone. Unresolved issues of policy development, infrastructure development, and safety will slow the penetration of hydrogen into the market even if the technical hurdles of production cost and energy efficiency are overcome. Significant industry investments in advance of market forces will not be made unless government creates a business environment that reflects societal priorities with respect to greenhouse gas emissions and oil imports. Second, the committee believes that a hydrogen economy will not result from a straightforward replacement of the present fossil-fuel-based economy. There are great uncertainties surrounding a transition period, because many innovations and technological breakthroughs will be required to address the cost, and energy-efficiency, distribution, and nontechnical issues. The hydrogen fuel for the very early transitional period, before distributed generation takes hold, would probably be supplied in the form of pressurized or liquefied molecular hydrogen, trucked from existing, centralized production facilities. But, as volume grows, such an approach may be judged too expensive and/or too hazardous. It seems likely that, in the next 10 to 30 years, hydrogen produced in distributed rather than centralized facilities will dominate. Distributed production of hydrogen seems most likely to be done with small-scale natural gas reformers or by electrolysis of water; however, new concepts in distributed production could be developed over this time period. Great uncertainties surround a number of fundamentally new innovations and technological breakthroughs that will have to happen in order to address costs, energy efficiency, distribution, and nontechnical issues. In its April 4, 2003, interim report to the Department of Energy, the committee recommended that exploratory research receive additional budgetary emphasis (see Appendix B).

Recommendation 9-5. There should be a shift in the hydrogen program away from some development areas and toward exploratory work—as has been done in the area of hydrogen storage. A hydrogen economy will require a number of technological and conceptual breakthroughs. The Department of Energy program calls for increased funding in some important exploratory research areas such as hydrogen storage and photoelectrochemical hydrogen production. However, the committee believes that much more exploratory research is needed. Other areas likely to benefit from an increased emphasis on exploratory research include delivery systems, pipeline materials, electrolysis, and materials science for many applications. The execution of such changes in emphasis would be facilitated by the establishment of DOE-sponsored academic energy research centers. These centers should focus on interdisciplinary areas of new science and engineering—such as materials research into nanostructures, and modeling for materials design—in which there are opportunities for breakthrough solutions to energy issues.

Industry Participation

The potential transition in the United States to the widespread replacement of gasoline by hydrogen for light-duty vehicles is essentially an alternative-fuel transition. As discussed in Chapter 3, "The Demand Side: Hydrogen End-Use Technologies," experience in the past with transitions to alternative fuels, such as ethanol, methanol, or compressed natural gas, has not resulted in the significant penetration of these alternative fuels into the light-duty-vehicle market. An essential consideration by the federal government for the transition to alternative fuels is to determine how to effectively involve the parts of the private sector that have experience in technology development, infrastructure, markets, financing, and other aspects required for market success.

For more than a decade there has been a significant level of participation with the automotive companies in the R&D directed toward advanced vehicles. For example, the three major U.S. automotive companies, Chrysler (now DaimlerChrysler), Ford, and General Motors, formed USCAR (United States Council for Automotive Research) and joined with the federal government in the Partnership for a New Generation of Vehicles (PNGV) program from 1993 to 2001. One of the goals of the PNGV program was to develop a midsize vehicle with up to three times the fuel economy of a comparable 1993 midsize vehicle. To achieve this goal, the focus was on the development of hybrid electric vehicles with diesel engines that required very-low-sulfur fuel in order to meet emissions requirements. Even though the fuel required was not completely new, such as for hydrogen or compressed natural gas, it was critical for the PNGV to involve the transportation fuels industry. In fact, the National Research Council's Standing Committee to Review the Research Program of the Partnership for a New Generation of Vehicles recommended that the PNGV propose ways to involve the transportation fuels industry in a partnership with government to help achieve the PNGV goals (NRC, 1998). The PNGV program, however, never developed an extensive partnership with the fuels industry equal in scope to what was done in the automotive sector.

The partnership between the federal government and USCAR continues, with the FreedomCAR program, an important part of which is focused on developing vehicle component technologies for future fuel cell vehicles. The transition to a completely different fuel, such as hydrogen, is obviously a much more significant change to the fuel system than what was required by the PNGV effort. The transition to a possible hydrogen future will require private sector investment to produce and distribute the hydrogen and to address, together with the government, the "chicken and egg" infrastructure problems that are outlined in the previous chapters of this report. But the "fuels" industry in this case may not only be the conventional petroleum and natural gas companies that would see an opportunity to supply hydrogen; also involved would be companies that produce electrolyzers, as well as the electric power industry that would supply electricity if hydrogen were produced by electrolysis. Since the committee believes that in the transition to a possible hydrogen economy the distributed production route would be the most likely strategy, the electric power sector may have a critical role, because electric power producers would be supplying the distributed generators.

An important partnership that is also a model of what will be needed to introduce hydrogen fuel cell vehicles is the California Fuel Cell Partnership (CaFCP). This partnership, headquartered in West Sacramento, was organized under the leadership of the California Air Resources Board in 1999. Its membership initially included oil and auto companies, but has expanded to include other energy suppliers, transit agencies, and other government agencies (including the DOE). In 2003, the partnership was renewed for another 4 years.

The goal of the California Fuel Cell Partnership is to promote further progress toward fuel cell vehicle commercialization. It is working with its members to accomplish the following goals over the 2004–2007 period (CaFCP, 2004):

1. **Conduct Fleet Demonstrations**. The CaFCP will facilitate members' placement of up to 300 fuel cell cars and buses in independent, fleet demonstration projects within the state during this phase. CaFCP members plan to focus these vehicles primarily in two main areas—the greater Los Angeles region, and the Sacramento-San Francisco area.
2. **Conduct Fuel Demonstrations.** The CaFCP members plan to construct fuel stations to support the independent demonstration projects. By concentrating vehicles and supporting refueling stations in defined regions the members will be able to focus resources more effectively in these early deployments. Fuel station interoperability—"common fit" fueling protocols—will allow all vehicles to utilize a growing network of fuel stations.
3. **Facilitate the Path to Commercialization.** The CaFCP and its members plan to work together to help prepare local communities for fuel cell vehicles and fueling by training local officials, facilitating permit processes and sharing lessons learned. The CaFCP and its members also plan to promote the development of practical codes and standards for FCVs and fueling stations, and help to obtain financial and other support where needed.
4. **Enhance Public Awareness, Education and Support.** The CaFCP plans to continue working together to raise public awareness through general outreach to public and media, consistent with pace of technology development. Increased focus will be placed on coordination with stakeholders and other fuel cell vehicle programs worldwide, sharing resource documents and lessons learned to further progress toward commercialization.

Recommendation 9-6. The committee commends the Department of Energy for significantly increasing its efforts to bring the energy industry into the hydrogen program. Particularly noteworthy are FreedomCar and the Hydrogen Fuel Initiative and the broad input solicited for the 2002 *National Energy Hydrogen Roadmap* and the 2003 *Hydrogen Posture Plan: An Integrated Research, Development, and Demonstration Plan.* The committee encourages the DOE to continue to seek broad input from the energy industry—which includes not only broad, multinational energy companies, but utilities and small companies. In particular, the committee believes that research and development partnerships that would enhance the energy industry and the objectives of the hydrogen program should be encouraged at the level of precompetitive technology.

INTERNATIONAL PARTNERSHIPS

Achieving some of the benefits of a hydrogen economy for the United States can also be facilitated if the DOE considers the role that hydrogen production and end-use technologies may play not only in the United States but also in other countries. For example, if future reductions of carbon dioxide become necessary to achieve, facilitating the development of low-carbon-emitting hydrogen production and end-use technologies that can be used around the world—especially in those developing countries where projected emissions of carbon dioxide are growing very rapidly—can help accelerate efforts to meet a goal of reducing global emissions of carbon dioxide. The successful development and use of hydrogen-related technologies in the transportation sectors of other countries can also substitute for oil, helping to ease any future supply-and-demand imbalances that may develop, and can help exert downward pressure on oil prices, which would benefit the U.S. economy. And U.S. companies, if successful in developing hydrogen technologies, can market those technologies not only in U.S. markets but in those of other countries, in both the developed and the developing world. Such activities would benefit U.S. companies and the economy. In addition, in other countries there are a number of efforts under way in the development of hydrogen-related technologies—efforts that may be important for the DOE to work with. Thus, it is important for a number of reasons for the DOE to adopt an international perspective as it sorts through its R&D priorities.

The DOE is already beginning to take some steps toward an international perspective on hydrogen technology development. On June 16, 2003, DOE Secretary Spencer Abraham delivered a keynote address to the European Union Conference on Hydrogen in which he noted that working together with international partners can leverage scarce resources and advance the schedule for research, development, and demonstration. He also stressed the value of international partnerships in achieving progress in the advance of scientific knowledge and technology applications in the energy area.

Secretary Abraham, joined by ministers representing 14 nations and the European Commission, signed an agreement on November 20, 2003, to formally establish the International Partnership for the Hydrogen Economy.[3]

STUDY OF ENVIRONMENTAL IMPACTS

An important public goal that has been expressed for an envisioned hydrogen energy system is that it could improve environmental quality. The committee in its report has limited its formal, quantitative analysis to one aspect of environmental quality—emissions of the greenhouse gas carbon dioxide. Other effects of switching to hydrogen fuels and fuel cell vehicles are discussed in various chapters. The committee believes that it will be important to consider environmental impacts in addition to those formally analyzed or discussed in this report if the goal of environmental quality is to be successfully achieved.

Following is a brief discussion of some known environmental impacts associated with today's energy sector that could be expected to be associated with a hydrogen economy. While not meant to be exhaustive, this discussion seeks to establish the scope of a proper analysis of a future hydrogen energy system. An analysis of environmental impacts should, in addition, consider the cumulative or delayed consequences of individual environmental threats, the difficulty in reversing them, and their potential to interact with one another (NRC, 1999a).

Carbon Dioxide Emissions

The committee examined the effect of the substitution of hydrogen fuel for petroleum in the transportation sector and its possible effects in reducing annual carbon emissions. These results are presented in Figures 6-7 through 6-10 in Chapter 6 and are discussed there in detail. This analysis showed that reductions in annual carbon emissions could be achieved[4] but that they would vary greatly depending, for example, on whether hydrogen fuel was generated from fossil fuel resources, whether carbon capture and storage were employed, or—in the case of distributed generation—whether electrolysis was used and powered by renewable energy sources, among other factors and choices.

Hydrogen Fuel Cycle

Production of hydrogen necessitates utilization of primary resources either as feedstock (e.g., natural gas for reforming, or coal or biomass for gasification) or as an energy source for electricity (e.g., coal or natural gas, or uranium leading to one of a few possible nuclear-related processes). From extraction and reclamation to end use—the entire fuel cycle—the use of these resources can have direct and indirect impacts on environmental quality. The quantity of a few of these resources that would be required to meet the needs of a hydrogen energy system was estimated by the committee in Chapter 6. For natural gas, coal, and land area for growing biomass, respectively, Figures 6-11, 6-14, and 6-15 present the committee's estimates of the use of these resources to the year 2050, assuming in each case that all hydrogen demand is met by one and only one type of resource (e.g., biomass).

Extraction of the primary resources such as coal, natural gas, or uranium that could be used for the production of hydrogen has created associated environmental impacts. The mining of uranium and coal can affect landscape, water quality, vegetation, and aquatic biota, often extending beyond the immediate area of the mine site (NRC, 1999b). The extraction of natural gas from subsurface deposits can degrade surface habitats and subsurface water resources (NRC, 2003a). Technologies for the extraction of oil and natural gas resources continue to evolve, however, and associated impacts such as the environmental footprint left by exploration and production have declined, according to some analyses (DOE, 1999).

Production of hydrogen can result in the release of various criteria pollutants[5] or their precursors. The types of emissions and the amounts depend on the primary feedstock and the technology used to convert the feedstock into hydrogen. Certain production technologies, such as nuclear energy methods and wind energy tied to electrolysis, will be inherently more favorable with respect to this issue. To take one example: hydrogen can be produced from coal gasification or natural gas reforming, which utilize two of the same primary resources used by the U.S. electric power system. Coal-fired and natural-gas-fired power plants emit criteria pollutants or their precursors, suggesting that hydrogen production from these resources may be subject to regulatory controls similar to those for coal-fired or natural-gas-fired power plants. The location of the emissions will depend on the type of distribution system, which in turn depends on the scale of production. A distributed production system could have thousands of small sites, each with some emissions, whereas

[3]Department of Energy Press Release, "Secretary of Energy Abraham Joins International Community to Establish International Partnership for the Hydrogen Economy," November 20, 2003.

[4]As a point of comparison, the analysis included comparison with emissions from hybrid electric vehicles and conventional gasoline internal combustion engines on a sector-wide basis.

[5]Criteria pollutants are air pollutants emitted from numerous or diverse stationary or mobile sources for which National Ambient Air Quality Standards have been set to protect human health and public welfare. The original list of criteria pollutants, adopted in 1971, consisted of carbon monoxide, total suspended particulate matter, sulfur dioxide, photochemical oxidants, hydrocarbons, and nitrogen oxides. Lead was added to the list in 1976, ozone replaced photochemical oxidants in 1979, and hydrocarbons were dropped in 1983. Total suspended particulate matter was revised in 1987 to include only particles with an equivalent aerodynamic particle diameter of less than or equal to 10 micrometers (PM_{10}). A separate standard for particles with an equivalent aerodynamic particle diameter of less than or equal to 2.5 micrometers ($PM_{2.5}$) was adopted in 1997.

a system based on central station production will have a much smaller number of sites but with larger emissions. In summary, the extent to which criteria pollutants would be an issue in a hydrogen energy system would depend on the specific production technologies deployed, the extent of their deployment, and the pollution control equipment and regulatory regimes that are implemented.

End use of hydrogen in fuel-cell-powered vehicles will result in a much different mix of types of emissions compared with those from today's gasoline vehicles, and also a much different profile of where in the life cycle (i.e., at resource extraction, production, distribution, and/or end use) the emissions will occur. Today's gasoline or diesel-powered car is a major source of criteria pollutant emissions in the United States, whereas the hydrogen-powered fuel cell vehicle will not emit any criteria pollutants. The only significant emission will be water in the form of vapor or liquid. Small amounts of hydrogen and nitrogen dioxide may be emitted from combusting the tail gas that passes through the fuel cell unreacted. The widespread use of hydrogen-powered fuel cell vehicles will have a positive impact on air quality in many urban areas of the United States, where cars currently are responsible for large amounts of emissions. However, as noted above, it is during the production phase of the fuel cycle of hydrogen that the potential for the emission of criteria pollutants or greenhouse gas emissions exists.

In addition to regulated environmental toxicants, the requirements of a hydrogen energy system with respect to resources such as water and land should be considered. In all of the production processes mentioned in this report, water is used as a source for at least a portion of the hydrogen production—one-third by mass of the hydrogen from biomass gasification comes from water.[6] In the hydrocarbon and coal-based processes, a significant portion of the hydrogen comes from water used in the water-gas-shift reaction. In the electrolytic processes, water is split using electricity. In the nuclear processes, water is split using high temperatures. Water is also used as a coolant in many of the processes, and large amounts are needed to grow biomass efficiently. The fuel cell, however, produces water from hydrogen and oxygen. The net balance and also the location of water needs were not reviewed by this committee.

Similarly, study is needed of the impact of large-scale biomass growth for feedstock for its impact on land use and any effect on nutrient runoff and eutrophication secondary to fertilizer demand (NRC, 2000).

Molecular Hydrogen

Molecular hydrogen is a short-lived trace atmospheric gas (with approximately a 2-year lifetime), having tropospheric concentrations of approximately 0.5 part per million. Its global distribution favors slightly higher concentrations in the Northern Hemisphere (about +5 percent), and well-defined seasonal cycles are observed (Novelli et al., 1999). A percentage of today's atmospheric burden of molecular hydrogen is believed to be secondary to biomass burning and technological processes such as motor vehicle use (Novelli et al., 1999). Hydrogen is removed from the troposphere by surface deposition and by chemical destruction via oxidation with hydroxyl (OH) (Novelli et al., 1999). Various authors have noted that hydrogen is one of many gases that are removed from the troposphere by OH, and that, furthermore, the resulting decreases in concentrations of OH could lead to higher concentrations of methane and tropospheric ozone (Derwent et al., 2001), both of which are established climate forcing agents (NRC, 2001b).

Finding 9-6. Any future hydrogen energy system, if based on coal, natural gas, or uranium, will likely imply some of the same environmental consequences that the use of those same resources has caused in today's energy system. The scope and magnitude of these consequences will depend on the nature of the hydrogen technologies deployed, on the portfolio mix of primary resources on which these technologies are based, and on the pollution control equipment and regulatory regimes that are implemented.

Recommendation 9-7. The committee recommends that the Department of Energy initiate a comprehensive assessment of the suite of environmental issues anticipated to arise secondary to deployment of a hydrogen energy system, and that the DOE develop a quantitative understanding of the tradeoffs and impacts.

DEPARTMENT OF ENERGY PROGRAM

As part of its effort, the committee reviewed the June 3, 2003, draft of "Hydrogen, Fuel Cells and Infrastructure Technologies Program: Multi-Year Research, Development and Demonstration Plan" (DOE, 2003b). Although very impressed by the plan and its thoroughness, the committee believes that several general aspects of the plan need to be addressed in greater detail. Comments on the individual technology sections of the plan are contained in Chapter 8.

First, the plan is focused primarily on the activities within the Office of Hydrogen, Fuel Cells and Infrastructure Technologies in the Office of Energy Efficiency and Renewable Energy, and it only casually mentions activities in the Office of Fossil Energy; the Office of Nuclear Energy, Science and Technology; the Office of Science; and activities related to CO_2 management. Development of an RD&D plan for the totality of the DOE's hydrogen program will require a plan with better balance and integration.

Second, it is very difficult to identify priorities within the myriad of activities that are proposed. A general budget is

[6]Margaret Mann and Ralph Overend, National Renewable Energy Laboratory, "Hydrogen from Biomass: Prospective Resources, Technologies, and Economics," presentation to the committee, January 22, 2003.

given in Appendix C in this report, but when discussing the various activities, no dollar amounts are given even for existing projects and programs. The committee found it difficult to judge the plans and priorities for each of the R&D areas. And finally, the plan needs to incorporate to a greater extent a set of "go/no go" decision points in the various development time lines.

Recommendation 9-8. The Department of Energy should continue to develop its hydrogen research, development, and demonstration (RD&D) plan to improve the integration and balance of activities within the Office of Energy Efficiency and Renewable Energy; the Office of Fossil Energy (including programs related to carbon sequestration); the Office of Nuclear Energy Science and Technology; and the Office of Science. The committee believes that, overall, the production, distribution, and dispensing portion of the program is probably underfunded, particularly because a significant fraction of appropriated funds is already earmarked. The committee understands that of the $78 million appropriated for hydrogen technology for FY 2004 in the Energy and Water appropriations bill (Public Law 108-137), $37 million is earmarked for activities that will not particularly advance the hydrogen initiative. The committee also believes that the hydrogen program, in an attempt to meet the extreme challenges set by senior government and DOE leaders, has tried to establish RD&D activities in too many areas, creating a very diverse, somewhat unfocused program. Thus, prioritizing the efforts both within and across program areas, establishing milestones and go/no-go decisions, and adjusting the program on the basis of results are all extremely important in a program with so many challenges. This approach will also help determine when it is appropriate to take a program to the demonstration stage. And finally, the committee believes that the probability of success in bringing the United States to a hydrogen economy will be greatly increased by partnering with a broader range of academic and industrial organizations—possibly including an international focus—and by establishing an independent program review process and board.

Recommendation 9-9. As a framework for recommending and prioritizing the Department of Energy program, the committee considered the following:

- Technologies that could significantly impact U.S. energy security and carbon dioxide emissions,
- The timescale for the evolution of the hydrogen economy,
- Technology developments needed for both the transition period and the steady state,
- Externalities that would decelerate technology implementation, and
- The comparative advantage of the DOE in research and development of technologies at the pre-competitive stage.

The committee recommends that the following areas receive increased emphasis:

- *Fuel cell vehicle development.* Increase research and development (R&D) to facilitate breakthroughs in fuel cell costs and in durability of fuel cell materials, as well as breakthroughs in on-board hydrogen storage systems;
- *Distributed hydrogen generation.* Increase R&D in small-scale natural gas reforming, electrolysis, and new concepts for distributed hydrogen production systems;
- *Infrastructure analysis.* Accelerate and increase efforts in systems modeling and analysis for hydrogen delivery, with the objective of developing options and helping guide R&D in large-scale infrastructure development;
- *Carbon sequestration and FutureGen.* Accelerate development and early evaluation of the viability of carbon capture and storage (sequestration) on a large scale because of its implications for the long-term use of coal for hydrogen production. Continue the FutureGen Project as a high-priority task; and
- *Carbon dioxide-free energy technologies.* Increase emphasis on the development of wind-energy-to-hydrogen as an important technology for the hydrogen transition period and potentially for the longer term. Increase exploratory and fundamental research on hydrogen production by photobiological, photoelectrochemical, thin-film solar, and nuclear heat processes.

10

Major Messages of This Report

BASIC CONCLUSIONS

As described below, the committee's basic conclusions address four topics: implications for national goals, priorities for research and development (R&D), the challenge of transition, and the impacts of hydrogen-fueled light-duty vehicles on energy security and CO_2 emissions.

Implications for National Goals

A transition to hydrogen as a major fuel in the next 50 years could fundamentally transform the U.S. energy system, creating opportunities to increase energy security through the use of a variety of domestic energy sources for hydrogen production while reducing environmental impacts, including atmospheric CO_2 emissions and criteria pollutants.[1] In his State of the Union address of January 28, 2003, President Bush moved energy, and especially hydrogen for vehicles, to the forefront of the U.S. political and technical debate. The President noted: "A simple chemical reaction between hydrogen and oxygen generates energy, which can be used to power a car producing only water, not exhaust fumes. With a new national commitment, our scientists and engineers will overcome obstacles to taking these cars from laboratory to showroom so that the first car driven by a child born today could be powered by hydrogen, and pollution-free."[2] This committee believes that investigating and conducting RD&D activities to determine whether a hydrogen economy might be realized are important to the nation. There is a potential for replacing essentially all gasoline with hydrogen over the next half century using only domestic resources. And there is a potential for eliminating almost all CO_2 and criteria pollutants from vehicular emissions. However, there are currently many barriers to be overcome before that potential can be realized.

Of course there are other strategies for reducing oil imports and CO_2 emissions, and thus the DOE should keep a balanced portfolio of R&D efforts and continue to explore supply-and-demand alternatives that do not depend upon hydrogen. If battery technology improved dramatically, for example, all-electric vehicles might become the preferred alternative. Furthermore, hybrid electric vehicle technology is commercially available today, and benefits from this technology can therefore be realized immediately. Fossil-fuel-based or biomass-based synthetic fuels could also be used in place of gasoline.

Research and Development Priorities

There are major hurdles on the path to achieving the vision of the hydrogen economy; the path will not be simple or straightforward. Many of the committee's observations generalize across the entire hydrogen economy: the hydrogen system must be cost-competitive, it must be safe and appealing to the consumer, and it would preferably offer advantages from the perspectives of energy security and CO_2 emissions. Specifically for the transportation sector, dramatic progress in the development of fuel cells, storage devices, and distribution systems is especially critical. Widespread success is not certain.

The committee believes that for hydrogen-fueled transportation, the four most fundamental technological and economic challenges are these:

1. *To develop and introduce cost-effective, durable, safe, and environmentally desirable fuel cell systems and hydrogen storage systems.* Current fuel cell lifetimes are much too short and fuel cell costs are at least an order of magnitude too high. An on-board vehicular hydrogen storage system

[1]Criteria pollutants are air pollutants (e.g., lead, sulfur dioxide, and so on) emitted from numerous or diverse stationary or mobile sources for which National Ambient Air Quality Standards have been set to protect human health and public welfare.

[2]*Weekly Compilation of Presidential Documents.* Monday, February 3, 2003. Vol. 39, No. 5, p. 111. Washington, D.C.: Government Printing Office.

that has an energy density approaching that of gasoline systems has not been developed. Thus, the resulting range of vehicles with existing hydrogen storage systems is much too short.

2. *To develop the infrastructure to provide hydrogen for the light-duty-vehicle user.* Hydrogen is currently produced in large quantities at reasonable costs for industrial purposes. The committee's analysis indicates that at a future, mature stage of development, hydrogen (H_2) can be produced and used in fuel cell vehicles at reasonable cost. The challenge, with today's industrial hydrogen as well as tomorrow's hydrogen, is the high cost of distributing H_2 to dispersed locations. This challenge is especially severe during the early years of a transition, when demand is even more dispersed. The costs of a mature hydrogen pipeline system would be spread over many users, as the cost of the natural gas system is today. But the transition is difficult to imagine in detail. It requires many technological innovations related to the development of small-scale production units. Also, nontechnical factors such as financing, siting, security, environmental impact, and the perceived safety of hydrogen pipelines and dispensing systems will play a significant role. All of these hurdles must be overcome before there can be widespread hydrogen use. An initial stage during which hydrogen is produced at small scale near the small user seems likely. In this case, production costs for small production units must be sharply reduced, which may be possible with expanded research.

3. *To reduce sharply the costs of hydrogen production from renewable energy sources, over a time frame of decades.* Tremendous progress has been made in reducing the cost of making electricity from renewable energy sources. But making hydrogen from renewable energy through the intermediate step of making electricity, a premium energy source, requires further breakthroughs in order to be competitive. Basically, these technology pathways for hydrogen production make electricity, which is converted to hydrogen, which is later converted by a fuel cell back to electricity. These steps add costs and energy losses that are particularly significant when the hydrogen competes as a commodity transportation fuel—leading the committee to believe that most current approaches—except possibly that of wind energy—need to be redirected. The committee believes that the required cost reductions can be achieved only by targeted fundamental and exploratory research on hydrogen production by photobiological, photochemical, and thin-film solar processes.

4. *To capture and store ("sequester") the carbon dioxide by-product of hydrogen production from coal.* Coal is a massive domestic U.S. energy resource that has the potential for producing cost-competitive hydrogen. However, coal processing generates large amounts of CO_2. In order to reduce CO_2 emissions from coal processing in a carbon-constrained future, massive amounts of CO_2 would have to be captured and safely and reliably sequestered for hundreds of years. Key to the commercialization of a large-scale, coal-based hydrogen production option (and also for natural-gas-based options) is achieving broad public acceptance, along with additional technical development, for CO_2 sequestration.

For a viable hydrogen transportation system to emerge, all four of these challenges must be addressed.

The Challenge of Transition

There will likely be a lengthy transition period during which fuel cell vehicles and hydrogen are not competitive with internal combustion engine vehicles, including conventional gasoline and diesel fuel vehicles, and hybrid gasoline electric vehicles. The committee believes that the transition to a hydrogen fuel system will best be accomplished initially through distributed production of hydrogen, because distributed generation avoids many of the substantial infrastructure barriers faced by centralized generation. Small hydrogen-production units located at dispensing stations can produce hydrogen through natural gas reforming or electrolysis. Natural gas pipelines and electricity transmission and distribution systems already exist; for distributed generation of hydrogen, these systems would need to be expanded only moderately in the early years of the transition. During this transition period, distributed renewable energy (e.g., wind or solar energy) might provide electricity to onsite hydrogen production systems, particularly in areas of the country where electricity costs from wind or solar energy are particularly low. A transition emphasizing distributed production allows time for the development of new technologies and concepts capable of potentially overcoming the challenges facing the widespread use of hydrogen. The distributed transition approach allows time for the market to develop before too much fixed investment is set in place. While this approach allows time for the ultimate hydrogen infrastructure to emerge, the committee believes that it cannot yet be fully identified and defined.

Impacts of Hydrogen-Fueled Light-Duty Vehicles

Several findings from the committee's analysis (see Chapter 6) show the impact on the U.S. energy system if successful market penetration of hydrogen fuel cell vehicles is achieved. In order to analyze these impacts, the committee posited that fuel cell vehicle technology would be developed successfully and that hydrogen would be available to fuel light-duty vehicles (cars and light trucks). These findings are as follows:

- The committee's upper-bound market penetration case for fuel cell vehicles, premised on hybrid vehicle experience, assumes that fuel cell vehicles enter the U.S. light-duty vehicle market in 2015 in competition with conventional and hybrid electric vehicles, reaching 25 percent of light-duty vehicle sales around 2027. The demand for hydrogen in about 2027 would be about equal to the current production of 9 million short tons (tons) per year, which would be only a small fraction of the 110 million tons required for full replacement of gasoline light-duty vehicles with hydrogen vehicles, posited to take place in 2050.

- If coal, renewable energy, or nuclear energy is used to produce hydrogen, a transition to a light-duty fleet of vehicles fueled entirely by hydrogen would reduce total energy imports by the amount of oil consumption displaced. However, if natural gas is used to produce hydrogen, and if, on the margin, natural gas is imported, there would be little if any reduction in total energy imports, because natural gas for hydrogen would displace petroleum for gasoline.
- CO_2 emissions from vehicles can be cut significantly if the hydrogen is produced entirely from renewables or nuclear energy, or from fossil fuels with sequestration of CO_2. The use of a combination of natural gas without sequestration and renewable energy can also significantly reduce CO_2 emissions. However, emissions of CO_2 associated with light-duty vehicles contribute only a portion of projected CO_2 emissions; thus, sharply reducing overall CO_2 releases will require carbon reductions in other parts of the economy, particularly in electricity production.
- Overall, although a transition to hydrogen could greatly transform the U.S. energy system in the long run, the impacts on oil imports and CO_2 emissions are likely to be minor during the next 25 years. However, thereafter, if R&D is successful and large investments are made in hydrogen and fuel cells, the impact on the U.S. energy system could be great.

MAJOR RECOMMENDATIONS

Systems Analysis of U.S. Energy Options

The U.S. energy system will change in many ways over the next 50 years. Some of the drivers for such change are already recognized, including at present the geology and geopolitics of fossil fuels and, perhaps eventually, the rising CO_2 concentration in the atmosphere. Other drivers will emerge from options made available by new technologies. The U.S. energy system can be expected to continue to have substantial diversity; one should expect the emergence of neither a single primary energy source nor a single energy carrier. Moreover, more-energy-efficient technologies for the household, office, factory, and vehicle will continue to be developed and introduced into the energy system. The role of the DOE hydrogen program[3] in the restructuring of the overall national energy system will evolve with time.

[3]The words "hydrogen program" refer collectively to the programs concerned with hydrogen production, distribution, and use within DOE's Office of Energy Efficiency and Renewable Energy, Office of Fossil Energy, Office of Science, and Office of Nuclear Energy, Science and Technology. There is no single program with this title.

To help shape the DOE hydrogen program, the committee sees a critical role for systems analysis. Systems analysis will be needed both to coordinate the multiple parallel efforts within the hydrogen program and to integrate the program within a balanced, overall DOE national energy R&D effort. Internal coordination must address the many primary sources from which hydrogen can be produced, the various scales of production, the options for hydrogen distribution, the crosscutting challenges of storage and safety, and the hydrogen-using devices. Integration within the overall DOE effort must address the place of hydrogen relative to other secondary energy sources—helping, in particular, to clarify the competition between electricity, liquid-fuel-based (e.g., cellulosic ethanol), and hydrogen-based transportation. This is particularly important as clean alternative fuel internal combustion engines, fuel cells, and batteries evolve. Integration within the overall DOE effort must also address interactions with end-use energy efficiency, as represented, for example, by high-fuel-economy options such as hybrid vehicles. Implications of safety, security, and environmental concerns will need to be better understood. So will issues of timing and sequencing: depending on the details of system design, a hydrogen transportation system initially based on distributed hydrogen production, for example, might or might not easily evolve into a centralized system as density of use increases.

Recommendation 10-1. The Department of Energy should continue to develop its hydrogen initiative as a potential long-term contributor to improving U.S. energy security and environmental protection. The program plan should be reviewed and updated regularly to reflect progress, potential synergisms within the program, and interactions with other energy programs and partnerships (e.g., the California Fuel Cell Partnership). In order to achieve this objective, the committee recommends that the DOE develop and employ a systems analysis approach to understanding full costs, defining options, evaluating research results, and helping balance its hydrogen program for the short, medium, and long term. Such an approach should be implemented for all U.S. energy options, not only for hydrogen.

As part of its systems analysis, the DOE should map out and evaluate a transition plan consistent with developing the infrastructure and hydrogen resources necessary to support the committee's hydrogen vehicle penetration scenario or another similar demand scenario. The DOE should estimate what levels of investment over time are required—and in which program and project areas—in order to achieve a significant reduction in carbon dioxide emissions from passenger vehicles by midcentury.

Fuel Cell Vehicle Technology

The committee observes that the federal government has been active in fuel cell research for roughly 40 years, while proton exchange membrane (PEM) fuel cells applied to hydrogen vehicle systems are a relatively recent development (as of the late 1980s). In spite of substantial R&D spending by the DOE and industry, costs are still a factor of 10 to 20 times too expensive, these fuel cells are short of required durability, and their energy efficiency is still too low for light-duty-vehicle applications. Accordingly, the challenges of developing PEM fuel cells for automotive applications are large, and the solutions to overcoming these challenges are uncertain.

The committee estimates that the fuel cell system, including on-board storage of hydrogen, will have to decrease in cost to less than $100 per kilowatt (kW)[4] before fuel cell vehicles (FCVs) become a plausible commercial option, and that it will take at least a decade for this to happen. In particular, if the cost of the fuel cell system for light-duty vehicles does not eventually decrease to the $50/kW range, fuel cells will not propel the hydrogen economy without some regulatory mandate or incentive.

Automakers have demonstrated FCVs in which hydrogen is stored on board in different ways, primarily as high-pressure compressed gas or as a cryogenic liquid. At the current state of development, both of these options have serious shortcomings that are likely to preclude their long-term commercial viability. New solutions are needed in order to lead to vehicles that have at least a 300 mile driving range; are compact, lightweight, and inexpensive; and meet future safety standards.

Given the current state of knowledge with respect to fuel cell durability, on-board storage systems, and existing component costs, the committee believes that the near-term DOE milestones for FCVs are unrealistically aggressive.

Recommendation 10-2. Given that large improvements are still needed in fuel cell technology and given that industry is investing considerable funding in technology development, increased government funding on research and development should be dedicated to the research on breakthroughs in on-board storage systems, in fuel cell costs, and in materials for durability in order to attack known inhibitors to the high-volume production of fuel cell vehicles.

Infrastructure

A nationwide, high-quality, safe, and efficient hydrogen infrastructure will be required in order for hydrogen to be used widely in the consumer sector. While it will be many years before hydrogen use is significant enough to justify an integrated national infrastructure—as much as two decades in the scenario posited by the committee—regional infrastructures could evolve sooner. The relationship between hydrogen production, delivery, and dispensing is very complex, even for regional infrastructures, as it depends on many variables associated with logistics systems and on many public and private entities. Codes and standards for infrastructure development could be a significant deterrent to hydrogen advancement if not established well ahead of the hydrogen market. Similarly, since resilience to terrorist attack has become a major performance criterion for any infrastructure system, the design of future hydrogen infrastructure systems may need to consider protection against such risks.

In the area of infrastructure and delivery there seem to be significant opportunities for making major improvements. The DOE does not yet have a strong program on hydrogen infrastructures. DOE leadership is critical, because the current incentives for companies to make early investments in hydrogen infrastructure are relatively weak.

Recommendation 10-3a. The Department of Energy program in infrastructure requires greater emphasis and support. The Department of Energy should strive to create better linkages between its seemingly disconnected programs in large-scale and small-scale hydrogen production. The hydrogen infrastructure program should address issues such as storage requirements, hydrogen purity, pipeline materials, compressors, leak detection, and permitting, with the objective of clarifying the conditions under which large-scale and small-scale hydrogen production will become competitive, complementary, or independent. The logistics of interconnecting hydrogen production and end use are daunting, and all current methods of hydrogen delivery have poor energy-efficiency characteristics and difficult logistics. Accordingly, the committee believes that exploratory research focused on new concepts for hydrogen delivery requires additional funding. The committee recognizes that there is little understanding of future logistics systems and new concepts for hydrogen delivery—thus making a systems approach very important.

Recommendation 10-3b. The Department of Energy should accelerate work on codes and standards and on permitting, addressing head-on the difficulties of working across existing and emerging hydrogen standards in cities, counties, states, and the nation.

Transition

The transition to a hydrogen economy involves challenges that cannot be overcome by research and development and demonstrations alone. Unresolved issues of policy development, infrastructure development, and safety will slow the penetration of hydrogen into the market even if the technical hurdles of production cost and energy efficiency are overcome. Significant industry investments in advance of market forces will not be made unless government creates a busi-

[4]The cost includes the fuel cell module, precious metals, the fuel processor, compressed hydrogen storage, balance of plant, and assembly, labor, and depreciation.

ness environment that reflects societal priorities with respect to greenhouse gas emissions and oil imports.

Recommendation 10-4. The policy analysis capability of the Department of Energy with respect to the hydrogen economy should be strengthened, and the role of government in supporting and facilitating industry investments to help bring about a transition to a hydrogen economy needs to be better understood.

The committee believes that a hydrogen economy will not result from a straightforward replacement of the present fossil-fuel-based economy. There are great uncertainties surrounding a transition period, because many innovations and technological breakthroughs will be required to address the costs and energy-efficiency, distribution, and nontechnical issues. The hydrogen fuel for the very early transitional period, before distributed generation takes hold, would probably be supplied in the form of pressurized or liquefied molecular hydrogen, trucked from existing, centralized production facilities. But, as volume grows, such an approach may be judged too expensive and/or too hazardous. It seems likely that, in the next 10 to 30 years, hydrogen produced in distributed rather than centralized facilities will dominate. Distributed production of hydrogen seems most likely to be done with small-scale natural gas reformers or by electrolysis of water; however, new concepts in distributed production could be developed over this time period.

Recommendation 10-5. Distributed hydrogen production systems deserve increased research and development investments by the Department of Energy. Increased R&D efforts and accelerated program timing could decrease the cost and increase the energy efficiency of small-scale natural gas reformers and water electrolysis systems. In addition, a program should be initiated to develop new concepts in distributed hydrogen production systems that have the potential to compete—in cost, energy efficiency, and safety—with centralized systems. As this program develops new concepts bearing on the safety of local hydrogen storage and delivery systems, it may be possible to apply these concepts in large-scale hydrogen generation systems as well.

Safety

Safety will be a major issue from the standpoint of commercialization of hydrogen-powered vehicles. Much evidence suggests that hydrogen can be manufactured and used in professionally managed systems with acceptable safety, but experts differ markedly in their views of the safety of hydrogen in a consumer-centered transportation system. A particularly salient and underexplored issue is that of leakage in enclosed structures, such as garages in homes and commercial establishments. Hydrogen safety, from both a technological and a societal perspective, will be one of the major hurdles that must be overcome in order to achieve the hydrogen economy.

Recommendation 10-6. The committee believes that the Department of Energy program in safety is well planned and should be a priority. However, the committee emphasizes the following:

- Safety policy goals should be proposed and discussed by Department of Energy with stakeholder groups early in the hydrogen technology development process.
- The Department of Energy should continue its work with standards development organizations and ensure increased emphasis on distributed production of hydrogen.
- Department of Energy systems analysis should specifically include safety, and it should be understood to be an overriding criterion.
- The goal of the physical testing program should be to resolve safety issues in advance of commercial use.
- The Department of Energy's public education program should continue to focus on hydrogen safety, particularly the safe use of hydrogen in distributed production and in consumer environments.

Carbon Dioxide-Free Hydrogen

The long timescale associated with the development of viable hydrogen fuel cells and hydrogen storage provides a time window for a more intensive DOE program to develop hydrogen from electrolysis, which, if economic, has the potential to lead to major reductions in CO_2 emissions and enhanced energy security. The committee believes that if the cost of fuel cells can be reduced to $50 per kilowatt, with focused research a corresponding dramatic drop in the cost of electrolytic cells to electrolyze water can be expected (to ~$125/kW). If such a low electrolyzer cost is achieved, the cost of hydrogen produced by electrolysis will be dominated by the cost of the electricity, not by the cost of the electrolyzer. Thus, in conjunction with research to lower the cost of electrolyzers, research focused on reducing electricity costs from renewable energy and nuclear energy has the potential to reduce overall hydrogen production costs substantially.

Recommendation 10-7. The Department of Energy should increase emphasis on electrolyzer development, with a target of $125 per kilowatt and a significant increase in efficiency toward a goal of over 70 percent (lower heating value basis). In such a program, care must be taken to properly account for the inherent intermittency of wind and solar energy, which can be a major limitation to their wide-scale use. In parallel, more aggressive electricity cost targets should be set for unsubsidized nuclear and renewable energy that might be used directly to generate electricity. Success in these areas would greatly increase the potential for carbon dioxide-free hydrogen production.

Carbon Capture and Storage

The DOE's various efforts with respect to hydrogen and fuel cell technology will benefit from close integration with carbon capture and storage (sequestration) activities and programs in the Office of Fossil Energy. If there is an expanded role for hydrogen produced from fossil fuels in providing energy services, the probability of achieving substantial reductions in net CO_2 emissions through sequestration will be greatly enhanced through close program integration. Integration will enable the DOE to identify critical technologies and research areas that can enable hydrogen production from fossil fuels with CO_2 capture and storage. Close integration will promote the analysis of overlapping issues such as the co-capture and co-storage with CO_2 of pollutants such as sulfur produced during hydrogen production.

Many early carbon capture and storage projects will not involve hydrogen, but rather will involve the capture of the CO_2 impurity in natural gas, the capture of CO_2 produced at electric plants, or the capture of CO_2 at ammonia and synfuels plants. All of these routes to capture, however, share carbon storage as a common component, and carbon storage is the area in which the most difficult institutional issues and the challenges related to public acceptance arise.

Recommendation 10-8. The Department of Energy should tighten the coupling of its efforts on hydrogen and fuel cell technology with the DOE Office of Fossil Energy's programs on carbon capture and storage (sequestration). Because of the hydrogen program's large stake in the successful launching of carbon capture and storage activity, the hydrogen program should participate in all of the early carbon capture and storage projects, even those that do not directly involve carbon capture during hydrogen production. These projects will address the most difficult institutional issues and the challenges related to issues of public acceptance, which have the potential of delaying the introduction of hydrogen in the marketplace.

The Department of Energy's Hydrogen Research, Development, and Demonstration Plan

As part of its effort, the committee reviewed the DOE's draft "Hydrogen, Fuel Cells & Infrastructure Technologies Program: Multi-Year Research, Development and Demonstration Plan," dated June 3, 2003 (DOE, 2003b). The committee's deliberations focused only on the hydrogen production and demand portion of the overall DOE plan. For example, while the committee makes recommendations on the use of renewable energy for hydrogen production, it did not review the entire DOE renewables program in depth. The committee is impressed by how well the hydrogen program has progressed. From its analysis, the committee makes two overall observations about the program:

- First, the plan is focused primarily on the activities in the Office of Hydrogen, Fuel Cells and Infrastructure Technologies Program within the Office of Energy Efficiency and Renewable Energy, and on some activities in the Office of Fossil Energy. The activities related to hydrogen in the Office of Nuclear Energy, Science and Technology, and in the Office of Science, as well as activities related to carbon capture and storage in the Office of Fossil Energy, are important, but they are mentioned only casually in the plan. The development of an overall DOE program will require better integration across all DOE programs.
- Second, the plan's priorities are unclear, as they are lost within the myriad of activities that are proposed. The general budget for DOE's hydrogen program is contained in the appendix of the plan, but the plan provides no dollar numbers at the project level, even for existing projects and programs. The committee found it difficult to judge the priorities and the go/no-go decision points for each of the R&D areas.

Recommendation 10-9. The Department of Energy should continue to develop its hydrogen research, development, and demonstration plan to improve the integration and balance of activities within the Office of Energy Efficiency and Renewable Energy; the Office of Fossil Energy (including programs related to carbon sequestration); the Office of Nuclear Energy, Science, and Technology; and the Office of Science. The committee believes that, overall, the production, distribution, and dispensing portion of the program is probably underfunded, particularly because a significant fraction of appropriated funds is already earmarked. The committee understands that of the $78 million appropriated for hydrogen technology for FY 2004 in the Energy and Water appropriations bill (Public Law 108-137), $37 million is earmarked for activities that will not particularly advance the hydrogen initiative. The committee also believes that the hydrogen program, in an attempt to meet the extreme challenges set by senior government and DOE leaders, has tried to establish RD&D activities in too many areas, creating a very diverse, somewhat unfocused program. Thus, prioritizing the efforts both within and across program areas, establishing milestones and go/no-go decisions, and adjusting the program on the basis of results are all extremely important in a program with so many challenges. This approach will also help determine when it is appropriate to take a program to the demonstration stage. And finally, the committee believes that the probability of success in bringing the United States to a hydrogen economy will be greatly increased by partnering with a broader range of academic and industrial organizations—possibly including an international focus[5]—and

[5]Secretary of Energy Spencer Abraham, joined by ministers representing 14 nations and the European Commission, signed an agreement on November 20, 2003, to formally establish the International Partnership for the Hydrogen Economy.

by establishing an independent program review process and board.

Recommendation 10-10. There should be a shift in the hydrogen program away from some development areas and toward exploratory work—as has been done in the area of hydrogen storage. A hydrogen economy will require a number of technological and conceptual breakthroughs. The Department of Energy program calls for increased funding in some important exploratory research areas such as hydrogen storage and photoelectrochemical hydrogen production. However, the committee believes that much more exploratory research is needed. Other areas likely to benefit from an increased emphasis on exploratory research include delivery systems, pipeline materials, electrolysis, and materials science for many applications. The execution of such changes in emphasis would be facilitated by the establishment of DOE-sponsored academic energy research centers. These centers should focus on interdisciplinary areas of new science and engineering—such as materials research into nanostructures, and modeling for materials design—in which there are opportunities for breakthrough solutions to energy issues.

Recommendation 10-11. As a framework for recommending and prioritizing the Department of Energy program, the committee considered the following:

- Technologies that could significantly impact U.S. energy security and carbon dioxide emissions,
- The timescale for the evolution of the hydrogen economy,
- Technology developments needed for both the transition period and the steady state,
- Externalities that would decelerate technology implementation, and
- The comparative advantage of the DOE in research and development of technologies at the pre-competitive stage.

The committee recommends that the following areas receive increased emphasis:

- *Fuel cell vehicle development.* Increase research and development to facilitate breakthroughs in fuel cell costs and in durability of fuel cell materials, as well as breakthroughs in on-board hydrogen storage systems;
- *Distributed hydrogen generation.* Increase R&D in small-scale natural gas reforming, electrolysis, and new concepts for distributed hydrogen production systems;
- *Infrastructure analysis.* Accelerate and increase efforts in systems modeling and analysis for hydrogen delivery, with the objective of developing options and helping guide R&D in large-scale infrastructure development;
- *Carbon sequestration and FutureGen.* Accelerate development and early evaluation of the viability of carbon capture and storage (sequestration) on a large scale because of its implications for the long-term use of coal for hydrogen production. Continue the FutureGen Project as a high-priority task; and
- *Carbon dioxide-free energy technologies.* Increase emphasis on the development of wind-energy-to-hydrogen as an important technology for the hydrogen transition period and potentially for the longer term. Increase exploratory and fundamental research on hydrogen production by photobiological, photoelectrochemical, thin-film solar, and nuclear heat processes.

References

Air Products and Chemicals. 2003. Information available online at http://www.airproducts.com/Products/LiquidBulkGases/HydrogenEnergyFuelCells/FrequentlyAskedQuestions.htm. Accessed December 10, 2003.

ANL. Argonne National Laboratory. 2003. Information available online at http://www.cmt.anl.gov/science-technology/lowtempthermochemical.shtml. Accessed December 10, 2003.

Archer, M.D., and R. Hill. 2001. *Clean Electricity from Photovoltaics.* London: Imperial College Press.

Arthur D. Little, Inc. 2001. *Guidance for Transportation Technologies: Fuel Choice for Fuel Cell Vehicles: Phase Two Final Report to DOE.* Cambridge, Mass.: Arthur D. Little.

AWEA. American Wind Energy Association. 2003. "Global Wind Energy Market Report 2003" and other reports. Available online at http://www.awea.org/. Accessed December 10, 2003.

Bailie, A., S. Bernow, B. Castelli, P. O'Connor, and J. Romm. 2003. "The Path to Carbon-Free Power: Switching to Clean Energy in the Utility Sector," April. Available online at http://www.worldwildlife.org/powerswitch/power_switch.pdf. Accessed December 10, 2003.

Bak, P.E. 2003. "Daimler Chrysler Minivan, Powered by Unique Fuel-Cell System, to Be Featured in Pentagon Display," April 20. Available online at http://www.h2cars.biz/artman/publish/article_144.shtml. Accessed December 10, 2003.

Besenbruch, G.E., L.C. Brown, J.F. Funk, and S.K. Showalter. 2001. "High Efficiency Generation of Hydrogen Fuels Using Nuclear Power." In *Nuclear Production of Hydrogen: First Information Exchange Meeting—Paris, France 2–3 October 2000.* Paris, France. October. Organization for Economic Cooperation and Development.

Boessel, Ulf, Baldur Eliasson, and Gordon Taylor. 2003. "The Future of the Hydrogen Economy: Bright or Bleak?" Available online at http://www.efcf.com/reports/. Accessed December 10, 2003.

Brown, L.C., R.D. Lentsch, G.E. Besenbruch, and K.R. Schultz. 2003. "Alternative Flowsheets for the Sulfur-Iodine Thermochemical Hydrogen Cycle." Paper presented at Spring 2003 National Meeting, American Institute of Chemical Engineers, New Orleans, La., March 30–April 3.

CaFCP. California Fuel Cell Partnership. 2004. "2004 to 2007 Plans." Available online at http://www.fuelcellpartnership.org/factsheet_04-07.html. Accessed January 15, 2004.

Corey, K., S. Bernow, W. Dougherty, S. Kartha, and E. Williams. 1999. "Analysis of Wind Turbine Cost Reductions: The Role of Research and Development and Cumulative Production." Paper presented at American Wind Energy Association (AWEA) Windpower '99 Conference, June. Available online at http://www.tellus.org/energy/publications/awea9_amy.pdf. Accessed November 17, 2003.

De la Torre Ugarte, D., M.E. Walsh, H. Shapouri, and S.P. Slinsky. 2003. *The Economic Impacts of Bioenergy Crop Production on U.S. Agriculture.* USDA Office of the Chief Economist Agricultural Economic Report 816. Washington, D.C.: U.S. Department of Agriculture.

Derwent, R.G., W.J. Collins, C.E. Johnson, and D.S. Stevenson. 2001. "Transient Behaviour of Tropospheric Ozone Precursors in a Global 3-D CTM and Their Indirect Effects." *Climatic Change,* Vol. 49, pp. 463–487.

Deutch, J., and E. Moniz, co-chairs. 2003. *The Future of Nuclear Power: An Interdisciplinary MIT Study.* Cambridge, Mass.: Massachusetts Institute of Technology.

Deyette, Jeff, Steve Clemmer, and Deborah Donovan. 2003. *Plugging in Renewable Energy: Grading the States.* Washington, D.C.: Union of Concerned Scientists, May.

DiPietro, P. 1997. "Incorporating Carbon Dioxide Sequestration into Hydrogen Energy Systems." In *Proceedings of the 1997 DOE Hydrogen Program Review,* Vol. II. Golden, Colo., National Renewable Energy Laboratory.

Doctor, R.D., D.C. Wade, and M.H. Mendelsohn. 2002. "STAR-H2: A Calcium-Bromine Hydrogen Cycle Using Nuclear Heat." Paper presented at Spring 2003 National Meeting, American Institute of Chemical Engineers, New Orleans, La., March 30–April 3.

DOE. Department of Energy. 1999. *Environmental Benefits of Advanced Oil and Gas Exploration and Production Technology.* DOE-FE-0385. Washington, D.C.: U.S. Department of Energy.

DOE. 2002a. *National Hydrogen Energy Roadmap.* Washington, D.C.: U.S. Department of Energy, November.

DOE. 2002b. *Proceedings of the Hydrogen Storage Workshop.* August 14-15. Argonne National Laboratory. Available online at http://www.eere.energy.gov/hydrogenandfuelcells/wkshp_h2_storage.html. Accessed December 9, 2003.

DOE. 2003a. *Hydrogen Posture Plan: An Integrated Research, Development, and Demonstration Plan.* Washington, D.C.: U.S. Department of Energy. July.

DOE. 2003b. "Hydrogen, Fuel Cells and Infrastructure Technologies Program: Multi-Year Research, Development and Demonstration Plan (Draft, June 3, 2003)." Washington, D.C.: U.S. Department of Energy, Office of Energy Efficiency and Renewable Energy.

DOE. 2003c. *Office of Fossil Energy—Hydrogen Program Plan, Hydrogen from Natural Gas and Coal: The Road to a Sustainable Energy Future.* Washington, D.C.: U.S. Department of Energy, Office of Fossil Energy, June.

DOE. 2003d. *Opportunities for Hydrogen Production and Use in the Industrial Sector.* Washington, D.C.: U.S. Department of Energy, May.

DOE. 2003e. *Basic Research Needs for the Hydrogen Economy: Report of the Basic Energy Sciences Workshop on Hydrogen Production Storage and Use*. Washington, D.C.: U.S. Department of Energy.

DOE. 2003f. Alternative Fuels Data Center, "Properties of Fuels." Available online at www.afdc.doe.gov/pdfs/fueltable.pdf. Accessed December 9, 2003.

DOE. 2003g. *A Plan for the Development of Fusion Energy: Final Report to FESAC*. Washington, D.C.: U.S. Department of Energy.

Doenitz, W., G. Dietrich, E. Erdle, and R. Streicher. 1988. "Electrochemical High Temperature Technology for Hydrogen Production or Direct Electricity Generation." *International Journal of Hydrogen Energy*, Vol. 13, No. 5, pp. 283–287.

Dostal, V., M.J. Driscoll, P. Hejzlar, and N.E. Todreas. 2002. *CO_2 Brayton Cycle Design and Optimization*. Report MIT-ANP-TR-09. Cambridge, Mass.: Massachusetts Institute of Technology, November.

East-West Center. 2000. *Interfuel Competition in the Power Sector: Fuel Prices Comparisons in the Asia-Pacific Region*. Report to the Energy Information Administration. Honolulu: East-West Center, May.

Edison Electric Institute. 2002. *Energy Infrastructure: Electricity Transmission Lines*. Washington, D.C.: Edison Electric Institute. February.

EIA. Energy Information Administration. 2000. *Long Term World Oil Supply*. Washington, D.C.: Energy Information Administration.

EIA. 2001. *Emissions of Greenhouse Gases in the United States 2000*. Washington, D.C.: U.S. Department of Energy, November.

EIA. 2002. *Emissions of Greenhouse Gases in the United States 2001*. Washington, D.C.: U.S. Department of Energy, December.

EIA. 2003. *Annual Energy Outlook 2003 with Projections to 2025*. Washington, D.C.: U.S. Department of Energy.

Elliott, D.L., and M.N. Schwartz. 1993. "Wind Energy Potential in the United States," Pacific Northwest Laboratory. Available online at http://www.nrel.gov/wind/wind_potential.html. Accessed November 17, 2003.

EPA. Environmental Protection Agency. 2002. *Inventory of U.S. Greenhouse Gas Emissions and Sinks: 1990–2000*. Washington, D.C.: EPA, April 15.

EPA. 2003. *Light-Duty Automotive Technology and Fuel Economy Trends: 1975 Through 2003: Executive Summary*. EPA420-S-03-004. Washington, D.C.: EPA.

Feder, B.J. 2003. "For Far Smaller Fuel Cells, a Far Shorter Wait." *New York Times*, March 16.

Fingersh, L.J. 2003. *Optimized Hydrogen and Electricity Generation*. NREL Technical Report NREL/TP-500-34364. Golden, Colo.: National Renewable Energy Laboratory, June.

Forsberg, C.W. 2003. "Hydrogen Production Using the Advanced High-Temperature Reactor," In *Proceedings of the 14th Annual U.S. Hydrogen Meeting*, Washington, D.C.: National Hydrogen Association, March.

Grätzel, M. 2001. "Photoelectrochemical Cells." *Nature*, Vol. 414, pp. 338–344.

Hall, D.O., and K.K. Rao. 1999. *Photosynthesis*. 6th ed. West Nyack, N.Y.: Cambridge University Press.

Herring, S. 2002. "High Temperature Electrolysis Using Solid Oxide Fuel Cell Technology," Presentation at Technical Workshop on Large Scale Production of Hydrogen Using Nuclear Power, San Diego, Calif., May 14–15.

Hill, Gardiner. 2003. "Using Carbon Dioxide to Recover Natural Gas and Oil." In *The Carbon Dioxide Dilemma*. Washington, D.C.: National Academies Press.

Holloway, S. 2001. "Storage of Fossil Fuel-Derived Carbon Dioxide Beneath the Surface of the Earth." *Annual Reviews of Energy and Environment*, Vol. 26, pp. 145–166.

Holt, N. 2003. "Hydrogen Production Options." Paper presented at Hydrogen-Electric Economy Workshop, Electric Power Research Institute, Palo Alto, Calif., July 17–18.

Huber, G.W., J.W. Shabaker, and J.A. Dumesic. 2003. "Raney Ni-Sn Catalyst for H_2 Production from Biomass-Derived Hydrocarbons." *Science*, Vol. 300, pp. 2075–2077.

IPCC. Intergovernmental Panel on Climate Change. 1995. *Second Assessment Report: Climate Change 1995*. Cambridge, United Kingdom: Cambridge University Press.

IPCC. 2001a. *IPCC Third Assessment Report—Climate Change 2001: The Scientific Basis*. Cambridge, United Kingdom: Cambridge University Press.

IPCC. 2001b. *IPCC Third Assessment Report—Climate Change 2001: Mitigation*. Cambridge, United Kingdom: Cambridge University Press.

JAERI. Japan Atomic Energy Research Institute. 2001. "Tests on Hydrogen Production from Water Using Thermal Energy Started." Available online at www.jaeri.go.jp/english/press/2001/010515.

Johansson, T.B. 1993. *Renewable Energy-Sources for Fuels and Electricity*. Washington, D.C.: Island Press.

Jones, R.M., and N.Z. Shilling. 2002. "IGCC Gas Turbines for Refining Applications." Paper presented at Gasification Technologies Conference, San Francisco, Calif., October 27–30.

Kuuskraa, V.A., and L.J. Pekot. 2003. "Defining Optimum CO_2 Sequestration Sites for Power and Industrial Plants." Pp. 609–614 in online *Proceedings of the 6th International Conference on Greenhouse Gas Control Technologies*, October 1–4, 2002, Kyoto, Japan, J. Gale and Y. Kaya, eds. Elsevier Science.

LaBar, M.P. 2002. "The Gas-Turbine Modular Helium Reactor: A Promising Option for Near-Term Deployment." In *Proceedings of International Congress on Advanced Nuclear Power Plants*, June, Hollywood, Fla. La Grange Park, Ill.: American Nuclear Society.

Lipman, T., and D. Sperling. 2003. "Market Concepts, Competing Technologies and Cost Challenges for Automotive and Stationary Applications." In *Handbook of Fuel Cells*, Vol . 4, Pt. 13, pp. 1318–1328. New York: John Wiley.

Lutsey, Nic, C.J. Brodrick, D. Sperling, and H. Dwyer. 2003. "Markets for Fuel Cell Auxiliary Power Units in Vehicles: A Preliminary Assessment." *Transportation Research Record*, No. 1842, Washington, D.C.: Transportation Research Board.

Milne T.A., C.C. Elam, and R.J. Evans. 2002. *Hydrogen from Biomass—State of the Art and Challenges*. IEA/H2/TR-02/001. Golden, Colo.: National Renewable Energy Laboratory.

Mintzer, Irving, J. Amber Leonard, and Peter Schwartz. 2003. "Global Business Network." In *U.S. Energy Scenarios for the 21st Century*. Arlington, Va.: Pew Center for Global Climate Change.

NASS. National Agricultural Statistics Service. 2003. *Agricultural Statistics 2003*. Washington, D.C.: U.S. Department of Agriculture.

National Energy Supergrid Workshop Report. 2002. Sponsored by University of Illinois at Urbana-Champaign and Richard Lounsbery Foundation, Palo Alto, Calif., November 6–8.

NIST. National Institute of Standards and Technology. 2003. Standard Reference Database Number 69 - March, 2003 Release. Gaithersburg, Md.: NIST.

National Mining Association. 2003. "U.S. Coal Production by State, Region and Method of Mining—2002." Available online at http://www.nma.org/pdf/c_production_method.pdf. Accessed January 3, 2004.

NPC. National Petroleum Council. 2001. *Securing Oil and Natural Gas Infrastructures in the New Economy*. Washington, D.C.: U.S. Department of Energy. June.

National Renewable Energy Laboratory. 2003. *Gas-Fired Distributed Energy Resource Technology Characterizations*. NREL/TP-620-34783. Golden, Colo.: National Renewable Energy Laboratory, November.

NRC. National Research Council. 1990. *Fuels to Drive Our Future*. Washington, D.C.: National Academy Press.

NRC. 1998. *Review of the Research Program of the Partnership for a New Generation of Vehicles, Fourth Report*. Washington, D.C.: National Academy Press.

NRC. 1999a. *Our Common Journey: A Transition Toward Sustainability*. Washington, D.C.: National Academy Press.

NRC. 1999b. *Hardrock Mining on Federal Lands*. Washington, D.C.: National Academy Press.

NRC. 2000. *Clean Coastal Waters: Understanding and Reducing the Effects of Nutrient Pollution.* Washington, D.C.: National Academy Press.

NRC. 2001a. *Energy Research at DOE: Was It Worth It? Energy Efficiency and Fossil Energy Research 1978–2000.* Washington, D.C.: National Academy Press.

NRC. 2001b. *Climate Change Science: An Analysis of Some Key Questions.* Washington, D.C.: National Academy Press.

NRC. 2002. *Making the Nation Safer: The Role of Science and Technology in Countering Terrorism.* Washington, D.C.: The National Academies Press.

NRC. 2003a. *Cumulative Environmental Effects of Oil and Gas Activities on Alaska's North Slope.* Washington, D.C.: The National Academies Press.

NRC. 2003b. *Review of DOE's Vision 21 Research and Development Program—Phase I.* Washington, D.C.: The National Academies Press.

Ness, H.M., S.S. Kim, and R. Massood. 1999. "Status of Advanced Coal-Fired Power Generation Technology Development in the US." Paper presented at 13th U.S./Korea Joint Workshop on Energy and Environment, Reno, Nev., September.

Novelli, P.C., P.M. Lang, K.A. Masarie, D.F. Hurst, R. Myers, and J.W. Elkins. 1999. "Molecular Hydrogen in the Troposphere: Global Distribution and Budget." *Journal of Geophysical Research,* Vol. 104, No. D23, pp. 30427–30444.

Nyborg, E.O., R.D. Hay, and P. Benard. 2003. "Clearance Distances and Hazardous Zone Issues for Hydrogen Systems." Paper presented at the National Hydrogen Association's 14th Annual U.S. Hydrogen Meeting, Washington, D.C., March 4–6, 2003.

Ogden, J. 1999. "Prospects for Building a Hydrogen Energy Infrastructure." *Annual Review of Energy and the Environment,* Vol. 24, pp. 227–279.

Ogden, J. 2003. "Modeling Infrastructure for a Fossil Hydrogen Energy System with CO_2 Sequestration." Pp. 1069–1074 in online *Proceedings of the 6th International Conference on Greenhouse Gas Control Technologies,* October 1–4, 2002, Kyoto, Japan, J. Gale and Y. Kaya, eds. Elsevier Science.

O'Regan, B., and M.A. Grätzel. 1991. "A Low-Cost, High Efficiency Solar Cell Based on Dye-Sensitized Colloidal TiO_2 Films." *Nature,* Vol. 353, pp. 737–740.

ORNL. Oak Ridge National Laboratory. 2003. *Transportation Energy Data Book: Edition 23.* Center for Transportation Analysis, Oak Ridge National Laboratory. ORNL-6970. Oak Ridge, Tenn.: UT-Battelle, LLC.

PCAST. President's Committee of Advisors on Science and Technology. 1999. *Powerful Partnerships: The Federal Role in International Cooperation on Energy.* Washington, D.C.: Executive Office of the President.

Pham, A.Q. 2000. "High Efficiency Steam Electrolyzer." In *Proceedings of 2000 Hydrogen Program Review.* NREL/CP-570-28890. Golden, Colo.: National Renewable Energy Laboratory.

Reeves, A. 2003. "Wind Electricity for Electric Power." A Renewable Energy Policy Project Issue Brief (REPP), July. Available online at http://www.repp.org/articles/static/1/binaries/wind%20issue%20brief_FINAL.pdf. Accessed December 9, 2003.

Resource Dynamics Corporation. 2003. *Installed Base of US Distributed Generation.* 2003 ed. Vienna, Va.: Resource Dynamics Corporation.

Sandell, L. 2003. *High Temperature Gas-Cooled Reactors for the Production of Hydrogen: An Assessment in Support of the Hydrogen Economy.* Report No. 1007802. Palo Alto, Calif.: Electric Power Research Institute.

Santini, Danilo J., Anant Vyas, and Margaret Singh. 2003. "The Scenario for Rate of Replacement of Oil by H_2 FCVs in Historical Context." April. Review draft. Argonne National Laboratory.

Schrattenholzer, L. 1998. "Energy Demand and Supply, 1900–2100." RR-98-4. International Institute for Applied Systems Analysis, Laxenburg, Austria. Reprinted from *Science Vision,* Vol. 3, No. 1, pp. 39–57.

Schroeder, B. 2003. "Status Report: Solar Cell Related Research and Development Using Amporphous and Microcrystalline Silicon Deposited by HW (Cat) CVD." *Thin Solid Films,* Vol. 430, pp. 1–6.

Schultz, K.R., L.C. Brown, G.E. Besenbruch, and C.J. Hamilton. 2002. "Production of Hydrogen by Nuclear Energy: The Enabling Technology for the Hydrogen Economy." In *Proceedings of The Americas Nuclear Energy Symposium (Simposio de Energía Nuclear de las Américas),* Coral Gables, Fla., 2002. Available online at http://anes2002.hcet.fiu.edu/proceedings_html.asp.

Shah, A., P. Torres, R. Tscharner, N. Wyrsch, and H. Keppner. 1999. "Photovoltaic Technology: The Case for Thin-Film Solar Cells." *Science,* Vol. 285, pp. 692–698.

Shipley, A., and R.N. Elliot. 2003. "Fuel Cells—Future Promise, Current Hype." American Council for an Energy-Efficient Economy (ACEEE), draft, May. Washington, D.C.: ACEEE.

Socolow, R. 2003. "The Century-Scale Problem of Carbon Management." *The Carbon Dioxide Dilemma: Promising Technologies and Policies. Proceedings of a Symposium,* April 23–24, 2002, pp. 11–14. Washington, D.C.: The National Academies Press.

Spath, P.L., J.M. Lane, M.K. Mann, and W.A. Amos. 2000. *Update of Hydrogen from Biomass—Determination of the Delivered Cost of Hydrogen.* April. Milestone Report for the U.S. Department of Energy's Hydrogen Program. Process Analysis Task, Milestone Type C.

Staebler, D.L. and C. Wronski. 1977. *Applied Physics Letters,* Vol. 70, p. 295.

Thauer, R.K., K. Jungermann, and K. Decker. 1977. "Energy Conversion in Chemotrophic Anaerobic Bacteria." *Bacteriological Reviews,* Vol. 41, pp. 100–180.

Thomas, George. 2003. "Hydrogen Storage: State of the Art." Presentation to the Basic Energy Sciences Workshop on Basic Research Needs for the Hydrogen Economy, Rockville, Md., May 13–15.

TIAX, LLC. 2003. *Aggressive Growth in the Use of Bioderived Energy and Products in the United States by 2010.* June. Final report Arthur D. Little, Inc.

Turner, J.A. 1999. "A Realizable Renewable Energy Future." *Science.* Vol. 285, pp. 687–689, July 30.

Ullal, H.S., K. Zweibel, and B. von Roedern. 2002. "Polycrystalline Thin Film Photovoltaics: Research, Development, and Technologies." Paper presented at 29th Institute of Electrical and Electronics Engineers Photovoltaic Specialists Conference (IEEE PVSC), New Orleans, La., May 20–24.

USNRC. U.S. Nuclear Regulatory Commission. 1996. Generic Environmental Impact Statement for License Renewal of Nuclear Plants (NUREG-1437 Vol. 1). Washington D.C.: USNRC.

Vesterby, Marlow, and Kenneth S. Krupa. 1997. *Major Uses of Land in the United States.* Statistical Bulletin No. 973. Washington, D.C.: U.S. Department of Agriculture, Economic Research Service, Resource Economics Division.

Walsh, M.E., D.G. de la Torre Ugarte, H. Shapouri, and S.P. Slinsky. 2000. "The Economic Impact of Bioenergy Crop Production on U.S. Agriculture." Paper presented at Conference on Sustainable Energy: New Challenges for Agriculture and Implications for Land Use, Wageningen, The Netherlands, May 18–20.

Wang, M. 2002. "Fuel Choices for Fuel Cell Vehicles: Well-to-Wheels Energy and Emissions Impacts." *Journal of Power Sources,* Vol. 112, pp. 307–321.

Ward's Communication. 2002. *Ward's World Motor Vehicle Data 2002.* Southfield, Michigan: Ward's Communication.

WEA. World Energy Assessment. 2000. *Energy and the Challenge of Sustainability.* Report prepared by United Nations Development Programme, United Nations Department of Economic and Social Affairs and World Energy Council, September. New York: United Nations Publications. Available online at http://www.undp.org/seed/eap/activities/wea/. Accessed December 10, 2003.

Wieting, R.D. 2002. "Why Thin Film Technology for Photovoltaics?" Paper presented at 29th Electrical and Electronics Engineers Photovoltaic Specialists Conference, New Orleans, La., May 20–24.

Winter, D., and K. Kelly. 2003. "Hybrid Heartburn." *Ward's Auto World,* March 1.

World Nuclear Association. 2002. "Supply of Uranium." Available online at http://world-nuclear.org/info/inf75.htm. Accessed November 17, 2003.

Worldwatch Institute. 1999. *Vital Signs 1999: The Environmental Trends That Are Shaping Our Future*. London: W.W. Norton.

Yang, J., A. Banerjee, and S. Guha. 1997. "Triple-Junction Amorphous Silicon Alloy Solar Cells with 14.6% Initial and 13.0% Stable Conversion Efficiencies." *Applied Physics Letters,* Vol. 70, No. 22, June 2, pp. 2975–2977.

Yildiz, B., and M. Kazimi. 2003. *Nuclear Options for Hydrogen and Hydrogen-Based Liquid Fuels Production.* MIT-NES-TR-001. Center for Advanced Nuclear Energy Systems, Cambridge, Mass.: Massachusetts Institute of Technology.

Appendixes

Appendix A

Biographies of Committee Members

Michael P. Ramage (NAE) (*Chair*) is retired executive vice president, ExxonMobil Research and Engineering Company. Previously he was executive vice president and chief technology officer, Mobil Oil Corporation. Dr. Ramage held a number of positions at Mobil, including those of research associate, manager of Process Research and Development, general manager of Exploration and Producing Research, vice president of Engineering, and president of Mobil Technology Company. He has broad experience in many aspects of the petroleum and chemical industries. He serves on a number of university visiting committees and is a member of the Government-University Industrial Research Roundtable. He is a director of the American Institute of Chemical Engineers and a member of several other professional organizations. Dr. Ramage is a member of the National Academy of Engineering (NAE) and serves on the NAE Council. He has a B.S., M.S., and Ph.D. in chemical engineering from Purdue University.

Rakesh Agrawal (NAE) is Air Products Fellow at Air Products and Chemicals, Inc., where he has worked since 1980. His research interests include basic and applied research in gas separations, process development, synthesis of distillation column configurations, adsorption and membrane separation processes, novel separation processes, gas liquefaction processes, cryogenics, and thermodynamics. He has broad experience in hydrogen production and purification technologies. His current interest is in energy production issues, especially those related to renewable sources such as solar. He holds more than 100 U.S. and 300 foreign patents. He has authored 61 technical papers and given many lectures and presentations. He chaired the Separations Division and the Chemical Technology Operating Council of the American Institute of Chemical Engineers and also a Gordon Conference on Separations. Dr. Agrawal received a B.Tech. from the Indian Institute of Technology, in Kanpur, India; an M.Ch.E. from the University of Delaware; and an Sc.D. in chemical engineering from the Massachusetts Institute of Technology.

David L. Bodde holds the Charles N. Kimball Chair in Technology and Innovation at the Henry W. Bloch School of Business and Public Administration, University of Missouri, Kansas City. He has extensive experience in energy policy and technology assessment, and his current work focuses on the role of entrepreneurs in the innovation and commercialization of energy technologies. He has served as corporate vice president, Midwest Research Institute (MRI), and president of MRI's for-profit subsidiary, MRI Ventures. He was executive director of the National Research Council's (NRC's) Commission on Engineering and Technical Systems; assistant director, Congressional Budget Office; deputy assistant secretary, Department of DOE (DOE); and manager, Engineering Analysis Office, Energy Systems Planning Division, TRW, Inc. He has worked on numerous studies involving nuclear energy, coal, synthetic fuels, electric utilities, renewable energy technologies, and commercialization. He recently served as chair of the Environmental Management Board, advising the DOE on the cleanup of the U.S. nuclear weapons complex, and is a member of the NRC Board on Energy and Environmental Systems. His current work includes research and teaching in the strategic use of technology to create new ventures in energy, the environment, and education. He holds the Doctor of Business Administration, Harvard University (1976); M.S. degrees in nuclear engineering (1972) and management (1973); and a B.S. from the United States Military Academy (1965).

Robert Epperly is a consultant. From 1994 to 1997, he was president of Catalytica Advanced Technologies, Inc., a company developing new catalytic technologies for the petroleum and chemical industries. Before joining Catalytica, he was chief executive officer of Fuel Tech N.V., a company specializing in new products for combustion and air pollution control. Earlier, he was general manager of Exxon Corporate Research. While at Exxon Research and Engineering Company, he was also general manager of the Synthetic Fuels Department and manager of the Baytown Research and Development Division. He is a fellow in the American Institute of Chemical Engineers (AIChE) and a past recipient of the AIChE's National Award in Chemical Engineering Practice. He has authored or co-authored more than 50 publications, including 2 books, and has 38 U.S. patents. He has extensive experience in the conversion of fossil feedstocks to alternative gaseous and liquid fuels, petroleum fuels, engines, catalysis, air pollution control, and R&D management. Since 1981, he has participated in seven committees at the National Research Council. He received B.S. and M.S. degrees in chemical engineering from Virginia Polytechnic Institute and State University.

Antonia V. Herzog is staff scientist in the Climate Center at the Natural Resources Defense Council, where she analyzes climate change issues and provides information to decision makers and the public. She had been a Congressional Legislative Science Fellow and a postdoctoral fellow at the University of California, Berkeley. She received a B.A. in physics from Vassar College, a B.Eng. from Dartmouth College, an M.S. in applied physics from Columbia University, and a Ph.D. in physics from the University of California, San Diego.

Robert L. Hirsch is currently a senior energy program advisor at Scientific Applications International Corporation. His past positions include those of senior energy analyst at the RAND Corporation; executive advisor to the president of Advanced Power Technologies, Inc.; vice president, Washington office, Electric Power Research Institute; vice president and manager, Research and Technical Services Department, ARCO Oil and Gas Company; chief executive officer of ARCO Power Technologies, a company that he founded; manager, Baytown Research and Development Division, and general manager, Exploratory Research, Exxon Research and Engineering Company. He was assistant administrator for Solar, Geothermal, and Advanced Energy Systems (presidential appointment); and director, Division of Magnetic Fusion Energy Research, U.S. Energy Research and Development Administration. He has served on numerous advisory committees, including as a member of the DOE Energy Research Advisory Board and a number of DOE national laboratory advisory boards. He has served on several National Research Council committees, including the one that wrote the report *Fuels to Drive Our Future* (1990), which examined the economics and technologies for producing transportation fuels from U.S. domestic resources, and he was chair of the Committee to Examine the Research Needs of the Advanced Extraction and Process Technology Program. He was formerly chair of the Board on Energy and Environmental Systems. He brings expertise in a number of areas of science, technology, and business related to energy production and consumption, research and development, and public policy. He received a Ph.D. in engineering and physics from the University of Illinois.

Mujid S. Kazimi is director of the Center for Advanced Nuclear Energy Systems and professor of nuclear engineering, Massachusetts Institute of Technology (MIT). He has been on the faculty at MIT since 1976, and previously served as head of the department. He also held positions at Brookhaven National Laboratory and the Westinghouse Electric Corporation before joining the MIT faculty. He has extensive expertise in advanced nuclear energy systems, in reactors, the nuclear fuel cycle, and nuclear research. He has served on numerous review committees and panels, and currently serves as a member of the Technical Review Committee of the Division of Nuclear and Energy Systems, Idaho National Engineering Laboratory, and member of the Organizing Committee of the International Congress on Advanced Power Plants, American Nuclear Society. He is coauthor of *Nuclear Systems*, a two-volume book on the thermal analysis and design of nuclear fission reactors. He served on the NRC Panel on Separations Technology and Transmutation Systems and is a fellow of the American Nuclear Society. He has a B.Eng. (Alexandria University), M.S. (MIT), and Ph.D. (MIT) in nuclear engineering.

Alexander MacLachlan (NAE) retired at the end of 1993 from E.I. du Pont de Nemours & Company after more than 36 years of service. He had been senior vice president for research and development and chief technical officer since 1986. In late 1994, he joined the U.S. Department of Energy as deputy undersecretary for technology partnerships and in 1995 was made deputy undersecretary for R&D management. He left the DOE in 1996, but remained on its Secretary of Energy Advisory Board, Laboratory Operations Board, Sandia President's Advisory Council, and the National Renewable Energy Laboratory's Advisory Council until 2003. He has participated in several studies for the National Research Council, including *Containing the Threat from Illegal Bombings* (1998); *Technology Commercialization: Russian Challenges, American Lessons* (1998); and *Building an Effective Environmental Management Science Program* (1997). Recently he was chair for the Committee to Review the Department of Transportation's Intelligent Vehicle Initiative. He currently serves on the NRC's Board on Radioactive Waste Management and is liaison to one of the board's current studies. He is a member of Phi Beta Kappa and a member of the National Academy of Engineering. He is a graduate of Tufts University with a B.S. in chemistry (1954) and of MIT with a Ph.D. in physical organic chemistry (1957).

Gene Nemanich is an independent consultant and chairman of the National Hydrogen Association. Prior to retiring from ChevronTexaco in late 2003, he was the vice president of hydrogen systems for ChevronTexaco Technology Ventures, where he was responsible for hydrogen supply and for developing and commercializing new hydrogen storage technologies. In 2000, he formed Texaco Ovonic Hydrogen Systems LLC, a joint venture between Texaco and Energy Conversion Devices to commercialize metal hydride hydrogen storage systems, and he was Texaco's managing director for this joint venture through 2003. He represented Texaco in the California Fuel Cell Partnership in 2000–2001. Mr. Nemanich was one of seven industry leaders to prepare the DOE-sponsored Hydrogen Roadmap in 2002. He has 32 years of experience with integrated oil companies, including Exxon, Cities Service, Texaco, and ChevronTexaco, working in the areas of refining, clean coal technology, oil supply and trading, and hydrogen systems. He was responsible for Texaco's worldwide oil products trading and supply business from 1987 to 1996 and was executive vice president of Tennessee Synfuels Associates, a company formed to build coal-to-gasoline plants, in 1980–1981. He has a B.S. in chemical engineering from the University of Illinois and an MBA from the University of Houston.

William F. Powers (NAE) is retired vice president, research, Ford Motor Company. His approximately 20 years at Ford included positions as director, Vehicle, Powertrain and Systems Research; director, Product and Manufacturing Systems; program manager, Specialty Car Programs; and executive director, Ford Research Laboratory and Information Technology. Prior positions also include those of professor, Department of Aerospace Engineering, University of Michigan, during which time he consulted with NASA, Northrop, Caterpillar, and Ford; research engineer, University of Texas; and mathematician and aerospace engineer, NASA Marshall Space Flight Center. He is a fellow, Institute of Electrical and Electronics Engineers; member, National Academy of Engineering; and foreign member, Royal Swedish Academy of Engineering Sciences. He has extensive expertise in advanced research and development of automotive technology. He has a B.S. in aerospace engineering, University of Florida, and a Ph.D. in engineering mechanics, University of Texas, Austin.

Maxine L. Savitz (NAE) is currently a consultant. She recently retired as general manager, Ceramic Components, AlliedSignal, Inc. She has held a number of positions in the federal government and private sector managing large R&D programs, especially with respect to the development of energy technologies. Some of her positions include those of chief, Buildings Conservation Policy Research, Federal Energy Administration; professional manager, Research Applied to National Needs, National Science Foundation; division director, Buildings and Industrial Conservation, Energy Research and Development Administration; deputy assistant secretary for conservation, U.S. Department of Energy; and president, Lighting Research Institute. She has extensive technical experience in materials, fuel cells, batteries and other storage devices, energy efficiency, and R&D management. She is a member of the National Academy of Engineering and has been or is serving as a member of numerous public and private sector boards, has served on many energy-related and other NRC committees, and is currently a member of the NRC Board on Energy and Environmental Systems. She recently served on the NRC's Committee on DOE R&D on Energy Efficiency and Fossil Energy. She has a Ph.D. in organic chemistry from the Massachusetts Institute of Technology.

Walter W. (Chip) Schroeder is a founder of Proton Energy Systems, Inc., and has served as the company's president and chief executive officer since its inception in 1996. Proton is involved in applications of proton exchange membrane fuel cell technology for energy conversion, storage, and power quality requirements. Mr. Schroeder has held executive positions with a number of energy and financial entities, including that of president, AES Corporation—Sonat Power; vice president, Investment Banking, Goldman Sachs & Company; and president, MidCon Corporation. Mr. Schroeder's energy background began in 1975 when he joined the staff of the Interstate and Foreign Commerce Committee of the U.S. House of Representatives; later he served as director of the Office of Regulatory Analysis at the Federal Energy Regulatory Commission. Mr. Schroeder received an S.M., Sloan School of Management, and a joint S.B., management and engineering, Massachusetts Institute of Technology.

Robert H. Socolow is a professor of mechanical and aerospace engineering at Princeton University, where he has been on the faculty since 1971. He was previously an assistant professor of physics at Yale University. Professor Socolow is a fellow of the American Physical Society and the American Association for the Advancement of Science. He currently codirects Princeton University's Carbon Mitigation Initiative, a multidisciplinary investigation of fossil fuels in a future carbon-constrained world. From 1979 to 1997, Professor Socolow directed Princeton University's Center for Energy and Environmental Studies. He has served on many NRC boards and committees, including the Committee on R&D Opportunities for Advanced Fossil-Fueled Energy Complexes, the Committee on Review of DOE's Vision 21 R&D Program, and the Board on Energy and Environmental Systems. He has a B.A., M.A., and Ph.D. in physics from Harvard University.

Daniel Sperling is director, Institute of Transportation Studies, University of California, Davis; professor of civil and environmental engineering; and professor of environmental science and policy. He has served on numerous Transportation Research Board committees, including as chair of the Alternative Fuels Committee and member of the Committee on Energy and the Committee on Transportation and a Sustainable Environment. He has also served on several NRC committees, including R&D Strategies for Biomass-Based Ethanol and Biodiesel Transportation Fuels and the NRC committee that wrote the report *Fuels to Drive Our Future* (1990), which examined the economics and technologies for producing transportation fuels from U.S. domestic resources. Professor Sperling has done extensive studies on alternative transportation fuels, fuel cell vehicles, and sustainable transportation, and is currently codirecting a research program at the University of California, Davis, on Hydrogen Pathways: Transportation and the Hydrogen Economy. He has a B.S. in civil engineering from Cornell University and a Ph.D. in transportation engineering from the University of California, Berkeley.

Alfred M. Spormann is a microbial physiologist and biochemist at Stanford University. His research interests include the microbial degradation of environmental pollutants, microbial interactions in biofilms, and biological production of molecular hydrogen. He employs biochemical, molecular, genomic, and advanced microscopic techniques to investigate fundamental aspects of microbial metabolism and physiology. He is associate professor of civil and environmental engineering and also has an appointment in the Department of Biological Sciences. He is director of the Stanford Biofilm Research Center. He serves as the editor of *Archives of Microbiology* and serves on the editorial board/committee of three publications: *Applied and Environmental Microbiology*, *Biodegradation*, and *Annual Review of Microbiology*. Professor Spormann received his bachelor's and doctorate degrees from Philipps-University in Marburg, Germany, and did postdoctoral work in the Department of Biochemistry at the University of Minneapolis.

James L. Sweeney is professor of management science and engineering, Stanford University, and senior fellow, Stanford Institute for Economic Policy Research. He has been director of the Office of Energy Systems, director of the Office of Quantitative Methods, and director of the Office of Energy Systems Modeling and Forecasting, Federal Energy Administration. At Stanford University, he has been chairman, Institute of Energy Studies; director, Center for Economic Policy Research; director, Energy Modeling Forum; chairman, Department of Engineering-Economic Systems; and chairman, Department of Engineering-Economic Systems and Operations Research. Professor Sweeney has served on several NRC committees, including the Committee on the National Energy Modeling System, the Committee on Effectiveness and Impact of Corporate Average Fuel Economy Standards, the Committee on the Human Dimensions of Global Climate Change, and the Committee on Benefits of DOE's R&D in Energy Efficiency and Fossil Energy, and has been a member of the Board on Energy and Environmental Systems. He also served on the NRC committee that issued the report *Fuels to Drive Our Future* (1990), which examined the economics and technologies for producing transportation fuels from U.S. domestic resources. He is a fellow of the California Council on Science and Technology. Professor Sweeney's research and writings address economic and policy issues important for natural resource production and use; energy markets including oil, natural gas, and electricity; environmental protection; and the use of mathematical models to analyze energy markets. He has a B.S. degree from MIT and a Ph.D. in engineering-economic systems from Stanford University.

Appendix B

Letter Report

April 4, 2003

Steve Chalk, Director
Office of Hydrogen, Fuel Cells, and Infrastructure Technologies
U.S. Department of Energy
1000 Independence Avenue, SW
Washington, DC 20585

Dear Mr. Chalk:

In response to a requirement in its statement of task item 5,[1] the National Research Council's (NRC's) Committee on Alternatives and Strategies for Future Hydrogen Production and Use submits this interim letter report.[2] This letter report is part of a larger project initiated at the request of the U.S. Department of Energy (DOE) by the NRC's Board on Energy and Environmental Systems and the National Academy of Engineering Program Office. The purpose of the project is to evaluate the cost and status of technologies for production, transportation, storage, and end-use of hydrogen and to review DOE's hydrogen research, development, and deployment (RD&D) strategy. The committee's observations and findings in this report are based on presentations made by various DOE representatives and others at the committee's first two meetings on December 2–4, 2002, and January 22–24, 2003, in Washington, D.C. This letter report provides some early feedback and recommendations that may be of help as you and your colleagues evolve your strategic directions for the fiscal year 2005 hydrogen research and development (R&D) programs.

INTRODUCTION

Hydrogen is a flexible energy carrier that can be produced from a variety of domestic energy resources and used in all sectors of the economy. An energy system based on domestic energy resources, using hydrogen as a carrier and deployed on a large scale, if accomplished, could improve energy security, air quality, and greenhouse gas management. Such a system will require development across a spectrum of complementary technologies for hydrogen production, transportation, storage, and use.

Today, hydrogen is generated for use in a variety of applications, the most significant of which are the refining of crude oil into commercial liquid fuels and the production of fertilizers and high-value chemicals. Accordingly, a great deal of practical commercial experience exists for producing, transporting, and using hydrogen.

The DOE's Office of Energy Efficiency and Renewable Energy has created a new program office, the Office of Hydrogen, Fuel Cells, and Infrastructure Technologies. The

[1] Item 5 of the statement of task asks for a letter report on the committee's interim findings. The balance of the statement of task (see Appendix B) will be addressed in the committee's final report in late 2003.

[2] This report has been reviewed in draft form by individuals chosen for their diverse perspectives and technical expertise, in accordance with procedures approved by the NRC's Report Review Committee. The purpose of this independent review is to provide candid and critical comments that will assist the institution in making its published report as sound as possible and to ensure that the report meets institutional standards for objectivity, evidence, and responsiveness to the study charge. The review comments and draft manuscript remain confidential to protect the integrity of the deliberative process. We wish to thank the following individuals for their review of this report: Daniel Arvizu, CH2M Hill; Alan Bard, NAS, University of Texas, Austin; H.M. Hubbard, Retired President and CEO, Pacific International Center for High Technology Research; James Katzer, NAE, ExxonMobil Research and Engineering Company; and Robert Shaw, Jr., Aretê Corporation.

Although the reviewers listed above have provided many constructive comments and suggestions, they were not asked to endorse the conclusions or recommendations, nor did they see the final draft of the report before its release. The review of this report was overseen by William Agnew, NAE, General Motors (retired). Appointed by the NRC, he was responsible for making sure that an independent examination of this report was carried out in accordance with institutional procedures and that all review comments were carefully considered. Responsibility for the final content of this report rests entirely with the authoring committee and the institution.

committee applauds DOE for providing one office as the focus for the hydrogen-related programs conducted under different DOE organizations. The purpose of this office is to facilitate overall strategic program direction, coordinate individual hydrogen-related activities across various DOE organizations, promote outreach to the public and private sectors, and coordinate with stakeholder partners.[3] An example of a coordination activity is the National Hydrogen Energy Roadmap (November 2002) [1].

The committee offers four recommendations based on its information gathering and deliberations thus far. Reflecting serious needs in DOE's program identified in an initial assessment by the committee, these recommendations may be refined and expanded upon in the committee's final report. They address a systems approach to hydrogen energy RD&D, exploratory research as the foundation for breakthroughs in technology, safety issues, and coordination of R&D strategy and programs.

SYSTEMS ANALYSIS

In its program overview, DOE personnel presented various R&D targets for a variety of possible future hydrogen energy system components. From its collective experience, the committee deems it essential that the DOE treat hydrogen energy development as a system ranging from hydrogen creation and production to transportation, storage, and end use. It is important that all aspects of the various conceivable hydrogen system pathways be adequately modeled to understand the complex interactions between components, system costs, environmental impacts of individual components and the system as a whole, societal impacts (e.g., offsets of imported oil per year), and possible system trade-offs. Indeed, such an analysis function is an essential tool for DOE personnel to optimally prioritize areas for R&D as well as to understand the ramifications of future R&D successes and disappointments. A competent, independent systems analysis group not only will help DOE program managers make better program decisions in the future, but also will help:

- Establish a high standard for assessments performed by program contractors,
- Provide a greater degree of confidence in program integrity,
- Enhance the private sector's willingness to participate in the hydrogen program, and
- Minimize the occurrence of unwarranted claims within DOE.

Indeed, in its recent review of the DOE Vision 21 Program [2], the NRC urged the establishment of such a systems analysis function for the coal-based energy program.

[3] Chalk, S. Overview of DOE Hydrogen Technology Activities. Presentation to the committee on December 2, 2002.

Recommendation 1: An independent systems engineering and analysis group should be established within the hydrogen program to identify the impacts of various technology pathways, to assess associated cost elements and drivers, to identify key cost and technological gaps, and to assist in the prioritization of R&D directions. The committee understands that DOE recognizes the importance of systems integration and suggests that its current analytical capabilities could be expanded into an in-house systems analysis group.

EXPLORATORY RESEARCH

A hydrogen economy[4] will not come about without significant improvements in technology. This in turn requires that DOE provide significant funding for fundamental, exploratory research supported by organizations and investigators that propose credible, promising, high-risk new concepts for technologies for hydrogen storage, production, transportation, and end-use. The cost reductions (e.g., fuel cell cost per kilowatt) and infrastructure necessary to bring about a hydrogen economy are indeed challenging. While progress will certainly result from further development and demonstrations of existing technologies, some hydrogen system components will require major scientific breakthroughs that development will not address. Such advances will require entirely new approaches and thinking, which can come about only through relatively fundamental, directed exploratory research aimed at identifying technologies that will achieve cost reduction and technology goals (e.g., weight percentage of stored hydrogen).

Demonstrations also have a place in a balanced research program because they can lead to cost reductions and accelerate the development of codes, standards, environmental permitting, and strategies for inspection and monitoring. But demonstrations can also distort budgets and divert effort toward technology with limited potential. Development of a careful plan for funding and evaluating demonstrations will serve the public interest.

Recommendation 2: Fundamental and exploratory research should receive additional budgetary emphasis, and the DOE should develop a careful plan for evaluating, funding, and validating emerging technologies for hydrogen production, transportation, storage, and end-use.

SAFETY

The nation's current hydrogen production, transportation, and utilization system is very safely managed [3]. The introduction of hydrogen into the commercial supply and con-

[4] The hydrogen economy has been envisioned as the large-scale use of hydrogen as an energy carrier, generated from any of a variety of fuels or feedstocks, to be used in the transportation, industrial, and building sectors, and requiring an infrastructure for its transmission and delivery.

sumption sectors of the economy, however, will present a number of new safety issues, due to hydrogen's wide explosive range and extremely low ignition energy. In addition, safety considerations can affect the choice of technology pathway. Accordingly, safety considerations must be an integral part of DOE's hydrogen program.

Recommendation 3: The committee recommends that DOE make a significant effort to address safety issues, and it supports DOE's plans to incorporate safety considerations into its various hydrogen research, development, and deployment programs.

ORGANIZATION

A transition from the current U.S. energy system to one based on hydrogen will be extremely difficult and challenging and will require a national coordinated effort across DOE's programs and the private sector. The private sector players in that new system will likely include a number of existing industries along with some entirely new companies. Considerable benefit can be gained from the experience and potential contributions of existing industry as well as new companies that may come into being along the way.

Recommendation 4: The committee strongly supports the DOE in its efforts to integrate its various hydrogen-related RD&D programs across the applied energy programs, the Office of Science, and appropriate private sector participants in the planning and development of hydrogen technologies and systems, and it recommends that DOE continue to leverage the knowledge and capabilities of the private sector. The committee further recommends that the Office of Science be integrated better into hydrogen program planning to help facilitate the needed exploratory research mentioned in Recommendation 2.

Michael P. Ramage, *Chair*
Committee on Alternatives and Strategies for Future Hydrogen Production and Use

REFERENCES

[1] U.S. Department of Energy. National Hydrogen Energy Roadmap. November 2002.

[2] National Research Council. Review of DOE's Vision 21 Research and Development Program: Phase 1. National Academies Press, Washington, D.C. 2003.

[3] Thomas, C.E. Direct-Hydrogen-Fueled Proton-Exchange Membrane Fuel Cell System for Transportation Applications: Hydrogen Vehicle Safety Report, DOE/CE/5039-502. U.S. Department of Energy, Washington, D.C. 1997.

Appendix C

DOE Hydrogen Program Budget

TABLE C-1 DOE Hydrogen Program Planning Levels, FY02–FY04 ($000)

	FY02		FY03(1)		FY04(2)	
	Direct	Associated	Direct	Associated	Direct	Associated
EERE						
Hydrogen Technology						
Production and Delivery	11,148		11,329		23,000	
Storage	6,125		10,921		30,000	
Infrastructure Validation	5,696		9,748		13,160	
Safety, Codes & Standards, and Utilization	4,486		4,611		16,000	
Education and Cross-cutting Analysis	1,437		1,926		5,822	
Fuel Cell Technology						
Transportation Systems	7,466		6,160		7,600	
Distributed Energy Systems	5,500		7,451		7,500	
Stack Component R&D	12,595		14,803		28,000	
Fuel Processor R&D	20,921		24,539		19,000	
Technology Validation	0		1,788		15,000	
Technical/Program Support	200		398		400	
FreedomCAR and Vehicle Technologies						
Vehicle Systems		5,100		3,656		3,800
Innovative Concepts		500		993		500
Hybrid and Electric Propulsion		42,180		38,700		45,525
Advanced Combustion R&D		15,339		18,625		12,799
Materials Technologies		18,326		16,491		20,840
Fuels Technologies		6,444		5,364		4,300
Technology Introduction		800		894		1,000
Technical/Programmatic Management Support		1,306		865		865
FreedomCAR Peer Review						1,500
Biomass and Biorefinery Systems R&D						
Advanced Biomass Technology R& D		14,486		15,950		14,000
Systems Integration and Production		17,140		14,585		0
TOTAL EERE	75,574	121,621	93,674	116,123	165,482	105,129
FE						
Natural Gas Technology, Hydrogen from Gas	0		0		6,555	
Fuels, Hydrogen from Coal	0		0		5,000	
Central Systems, IGCC		41,990		44,360		51,000
Distributed Generation Systems, Fuel Cells (3)		56,678		63,608		47,000
Sequestration		31,486		39,939		62,000
TOTAL FE	0	130,154	0	147,907	11555	160,000
NE						
Generation IV Nuclear Systems		4,000		3,800		8,200
NERI		1,530		1,291		1,268
I-NERI (International)		750		750		750
Nuclear Hydrogen Initiative	0		2,000		4,000	
TOTAL NE	0	6,280	2,000	5,841	4,000	10,218
SC						
Chemical Sci., Geosci. and Energy Biosci. (4)		2,910		9,370		2,930
Materials Sci. and Engineering (4)		3,063		2,895		3,063
Chemical Sci., Geosci. and Energy Biosci. (5)		1,744		3,023		1,744
Biological and Environmental Research (6)		0		1,722		17,710
TOTAL SC	0	7,717	0	17,010	0	25,447
Dept. of Transportation (7)					674	
GRAND TOTAL	75,574	265,772	95,674	286,881	181,711	300,794

NOTES:
(1) FY 2003 Appropriation
(2) FY 2004 Administration Budget Request
(3) FE fuel cells for FY04 includes SECA, Advanced Research and Vision 21 funding
(4) for research in hydrogen storage in nanotubes, hydrogen combustion, and catalysts for combustion
(5) basic science to improve materials for fuel cells
(6) for research in the biological production of hydrogen
(7) this planning level has not yet been coordinated with DOT
EE = Office of Energy Efficiency and Renewable Energy
FE = Office of Fossil Energy
NE = Office of Nuclear Energy Science and Technology
SC = Office of Science
Direct = Funding that would not be requested if there were no DOE Hydrogen activities.
Associated = Efforts that are necessary for a Hydrogen pathway, i.e., hybrid electric components of FreedomCAR, high-temperature stationary fuel cells, sequestration, etc.
SOURCE: DOE, Office of Energy Efficiency and Renewable Energy.

Appendix D

Presentations and Committee Meetings

1. COMMITTEE MEETING, THE NATIONAL ACADEMIES, WASHINGTON, D.C., DECEMBER 2–3, 2002

DOE Expectations from the Study
David Garman, Office of Energy Efficiency and Renewable Energy (EERE), DOE

DOE Hydrogen Program Overview
Steve Chalk, Office of Hydrogen, Fuel Cells and Infrastructure Technologies, DOE

Hydrogen from Natural Gas
Pete Devlin, Office of Hydrogen, Fuel Cells and Infrastructure Technologies, DOE

Hydrogen from Coal
Lowell Miller, Office of Fossil Energy, DOE

Hydrogen from Biomass
Roxanne Danz, Office of Hydrogen, Fuel Cells and Infrastructure Technologies, DOE

Hydrogen from Other Renewable Resources
Cathy Grégoire-Padró, National Renewable Energy Laboratory (NREL)

Hydrogen from Nuclear
Dave Henderson, Office of Nuclear Energy, Science and Technology, DOE

Hydrogen Storage
JoAnn Milliken, Office of Hydrogen, Fuel Cells and Infrastructure Technologies, DOE

DOE Wrap-up
Steve Chalk, Office of Hydrogen, Fuel Cells and Infrastructure Technologies, DOE

2. COMMITTEE MEETING, THE NATIONAL ACADEMIES, WASHINGTON, D.C., JANUARY 22–24, 2003

Electrolytic Hydrogen Technology and Economics
Chip Schroeder, Proton Energy Systems

Production of Hydrogen Using Nuclear Energy
Charles Forsberg, Oak Ridge National Laboratory (ORNL)

Carbon Management in a Greenhouse-Constrained World
Robert Socolow, Princeton University

Introduction to Gasification
Neal Richter, Chevron-Texaco

Hydrogen from Biomass: Prospective Resources, Technologies, and Economics
Margaret Mann and Ralph Overend, NREL

Hydrogen Production and Infrastructure Economics: The Biggest Challenge for Fuel Cell Vehicle Commercialization
Dale Simbeck and Elaine Chang, SFA Pacific

Economics of Hydrogen Production and Use
Joan Ogden, Princeton University

3. COMMITTEE MEETING, THE NATIONAL ACADEMIES, WASHINGTON, D.C., APRIL 23–25, 2003

Sequestration
Lynn Orr, Jr., Stanford University

Sequestration
Gardiner Hill, BP

SOFCs, Direct Firing, Wind
Jon Ebacher, GE Power Systems

On-Board Hydrogen Storage
Scott Jorgensen, General Motors

Fuel Cell Commercialization
John Cassidy, UTC, Inc.

Renewable Energy
David Pimentel, Cornell University

Hydrogen from Coal
David Gray and Glen Tomlinson, Mitretek

Solid Oxide Fuel Cells
Joseph Strakey, DOE/National Energy Technology Laboratory

Hydrogen Production from Solar Energy
Nathan Lewis, California Institute of Technology

Methane Conversion
Alex Bell, University of California, Berkeley

Methane Conversion
Jens Rostrup-Nielsen, Haldor Topsoe

4. COMMITTEE MEETING, THE NATIONAL ACADEMIES, WASHINGTON, D.C., JUNE 10–12, 2003

Hydrogen, Fuel Cell Vehicles and the Transportation Sector
David Friedman, Union of Concerned Scientists

Hydro-Chem
Dennis Norton, Hydro-Chem

Small Hydrogen Plants for the Hydrogen Economy
Marvin A. Crews, Howe Baker

DOE's Hydrogen RD&D Plan
Steve Chalk, Pete Devlin, David Henderson, Mark Paster, JoAnn Milliken, Pat Davis, Sig Gronich, Christy Cooper, and Neil Rossmeissl, DOE

Hydrogen: Opportunities and Challenges
Dan Reicher, Northern Power Systems and New Energy Capital

DOE's Hydrogen Feedstock Strategy
Mark Pastor, Office of Hydrogen, Fuel Cells and Infrastructure Technologies, DOE

Fuel Cell Vehicles and the Hydrogen Economy
Larry Burns, General Motors

Issues Confronting Future Hydrogen Production and Use for Transportation
Bill Innes, ExxonMobil Research and Engineering

5. COMMITTEE MEETING, THE NATIONAL ACADEMIES, WASHINGTON, D.C., AUGUST 13–15, 2003

Closed Session

Appendix E

Spreadsheet Data from Hydrogen Supply Chain Cost Analyses

INTRODUCTION

Following are the hydrogen production spreadsheets that are the basis for Chapter 5 in this report. As noted there, these charts are for different combinations of feedstock, status of technology (current versus possible future), and whether or not sequestration of carbon dioxide is required at facilities processing hydrocarbon feedstock. A modified version of Table 5-2, with additional pathways, is included here as Table E-1 for convenience in following the charts. This table lists the code for each pathway as used for identification in the charts.

The first spreadsheets are for the central station pathways, starting with a summary of the results in Table E-2. For each pathway, the hydrogen cost to the end user is listed as the sum of the production, distribution, and dispensing costs. The capital investment requirements and carbon dioxide emissions are also listed. Table E-3 is a summary of the design basis inputs for central plants. The central plants considered here can produce 1,200,000 kilograms of hydrogen per day (90 percent of the year), which will fuel more than 2 million fuel cell vehicles via four main transmission lines of 150 kilometers and 438 dispensing stations. The details for each pathway are then shown in Tables E-4 through E-12: first for natural gas, then coal, and nuclear. Finally, the associated transmission and dispensing analyses are shown in Tables E-13 through E-16.

To illustrate the detailed production spreadsheets, take CS NG-C (Table E-4) as an example. The size of the plant (1,102,041 kg/day of hydrogen) and annual load factor (98 percent) are entered at the top. In the top right corner, the spreadsheet calculates the vehicles that this plant could fuel for the assumed fleet (2,135,250 vehicles equivalent to 65 miles-per-gallon hybrids). Then characteristics of the natural gas and electricity inputs are entered (the latter to calculate overall carbon dioxide emissions). Capital costs are calculated on a unit basis and then multiplied by the capacity assumed, for a total of $453 million. Finally, the hydrogen costs are calculated for the assumed fuel, other variable, and capital costs. The final figure is $1.03/kg at the plant, not including any charge for carbon emissions. Delivering the hydrogen to the filling station would add $0.42 (Table E-13), and the costs of the filling station itself would add another $0.54 (Table E-15) for a total of $1.99. The assumed charge for carbon emissions at the production plant and indirectly from electric power generation would add $0.13, for a grand total of $2.11 ("total H_2 costs" for "CS NG-C" in Table E-2).

Midsize plants are next, starting with the summary and design basis inputs in Tables E-17 and E-18. Natural gas and electrolysis plants are included here, unlike in Chapter 5. The analyses are shown in Tables E-19 to E-28. Also included is the distribution cost via tanker truck and dispensing costs, in Tables E-29 to E-32.

Distributed plants start with a summary and design basis inputs (Tables E-33 and E-34), then natural gas (Tables E-35 and E-36) and grid-based electrolysis (Tables E-37 to E-39) Included under the latter is a combination of natural-gas-assisted steam electrolysis case, for future technology only. Wind and photovoltaics are shown for both stand-alone units (Tables E-40 to E-43) and in combination with the power grid (Tables E-44 to E-47). Tables E-48 and E-49 show the detailed buildup of the cost of electricity from photovoltaics, which is an input to the electrolysis calculations.

TABLE E-1
Hydrogen Supply Chain Pathways Examined

Scale	Primary Energy Source						
	Natural Gas	Coal	Nuclear	Biomass	Grid-Based Electricity (from any source)	Wind	Photovoltaics
Central station	*CN:* CS NG-C *FN:* CS NG-F *CY:* CS NG-C-Seq *FY:* CS NG-F-Seq	*CN:* CS Coal-C *FN:* CS Coal-F *CY:* CS Coal-C-Seq *FY:* CS Coal-F-Seq	*F:* CS Nu-F				
Midsize				*CN:* MS Bio-C *FN:* MS Bio-F *CY:* MS Bio-C-Seq *FY:* MS Bio-F-Seq			
Distributed	*C:* Dist NG-C *F:* Dist NG-F				*C:* Dist Elec-C *F:* Dist Elec-F	*C:* Dist WT-Gr Elec-C *F:* Dist WT-Gr Elec-F	*C:* Dist PV-Gr Elec-C *F:* Dist PV-Gr-Elec-F

NOTES: *C* = current technology; *Y* = sequestration (hydrocarbon feedstock in central station and midsize plants only); *F* = future technology; *N* = no sequestration. The abbreviations for the hydrogen supply chain pathways (e.g., CS NG-C) are from the detailed spreadsheets that follow.

TABLE E-2
Central Plant Summary of Results

Pathway Table E-	CS NG-C 4	CS NG-F 5	CS NG-C-Seq 6	CS NG-F- Seq 7	CS Coal-C 8	CS Coal-F 9	CS Coal-C-Seq 10	CS Coal-F-Seq 11	CS Nu-F 12
Capital investment, milion $									
H_2 production	453.39	326.85	623.75	425.54	1,151.92	868.18	1,177.33	889.97	2,468.19
Distribution	724.75	532.98	724.75	532.98	724.75	532.98	724.75	532.98	532.98
Dispensing	714.51	507.42	714.51	507.42	714.51	507.42	714.51	507.42	507.42
Total Capital Investment	1,892.66	1,367.25	2,063.01	1,465.95	2,591.18	1,908.58	2,616.59	1,930.38	3,508.59
Production costs, $/kg H_2									
Variable costs									
Feed	0.75	0.71	0.79	0.73	0.21	0.19	0.21	0.19	0.20
Electricity	0.03	0.03	0.08	0.06	0.11	0.04	0.17	0.08	
Decommission fund									0.06
Non-fuel O&M, 1%/yr of capital	0.01	0.01	0.02	0.01	0.03	0.02	0.03	0.02	0.06
Total variable costs	0.79	0.74	0.89	0.80	0.35	0.25	0.41	0.30	0.32
Fixed costs, 5%/yr of capital	0.06	0.04	0.08	0.05	0.15	0.11	0.15	0.11	0.31
Capital charges, 18%/yr of capital	0.18	0.13	0.25	0.17	0.46	0.35	0.47	0.36	1.00
Total Production Costs	1.03	0.92	1.22	1.02	0.96	0.71	1.03	0.77	1.63
Distribution costs, $/kg H_2									
Variable costs									
Labor	0	0	0	0	0	0	0	0	0
Fuel	0	0	0	0	0	0	0	0	0
Electricity	0.01	0.01	0.01	0.01	0.01	0.01	0.01	0.01	0.01
Non-fuel O&M, 1%/yr of capital	0.02	0.01	0.02	0.01	0.02	0.01	0.02	0.01	0.01
Total variable costs	0.03	0.02	0.03	0.02	0.03	0.02	0.03	0.02	0.02
Fixed costs, 5%/yr of capital	0.09	0.07	0.09	0.07	0.09	0.07	0.09	0.07	0.07
Capital charges, 18%/yr of capital	0.29	0.21	0.29	0.21	0.29	0.21	0.29	0.21	0.21
Total Distribution Costs	0.42	0.31	0.42	0.31	0.42	0.31	0.42	0.31	0.31

continued

TABLE E-2 *continued*

Pathway Table E-	CS NG-C 4	CS NG-F 5	CS NG-C-Seq 6	CS NG-F- Seq 7	CS Coal-C 8	CS Coal-F 9	CS Coal-C-Seq 10	CS Coal-F-Seq 11	CS Nu-F 12
Dispensing costs, $/kg H_2									
Variable costs									
Electricity	0.14	0.11	0.14	0.11	0.14	0.11	0.14	0.11	0.11
Non-fuel O&M, 1%/yr of capital	0.09	0.06	0.09	0.06	0.09	0.06	0.09	0.06	0.06
Total variable costs	0.23	0.17	0.23	0.17	0.23	0.17	0.23	0.17	0.17
Fixed costs, 5%/yr of capital	0.05	0.04	0.05	0.04	0.05	0.04	0.05	0.04	0.04
Capital charges, 18%/yr of capital	0.25	0.18	0.25	0.18	0.25	0.18	0.25	0.18	0.18
Total Dispensing Costs	0.54	0.39	0.54	0.39	0.54	0.39	0.54	0.39	0.39
H_2 costs, $/kg									
Production	1.03	0.92	1.22	1.02	0.96	0.71	1.03	0.77	1.63
Distribution	0.42	0.31	0.42	0.31	0.42	0.31	0.42	0.31	0.31
Dispensing	0.54	0.39	0.54	0.39	0.54	0.39	0.54	0.39	0.39
CO_2 disposal	0	0	0.09	0.08	0	0	0.16	0.15	
Carbon tax	0.13	0.12	0.02	0.02	0.26	0.23	0.04	0.03	
Total H_2 Costs	2.11	1.73	2.28	1.82	2.17	1.63	2.19	1.64	2.33
Carbon dioxide vented to atmosphere									
kg carbon/kg H_2	2.51	2.39	0.42	0.35	5.12	4.56	0.82	0.60	
Direct use	2.45	2.34	0.26	0.24	4.90	4.50	0.49	0.45	
Indirect use	0.06	0.05	0.16	0.12	0.21	0.07	0.33	0.15	
kg CO_2/kg H_2	9.22	8.75	1.53	1.30	18.76	16.73	3.00	2.21	
Direct use	8.99	8.56	0.95	0.88	17.98	16.49	1.80	1.65	
Indirect use	0.23	0.18	0.58	0.42	0.77	0.25	1.20	0.56	
Carbon charge ($/kg)	0.13	0.12	0.02	0.02	0.26	0.23	0.04	0.03	
Carbon dioxide sequestered									
Carbon dioxide (kg CO_2/kg H_2)			8.56	7.91			16.19	14.84	
Carbon			2.34	2.16			4.41	4.05	

TABLE E-3
Central Hydrogen Plant Summary of Inputs

Inputs | Boxed | are the key input variables you must choose, **current inputs are just an example**

design basis

Central Hydrogen Plant Production Inputs

Key Variables Inputs		Value	Unit	Notes
				1 kg H2 is the same energy content as 1 gallon of gasoline
Design hydrogen production		1,200,000	**kg/d H2 design**	497,400,000 **scf/d H2** CS range 500,000 to 2,500,000 kg/d
Annual average load factor		90%	/yr of design 32,850,000	kg/month actual or **394,200,000** kg/yr actual H2 production
Use in Vehicles				
FC Vehicle gasoline equiv mileage		65	mpg (U.S. gallons) or	28 km/liter
FC Vehicle miles per year		12,000	mile/yr thereby requires	185 kg/yr H2 for each FCV or 2,135,250 FCV
Typical gasoline sales/month per station		150,000	gallons/month per station	100,000 - 250,000 gallons/month is typical or 4,932 gal/d
Hydrogen as % of gasoline at each station		50.0%	of gasoline/station or	75,000 kg H2/month per stations or 2,466 kg/d/station
Capital Cost Buildup Inputs from process unit costs				**All major utilities included as process units**
General Facilities		20%	of process units	20-40% typical for SMR + 10% more for gasification
Engineering, Permitting & Startup		10%	of process units	10-20% typical
Contingencies		10%	of process units	10-20% typical, should go down as many units are built
Working Capital, Land & Misc.		7%	of process units	5-10% typical
Site specific factor		110%	above US Gulf Coast	90-130% typical; sales tax, labor rates & weather issues
Product Cost Buildup Inputs				
Non-fuel Variable O&M		1.0%	/yr of capital	0.5-1.5% is typical
Fuels	Natural Gas	$ 4.50	/MM Btu HHV	$2.50-4.50/MM Btu typical **industrial** rate, see www.eia.doe.gov
	Electricity	$ 0.045	/kWh	$0.04-0.05/kWh typical **industrial** rate, see www.eia.doe.gov
	Electricity generation eff	50%		50% 65% future: incremental efficiency of new plants
	Electricity CO2 emissions	0.32	kg/hr CO2/kWe	Based on NGCC, coal and current grid at 0.75 kg/hr CO2/kWe
	Biomass production costs	$ 500	/ha/yr gross revenues	$400-600/ha/yr typical in U.S. .lower in developing nations or wastes
	Biomass yield	10	tonne/ha/yr bone dry	8-12 ton/hr/yr typical if farmed, 3-5 ton/hr/yr if forestation or wastes
	Coal	$ 1.22	/million Btu dry HHV	$0.75-1.25/million Btu coal utility delivered see www.eia.doe.gov
Carbon tax		$ 50.00	$/tonne C	
Carbon Price for Carbon Vented		$50	$/tonne C	
CO2 disposal cost		$ 10.00	/tonne CO2	From plant gate at high pressure to injection
Fixed O&M Costs		5.0%	/yr of capital	4-7% typical for refiners: labor, overhead, insurance, taxes, G&A
Capital Charges		15.9%	/yr of capital	20-25%/yr CC typical for refiners & 14-20%/yr CC typical for utilities
				20%/yr CC is about 12% IRR DCF on 100% equity where as
				15%/yr CC is about 12% IRR DCF on 50% equity & debt at 7%
				Above annual CC rates assume quick ramp-up to design capacity

Hydrogen Distribution to Forecourt Inputs

Key Variables Inputs	Value	Unit	Notes	
Maximum pipeline distance to forecourt	150	km, key assumption for pipeline	70,650 sq km area	
Number of pipeline arms	4	directions	30 FCV per sq km	
Days of H2 storage for high availability	5	days at design H2 rate		
Pipeline	600,000	$/km equivalent to	$ 965,562 per mile	
Capital Cost Buildup Inputs from process unit costs				
General Facilities	20%		20-40% typical	assume low for pipeline
Engineering, Permitting & Startup	10%		10-20% typical	assume low for pipeline
Contingencies	7%	of process units	10-20% typical, should be low after the first few	
Working Capital, Land & Misc.	7%	of process units	5-10% typical	
Site specific factor	110%	above US Gulf Coast	90-130% typical; sales tax, labor rates & weather issues	
Operating Cost Buildup Inputs				
Non-fuel Variable O&M	1.0%	/yr of capital	0.5-1.5% typical but could be lower for pipeline	
Electricity	$ 0.045	/kWh	$0.04-0.05/kWh typical **industrial** rate, see www.eia.doe.gov	
Fixed O&M Costs	5.0%	/yr of capital	4-7% typical for refiners: labor, overhead, insurance, taxes, G&A	
Capital Charges	15.9%	/yr of capital	20-25%/yr CC typical for refiners for 12% DCF IRR	

Hydrogen Fueling Station Inputs

Key Variables Inputs	Value	Unit	Notes
			7.2 km radius per station
Hydrogen sales/month per station from above	75,000	kg/month	thereby supplying 438 stations
Average hydrogen sales/d per station from above	2,466	kg/d actual	2,740 kg/d design 161 sq km per station
Forecourt loading factor	90%	/yr of design	"plug & play" 24 hr replacements for reasonable availability
High pressure gas storage buffer	3	hours at peak surge rate due to higher demand before & after work typical fill-up times	
Capital Cost Buildup Inputs from process unit costs			assuming peak is 3 times the daily average H2 flow
General Facilities	25%		
Engineering, Permitting & Startup	10%	of process units	10-20% typical Engineering costs spread over multiple stations
Contingencies	10%	of process units	10-20% typical, should be low after the first few
Working Capital, Land & Misc.	7%	of process units	5-10% typical
Site specific factor	100%	above US Gulf Coast	90-130% typical; sales tax, labor rates & weather issues
Operating Cost Buildup Inputs			
Road tax or (subsidy)	$ -	/gal gaso equiv.	may need subsidy like EtOH to get it going
Gas Station mark-up	$ -	/gal gaso equiv.	may be needed if H2 sales drops total station revenues
Electricity	$ 0.070	/kWh	$0.06-.0.09/kWh typical **commercial** rate, see www.eia.doe.gov
Non-fuel Variable O&M	5.0%	/yr of capital	0.5-1.5% is typical, assumed low here for "plug & play"
Fixed O&M Costs	3.0%	/yr of capital	4-7% typical for refiners: labor, overhead, insurance, taxes, G&A
Capital Charges	14.0%	/yr of capital	20-25%/yr CC typical for refiners & 14-20%/yr CC typical for utilities

TABLE E-4
CS NG-C
CS Size Hydrogen Steam Reforming of Natural Gas with Current Technology

Color codes variables via summary inputs key outputs

CS-Size Plant Design	Hydrogen	Design LHV energy equivalent			
Size range	kg/d H2	gasoline gal/d	million Btu/hr	scf/d H2	MW t
Maximum	2,500,000	2,500,000	11,855	1,036,250,000	3,474
This run	1,102,041	1,102,041	5,226	456,795,918	1,531
Minimum	500,000	500,000	2,371	207,250,000	695

	gasoline equivalent	
Assuming	65	mpg and 12,000 mile/yr
requires	185	kg/yr H2/vehicle or gal/yr gaso equiv
Assuming	98%	annual load factor at
actual H2	394,200,000	kg/y H2 /station or gal/y gaso equiv
or	32,850,000	kg/month H2 or gal/mo. gaso equiv
thereby	2,135,250	vehicles can be serviced at
	305,036	fill-ups/d @ 4.2 kg or gal equiv/fill-up
		or each vehicle fills up one a week

Electric Power

Compress	22,959
SMR & misc.	9,517
Total	**32,476 kW**

H2 Compress 0.5 kW/kg/h — 45,918 kg/hr H2, 75 atm; 2.5 compression ratio

19,033,163 scf/hr H2, 30 atm

Natural Gas → SMR 76.2% LHV effic

6,858 MM Btu/h LHV
7,613 MM Btu/h HHV
7,612,798 scf/hr @ 1,000 Btu/scf
150,137 kg/hr @23,000 Btu/lb

360 Btu LHV/scf H2

at 0.32 kg CO2/kWh from input

CO2 emission to atmosphere

Source	kg CO2/kg H2	kg C/kg H2
From NG	8.99	2.45
From Ele Gen	0.23	0.06
Total	**9.22**	**2.51**

74% NG + fuel for power to H2 efficiency

Hydrogen in Gas Pipeline @ 75 atm
1,102,041 design kg/d H2 or gal/d gasoline equivalent
1,080,000 actual kg/d annual ave.

412,876 kg/hr CO2, however in dilute N2 rich SMR flue gas
10,392 kg/hr CO2 at power plants

Capital Costs	Unit cost basis at 500,000 kg/d H2		cost/size factors	Unit cost at 1,102,041 kg/d H2		millions of $ for 1 plant	Notes
SMR	$ 0.65	/scf/d	75%	$ 0.53	/scf/d	244	$ 221 /kg/d H2
H2 Compressor	$ 1,800	/kW	85%	$ 1,599	/kW	37	$ 33 /kg/d H2
				Total process units		280	
General Facilities	20%	of process units				56	20-40% typical
Engineering Permitting & Startup	10%	of process units				28	10-20% typical
Contingencies	10%	of process units				28	10-20% typical, low after the first few
Working Capital, Land & Misc.	7%	of process units				20	5-10% typical
				U.S. Gulf Coast Capital Costs		412	
Site specific factor	110%	of US Gulf Coast costs		**Total Capital Costs**		**453**	
Unit Capital Costs	**0.99**	**/scf/d H2 or**	411	/kg/d H2 or		411	/gal/d gaso equiv

Hydrogen Costs at	98%	ann load factor	million $/yr of 1 plant	$/million Btu LHV	$/1,000 scf H2	$/kg H2 or $/gal gaso equiv	Notes
Variable Non-fuel O&M	1.0%	/yr of capital	5	0.10	0.03	0.01	0.5-1.5% typical
Natural Gas	$ 4.50	/MM Btu HHV	294	6.56	1.80	0.75	$2.50-4.50/MM Btu **industrial** rate
Electricity	$ 0.045	/kWh	13	0.28	0.08	0.03	$0.04-0.05/kWh **industrial** rate
Variable Operating Cost			**311**	**6.94**	**1.90**	**0.79**	
Fixed Operating Cost	5.0%	/yr of capital	**23**	**0.51**	**0.14**	**0.06**	4-7% typical for refining
Capital Charges	16%	/yr of capital	**72**	**1.61**	**0.44**	**0.18**	20-25% typical for refining
Total Gaseous Hydrogen Costs from Natural Gas			**406**	**9.05**	**2.48**	**1.03**	including return on investment
Carbon Tax	$ 50.00	/ton	50	1.10	0.30	0.13	include indirect C from power

notes: Costs are for production at the plant only and do not include distribution costs. Assume no plant storage but compression of hydrogen to 75 atm for the pipeline.

TABLE E-5
CS NG-F
CS Size Hydrogen via Steam Reforming of Natural Gas with Future Optimism

Color codes variables via summary inputs key outputs

CS-Size Plant Design		Design LHV energy equivalent			
Size range	Hydrogen kg/d H2	gasoline gal/d	million Btu/hr	scf/d H2	MW t
Maximum	2,500,000	2,500,000	11,855	1,036,250,000	3,474
This run	1,102,041	1,102,041	5,226	456,795,918	1,531
Minimum	500,000	500,000	2,371	207,250,000	695

gasoline equivalent
Assuming 65 mpg and 12,000 mile/yr
requires 185 kg/yr H2/vehicle or gal/yr gaso equiv
Assuming 98% annual load factor at
actual H2 394,200,000 kg/y H2 /station or gal/y gaso equiv
or 32,850,000 kg/month H2 or gal/mo. gaso equiv
thereby 2,135,250 vehicles can be serviced at
305,036 fill-ups/d @ 4.2 kg or gal equiv/fill-up
or each vehicle fills up one a week

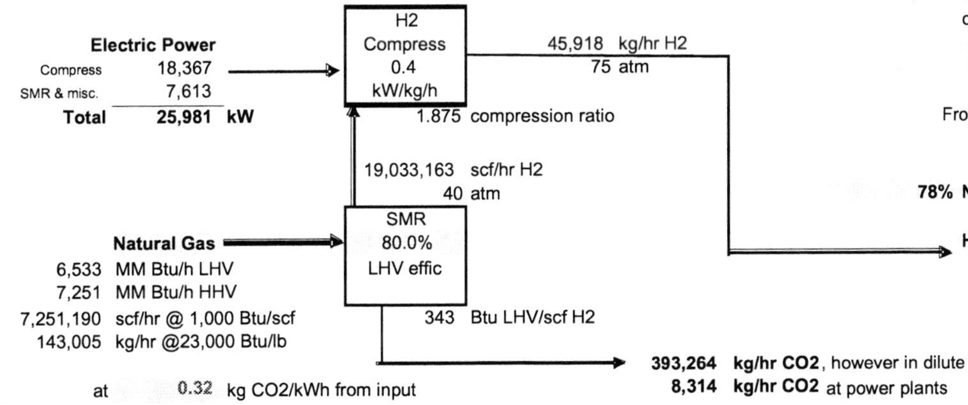

CO2 emission to atmosphere

Source	kg CO2/kg H2	kg C/kg H2
From NG	8.56	2.34
From Ele Gen	0.18	0.05
Total	**8.75**	**2.39**

78% NG + fuel for power to H2 efficiency

Electric Power

Compress	18,367
SMR & misc.	7,613
Total	**25,981 kW**

H2 Compress 0.4 kW/kg/h
45,918 kg/hr H2
75 atm
1.875 compression ratio

19,033,163 scf/hr H2
40 atm

SMR 80.0% LHV effic

Natural Gas
6,533 MM Btu/h LHV
7,251 MM Btu/h HHV
7,251,190 scf/hr @ 1,000 Btu/scf
143,005 kg/hr @23,000 Btu/lb

343 Btu LHV/scf H2

Hydrogen in Gas Pipeline @ 75 atm
1,102,041 design kg/d H2 or gal/d gasoline equivalent
1,080,000 actual kg/d annual ave.

393,264 kg/hr CO2, however in dilute N2 rich SMR flue gas
8,314 kg/hr CO2 at power plants

at 0.32 kg CO2/kWh from input

Capital Costs	Unit cost basis at 500,000 kg/d H2		cost/size factors	Unit cost at 1,102,041 kg/d H2		millions of $ for 1 plant	Notes
Adv SMR or ITM	$ 0.50	/scf/d	75%	$ 0.41	/scf/d	187	$ 170 /kg/d H2
H2 Compressor	$ 900	/kW	85%	$ 799	/kW	15	$ 13 /kg/d H2
				Total process units		202	
General Facilities	20%	of process units				40	20-40% typical
Engineering Permitting & Startup	10%	of process units				20	10-20% typical
Contingencies	10%	of process units				20	10-20% typical, low after the first few
Working Capital, Land & Misc.	7%	of process units				14	5-10% typical
				U.S. Gulf Coast Capital Costs		297	
Site specific factor	110%	of US Gulf Coast costs		**Total Capital Costs**		**327**	
Unit Capital Costs	**0.72**	**/scf/d H2 or**	297	/kg/d H2 or	297	/gal/d gaso equiv	
	72%	of equivalent high estimate					

Hydrogen Costs at	98%	ann load factor	million $/yr of 1 plant	$/million Btu LHV	$/1,000 scf H2	$/kg H2 or $/gal gaso equiv	Notes
Variable Non-fuel O&M	1.0%	/yr of capital	3	0.07	0.02	0.01	0.5-1.5% typical
Natural Gas	$ 4.50	/MM Btu HHV	280	6.24	1.71	0.71	$2.50-4.50/MM Btu **industrial** rate
Electricity	$ 0.045	/kWh	10	0.22	0.06	0.03	$0.04-0.05/kWh **industrial** rate
Variable Operating Cost			**293**	**6.54**	**1.80**	**0.74**	
Fixed Operating Cost	5.0%	/yr of capital	**16**	**0.36**	**0.10**	**0.04**	4-7% typical for refining
Capital Charges	16%	/yr of capital	**52**	**1.16**	**0.32**	**0.13**	20-25% typical for refining
Total Gaseous Hydrogen Costs from Natural Gas			**362**	**8.06**	**2.21**	**0.92**	89% of equivalent high
Carbon Tax	$ 50.00	/ton	47	1.05	0.29	0.12	include indirect C from power

notes: Costs are for production at the plant only and do not include distribution costs. Assume no plant storage but compression of hydrogen to 75 atm for the pipeline.

TABLE E-6
CS NG-C Seq
CS Size Hydrogen via Steam Reforming of Natural Gas plus CO2 Capture with Current Technology

Color codes variables via summary inputs key outputs

CS-Size Plant Design		Design LHV energy equivalent			
	Hydrogen	gasoline	million		
Size range	kg/d H2	gal/d	Btu/hr	scf/d H2	MW t
Maximum	2,500,000	2,500,000	11,855	1,036,250,000	3,474
This run	1,200,000	**1,200,000**	**5,691**	**497,400,000**	**1,667**
Minimum	500,000	500,000	2,371	207,250,000	695

		gasoline equivalent
Assuming	65	mpg and 12,000 mile/yr
requires	185	kg/yr H2/vehicle or gal/yr gaso equiv
Assuming	90%	**annual load factor at**
actual H2	394,200,000	kg/y H2 /station or gal/y gaso equiv
or	32,850,000	kg/month H2 or gal/mo. gaso equiv
thereby	**2,135,250**	**vehicles can be serviced at**
	305,036	fill-ups/d @ 4.2 kg or gal equiv/fill-up
		or each vehicle fills up one a week

Electric Power

Compress	25,000
SMR & misc.	12,435
CO_2 Compres	53,099
Total	**90,534 kW**

H2 Compress 0.5 kW/kg/h

50,000 kg/hr H2
75 atm

2.5 compression ratio

67% NG + fuel for power to H2 efficiency

Hydrogen in Gas Pipeline @ 75 atm
1,200,000 design kg/d H2 or gal/d gasoline equivalent
1,080,000 actual kg/d annual ave.

20,725,000 scf/hr H2
30 atm

Natural Gas
7,904 MM Btu/h LHV
8,773 MM Btu/h HHV
8,773,045 scf/hr @ 1,000 Btu/scf
173,019 kg/hr @23,000 Btu/lb
90% Percent CO2 Separated
at 0.32 kg CO2/kWh from input

SMR 72.0% LHV effic

1 Atm

381 Btu LHV/scf H2

CO2 Compressor 53,099 kWe

135 atm
428,221 kg/hr CO2 to gelogic disposal
28,971 kg/hr CO2 at power plants

	CO2 emission to atmosphere		CO2 Sequestered	
Source	kg CO2/kg H2	kg C/kg H2	CO2/kg H2	kg C/kg H2
From NG is seq	0.95	0.26	8.56	2.34
From Ele gen	0.58	0.16		
Total	**1.53**	**0.42**		

47,580.11 kg/hr CO2 Vented to Atmosphere

Capital Costs		Unit cost basis at 500,000 kg/d H2		cost/size factors		Unit cost at 1,200,000 kg/d H2		millions of $ for 1 plant	Notes	
SMR with MEA scrubber	$	0.75	/scf/d	75%	$	0.60	/scf/d	300	$ 250	/kg/d H2
H2 Compressor	$	1,800	/kW	85%	$	1,578	/kW	39	$ 33	/kg/d H2
CO2 Compressor	$	1,000	/kW	85%	$	877	/kW	47	$ 39	/kg/d H2
						Total process units		386		
General Facilities		20%	of process units					77	20-40% typical	
Engineering Permitting & Startup		10%	of process units					39	10-20% typical	
Contingencies		10%	of process units					39	10-20% typical, low after the first few	
Working Capital, Land & Misc.		7%	of process units					27	5-10% typical	
						U.S. Gulf Coast Capital Costs		567		
Site specific factor		110%	of US Gulf Coast costs			**Total Capital Costs**		**624**		
Unit Capital Costs		**1.25**	**/scf/d H2 or**	520	/kg/d H2 or	520	/gal/d gaso equiv			
		126%	of equivalent without CO2 capture							

Hydrogen Costs at		90%	ann load factor	million $/yr of 1 plant	$/million Btu LHV	$/1,000 scf H2	$/kg H2 or $/gal gaso equiv	Notes
Variable Non-fuel O&M		1.0%	/yr of capital	6	0.14	0.04	0.02	0.5-1.5% typical
Natural Gas	$	4.50	/MM Btu HHV	311	6.94	1.90	0.79	$2.50-4.50/MM Btu **industrial** rate
Electricity	$	0.045	/kWh	32	0.72	0.20	0.08	$0.04-0.05/kWh **industrial** rate
Variable Operating Cost				**350**	**7.79**	**2.14**	**0.89**	
Fixed Operating Cost		5.0%	/yr of capital	**31**	**0.70**	**0.19**	**0.08**	4-7% typical for refining
Capital Charges		16%	/yr of capital	**99**	**2.21**	**0.61**	**0.25**	20-25% typical for refining
Total Gaseous Hydrogen Costs from Natural Gas				**480**	**10.70**	**2.94**	**1.22**	including return on investment
Carbon Tax	$	50.00	/ton	8	0.18	0.05	0.02	include indirect C from power
CO2 Disposal	$	10.00	/ton	34	0.75	0.21	0.09	From plant gate at high pressure

notes: Costs are for production at the plant only and do not include distribution costs. Assume no plant storage but compression of hydrogen to 75 atm for the pipeline.

TABLE E-7
CS NG-F Seq
CS Size Hydrogen via Steam Reforming of Natural Gas Plus CO2 Capture with Future Optimism

Color codes variables via summary inputs key outputs

CS-Size Plant Design — **Design LHV energy equivalent**

Size range	Hydrogen kg/d H2	gasoline gal/d	million Btu/hr	scf/d H2	MW t
Maximum	1,000,000	1,000,000	4,742	414,500,000	1,389
This run	1,200,000	**1,200,000**	**5,691**	**497,400,000**	**1,667**
Minimum	20,000	20,000	95	8,290,000	28

		gasoline equivalent
Assuming	65	mpg and 12,000 mile/yr
requires	185	kg/yr H2/vehicle or gal/yr gaso equiv
Assuming	90%	**annual load factor at**
actual H2	394,200,000	kg/y H2 /station or gal/y gaso equiv
or	32,850,000	kg/month H2 or gal/mo. gaso equiv
thereby	**2,135,250**	**vehicles can be serviced at**
	305,036	fill-ups/d @ 4.2 kg or gal equiv/fill-up
		or each vehicle fills up one a week

Electric Power

Compress	20,000
SMR & misc.	10,363
CO_2 Compres	35,575
Total	**65,938 kW**

H2 Compress 0.4 kW/kg/h

50,000 kg/hr H2

75 atm

1.875 compression ratio

73% NG + fuel for power to H2 efficiency

Hydrogen in Gas Pipeline @ 75 atm

1,200,000 design kg/d H2 or gal/d gasoline equivalent

1,080,000 actual kg/d annual ave.

20,725,000 scf/hr H2

40 atm

Natural Gas → ATR/SMR 78.0% LHV effic → 3 Atm → CO2 Compressor 35,575 kWe

7,296 MM Btu/h LHV

8,098 MM Btu/h HHV

8,098,195 scf/hr @ 1,000 Btu/scf

159,709 kg/hr @23,000 Btu/lb

90% Percent CO2 Separated

at 0.32 kg CO2/kWh from input

352 Btu LHV/scf H2

135 atm

395,281 kg/hr CO2 to gelogic disposal

21,100 kg/hr CO2 at power plants

	CO2 emission to atmosphere kg CO2/kg H2	kg C/kg H2	CO2 Sequestered kg CO2/kg H2	kg C/kg H2
From NG is seq	0.88	0.24	7.91	2.16
From Ele gen	0.42	0.12		
Total	**1.30**	**0.35**		
	43,920.10			

Capital Costs	Unit cost basis at 500,000 kg/d H2	cost/size factors	Unit cost at 1,200,000 kg/d H2	millions of $ for 1 plant	Notes
Adv heat exch or ITM	$ 0.58 /scf/d	75%	$ 0.47 /scf/d	232	$ 193 /kg/d H2
H2 Compressor	$ 900 /kW	85%	$ 789 /kW	16	$ 13 /kg/d H2
CO2 Compressor	$ 500 /kW	85%	$ 438 /kW	16	$ 13 /kg/d H2
			Total process units	263	
General Facilities	20% of process units			53	20-40% typical
Engineering Permitting & Startup	10% of process units			26	10-20% typical
Contingencies	10% of process units			26	10-20% typical, low after the first few
Working Capital, Land & Misc.	7% of process units			18	5-10% typical
			U.S. Gulf Coast Capital Costs	387	
Site specific factor	110% of US Gulf Coast costs		**Total Capital Costs**	**426**	
Unit Capital Costs	**0.86 /scf/d H2 or**	355 /kg/d H2 or	355 /gal/d gaso equiv		
	68% of equivalent high				

Hydrogen Costs at	90% ann load factor	million $/yr of 1 plant	$/million Btu LHV	$/1,000 scf H2	$/kg H2 or $/gal gaso equiv	Notes
Variable Non-fuel O&M	1.0% /yr of capital	4	0.09	0.03	0.01	0.5-1.5% typical
Natural Gas	$ 4.50 /MM Btu HHV	287	6.40	1.76	0.73	$2.50-4.50/MM Btu **industrial** rate
Electricity	$ 0.045 /kWh	23	0.52	0.14	0.06	$0.04-0.05/kWh **industrial** rate
Variable Operating Cost		**315**	**7.02**	**1.93**	**0.80**	
Fixed Operating Cost	5.0% /yr of capital	**21**	**0.47**	**0.13**	**0.05**	4-7% typical for refining
Capital Charges	16% /yr of capital	**68**	**1.51**	**0.41**	**0.17**	20-25% typical for refining
Total Gaseous Hydrogen Costs from Natural Gas		**404**	**9.00**	**2.47**	**1.02**	84% of equivalent high
Carbon Tax	$ 50.00 /ton	7	0.16	0.04	0.02	include indirect C from power
CO2 Disposal	$ 10.00 /ton	31	0.69	0.19	0.08	From plant gate at high pressure

notes: Costs are for production at the plant only and do not include distribution costs. Assume no plant storage but compression of hydrogen to 75 atm for the pipeline.

TABLE E-8
CS Coal-C
CS Size Hydrogen via Coal Gasification with Current Technology

Color codes variables via summary inputs key outputs

CS-Size Plant Design	Hydrogen kg/d H2	gasoline gal/d	Design LHV energy equivalent million Btu/hr	scf/d H2	MW t
Size range					
Maximum	2,500,000	2,500,000	11,855	1,036,250,000	3,474
This run	1,200,000	**1,200,000**	**5,691**	**497,400,000**	**1,667**
Minimum	500,000	500,000	2,371	207,250,000	695

gasoline equivalent
Assuming 65 mpg and 12,000 mile/yr
requires 185 kg/yr H2/vehicle or gal/yr gaso equiv
Assuming 90% **annual load factor at**
actual H2 394,200,000 kg/y H2 /station or gal/y gaso equiv
or 32,850,000 kg/month H2 or gal/mo. gaso equiv
thereby **2,135,250 vehicles can be serviced at**
305,036 fill-ups/d @ 4.2 kg or gal equiv/fill-up
or each vehicle fills up once a week

Coal → Coal gasifier 75.0% LHV effic
8,399 MM Btu/h LHV
8,651 MM Btu/h HHV
326,994 kg/hr @12,000 Btu/lb dry
7,848 tons/d dry bit coal
2% sulfur

6,299 MM Btu/hr hot raw syngas
58% CO/(H2+CO) syngas
75 atm

326,994 kg/hr O2

157 ton/d sulfur
CO shift cool & clean 3.0% PSA loses
899,233 kg/hr CO2
189 MM Btu/hr PSA fuel gas for steam
548 MM Btu/h CO to H2 shifting heat

CO2 emission to atmosphere

Source	kg CO2/kg H2	kg C/kg H2
From Coal	17.98	4.90
From Ele Gen	0.77	0.21
Total	**18.76**	**5.12**

50,000 kg/hr H2
5,691 MM Btu/hr H2
20,725,000 scf/hr H2 @ 75 atm

Electric Power
ASU 127,528 → ASU 0.39 kWh/kg O2
Misc. 33,347
Total gross 160,874 kW
Net import 120,656 kW if
assuming 25% from use of steam via cooling of hot raw syngas & shift heat
at 0.32 kg CO2/kWh from input

7,848 metric tons/d O2
1.00 tons O2/ton dry feed

Hydrogen in Gas Pipeline @ 75 atm
1,200,000 design kg/d H2 or gal/d gasoline equivalent
1,080,000 actual kg/d annual ave.

62% Coal + fuel for power to H2 efficiency
38,610 kg/hr CO2 at power plants

Capital Costs		**Unit cost basis at 500,000**	**kg/d H2**	**cost/size factors**		**Unit cost at 1,200,000**	**kg/d H2**	**millions of $ for 1 plant**	**Notes**
Coal handling & prep	$	15	/kg/d coal	75%	$	12	/kg/d coal	95	solids & slurry prep
Texaco coal gasifers	$	21	/kg/d coal	85%	$	18	/kg/d coal	173	20% spare unit HP quench
Air separation unit (ASU)	$	22	/kg/d oxygen	85%	$	19	/kg/d oxygen	151	$ 1,187 /kW ASU power
CO shift, cool & cleanup	$	12	/kg/d CO2	75%	$	10	/kg/d CO2	208	$ 0.4 /scf/d H2 MDEA & PSA
Sulfur recovery	$	300	/kg/d sulfur	80%	$	252	/kg/d sulfur	40	O2 Claus & tailgas treat
						Total process units		667	
General Facilities				30% of process units				200	20-40% typical, SMR + 10%
Engineering Permitting & Startup				10% of process units				67	10-20% typical
Contingencies				10% of process units				67	10-20% typical, low after the first few
Working Capital, Land & Misc.				7% of process units				47	5-10% typical
						U.S. Gulf Coast Capital Costs		1,047	
Site specific factor		110%	of US Gulf Coast costs			**Total Capital Costs**		**1,152**	691 $/kWt H2
Unit Capital Costs		**2.32**	**/scf/d H2 or**	960	/kg/d H2 or	960	/gal/d gaso equiv		1,309 $/kWe equiv
		233%	of equivalent NG case						assuming "H" 60% LHV minus import power & CC

Hydrogen Costs at		**90%**	ann load factor	**million $/yr of 1 plant**	**$/million Btu LHV**	**$/1,000 scf H2**	**$/kg H2 or $/gal gaso equiv**	**Notes**
Variable Non-fuel O&M		1.0%	/yr of capital	12	0.26	0.07	0.03	0.5-1.5% typical
Coal	$	1.22	/MM Btu HHV	83	1.85	0.51	0.21	$0.75-1.25/MM Btu typical
Electricity	$	0.045	/kWh	43	0.95	0.26	0.11	$0.04-0.05/kWh **industrial** rate
Variable Operating Cost				**138**	**3.07**	**0.84**	**0.35**	
Fixed Operating Cost		5.0%	/yr of capital	**58**	**1.28**	**0.35**	**0.15**	4-7% typical for refining
Capital Charges		16%	/yr of capital	**183**	**4.08**	**1.12**	**0.46**	20-25% typical for refining
Total Gaseous Hydrogen Costs from Coal				**378**	**8.43**	**2.32**	**0.96**	including return of investment
Carbon Tax	$	50.00	/ton	101	2.25	0.62	0.26	include indirect C from power

notes: **$ 32.28** /tonne coal price from above $/MM Btu input at 12,000 Btu/lb HHV

Costs are for production at the plant only and do not include distribution costs. Assume no plant storage but compression of hydrogen to 75 atm for the pipeline.

TABLE E-9
CS Coal-F
CS Size Hydrogen via Coal Gasification with Future Technology

Color codes variables via summary inputs key outputs

CS-Size Plant Design — **Design LHV energy equivalent**

Size range	Hydrogen kg/d H2	gasoline gal/d	million Btu/hr	scf/d H2	MW t
Maximum	2,500,000	2,500,000	11,855	1,036,250,000	3,474
This run	1,200,000	**1,200,000**	**5,691**	**497,400,000**	**1,667**
Minimum	500,000	500,000	2,371	207,250,000	695

gasoline equivalent
Assuming 65 mpg and 12,000 mile/yr
requires 185 kg/yr H2/vehicle or gal/yr gaso equiv
Assuming 90% **annual load factor at**
actual H2 394,200,000 kg/y H2 /station or gal/y gaso equiv
or 32,850,000 kg/month H2 or gal/mo. gaso equiv
thereby **2,135,250 vehicles can be serviced at**
305,036 fill-ups/d @ 4.2 kg or gal equiv/fill-up
144 ton/d sulfur or each vehicle fills up one a week

Coal → Coal gasifier 80.0% LHV effic
7,698 MM Btu/h LHV
7,929 MM Btu/h HHV
299,728 kg/hr @12,000 Btu/lb dry
7,193 tons/d dry bit coal
2% sulfur

Coal gasifier → 6,159 MM Btu/hr hot raw syngas, 58% CO/(H2+CO) syngas, 75 atm → CO shift cool & clean 1% PSA losses

CO shift → **824,251 kg/hr CO2**
62 MM Btu/hr PSA fuel gas
536 MM Bur/h CO to H2 shifting heat

CO2 emission to atmosphere

Source	kg CO2/kg H2	kg C/kg H2
From Coal	16.49	4.50
From Ele Gen	0.25	0.07
Total	**16.73**	**4.56**

269,755 kg/hr O2

50,000 kg/hr H2
5,691 MM Btu/hr H2
20,725,000 scf/hr H2 @ 75 atm

advanced ASU 0.30 kWh/kg O2
6,474 metric tons/d O2
0.90 tons O2/ton dry feed

Electric Power
ASU 80,926
Misc. 16,673
Total gross 97,600 kW
Net import 39,040 kW if
assuming 60% from use of steam via cooling of hot raw syngas & shift heat
at 0.32 kg CO2/kWh from input

Hydrogen in Gas Pipeline @ 75 atm
1,200,000 design kg/d H2 or gal/d gasoline equivalent
1,080,000 actual kg/d annual ave.

71% Coal + fuel for power to H2 efficiency
12,492.78 kg/hr CO2 equivalent for net import power

Capital Costs	**Unit cost basis at 500,000**	**kg/d H2**	**cost/size factors**	**Unit cost at 1,200,000**	**kg/d H2**	**millions of $ for 1 plant**	**Notes**
Coal handling & prep	$ 13	/kg/d coal	75%	$ 10	/kg/d coal	75	solids & slurry prep
Advanced HP slurry feed	$ 18	/kg/d coal	85%	$ 16	/kg/d coal	136	20% spare unit heat recovery
Advanced ASU	$ 18	/kg/d oxygen	85%	$ 16	/kg/d oxygen	102	$ 1,263 /kW ASU power
CO shift, cool & cleanup	$ 10	/kg/d CO2	75%	$ 8	/kg/d CO2	159	$ 0.3 /scf/d H2
Sulfur recovery	$ 250	/kg/d sulfur	80%	$ 210	/kg/d sulfur	30	O2 Claus & tailgas treat
					Total process units	503	
General Facilities			30% of process units			151	20-40% typical, SMR + 10%
Engineering Permitting & Startup			10% of process units			50	10-20% typical
Contingencies			10% of process units			50	10-20% typical, low after the first few
Working Capital, Land & Misc.			7% of process units			35	5-10% typical
					U.S. Gulf Coast Capital Costs	789	
Site specific factor	110%	of US Gulf Coast costs			**Total Capital Costs**	**868**	521 $/kWt H2
Unit Capital Costs	**1.75**	**/scf/d H2 or**	723 /kg/d H2 or		723	/gal/d gaso equiv	903 $/kWe equiv
	75%	of equivalent high					assuming "H" 60% LHV minus import power & CC

Hydrogen Costs at	90% ann load factor	**million $/yr of 1 plant**	**$/million Btu LHV**	**$/1,000 scf H2**	**$/kg H2 or $/gal gaso equiv**	**Notes**
Variable Non-fuel O&M	1.0% /yr of capital	9	0.19	0.05	0.02	0.5-1.5% typical
Coal	$ 1.22 /MM Btu HHV	76	1.70	0.47	0.19	$0.75-1.25/MM Btu typical
Electricity	$ 0.045 /kWh	13.85	0.31	0.08	0.04	$0.04-0.05/kWh **industrial** rate
Variable Operating Cost		**99**	**2.20**	**0.60**	**0.25**	
Fixed Operating Cost	5.0% /yr of capital	**43**	**0.97**	**0.27**	**0.11**	4-7% typical for refining
Capital Charges	16% /yr of capital	**138**	**3.08**	**0.84**	**0.35**	20-25% typical for refining
Total Gaseous Hydrogen Costs from Coal		**280**	**6.25**	**1.72**	**0.71**	74% of equivalent high
Carbon Tax	$ 50.00 /ton	90	2.01	0.55	0.23	include indirect C from power

notes: **$ 32.28** /tonne coal price from above $/MM Btu input at 12,000 Btu/lb HHV

Costs are for production at the plant only and do not include distribution costs. Assume no plant storage but compression of hydrogen to 75 atm for the pipeline.

TABLE E-10

CS Coal-C Seq

CS Size Hydrogen via Coal Gasification with CO2 Capture with Current Technology

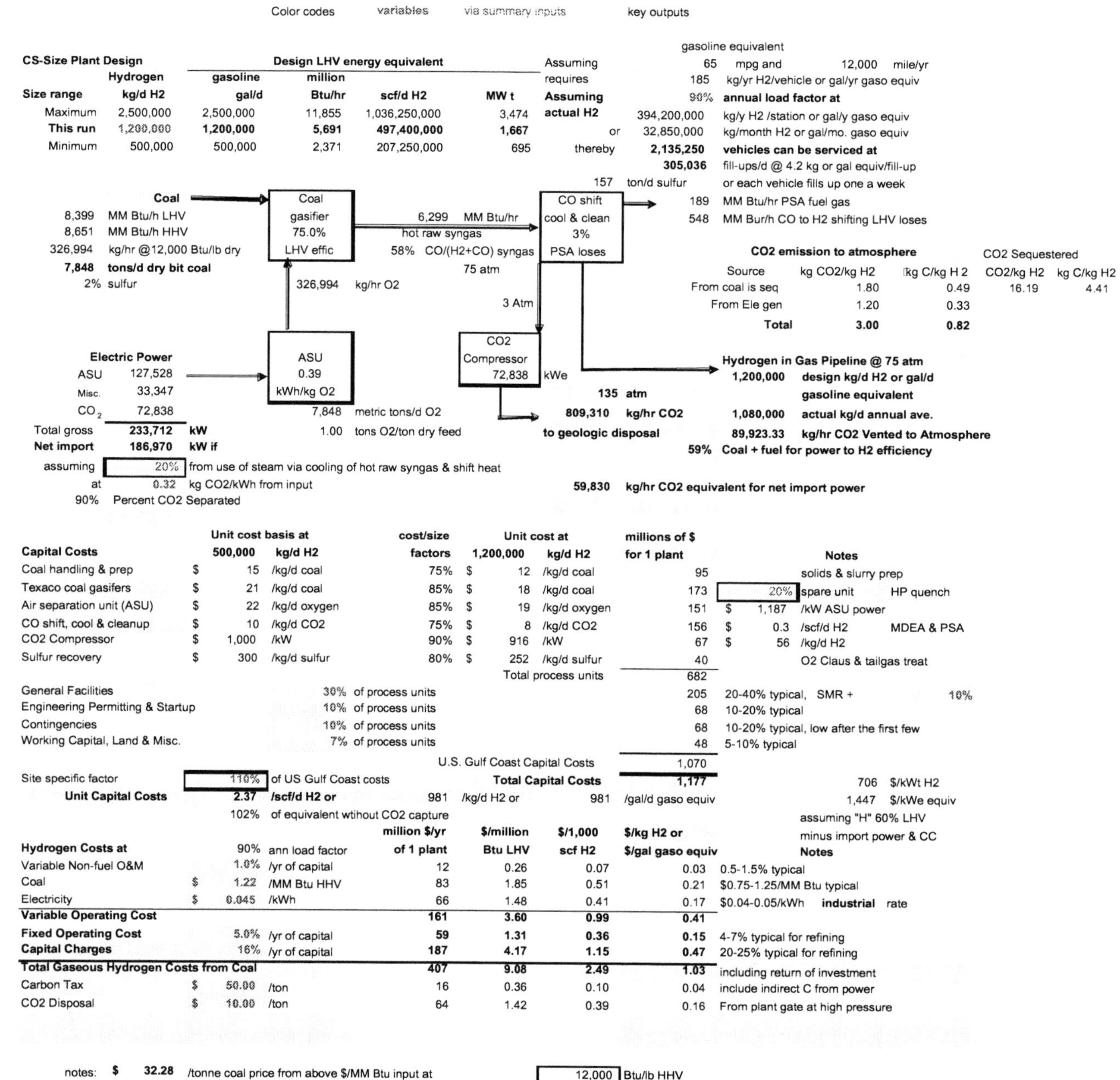

Color codes variables via summary inputs key outputs

CS-Size Plant Design	Hydrogen	gasoline	Design LHV energy equivalent million		
Size range	kg/d H2	gal/d	Btu/hr	scf/d H2	MW t
Maximum	2,500,000	2,500,000	11,855	1,036,250,000	3,474
This run	1,200,000	1,200,000	5,691	497,400,000	1,667
Minimum	500,000	500,000	2,371	207,250,000	695

Assuming 65 mpg and 12,000 mile/yr (gasoline equivalent)
requires 185 kg/yr H2/vehicle or gal/yr gaso equiv
Assuming 90% annual load factor at
actual H2 394,200,000 kg/y H2 /station or gal/y gaso equiv
or 32,850,000 kg/month H2 or gal/mo. gaso equiv
thereby 2,135,250 vehicles can be serviced at
305,036 fill-ups/d @ 4.2 kg or gal equiv/fill-up
or each vehicle fills up one a week

Coal: 8,399 MM Btu/h LHV; 8,651 MM Btu/h HHV; 326,994 kg/hr @12,000 Btu/lb dry; 7,848 tons/d dry bit coal; 2% sulfur

Coal gasifier 75.0% LHV effic — 6,299 MM Btu/hr hot raw syngas; 58% CO/(H2+CO) syngas; 75 atm; 326,994 kg/hr O2

CO shift cool & clean 3% PSA loses — 157 ton/d sulfur; 189 MM Btu/hr PSA fuel gas; 548 MM Bur/h CO to H2 shifting LHV loses; 3 Atm

Source	kg CO2/kg H2	kg C/kg H 2
CO2 emission to atmosphere		
From coal is seq	1.80	0.49
From Ele gen	1.20	0.33
Total	3.00	0.82

CO2 Sequestered: CO2/kg H2 16.19; kg C/kg H2 4.41

Electric Power: ASU 127,528; Misc. 33,347; CO_2 72,838; Total gross 233,712 kW; Net import 186,970 kW if

ASU 0.39 kWh/kg O2; 7,848 metric tons/d O2; 1.00 tons O2/ton dry feed

CO2 Compressor 72,838 kWe; 135 atm; 809,310 kg/hr CO2 to geologic disposal

Hydrogen in Gas Pipeline @ 75 atm
1,200,000 design kg/d H2 or gal/d gasoline equivalent
1,080,000 actual kg/d annual ave.
89,923.33 kg/hr CO2 Vented to Atmosphere
59% Coal + fuel for power to H2 efficiency

assuming 20% from use of steam via cooling of hot raw syngas & shift heat
at 0.32 kg CO2/kWh from input
90% Percent CO2 Separated

59,830 kg/hr CO2 equivalent for net import power

Capital Costs	Unit cost basis at 500,000 kg/d H2		cost/size factors	Unit cost at 1,200,000 kg/d H2		millions of $ for 1 plant	Notes
Coal handling & prep	$ 15	/kg/d coal	75%	$ 12	/kg/d coal	95	solids & slurry prep
Texaco coal gasifers	$ 21	/kg/d coal	85%	$ 18	/kg/d coal	173	20% spare unit HP quench
Air separation unit (ASU)	$ 22	/kg/d oxygen	85%	$ 19	/kg/d oxygen	151	$ 1,187 /kW ASU power
CO shift, cool & cleanup	$ 10	/kg/d CO2	75%	$ 8	/kg/d CO2	156	$ 0.3 /scf/d H2 MDEA & PSA
CO2 Compressor	$ 1,000	/kW	90%	$ 916	/kW	67	$ 56 /kg/d H2
Sulfur recovery	$ 300	/kg/d sulfur	80%	$ 252	/kg/d sulfur	40	O2 Claus & tailgas treat
				Total process units		682	
General Facilities	30%	of process units				205	20-40% typical, SMR + 10%
Engineering Permitting & Startup	10%	of process units				68	10-20% typical
Contingencies	10%	of process units				68	10-20% typical, low after the first few
Working Capital, Land & Misc.	7%	of process units				48	5-10% typical
				U.S. Gulf Coast Capital Costs		1,070	
Site specific factor	110%	of US Gulf Coast costs		Total Capital Costs		1,177	706 $/kWt H2
Unit Capital Costs	2.37	/scf/d H2 or	981 /kg/d H2 or		981	/gal/d gaso equiv	1,447 $/kWe equiv
	102%	of equivalent wtihout CO2 capture					assuming "H" 60% LHV minus import power & CC

Hydrogen Costs at	90%	ann load factor	million $/yr of 1 plant	$/million Btu LHV	$/1,000 scf H2	$/kg H2 or $/gal gaso equiv	Notes
Variable Non-fuel O&M	1.0%	/yr of capital	12	0.26	0.07	0.03	0.5-1.5% typical
Coal	$ 1.22	/MM Btu HHV	83	1.85	0.51	0.21	$0.75-1.25/MM Btu typical
Electricity	$ 0.045	/kWh	66	1.48	0.41	0.17	$0.04-0.05/kWh industrial rate
Variable Operating Cost			**161**	**3.60**	**0.99**	**0.41**	
Fixed Operating Cost	5.0%	/yr of capital	**59**	**1.31**	**0.36**	**0.15**	4-7% typical for refining
Capital Charges	16%	/yr of capital	**187**	**4.17**	**1.15**	**0.47**	20-25% typical for refining
Total Gaseous Hydrogen Costs from Coal			**407**	**9.08**	**2.49**	**1.03**	including return of investment
Carbon Tax	$ 50.00	/ton	16	0.36	0.10	0.04	include indirect C from power
CO2 Disposal	$ 10.00	/ton	64	1.42	0.39	0.16	From plant gate at high pressure

notes: $ 32.28 /tonne coal price from above $/MM Btu input at 12,000 Btu/lb HHV

Costs are for production at the plant only and do not include distribution costs. Assume no plant storage but compression of hydrogen to 75 atm for the pipeline.

TABLE E-11
CS Coal F Seq
CS Size Hydrogen via Coal Gasification Plus CO2 Capture with Future Optimism

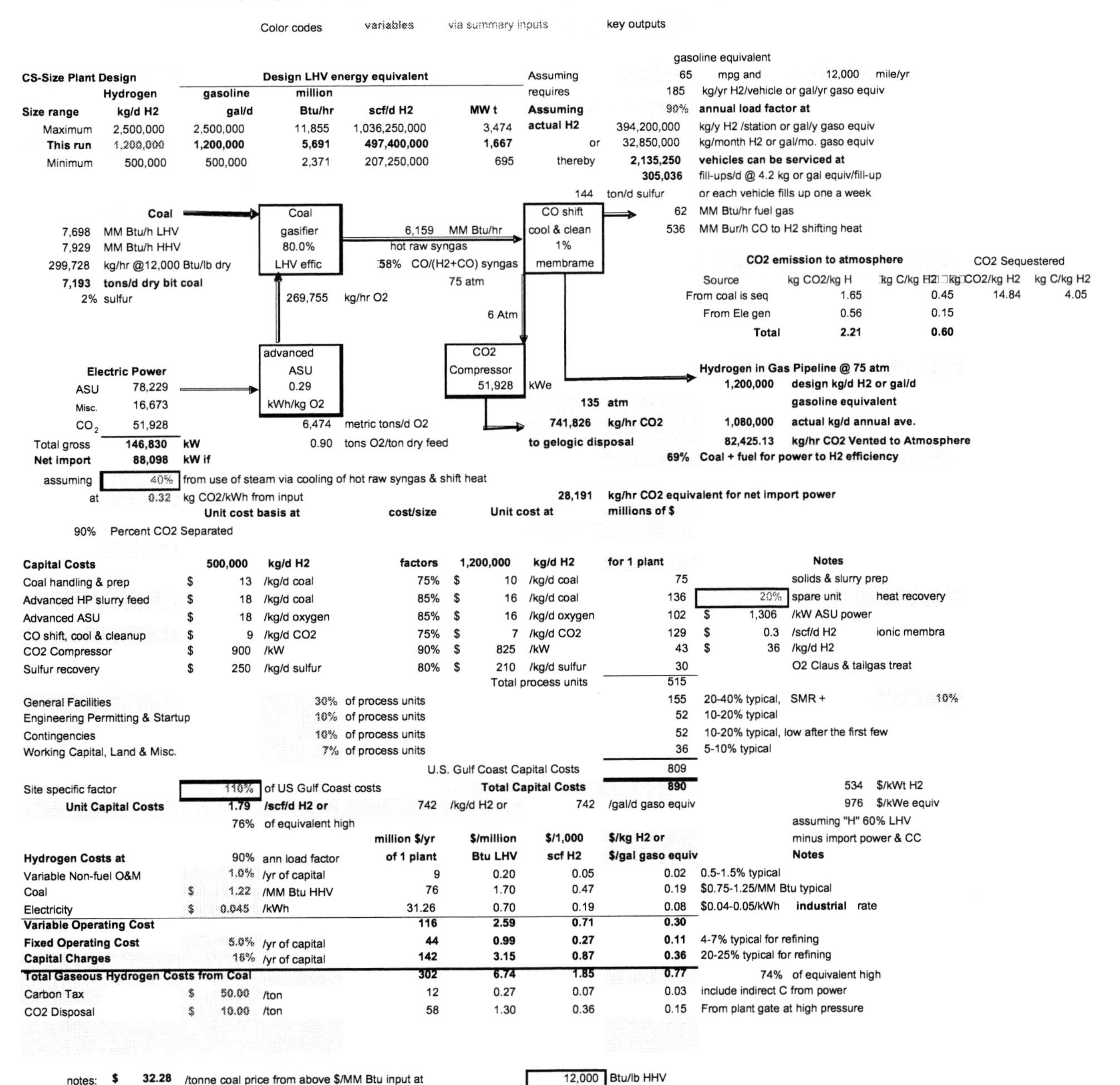

Color codes variables via summary inputs key outputs

CS-Size Plant Design	Hydrogen	gasoline	Design LHV energy equivalent million		
Size range	kg/d H2	gal/d	Btu/hr	scf/d H2	MW t
Maximum	2,500,000	2,500,000	11,855	1,036,250,000	3,474
This run	1,200,000	1,200,000	5,691	497,400,000	1,667
Minimum	500,000	500,000	2,371	207,250,000	695

gasoline equivalent
Assuming 65 mpg and 12,000 mile/yr
requires 185 kg/yr H2/vehicle or gal/yr gaso equiv
Assuming 90% annual load factor at
actual H2 394,200,000 kg/y H2 /station or gal/y gaso equiv
or 32,850,000 kg/month H2 or gal/mo. gaso equiv
thereby 2,135,250 vehicles can be serviced at
305,036 fill-ups/d @ 4.2 kg or gal equiv/fill-up
or each vehicle fills up one a week
144 ton/d sulfur

Coal → Coal gasifier 80.0% LHV effic
7,698 MM Btu/h LHV
7,929 MM Btu/h HHV
299,728 kg/hr @12,000 Btu/lb dry
7,193 tons/d dry bit coal
2% sulfur

6,159 MM Btu/hr hot raw syngas
58% CO/(H2+CO) syngas
75 atm
269,755 kg/hr O2
6 Atm

CO shift cool & clean 1% membrame
62 MM Btu/hr fuel gas
536 MM Bur/h CO to H2 shifting heat

CO2 emission to atmosphere			CO2 Sequestered	
Source	kg CO2/kg H	kg C/kg H2	kg CO2/kg H2	kg C/kg H2
From coal is seq	1.65	0.45	14.84	4.05
From Ele gen	0.56	0.15		
Total	2.21	0.60		

Electric Power		
ASU	78,229	
Misc.	16,673	
CO_2	51,928	
Total gross	146,830	kW
Net import	88,098	kW if

advanced ASU 0.29 kWh/kg O2
6,474 metric tons/d O2
0.90 tons O2/ton dry feed

CO2 Compressor 51,928 kWe
135 atm
741,826 kg/hr CO2
to gelogic disposal

Hydrogen in Gas Pipeline @ 75 atm
1,200,000 design kg/d H2 or gal/d gasoline equivalent
1,080,000 actual kg/d annual ave.
82,425.13 kg/hr CO2 Vented to Atmosphere
69% Coal + fuel for power to H2 efficiency

assuming 40% from use of steam via cooling of hot raw syngas & shift heat
at 0.32 kg CO2/kWh from input
28,191 kg/hr CO2 equivalent for net import power

90% Percent CO2 Separated

Capital Costs	Unit cost basis at 500,000 kg/d H2			cost/size factors	Unit cost at 1,200,000 kg/d H2			millions of $ for 1 plant	Notes	
Coal handling & prep	$	13	/kg/d coal	75%	$	10	/kg/d coal	75	solids & slurry prep	
Advanced HP slurry feed	$	18	/kg/d coal	85%	$	16	/kg/d coal	136	20% spare unit	heat recovery
Advanced ASU	$	18	/kg/d oxygen	85%	$	16	/kg/d oxygen	102	$ 1,306 /kW ASU power	
CO shift, cool & cleanup	$	9	/kg/d CO2	75%	$	7	/kg/d CO2	129	$ 0.3 /scf/d H2	ionic membra
CO2 Compressor	$	900	/kW	90%	$	825	/kW	43	$ 36 /kg/d H2	
Sulfur recovery	$	250	/kg/d sulfur	80%	$	210	/kg/d sulfur	30	O2 Claus & tailgas treat	
							Total process units	515		
General Facilities				30% of process units				155	20-40% typical, SMR +	10%
Engineering Permitting & Startup				10% of process units				52	10-20% typical	
Contingencies				10% of process units				52	10-20% typical, low after the first few	
Working Capital, Land & Misc.				7% of process units				36	5-10% typical	
							U.S. Gulf Coast Capital Costs	809		
Site specific factor		110%	of US Gulf Coast costs				Total Capital Costs	890	534 $/kWt H2	
Unit Capital Costs		1.79	/scf/d H2 or	742	/kg/d H2 or	742	/gal/d gaso equiv		976 $/kWe equiv	
		76%	of equivalent high						assuming "H" 60% LHV minus import power & CC	

Hydrogen Costs at		90%	ann load factor	million $/yr of 1 plant	$/million Btu LHV	$/1,000 scf H2	$/kg H2 or $/gal gaso equiv	Notes
Variable Non-fuel O&M		1.0%	/yr of capital	9	0.20	0.05	0.02	0.5-1.5% typical
Coal	$	1.22	/MM Btu HHV	76	1.70	0.47	0.19	$0.75-1.25/MM Btu typical
Electricity	$	0.045	/kWh	31.26	0.70	0.19	0.08	$0.04-0.05/kWh industrial rate
Variable Operating Cost				116	2.59	0.71	0.30	
Fixed Operating Cost		5.0%	/yr of capital	44	0.99	0.27	0.11	4-7% typical for refining
Capital Charges		18%	/yr of capital	142	3.15	0.87	0.36	20-25% typical for refining
Total Gaseous Hydrogen Costs from Coal				302	6.74	1.85	0.77	74% of equivalent high
Carbon Tax	$	50.00	/ton	12	0.27	0.07	0.03	include indirect C from power
CO2 Disposal	$	10.00	/ton	58	1.30	0.36	0.15	From plant gate at high pressure

notes: $ 32.28 /tonne coal price from above $/MM Btu input at 12,000 Btu/lb HHV

Costs are for production at the plant only and do not include distribution costs. Assume no plant storage but compression of hydrogen to 75 atm for the pipeline.

TABLE E-12
CS Nu-F
CS Size Hydrogen via Nuclear Thermal Splitting of Water with Future Optimism

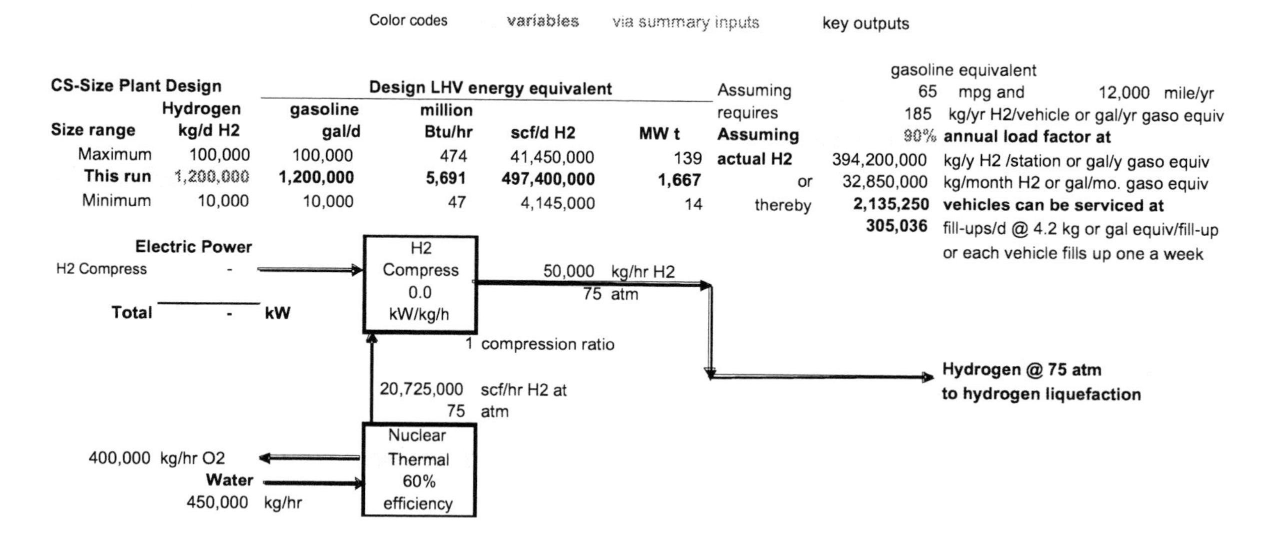

Color codes variables via summary inputs key outputs

CS-Size Plant Design	Hydrogen	Design LHV energy equivalent			
Size range	kg/d H2	gasoline gal/d	million Btu/hr	scf/d H2	MW t
Maximum	100,000	100,000	474	41,450,000	139
This run	1,200,000	1,200,000	5,691	497,400,000	1,667
Minimum	10,000	10,000	47	4,145,000	14

		gasoline equivalent
Assuming	65	mpg and 12,000 mile/yr
requires	185	kg/yr H2/vehicle or gal/yr gaso equiv
Assuming	90%	annual load factor at
actual H2	394,200,000	kg/y H2 /station or gal/y gaso equiv
or	32,850,000	kg/month H2 or gal/mo. gaso equiv
thereby	2,135,250	vehicles can be serviced at
	305,036	fill-ups/d @ 4.2 kg or gal equiv/fill-up
		or each vehicle fills up one a week

Capital Costs		Unit cost basis at 500,000 kg/d H2	cost/size factors		Unit cost at 1,200,000 kg/d H2	millions of $		Notes
NGAS	$	750 /kWt	75%	$	603 /kWt	1,674	$	3.4 /scf/d H2
H2 Compressor	$	1,000 /kW	85%	$	877 /kW	-	$	- /kg/d H2
					Total process units	1,674		
General Facilities			10% of process units			167		20-40% typical
Engineering Permitting & Startup			10% of process units			167		10-20% typical
Contingencies			7% of process units			117		10-20% typical, low after the first few
Working Capital, Land & Misc.			7% of process units			117		5-10% typical
					U.S. Gulf Coast Capital Costs	2,244		
Site specific factor		110% of US Gulf Coast costs			Total Capital Costs	2,468		1,480 $/kWt H2
Unit Capital Costs of		4.96 /scf/d H2 or	2,057 /kg/d H2 or		2,057 /gal/d gaso equiv			

Hydrogen Costs at		90% ann load factor	million $/yr of 1 plant	$/million Btu LHV	$/1,000 scf H2	$/kg H2 or $/gal gaso equiv	Notes
Non-fuel Variable O&M		1.0% /yr of capital	25	0.55	0.15	0.06	0.5-1.5% typical
Oxygen byproduct	$	(10) /ton O2	(32)	(0.70)	(0.19)	(0.08)	large amount could create min. value
Nuclear Fuel	$	5.0 /MWt h	110	2.44	0.67	0.28	
Decommission fund		1% /yr of capital	25	0.55	0.15	0.06	
Variable Operating Cost			**127**	**2.84**	**0.78**	**0.32**	
Fixed Operating Cost		5.0% /yr of capital	**123**	**2.75**	**0.76**	**0.31**	4-7% typical for refining
Capital Charges		15.9% /yr of capital	**392**	**8.75**	**2.40**	**1.00**	14-20%/yr CC typical for utilities
Total H2 Costs from Nuclear water splitting			**643**	**14.34**	**3.94**	**1.63**	including return on investment
Carbon Tax	$	50.00 /ton	-	-	-	-	include indirect C from power

notes: Costs are for production at the plant only and do not include distribution costs. Assume no plant storage but compression of hydrogen to 75 atm for the pipeline.

TABLE E-13
CS Pipe-C
Gaseous Hydrogen Distributed via Pipeline With Current Technology and Regulations

CS-Size Plant Design	Hydrogen	Design LHV energy equivalent			
Size range	kg/d H2	gasoline gal/d	million Btu/hr	scf/d H2	MW t
Maximum	2,500,000	2,500,000	11,855	1,036,250,000	3,474
This run	1,200,000	**1,200,000**	**5,691**	**497,400,000**	**1,667**
Minimum	500,000	500,000	2,371	207,250,000	695

	gasoline equivalent	
	65	mpg and 12,000 mile/yr
Assuming	185	kg/yr H2/vehicle or gal/yr gaso equiv
requires	90%	**annual load factor at**
Assuming	900,000	kg/y H2 /station or gal/y gaso equiv
actual H2	75,000	kg/month H2 or gal/mo. gaso equiv
or	**4,875**	**vehicles can be serviced at**
thereby	**587**	fill-ups/d @ 4.2 kg or gal equiv/fill-up
	438	station supported by this central faciltiy

5 Days liquid hydrogen storage at central plant gate to assure high availability

Note: Nothing else depends on above calculations

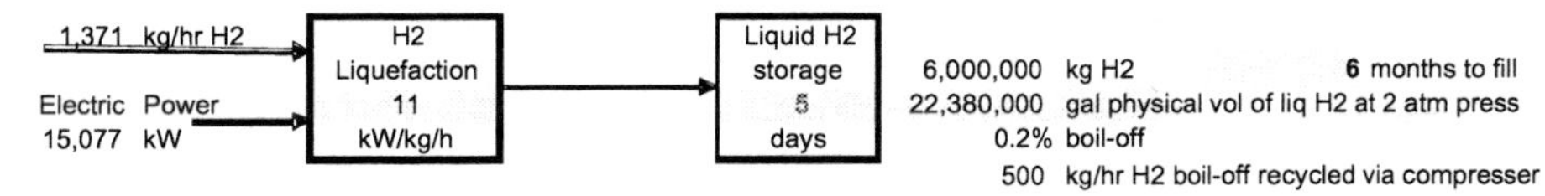

6,000,000	kg H2	**6** months to fill
22,380,000	gal physical vol of liq H2 at 2 atm press	
0.2%	boil-off	
500	kg/hr H2 boil-off recycled via compresser	

Key pipeline input assumption from summary sheet

Delivery distance	150	km	key input
Number of arms	4	key input	Radiate four directions or
Delivery pressure	440	psia	
Pipeline cost	600,000	$/km	includes right of way costs which is the key cost issue in urban areas
Electricity cost	0.045	$/kwh	if a booster compressor is required for long pipeline

	Unit cost 100,000	Basis at kg/d H2	cost/size factors	Unit cost at 32,895 Million $	kg/d H2	Million $		
Capital costs								
H2 Cryo Liquefaction	$ 700	/kg/d H2	75%	$ 924	/kg/d H2	30		
Liquid H2 storage	$ 3	/gal phy vol	100%	$ 3	/gal phy vol	67	$ 11	kg of H2 liquid storage
Pipeline from above assumptions				$ 600,000	/km	360		
Capital cost						458		
General Facilities & permitting		20% of unit cost				92		could be lower for pipelines
Eng. startup & contingencies		10% of unit cost				46		
Contingencies		7% of unit cost				32		
Working Capital, Land & Misc.		7% of unit cost				32		could be lower for pipelines
						659		
Location factor		110% of US Gulf Coast				**725**		
Unit Capital Costs	**1.46**	**/scf/d H2 or**	604	/kg/d H2 or	604	/gal/d gaso equiv		

			Million $/yr	$/million Btu LHV	$/k scf H2	$/kg H2 or $/gal gaso equiv	
Variable Operating Cost							
Electricity	$ 0.045	/kWh	5	0.12	0.03	0.01	
Variable non-fuel O&M	1%	/yr of capital	7	0.16	0.04	0.02	could be lower for pipelines
Total variable operating costs			**13**	**0.28**	**0.08**	**0.03**	
Fixed Operating Cost	5%	/yr of capital	**36**	**0.81**	**0.22**	**0.09**	could be lower for pipelines
Capital Charges	16%	/yr of capital	**115**	**2.57**	**0.71**	**0.29**	
Total operating costs			**164**	**3.66**	**1.00**	**0.42**	

TABLE E-14
CS Pipe-F
Gaseous Hydrogen Distributed via Pipeline with Future Optimism

CS-Size Plant Design	Hydrogen	Design LHV energy equivalent			
		gasoline	million		
Size range	kg/d H2	gal/d	Btu/hr	scf/d H2	MW t
Maximum	2,500,000	2,500,000	11,855	1,036,250,000	3,474
This run	1,200,000	**1,200,000**	**5,691**	**497,400,000**	**1,667**
Minimum	500,000	500,000	2,371	207,250,000	695

	gasoline equivalent	
	65	mpg and 12,000 mile/yr
Assuming	185	kg/yr H2/vehicle or gal/yr gaso equiv
requires	90%	**annual load factor at**
Assuming	900,000	kg/y H2 /station or gal/y gaso equiv
actual H2	75,000	kg/month H2 or gal/mo. gaso equiv
or	**4,875**	**vehicles can be serviced at**
thereby	**587**	fill-ups/d @ 4.2 kg or gal equiv/fill-up
	438	station supported by this central faciltiy

5 Days liquid hydrogen storage at central plant gate to assure high availability

Note: Nothing else depends on above calculations

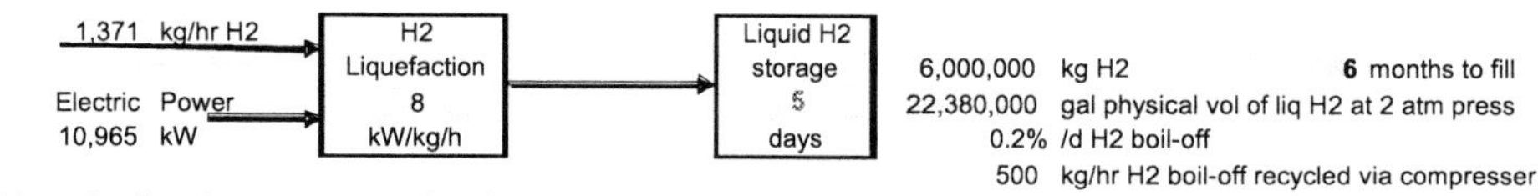

6,000,000	kg H2	**6** months to fill
22,380,000	gal physical vol of liq H2 at 2 atm press	
0.2%	/d H2 boil-off	
500	kg/hr H2 boil-off recycled via compresser	

Key pipeline input assumption from summary sheet

Delivery distance	150	km	key input
Number of arms	4	key input	Radiate four directions or
Delivery pressure	440	psia	
Pipeline cost	450,000	$/km	includes right of way costs which is the key cost issue in urban areas
Electricity cost	0.045	$/kwh	if a booster compressor is required for long pipeline

	Unit cost 100,000	Basis at kg/d H2	cost/size factors	Unit cost at 32,895 Million $	kg/d H2	Million $		
Capital costs								
H2 Cryo Liquefaction	$ 500	/kg/d H2	75%	$ 660	/kg/d H2	22		
Liquid H2 storage	$ 2	/gal phy vol	100%	$ 2	/gal phy vol	45	$ 7	kg of H2 liquid storage
Pipeline				$ 450,000	/km	270		
Capital cost						336		
General Facilities & permitting		20% of unit cost				67		could be lower for pipelines
Eng. startup & contingencies		10% of unit cost				34		
Contingencies		7% of unit cost				24		
Working Capital, Land & Misc.		7% of unit cost				24		could be lower for pipelines
						485		
Location factor		110% of US Gulf Coast				**533**		
Unit Capital Costs	**1.07**	**/scf/d H2 or**	444.15	/kg/d H2 or		444	/gal/d gaso equiv	
	74%	of equivalent high						

			Million $/yr	$/million Btu LHV	$/k scf H2	$/kg H2 or $/gal gaso equiv	
Variable Operating Cost							
Electricity	$ 0.045	/kWh	4	0.09	0.02	0.01	
Variable non-fuel O&M	1%	/yr of capital	5	0.12	0.03	0.01	could be lower for pipelines
Total variable operating costs			**9**	**0.21**	**0.06**	**0.02**	
Fixed Operating Cost	5%	/yr of capital	**27**	**0.59**	**0.16**	**0.07**	could be lower for pipelines
Capital Charges	16%	/yr of capital	**85**	**1.89**	**0.52**	**0.21**	
Total operating costs			**121**	**2.69**	**0.74**	**0.31**	74% of equivalent high

TABLE E-15
CS Disp-C
Gaseous Pipeline Hydrogen-Based Fueling Stations with Current Technology

CS-Size hydrogen production	1,200,000	kg/d
Annual average load factor	90%	/yr of design

Design per station	Hydrogen	Design LHV energy equivalent			
		gasoline	million		
Size range	kg/d/station H2	gal/d	Btu/hr	scf/d H2	MW t
This run	2,740	2,740	12.992	1,135,616	3.807
one of	438 stations				

Assuming	65	mpg and 12,000 mile/yr
requires	185	kg/yr H2/vehicle or gal/yr gaso equiv
Assuming	90%	Forecourt loading factor
or	75,000	kg/month H2 or gal/mo. gaso equiv
	4,875	**FC vehicles can be supported at**
	587	fill-ups/d @ 4.2 kg or gal equiv/fill-up

Note: Nothing else depends on above calculations

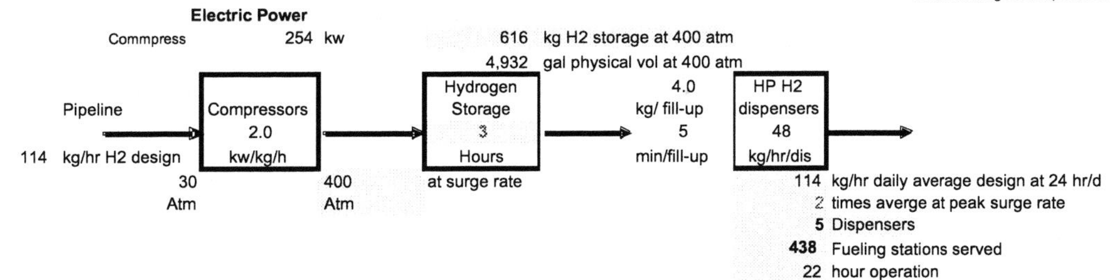

Capital Costs	**Unit cost basis at 1,000**	**kg/d H2**	**cost/size factors**	**Unit cost at 2,740**	**kg/d H2**	**millions of $ for 1 fueling station**		
H2 Compressors	$ 3,000	/kwh	80%	$ 2,452	/kg/d H2	0.62	$ 227	/kg/d H2
H2 buffer HP surge storage	$ 100	/gal phy vol	75%	$ 78	/gal phy vol	0.38	$ 622	/kg of HP H2 gas storage
Gaseous H2 dispenser	$ 15,000	/dispenser	90%	$ 13,562	/dispenser	0.07	$ 25	/kg/d dispenser design
					Unit cost	1.07		
General Facilities & permitting	25%					0.27		
Eng. startup & contingencies	10%					0.11		
Contingencies	10%					0.11		
Working Capital, Land & Misc.	7%					0.08		
					Capital Costs	**1.63**	**for 1 of**	**438 stations**
					Total Capital Costs	**715**	**for all**	**438 stations**
Unit Capital Costs	**1.44**	**/scf/d H2 or**	595	/kg/d H2 or	595	/gal/d gaso equiv		

Hydrogen Costs at		**90%** ann load factor	**$/yr of 1 station**	**$/million Btu LHV**	**$/k scf H2**	**$/kg H2 or $/gal gaso equiv**	
Road tax or (subsidy)	$ -	/gal gaso equiv.	-	-	-	-	can be subsidy like EtOH
Gas Station mark-up	$ -	/gal gaso equiv.	-	-	-	-	if H2 drops total station revenues
Variable Non-fuel O&M	5.0%	/yr of capital	81,565	0.80	0.22	0.09	0.5-1.5 typical many be low here
Electricity	$ 0.070	/kWh	126,000	1.23	0.34	0.14	0.06-0.09 typical commercial rates
Variable Operating Cost			**207,565**	**2.03**	**0.56**	**0.23**	
Fixed Operating Cost	3.0%	/yr of capital	**48,939**	**0.48**	**0.13**	**0.05**	3-5% typical, may be lower here
Capital Charges	14.0%	/yr of capital	**228,383**	**2.23**	**0.61**	**0.25**	20-25% typical for refiners
Fueling Station Cost including return of investment			**484,887**	**4.73**	**1.30**	**0.54**	

Total Hydrogen Fueling Costs of	**438 Stations**
	Million $/yr
Variable Operating Cost	91
Fixed Operating Cost	21
Capital Charges	100
Total Fueling Station Cost	**212**

TABLE E-16
CS Disp-F
Gaseous Pipeline Hydrogen-Based Fueling Stations with Future Optimism

CS-Size hydrogen production	1,200,000	kg/d
Annual average load factor	90%	/yr of design

Design per station		Design LHV energy equivalent			
	Hydrogen	gasoline	million		
Size range	kg/d/station H2	gal/d	Btu/hr	scf/d H2	MW t
This run	2,740	2,740	12.992	1,135,616	3.807
one of	438 stations				

Assuming	65	mpg and 12,000 mile/yr
requires	185	kg/yr H2/vehicle or gal/yr gaso equiv
Assuming	90%	Forecourt loading factor
or	75,000	kg/month H2 or gal/mo. gaso equiv
	4,875	FC vehicles can be supported at
	587	fill-ups/d @ 4.2 kg or gal equiv/fill-up

Note: Nothing else depends on above calculations

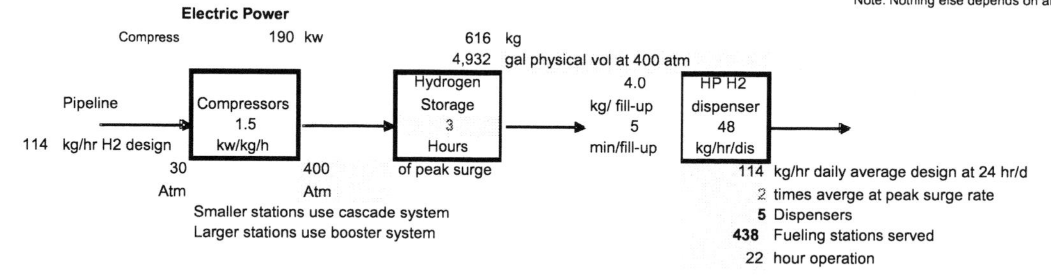

Smaller stations use cascade system
Larger stations use booster system

114	kg/hr daily average design at 24 hr/d
2	times averge at peak surge rate
5	Dispensers
438	Fueling stations served
22	hour operation

	Unit cost basis at		cost/size	Unit cost at		millions of $		
Capital Costs	1,000	kg/d H2	factors	2,740	kg/d H2	for 1 fueling station		
H2 Compressors	$ 2,000	/kwh	80%	$ 1,635	/kg/d H2	0.31	$ 114	/kg/d H2
H2 buffer storage	$ 100	/gal phy vol	75%	$ 78	/gal phy vol	0.38	$ 622	/kg of HP H2 gas storage
Gaseous H2 dispenser	$ 15,000	/dispenser	90%	$ 13,562	/dispenser	0.07	$ 25	/kg/d dispenser design
					Unit cost	0.76		
General Facilities & permitting	25%					0.19		
Eng. startup & contingencies	10%					0.08		
Contingencies	10%					0.08		
Working Capital, Land & Misc.	7%					0.05		
					Capital Costs	1.16	for 1 of	438 stations
					Total Capital Costs	507	for all	438 stations
Unit Capital Costs	1.02	/scf/d H2 or	423	/kg/d H2 or	423	/gal/d gaso equiv		
	71%	of equivalent high						

Hydrogen Costs at	90%	ann load factor	$/yr of 1 station	$/million Btu LHV	$/k scf H2	$/kg H2 or $/gal gaso equiv	
Road tax or (subsidy)	$ -	/gal gaso equiv.	-	-	-	-	can be subsidy like EtOH
Gas Station mark-up	$ -	/gal gaso equiv.	-	-	-	-	if H2 drops total station revenues
Variable Non-fuel O&M	5.0%	/yr of capital	57,925	0.57	0.16	0.06	0.5-1.5 typical many be low here
Electricity	$ 0.070	/kWh	94,500	0.92	0.25	0.11	0.06-0.09 typical commercial rates
Variable Operating Cost			152,425	1.49	0.41	0.17	
Fixed Operating Cost	3.0%	/yr of capital	34,755	0.34	0.09	0.04	3-5% typical, may be lower here
Capital Charges	14.0%	/yr of capital	162,191	1.58	0.43	0.18	20-25% typical for refiners
Fueling Station Cost			349,371	3.41	0.94	0.39	72% of equivalent high

including return of investment

Total Hydrogen Fueling Costs of 438 Stations

	Million $/yr
Variable Operating Cost	67
Fixed Operating Cost	15
Capital Charges	71
Total Fueling Station Cost	153

TABLE E-17
Midsize Plants Summary of Results

Pathway	MS NG-C	MS NG-F	1S NG-C-Se	1S NG-F-Se	MS Bio-C	MS Bio-F	1S Bio-C-Se	1S Bio-F-Se	MS Ele-C	MS Ele-F
Table E-	19	20	21	22	23	24	25	26	27	28
Capital investment, MM										
H_2 production	21.52	17.12	29.26	23.06	121.04	59.04	123.91	60.53	84.78	9.71
Distribution	34.73	18.54	34.73	18.54	34.73	18.54	34.73	18.54	34.73	18.54
Dispensing	20.70	9.01	20.70	9.01	20.70	9.01	20.70	9.01	20.70	9.01
Total Capital Investment	76.95	44.67	84.69	50.61	176.46	86.59	179.33	88.08	140.20	37.26
Production costs, $/kg H_2										
Variable costs										
Feed	0.79	0.74	0.82	0.79	0.98	0.42	0.98	0.42	(0.08)	(0.08)
Electricity	0.04	0.03	0.10	0.08	0.29	0.14	0.41	0.21	2.43	2.11
Non-fuel O&M, 1%/yr of capital	0.03	0.02	0.03	0.03	0.15	0.07	0.16	0.08	0.11	0.01
Total variable costs	0.86	0.79	0.96	0.90	1.42	0.64	1.54	0.71	2.46	2.04
Fixed costs, 5%/yr of capital	0.13	0.10	0.17	0.13	0.77	0.37	0.79	0.38	0.54	0.06
Capital charges, 18%/yr of capital	0.40	0.32	0.54	0.43	2.44	1.19	2.50	1.22	1.71	0.20
Total Production Costs	1.38	1.21	1.67	1.46	4.63	2.21	4.82	2.32	4.70	2.30
Distribution costs, $/kg H_2										
Variable costs										
Labor	0.09	0.09	0.09	0.09	0.09	0.09	0.09	0.09	0.09	0.09
Fuel	0.01	0.01	0.01	0.01	0.01	0.01	0.01	0.01	0.01	0.01
Electricity	0.50	0.36	0.50	0.36	0.50	0.36	0.50	0.36	0.50	0.36
Non-fuel O&M, 1%/yr of capital	0.06	0.03	0.06	0.03	0.06	0.03	0.06	0.03	0.06	0.03
Total variable costs	0.65	0.49	0.65	0.49	0.65	0.49	0.65	0.49	0.65	0.49
Fixed costs, 5%/yr of capital	0.28	0.15	0.28	0.15	0.28	0.15	0.28	0.15	0.28	0.15
Capital charges, 18%/yr of capital	0.88	0.47	0.88	0.47	0.88	0.47	0.88	0.47	0.88	0.47
Total Distribution Costs	1.80	1.10	1.80	1.10	1.80	1.10	1.80	1.10	1.80	1.10
Dispensing costs, $/kg H_2										
Variable costs										
Road tax	-	-	-	-	-	-	-	-	-	-
Station markup	-	-	-	-	-	-	-	-	-	-
Electricity	0.06	0.06	0.06	0.06	0.06	0.06	0.06	0.06	0.06	0.06
Non-fuel O&M, 1%/yr of capital	0.13	0.06	0.13	0.06	0.13	0.06	0.13	0.06	0.13	0.06
Total variable costs	0.18	0.11	0.18	0.11	0.18	0.11	0.18	0.11	0.18	0.11
Fixed costs, 5%/yr of capital	0.08	0.03	0.08	0.03	0.08	0.03	0.08	0.03	0.08	0.03
Capital charges, 18%/yr of capital	0.36	0.16	0.36	0.16	0.36	0.16	0.36	0.16	0.36	0.16
Total Dispensing Costs	0.62	0.30	0.62	0.30	0.62	0.30	0.62	0.30	0.62	0.30
H_2 Costs, $/kg										
Production	1.38	1.21	1.67	1.46	4.63	2.21	4.82	2.32	4.70	2.30
Distribution	1.80	1.10	1.80	1.10	1.80	1.10	1.80	1.10	1.80	1.10
Dispensing	0.62	0.30	0.62	0.30	0.62	0.30	0.62	0.30	0.62	0.30
CO_2 disposal	0	0	0.09	0.09	0	0	0.26	0.17	0	0
Carbon Tax	0.13	0.12	0.02	0.02	0.03	0.01	(0.31)	(0.21)	0.24	0.20
Total H_2 Costs	3.94	2.74	4.20	2.97	7.07	3.63	7.19	3.68	7.36	3.91

TABLE E-18
Midsize Hydrogen Plant Summary of Inputs and Outputs

Inputs | Boxed | are the key input variables you must choose, **current inputs are just an example**
design basis

Mid-Size Hydrogen Production Inputs

Key Variables Inputs	Value	Unit		Notes
				1 kg H2 is the same energy content as 1 gallon of gasoline
Design hydrogen production	24,000	**kg/d H2**		9,948,000 **scf/d H2** size range of 10,000 to 100,000 kg/d
Annual average load factor	90%	/yr of design	657,000	kg/month actual or 7,884,000 kg/yr actual
Distribution distance to forecourt				17 miles average distance 25-200 miles is typical
FC Vehicle gasoline equiv mileage	65	mpg (U.S. gallons) or		28 km/liter
FC Vehicle miles per year	12,000	mile/yr thereby requires		185 kg/yr H2 for each FC vehicle **42,705 FCV**
Typical gasoline sales/month/station	150,000	gallons/month per station		100,000 - 250,000 gallons/month typical or 4,932 gal/d
Hydrogen as % of gasoline/station	50.0%	of gasoline/station or		75,000 kg H2/month per stations or 2,466 kg/d/station
Capital Cost Buildup Inputs from process unit costs				**All major utilities included as process units**
General Facilities	20%	of process units		20-40% typical for SMR + 10% more for gasification
Engineering, Permitting & Startup	10%	of process units		10-20% typical
Contingencies	10%	of process units		10-20% typical, should be low after the first few
Working Capital, Land & Misc.	7%	of process units		5-10% typical
Site specific factor	110%	above US Gulf Coast		90-130% typical; sales tax, labor rates & weather issues
Product Cost Buildup Inputs				
Non-fuel Variable O&M	1.0%	/yr of capital		0.5-1.5% is typical
Fuels — Natural Gas	$ 4.50	/MM Btu HHV		$2.50-4.50/MM Btu typical **industrial** rate, see www.eia.doe.gov
Electricity	$ 0.045	/kWh		$0.04-0.05/kWh typical **industrial** rate, see www.eia.doe.gov
Electricity generation eff	50%			Incremental grid efficiency
Electricity CO2 emissions	0.32	kg/hr CO2/kWe		Based on NGCC, coal and current grid at 0.75 kg/hr CO2/kWe
Biomass production costs	$ 500	/ha/yr gross revenues		$400-600/hr/yr typical in U.S. .lower in developing nations or wastes
Biomass yield	10	tonne/ha/yr bone dry		8-12 ton/hr/yr typical if farmed, 3-5 ton/hr/yr if forestation or wastes
Coal	$ 1.22	/million Btu dry HHV		$0.75-1.25/million Btu coal utility delivered go to www.eia.doe.gov
Carbon tax	$ 50.00	$/tonne C		
Carbon Price for Carbon Vented	$50	$/tonne C		
CO2 disposal cost	$ 10.00	/ton		From plant gate at high pressure
Fixed O&M Costs	5.0%	/yr of capital		4-7% typical for refiners: labor, overhead, insurance, taxes, G&A
Capital Charges	15.9%	/yr of capital		20-25%/yr CC typical for refiners & 14-20%/yr CC typical for utilities
				20%/yr CC is about 12% IRR DCF on 100% equity where as
				15%/yr CC is about 12% IRR DCF on 50% equity & debt at 7%

Liquid Hydrogen Distribution to Forecout Inputs

Average distance to forecourt	150	km, key assumption that favors liquid hydrogen over pipelines	70,650 sq km
Truck utilization	80%		0.60 FVC/sq km
Days of H2 storage for high availability	5	dyas at design H2 rate	
Capital Cost Buildup Inputs from process unit costs			
General Facilities	20%		
Engineering, Permitting & Startup	15%		
Contingencies	10%	of process units	10-20% typical, should be low after the first few
Working Capital, Land & Misc.	7%	of process units	5-10% typical
Site specific factor	110%	above US Gulf Coast	90-130% typical; sales tax, labor rates & weather issues
Operating Cost Buidup Inputs			
Labor	$ 28.75	/hr	
Truck Fuel price	$ 1.00	$/gal	
Electricity	$ 0.045	/kWh	
Non-fuel Variable O&M	1.0%	/yr of capital	0.5-1.5% is typical
Fixed O&M Costs	5.0%	/yr of capital	4-7% typical for refiners: labor, overhead, insurance, taxes, G&A
Capital Charges	15.9%	/yr of capital	20-25%/yr CC typical for refiners & 14-20%/yr CC typical for utilities

Hydrogen Fueling Station (Forecourt) Inputs

			50.7 km radius per station
Hydrogen sales/month per station from above	75,000	kg/month	thereby supplying 9 stations
Average hydrogen sales/d per station from above	2,466	kg/d acutal or	2,740 kg/d design 8,065 sq km per station
Forecourt loading factor	90%	/yr of design	"plug & play" 24 hr replacements for reasonable availability
High pressure gas storage buffer	3	hours at peak surge rate due to higher demand before & after work typical fill-up times	
Capital Cost Buildup Inputs from process unit costs			
General Facilities	25%		
Engineering, Permitting & Startup	10%	of process units	10-20% typical Engineering costs spread over multiple stations
Contingencies	10%	of process units	10-20% typical, should be low after the first few
Working Capital, Land & Misc.	7%	of process units	5-10% typical
Site specific factor	100%	above US Gulf Coast	90-130% typical; sales tax, labor rates & weather issues
Operating Cost Buildup Inputs			
Road tax or (subsidy)	$ -	/gal gaso equiv.	may need subsidy like EtOH to get it going
Gas Station mark-up	$ -	/gal gaso equiv.	may be needed if H2 sales drops total station revenues
Electricity	$ 0.070	/kWh	$0.06-.0.09/kWh typical **commercial** rate, see www.eia.doe.gov
Non-fuel Variable O&M	5.0%	/yr of capital	0.5-1.5% is typical, assumed low here for "plug & play"
Fixed O&M Costs	3.0%	/yr of capital	
Capital Charges	14.0%	/yr of capital	

TABLE E-19
MS NG-C
Midsize Hydrogen via Current Steam Methane Reforming Technology

Color codes variables via summary inputs key outputs

MS-Size Plant Design	Hydrogen kg/d H2	gasoline gal/d	million Btu/hr	scf/d H2	MW t
Size range		Design LHV energy equivalent			
Maximum	100,000	100,000	474	41,450,000	139
This run	24,000	**24,000**	**114**	**9,948,000**	**33**
Minimum	10,000	10,000	47	4,145,000	14

Assuming requires gasoline equivalent 65 mpg and 12,000 mile/yr
185 kg/yr H2/vehicle or gal/yr gaso equiv
Assuming 98% **annual load factor at**
actual H2 8,584,800 kg/y H2 /station or gal/y gaso equiv
or 715,400 kg/month H2 or gal/mo. gaso equiv
thereby **46,501** **vehicles can be serviced at**
6,643 fill-ups/d @ 4.2 kg or gal equiv/fill-up
or each vehicle fills up one a week

Electric Power	
Compress	700
SMR & misc.	290
Total	**990 kW**

H2 Compress 0.7 kW/kg/h
1,000 kg/hr liq H2, 75 atm
2.5 compression ratio
414,500 scf/hr H2
30 atm

SMR 72.0% LHV effic

Natural Gas
158 MM Btu/h LHV
175 MM Btu/h HHV
175,461 scf/hr @ 1,000 Btu/scf
3,460 kg/hr @23,000 Btu/lb

381 Btu LHV/scf H2

CO2 emission to atmosphere

Source	kg CO2/kg H2	kg C/kg H2
From NG	9.52	2.60
From Ele Gen	0.32	0.09
Total	**9.83**	**2.68**

69% NG + fuel for power to H2 efficiency

Hydrogen @ 75 atm to hydrogen liquefaction
24,000 design kg/d H2 or gal/d gasoline equivalent
23,520 actual kg/d annual ave.

9,516 kg/hr CO2, however in dilute N2 rich SMR flue gas

at 0.32 kg CO2/kWh incremental U.S. NG = **317 kg/hr CO2** equivalent at power plants

Capital Costs		**Unit cost basis at 100,000 kg/d H2**		**cost/size factors**		**Unit cost at 24,000 kg/d H2**		**millions of $ for 1 plant**		**Notes**	
SMR	$	0.75	/scf/d H2	70%	$	1.15	/scf/d H2	11.4	$	477	/kg/d H2
H2 Compressor	$	2,000	/kW	80%	$	2,661	/kW	1.9	$	78	/kg/d H2
						Total process units		13.3			
General Facilities				20%	of process units			2.7	20-40% typical		
Engineering Permitting & Startup				10%	of process units			1.3	10-20% typical		
Contingencies				10%	of process units			1.3	10-20% typical, low after the first few		
Working Capital, Land & Misc.				7%	of process units			0.9	5-10% typical		
						U.S. Gulf Coast Capital Costs		19.6			
Site specific factor		110%	of US Gulf Coast costs			**Total Capital Costs**		**21.5**			
Unit Capital Costs		**2.16**	**/scf/d H2 or**	897	/kg/d H2 or	897	/gal/d gaso equiv				

Hydrogen Costs at		98%	ann load factor	**million $/yr of 1 plant**	**$/million Btu LHV**	**$/1,000 scf H2**	**$/kg H2 or $/gal gaso equiv**	**Notes**
Variable Non-fuel O&M		1.0%	/yr of capital	0.2	0.22	0.06	0.03	0.5-1.5% typical
Natural Gas	$	4.50	/MM Btu HHV	6.8	6.94	1.90	0.79	$2.50-4.50/MM Btu **industrial** rate
Electricity	$	0.045	/kWh	0.4	0.39	0.11	0.04	$0.04-0.05/kWh **industrial** rate
Variable Operating Cost				**7.4**	**7.55**	**2.07**	**0.86**	
Fixed Operating Cost		5.0%	/yr of capital	**1.1**	**1.10**	**0.30**	**0.13**	4-7% typical for refining
Capital Charges		16%	/yr of capital	**3.4**	**3.50**	**0.96**	**0.40**	20-25% typical for refining
Total Liquid Hydrogen Production Costs from Natural Gas				**11.9**	**12.15**	**3.34**	**1.38**	including return on investment
Carbon Tax	$	50.00	/ton	1	1.18	0.32	0.13	include indirect C from power

notes: Costs are for production at the plant only and do not include distribution costs. Assume no plant storage and compression of hydrogen to 75 atm for distribution.

TABLE E-20

MS NG-F

Midsize Hydrogen via Steam Methane Reforming with Future Optimism

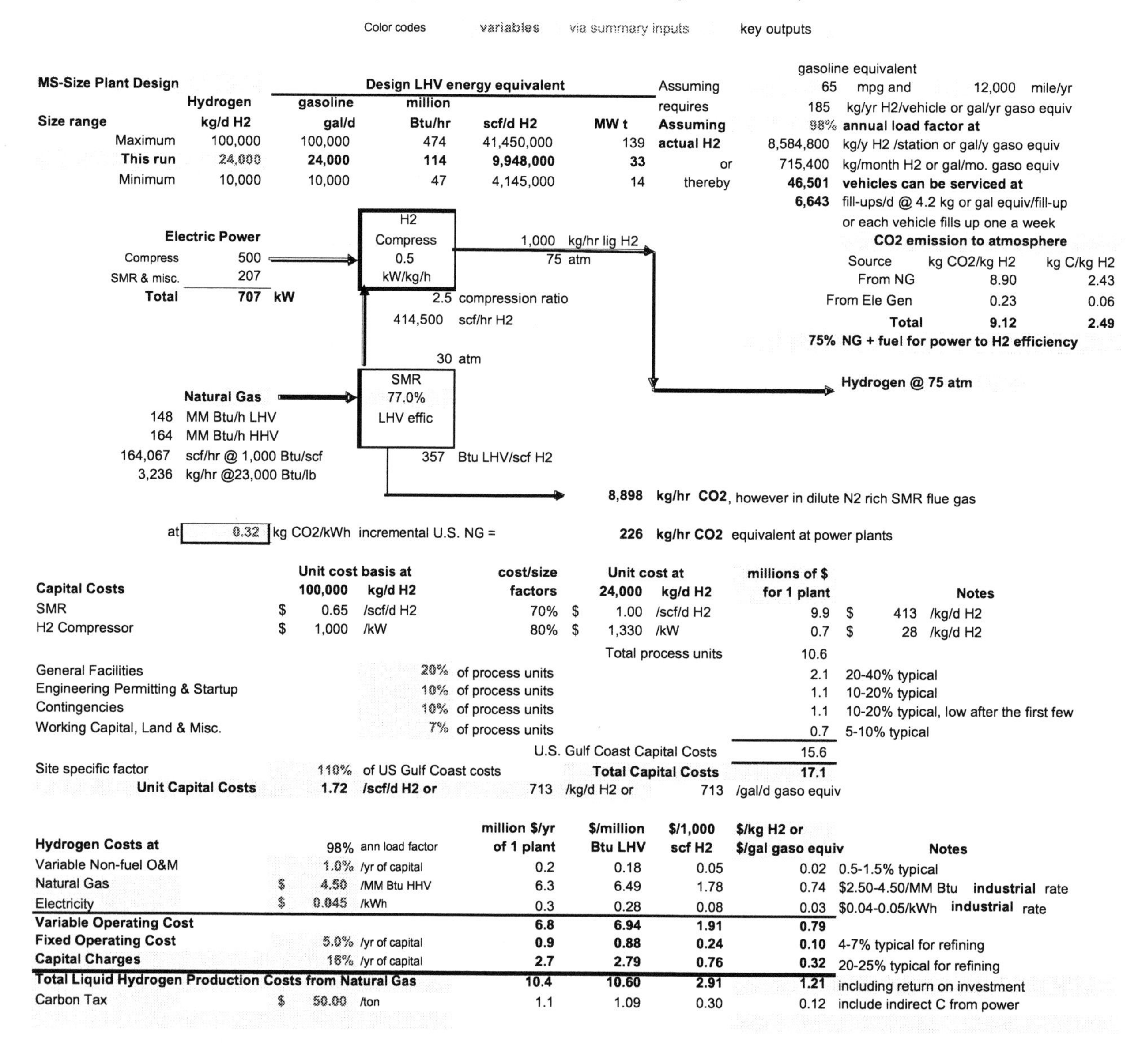

Color codes variables via summary inputs key outputs

MS-Size Plant Design

Size range	Hydrogen kg/d H2	Design LHV energy equivalent: gasoline gal/d	million Btu/hr	scf/d H2	MW t
Maximum	100,000	100,000	474	41,450,000	139
This run	24,000	**24,000**	**114**	**9,948,000**	**33**
Minimum	10,000	10,000	47	4,145,000	14

Assuming requires **Assuming actual H2** or thereby

gasoline equivalent
65 mpg and 12,000 mile/yr
185 kg/yr H2/vehicle or gal/yr gaso equiv
98% **annual load factor at**
8,584,800 kg/y H2 /station or gal/y gaso equiv
715,400 kg/month H2 or gal/mo. gaso equiv
46,501 vehicles can be serviced at
6,643 fill-ups/d @ 4.2 kg or gal equiv/fill-up
or each vehicle fills up one a week

CO2 emission to atmosphere

Source	kg CO2/kg H2	kg C/kg H2
From NG	8.90	2.43
From Ele Gen	0.23	0.06
Total	**9.12**	**2.49**

75% NG + fuel for power to H2 efficiency

Electric Power

Compress	500
SMR & misc.	207
Total	**707 kW**

148 MM Btu/h LHV
164 MM Btu/h HHV
164,067 scf/hr @ 1,000 Btu/scf
3,236 kg/hr @23,000 Btu/lb

8,898 kg/hr CO2, however in dilute N2 rich SMR flue gas

at 0.32 kg CO2/kWh incremental U.S. NG = **226 kg/hr CO2** equivalent at power plants

Capital Costs	Unit cost basis at 100,000 kg/d H2		cost/size factors	Unit cost at 24,000 kg/d H2		millions of $ for 1 plant	Notes
SMR	$ 0.65	/scf/d H2	70%	$ 1.00	/scf/d H2	9.9	$ 413 /kg/d H2
H2 Compressor	$ 1,000	/kW	80%	$ 1,330	/kW	0.7	$ 28 /kg/d H2
				Total process units		10.6	
General Facilities			20% of process units			2.1	20-40% typical
Engineering Permitting & Startup			10% of process units			1.1	10-20% typical
Contingencies			10% of process units			1.1	10-20% typical, low after the first few
Working Capital, Land & Misc.			7% of process units			0.7	5-10% typical
				U.S. Gulf Coast Capital Costs		15.6	
Site specific factor	110%	of US Gulf Coast costs		**Total Capital Costs**		**17.1**	
Unit Capital Costs	**1.72**	**/scf/d H2 or**	713	/kg/d H2 or	713	/gal/d gaso equiv	

Hydrogen Costs at	98% ann load factor	million $/yr of 1 plant	$/million Btu LHV	$/1,000 scf H2	$/kg H2 or $/gal gaso equiv	Notes
Variable Non-fuel O&M	1.0% /yr of capital	0.2	0.18	0.05	0.02	0.5-1.5% typical
Natural Gas	$ 4.50 /MM Btu HHV	6.3	6.49	1.78	0.74	$2.50-4.50/MM Btu **industrial** rate
Electricity	$ 0.045 /kWh	0.3	0.28	0.08	0.03	$0.04-0.05/kWh **industrial** rate
Variable Operating Cost		**6.8**	**6.94**	**1.91**	**0.79**	
Fixed Operating Cost	5.0% /yr of capital	**0.9**	**0.88**	**0.24**	**0.10**	4-7% typical for refining
Capital Charges	16% /yr of capital	**2.7**	**2.79**	**0.76**	**0.32**	20-25% typical for refining
Total Liquid Hydrogen Production Costs from Natural Gas		**10.4**	**10.60**	**2.91**	**1.21**	including return on investment
Carbon Tax	$ 50.00 /ton	1.1	1.09	0.30	0.12	include indirect C from power

notes: Costs are for production at the plant only and do not include distribution costs. Assume no plant storage and compression of hydrogen to 75 atm for distribution.

TABLE E-21
MS NG-C Seq
Midsize Hydrogen via Steam Methane Reforming Plus CO2 Capture with Current Technology

Color codes variables via summary inputs key outputs

MS-Size Plant Design	Hydrogen	gasoline	Design LHV energy equivalent million			
Size range	kg/d H2	gal/d	Btu/hr	scf/d H2	MW t	
Maximum	100,000	100,000	474	41,450,000	139	
This run	24,000	24,000	114	9,948,000	33	
Minimum	10,000	10,000	47	4,145,000	14	

gasoline equivalent

Assuming	65	mpg and 12,000 mile/yr
requires	185	kg/yr H2/vehicle or gal/yr gaso equiv
Assuming	98%	annual load factor at
actual H2	8,584,800	kg/y H2 /station or gal/y gaso equiv
or	715,400	kg/month H2 or gal/mo. gaso equiv
thereby	46,501	vehicles can be serviced at
	6,643	fill-ups/d @ 4.2 kg or gal equiv/fill-up or each vehicle fills up one a week

Electric Power	
Compressor	700
SMR & misc.	290
CO_2	1,251
Total	2,241 kW

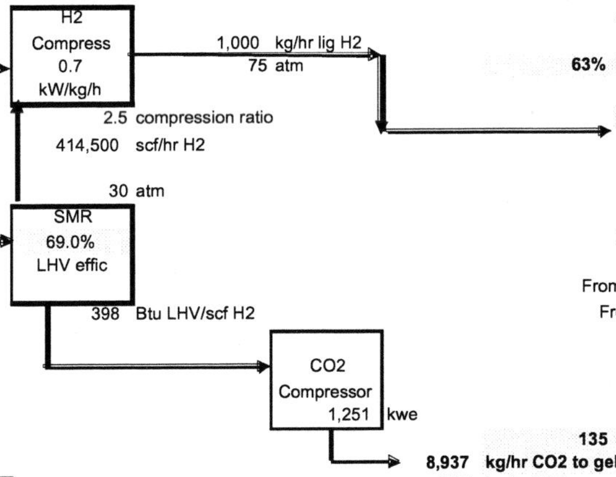

H2 Compress 0.7 kW/kg/h

1,000 kg/hr liq H2
75 atm

2.5 compression ratio
414,500 scf/hr H2

30 atm

SMR 69.0% LHV effic

Natural Gas
165 MM Btu/h LHV
183 MM Btu/h HHV
183,090 scf/hr @ 1,000 Btu/scf
3,611 kg/hr @23,000 Btu/lb

398 Btu LHV/scf H2

CO2 Compressor 1,251 kwe

63% NG + fuel for power to H2 efficiency

Hydrogen @ 75 atm to Liqufaction
24,000 design kg/d H2 or gal/d gasoline equivalent
23,520 actual kg/d annual ave.

	CO2 emission to atmosphere		CO2 Sequestered	
Source	kg CO2/kg H2	kg C/kg H2	CO2/kg H2	kg C/kg H2
From NG is seq	0.99	0.27	8.94	2.44
From Ele gen	0.72	0.20		
Total	1.71	0.47		

135 atm
8,937 kg/hr CO2 to gelogic storage 992.98 kg/hr CO2 Vented to Atmosphere
717 kg/hr CO2 equivalent at power plants

90% Percent CO2 Separated
at 0.32 kg CO2/kWh incremental U.S. NG =

Capital Costs	Unit cost basis at 100,000 kg/d H2	cost/size factors	Unit cost at 24,000 kg/d H2	millions of $ for 1 plant	Notes
SMR & MEA scrubber	$ 0.90 /scf/d H2	70%	$ 1.38 /scf/d H2	13.7	$ 572 /kg/d H2
H2 Compressor	$ 2,000 /kW	80%	$ 2,661 /kW	1.9	$ 78 /kg/d H2
CO2 Compressor	$ 1,500 /kW	80%	$ 1,995 /kW	2.5	
			Total process units	18.1	
General Facilities	20% of process units			3.6	20-40% typical
Engineering Permitting & Startup	10% of process units			1.8	10-20% typical
Contingencies	10% of process units			1.8	10-20% typical, low after the first few
Working Capital, Land & Misc.	7% of process units			1.3	5-10% typical
			U.S. Gulf Coast Capital Costs	26.6	
Site specific factor	110% of US Gulf Coast costs		Total Capital Costs	29.3	
Unit Capital Costs	2.94 /scf/d H2 or	1,219 /kg/d H2 or	1,219 /gal/d gaso equiv		

Hydrogen Costs at	98% ann load factor	million $/yr of 1 plant	$/million Btu LHV	$/1,000 scf H2	$/kg H2 or $/gal gaso equiv	Notes
Variable Non-fuel O&M	1.0% /yr of capital	0.3	0.30	0.08	0.03	0.5-1.5% typical
Natural Gas	$ 4.50 /MM Btu HHV	7.1	7.24	1.99	0.82	$2.50-4.50/MM Btu industrial rate
Electricity	$ 0.045 /kWh	0.9	0.89	0.24	0.10	$0.04-0.05/kWh industrial rate
Variable Operating Cost		8.2	8.42	2.31	0.96	
Fixed Operating Cost	5.0% /yr of capital	1.5	1.50	0.41	0.17	4-7% typical for refining
Capital Charges	16% /yr of capital	4.7	4.76	1.31	0.54	20-25% typical for refining
Total Liquid Hydrogen Costs from Natural Gas		14.3	14.68	4.03	1.67	including return on investment
Carbon Tax	$ 50.00 /ton	0.2	0.20	0.06	0.02	include indirect C from power
CO2 Disposal	$ 10.00 /ton	0.8	0.79	0.22	0.09	From plant gate at high pressure

note: Costs are for production at the plant only and do not include distribution costs. Assume no plant storage and compression of hydrogen to 75 atm for distribution.

TABLE E-22
MS NG-F Seq
Midsize Hydrogen via Steam Methane Reforming Plus CO2 Capture with Future Optimism

Color codes variables via summary inputs key outputs

MS-Size Plant Design

Size range	Hydrogen kg/d H2	gasoline gal/d	Design LHV energy equivalent million Btu/hr	scf/d H2	MW t
Maximum	1,000,000	1,000,000	4,742	414,500,000	1,389
This run	24,000	**24,000**	**114**	**9,948,000**	**33**
Minimum	20,000	20,000	95	8,290,000	28

gasoline equivalent
Assuming 65 mpg and 12,000 mile/yr
requires 185 kg/yr H2/vehicle or gal/yr gaso equiv
Assuming 98% **annual load factor at**
actual H2 8,584,800 kg/y H2 /station or gal/y gaso equiv
or 715,400 kg/month H2 or gal/mo. gaso equiv
thereby **46,501** **vehicles can be serviced at**
6,643 fill-ups/d @ 4.2 kg or gal equiv/fill-up
or each vehicle fills up one a week

Electric Power

Compressor	500	
SMR & misc.	207	
CO_2	1,113	
Total	**1,821**	**kW**

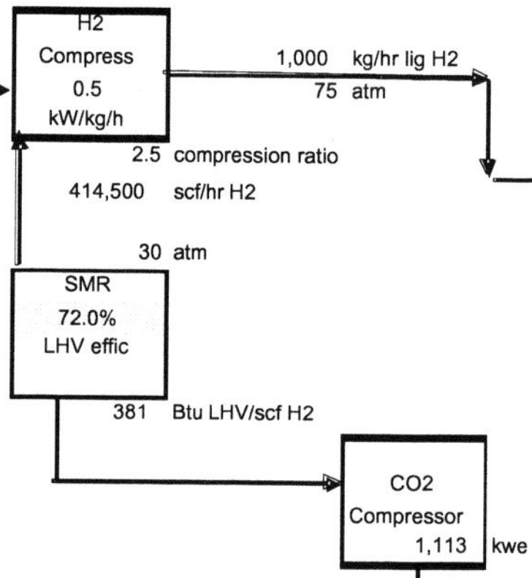

H2 Compress 0.5 kW/kg/h

1,000 kg/hr liq H2
75 atm

67% NG + fuel for power to H2 efficiency

2.5 compression ratio
414,500 scf/hr H2

Hydrogen @ 75 atm to Liqufaction
24,000 **design kg/d H2 or gal/d gasoline equivalent**
23,520 **actual kg/d annual ave.**

30 atm

SMR 72.0% LHV effic

Natural Gas
158 MM Btu/h LHV
175 MM Btu/h HHV
175,461 scf/hr @ 1,000 Btu/scf
3,460 kg/hr @23,000 Btu/lb

381 Btu LHV/scf H2

CO2 Compressor 1,113 kwe

CO2 emission to atmosphere

Source	kg CO2/kg H2	kg C/kg H2
From NG is seq	0.95	0.26
From Ele gen	0.58	0.16
Total	**1.53**	**0.42**

CO2 Sequestered

CO2/kg H2	kg C/kg H2
8.56	2.34

90% Percent CO2 Separated
at 0.32 kg CO2/kWh incremental U.S. NG =

8,564 **kg/hr CO2 to gelogic storage** **951.60** **kg/hr CO2 Vented to Atmosphere**
583 **kg/hr CO2** equivalent at power plants

Capital Costs		**Unit cost basis at 100,000**	**kg/d H2**	**cost/size factors**		**Unit cost at 24,000**	**kg/d H2**	**millions of $ for 1 plant**	**Notes**	
SMR	$	0.75	/scf/d H2	70%	$	1.15	/scf/d H2	11.4	$ 477	/kg/d H2
H2 Compressor	$	2,000	/kW	80%	$	2,661	/kW	1.3	$ 55	/kg/d H2
CO2 Compressor	$	1,000	/kW	80%	$	1,330	/kW	1.5		
						Total process units		14.3		
General Facilities				20%	of process units			2.9	20-40% typical	
Engineering Permitting & Startup				10%	of process units			1.4	10-20% typical	
Contingencies				10%	of process units			1.4	10-20% typical, low after the first few	
Working Capital, Land & Misc.				7%	of process units			1.0	5-10% typical	
						U.S. Gulf Coast Capital Costs		21.0		
Site specific factor		110%	of US Gulf Coast costs			**Total Capital Costs**		**23.1**		
Unit Capital Costs		**2.32**	**/scf/d H2 or**	961	/kg/d H2 or	961	/gal/d gaso equiv			

Hydrogen Costs at		98%	ann load factor	**million $/yr of 1 plant**	**$/million Btu LHV**	**$/1,000 scf H2**	**$/kg H2 or $/gal gaso equiv**	**Notes**
Variable Non-fuel O&M		1.0%	/yr of capital	0.2	0.24	0.06	0.03	0.5-1.5% typical
Natural Gas	$	4.50	/MM Btu HHV	6.8	6.94	1.90	0.79	$2.50-4.50/MM Btu **industrial** rate
Electricity	$	0.045	/kWh	0.7	0.72	0.20	0.08	$0.04-0.05/kWh **industrial** rate
Variable Operating Cost				**7.7**	**7.89**	**2.17**	**0.90**	
Fixed Operating Cost		5.0%	/yr of capital	**1.2**	**1.18**	**0.32**	**0.13**	4-7% typical for refining
Capital Charges		16%	/yr of capital	**3.7**	**3.75**	**1.03**	**0.43**	20-25% typical for refining
Total Liquid Hydrogen Costs from Natural Gas				**12.5**	**12.83**	**3.52**	**1.46**	including return on investment
Carbon Tax	$	50.00	/ton	0.2	0.18	0.05	0.02	include indirect C from power
CO2 Disposal	$	10.00	/ton	0.7	0.75	0.21	0.09	From plant gate at high pressure

notes: Costs are for production at the plant only and do not include distribution costs. Assume no plant storage and compression of hydrogen to 75 atm for distribution.

TABLE E-23
MS Bio-C
Midsize Hydrogen via Current Biomass Gasification Technology

Color codes variables via summary inputs key outputs

MS-Size Plant Design	Hydrogen	gasoline	Design LHV energy equivalent million		
Size range	kg/d H2	gal/d	Btu/hr	scf/d H2	MW t
Maximum	100,000	100,000	474	41,450,000	139
This run	24,000	24,000	114	9,948,000	33
Minimum	10,000	10,000	47	4,145,000	14

Assuming requires **Assuming actual H2** or thereby

gasoline equivalent

65	mpg and 12,000 mile/yr
185	kg/yr H2/vehicle or gal/yr gaso equiv
90%	**annual load factor at**
7,884,000	kg/y H2 /station or gal/y gaso equiv
657,000	kg/month H2 or gal/mo. gaso equiv
42,705	**vehicles can be serviced at**
6,101	fill-ups/d @ 4.2 kg or gal equiv/fill-up
	or each vehicle fills up one a week

Shell gasifier to avoid high CH4 & secondary SMR or ATR

Biomass → biomass gasifier 50.0% LHV effic → 153 MM Btu/hr hot raw syngas, 50% CO/(H2+CO) syngas, 35 atm → CO shift cool & clean 5% PSA loses

306	MM Btu/h LHV
324	MM Btu/h HHV
18,396	kg/hr @8,000 Btu/lb dry
442	**tons/d biomass bone dry**
145,036	tons/yr biomass bone dry
14,504	hectares of land for biomass
57	**square miles of land to grow biomass**

14,717 kg/hr O2

28,667	**kg/hr CO2** plus 15% from dryer
8	MM Btu/hr PSA fuel gas
11	MM Bur/h CO to H2 shifting LHV loses

CO2 emission to atmosphere

Note: Need to subtract carbon reduced from growing biomass

Source	kg CO2/kg H2	kg C/kg H2
From biomass	-	-
From Ele Gen	2.06	0.56
Total	**2.06**	**0.56**

33% Biomass + fuel for power to H2 efficiency

Electric Power		
ASU	5,445	
Misc.	1,000	
Total	**6,446**	**kW**

ASU 0.370 kWh/kg O2

353 metric tons/d O2

0.80 tons O2/ton dry feed

Hydrogen @ 75 atm to Liqufaction

24,000 design kg/d H2 or gal/d gasoline equivalent

15% of biomass fired in FBC to dry gasifier biomass feed

375 tons/day bone dry biomass to gasifier

at 0.32 kg CO2/kWh incremental U.S. NG =

1,902 Btu/lb water vaporized

1,500 Btu/lb water vaporized minimum

21,600 actual kg/d annual ave.

2,063 kg/hr CO2 equivalent at power plants

Capital Costs	**Unit cost basis at 100,000 kg/d H2**			**cost/size factors**	**Unit cost at 24,000 kg/d H2**			**millions of $ for 1 plant**	**Notes**		
Biomass handling & drying	$	25	/kg/d dry bio	75%	$	36	/kg/d dry bio	15.8	18	/kg/d green (wet) biomass	
Shell gasifer	$	20	/kg/d dry bio	80%	$	27	/kg/d dry bio	20.0	100%	spare unit	H2O quench
Air separation unit (ASU)	$	33	/kg/d oxygen	75%	$	47	/kg/d oxygen	16.7	$ 3,058	/kW power	
CO shift, cool & cleanup	$	18	/kg/d CO2	75%	$	26	/kg/d CO2	17.7	$ 1.8	/scf/d H2	MDEA & PSA
					Total process units			70.1			
General Facilities				30% of process units				21.0	20-40% typical,	SMR +	10%
Engineering Permitting & Startup				10% of process units				7.0	10-20% typical		
Contingencies				10% of process units				7.0	10-20% typical, low after the first few		
Working Capital, Land & Misc.				7% of process units				4.9	5-10% typical		
					U.S. Gulf Coast Capital Costs			110.0			
Site specific factor		110%	of US Gulf Coast costs		**Total Capital Costs**			**121.0**			
Unit Capital Costs		**12.17**	**/scf/d H2 or**	5,043	/kg/d H2 or		5,043	/gal/d gaso equiv			

Hydrogen Costs at				**million $/yr of 1 plant**	**$/million Btu LHV**	**$/1,000 scf H2**	**$/kg H2 or $/gal gaso equiv**	**Notes**
		90%	ann load factor					
Variable Non-fuel O&M		1.0%	/yr of capital	1.2	1.35	0.37	0.15	0.5-1.5% typical
Delivered biomass	$	**3.01**	/MM Btu HHV	7.7	8.57	2.35	0.98	**based on costs below**
Electricity	$	0.045	/kWh	2.3	2.55	0.70	0.29	0.04-0.05/kWh typical **industrial** rates
Variable Operating Cost				**11.2**	**12.47**	**3.42**	**1.42**	
Fixed Operating Cost		5.0%	/yr of capital	**6.1**	**6.74**	**1.85**	**0.77**	4-7% typical for refining
Capital Charges		16%	/yr of capital	**19.2**	**21.45**	**5.89**	**2.44**	20-25% typical for refining
Total Liquid Hydrogen Costs from Biomass				**36.5**	**40.66**	**11.16**	**4.63**	including return on investment
Carbon Tax	$	50.00	/ton	0.2	0.25	0.07	0.03	include indirect C from power

note: Costs are for production at the plant only and do not include distribution costs. Assume no plant storage and compression of hydrogen to 75 atm for distribution.

Delivered biomass @ $ 53.02 /bone dry ton (BDT) or $ 3.01 /million Btu LHV based on below:

$	500	/hectare per yr gross total revenues or	$	200	/acre per yr gross total revenues
	10	ton biomass/yr per ha - bone dry basic or		4.0	tons biomass/yr per acre - bone dry
	8,000	Btu/lb HHV bone dry and		50%	moisture of green biomass
$	2.08	/mile round trip for typical 25 ton truck hauling green biomass			
	18	miles round trip haul = $ 1.51 /ton green or	$	**3.02**	**/ton bone dry equivalent transportation**

If waste bio or coproduct lower gross revenue needs but much lower yield/ha

TABLE E-24
MS Bio-F
Midsize Hydrogen via Biomass Gasification with Future Optimism

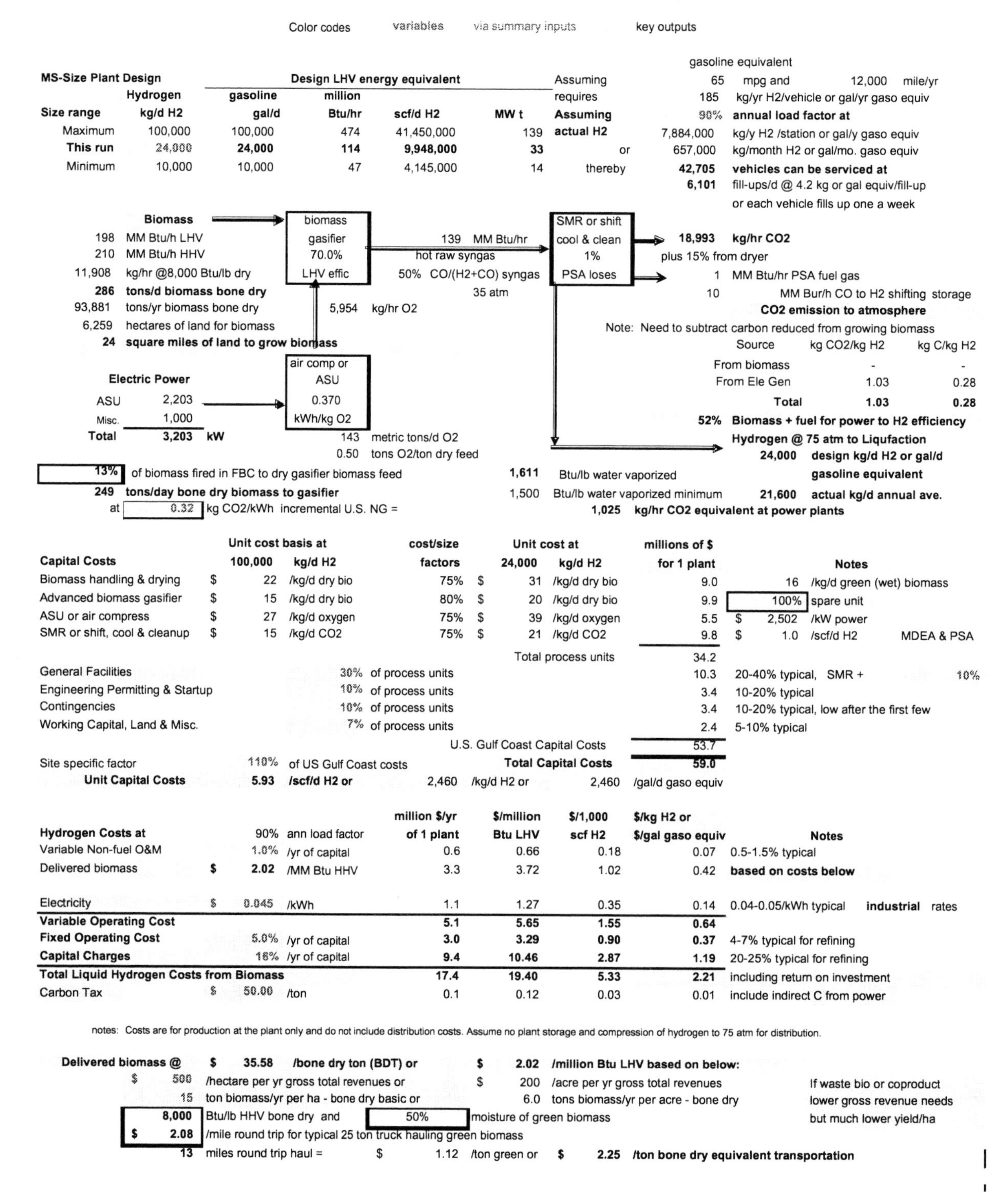

Color codes variables via summary inputs key outputs

MS-Size Plant Design	Hydrogen	Design LHV energy equivalent gasoline	million		
Size range	kg/d H2	gal/d	Btu/hr	scf/d H2	MW t
Maximum	100,000	100,000	474	41,450,000	139
This run	24,000	24,000	114	9,948,000	33
Minimum	10,000	10,000	47	4,145,000	14

Assuming requires: gasoline equivalent 65 mpg and 12,000 mile/yr; 185 kg/yr H2/vehicle or gal/yr gaso equiv
Assuming actual H2: 90% annual load factor at 7,884,000 kg/y H2 /station or gal/y gaso equiv or 657,000 kg/month H2 or gal/mo. gaso equiv
thereby 42,705 vehicles can be serviced at 6,101 fill-ups/d @ 4.2 kg or gal equiv/fill-up or each vehicle fills up one a week

Biomass → biomass gasifier 70.0% LHV effic
198 MM Btu/h LHV
210 MM Btu/h HHV
11,908 kg/hr @8,000 Btu/lb dry
286 tons/d biomass bone dry
93,881 tons/yr biomass bone dry
6,259 hectares of land for biomass
24 square miles of land to grow biomass

139 MM Btu/hr hot raw syngas → SMR or shift cool & clean 1% PSA loses
50% CO/(H2+CO) syngas
35 atm
5,954 kg/hr O2

→ 18,993 kg/hr CO2 plus 15% from dryer
→ 1 MM Btu/hr PSA fuel gas
10 MM Bur/h CO to H2 shifting storage
CO2 emission to atmosphere
Note: Need to subtract carbon reduced from growing biomass

Source	kg CO2/kg H2	kg C/kg H2
From biomass	-	-
From Ele Gen	1.03	0.28
Total	1.03	0.28

52% Biomass + fuel for power to H2 efficiency

Electric Power → air comp or ASU 0.370 kWh/kg O2

ASU	2,203	
Misc.	1,000	
Total	3,203	kW

143 metric tons/d O2
0.50 tons O2/ton dry feed

→ Hydrogen @ 75 atm to Liqufaction
24,000 design kg/d H2 or gal/d gasoline equivalent
21,600 actual kg/d annual ave.

13% of biomass fired in FBC to dry gasifier biomass feed
249 tons/day bone dry biomass to gasifier
at 0.32 kg CO2/kWh incremental U.S. NG =
1,611 Btu/lb water vaporized
1,500 Btu/lb water vaporized minimum
1,025 kg/hr CO2 equivalent at power plants

Capital Costs		Unit cost basis at 100,000	kg/d H2	cost/size factors		Unit cost at 24,000	kg/d H2	millions of $ for 1 plant	Notes
Biomass handling & drying	$	22	/kg/d dry bio	75%	$	31	/kg/d dry bio	9.0	16 /kg/d green (wet) biomass
Advanced biomass gasifier	$	15	/kg/d dry bio	80%	$	20	/kg/d dry bio	9.9	100% spare unit
ASU or air compress	$	27	/kg/d oxygen	75%	$	39	/kg/d oxygen	5.5	$ 2,502 /kW power
SMR or shift, cool & cleanup	$	15	/kg/d CO2	75%	$	21	/kg/d CO2	9.8	$ 1.0 /scf/d H2 MDEA & PSA
						Total process units		34.2	
General Facilities				30%	of process units			10.3	20-40% typical, SMR + 10%
Engineering Permitting & Startup				10%	of process units			3.4	10-20% typical
Contingencies				10%	of process units			3.4	10-20% typical, low after the first few
Working Capital, Land & Misc.				7%	of process units			2.4	5-10% typical
						U.S. Gulf Coast Capital Costs		53.7	
Site specific factor		110%	of US Gulf Coast costs			Total Capital Costs		59.0	

Unit Capital Costs 5.93 /scf/d H2 or 2,460 /kg/d H2 or 2,460 /gal/d gaso equiv

Hydrogen Costs at		90%	ann load factor	million $/yr of 1 plant	$/million Btu LHV	$/1,000 scf H2	$/kg H2 or $/gal gaso equiv	Notes
Variable Non-fuel O&M		1.0%	/yr of capital	0.6	0.66	0.18	0.07	0.5-1.5% typical
Delivered biomass	$	2.02	/MM Btu HHV	3.3	3.72	1.02	0.42	based on costs below
Electricity	$	0.045	/kWh	1.1	1.27	0.35	0.14	0.04-0.05/kWh typical industrial rates
Variable Operating Cost				5.1	5.65	1.55	0.64	
Fixed Operating Cost		5.0%	/yr of capital	3.0	3.29	0.90	0.37	4-7% typical for refining
Capital Charges		18%	/yr of capital	9.4	10.46	2.87	1.19	20-25% typical for refining
Total Liquid Hydrogen Costs from Biomass				17.4	19.40	5.33	2.21	including return on investment
Carbon Tax	$	50.00	/ton	0.1	0.12	0.03	0.01	include indirect C from power

notes: Costs are for production at the plant only and do not include distribution costs. Assume no plant storage and compression of hydrogen to 75 atm for distribution.

Delivered biomass @ $ 35.58 /bone dry ton (BDT) or $ 2.02 /million Btu LHV based on below:
$ 500 /hectare per yr gross total revenues or $ 200 /acre per yr gross total revenues
15 ton biomass/yr per ha - bone dry basic or 6.0 tons biomass/yr per acre - bone dry
8,000 Btu/lb HHV bone dry and 50% moisture of green biomass
$ 2.08 /mile round trip for typical 25 ton truck hauling green biomass
13 miles round trip haul = $ 1.12 /ton green or $ 2.25 /ton bone dry equivalent transportation

If waste bio or coproduct lower gross revenue needs but much lower yield/ha

TABLE E-25
MS Bio-C Seq

Midsize Hydrogen via Current Biomass Gasification Technology with CO_2 Capture

Color codes **variables** via summary inputs key outputs

MS-Size Plant Design		**Design LHV energy equivalent**			
	Hydrogen	**gasoline**	**million**		
Size range	**kg/d H2**	**gal/d**	**Btu/hr**	**scf/d H2**	**MW t**
Maximum	200,000	200,000	948	82,900,000	278
This run	**24,000**	**24,000**	**114**	**9,948,000**	**33**
Minimum	20,000	20,000	95	8,290,000	28

gasoline equivalent

Assuming 65 mpg and 12,000 mile/yr
requires 185 kg/yr H2/vehicle or gal/yr gaso equiv
Assuming 90% annual load factor at
actual H2 7,884,000 kg/y H2 /station or gal/y gaso equiv
or 657,000 kg/month H2 or gal/mo. gaso equiv
thereby **42,705 vehicles can be serviced at**
6,101 fill-ups/d @ 4.2 kg or gal equiv/fill-up
or each vehicle fills up one a week

Shell gasifier to avoid high CH4 & secondary SMR or ATR

Biomass → biomass gasifier 50.0% LHV effic → 153 MM Btu/hr hot raw syngas 50% CO/(H2+CO) 35 atm → CO shift cool & clean 5% syngas PSA loses

306 MM Btu/h LHV
324 MM Btu/h HHV
18,396 kg/hr @8,000 Btu/lb dry
442 tons/d biomass bone dry
145,036 tons/yr biomass bone dry
14,504 hectares of land for biomass
57 square miles of land to grow biomass

14,717 kg/hr O2

8 MM Btu/hr PSA fuel gas
11 MM Bur/h CO to H2 shifting LHV loses

CO2 emission to atmosphere

	Subtracting growth of biomass		CO2 Sequestered	
Source	kg CO2/kg H2	kg C/kg H2	kg CO2/kg H2	kg C/kg H2
From biomass	(25.80)	(7.04)	25.80	7.04
From Ele Gen	2.89	0.79		
Total	**(22.91)**	**(6.25)**		

Electric Power	
ASU	5,445
Misc.	1,000
CO2	2,580
Total	**9,026 kW**

ASU 0.370 kWh/kg O2
353 metric tons/d O2
0.80 t O2/ton dry feed

CO2 Compression 2,580 kwe
25,801 kg/hr CO2
33.2 kg CO2/kg H2

37% overall effic raw bio to H2
1,000 kg/hr H2 to hydrogen liquefaction
114 MM Btu/hr H2
414,500 scf/hr H2 @ 30 atm
31% Biomass + fuel for power to H2 efficiency
plus 15% from dryer

15% of biomass fired in FBC to dry gasifier biomass feed
375 tons/day bone dry biomass to gasifier
at **0.32** kg CO2/kWh incremental U.S. NG =
90% Percent CO2 Separated

1,902 Btu/lb water vaporized
1,500 Btu/lb water vaporized minimum
2,888 kg/hr CO2 equivalent at power plants

	Unit cost basis at		cost/size	Unit cost at		millions of $	Notes
Capital Costs	**100,000**	**kg/d H2**	**factors**	**24,000**	**kg/d H2**	**for 1 plant**	
Biomass handling & drying	$ 25	/kg/d dry bio	75%	$ 36	/kg/d dry bio	15.8	18 /kg/d green (wet) biomass
Shell gasifer	$ 20	/kg/d dry bio	80%	$ 27	/kg/d dry bio	20.0	100% spare unit H2O quench
Air separation unit (ASU)	$ 33	/kg/d oxygen	75%	$ 47	/kg/d oxygen	16.7	$ 3,058 /kW power
CO2 Compressor	$ 1,000	/kW	80%	$ 1,330	/kW	3.4	$ 143 /kg/d H2
CO shift, cool & cleanup	$ 18	/kg/d CO2	75%	$ 26	/kg/d CO2	15.9	$ 1.6 /scf/d H2 MDEA & PSA
				Total process units		71.7	
General Facilities		**30%** of process units				21.5	20-40% typical, SMR + **10%**
Engineering Permitting & Startup		**10%** of process units				7.2	10-20% typical
Contingencies		**10%** of process units				7.2	10-20% typical, low after the first few
Working Capital, Land & Misc.		**7%** of process units				5.0	5-10% typical
				U.S. Gulf Coast Capital Costs		112.6	
Site specific factor	**110%**	of US Gulf Coast costs		**Total Capital Costs**		**123.9**	
Unit Capital Costs	**12.46**	**/scf/d H2 or**	5,163 /kg/d H2 or			5,163 /gal/d gaso equiv	

Hydrogen Costs at	90%	ann load factor	million $/yr of 1 plant	$/million Btu LHV	$/1,000 scf H2	$/kg H2 or $/gal gaso equiv	Notes
Variable Non-fuel O&M	**1.0%**	/yr of capital	1.2	1.38	0.38	0.16	0.5-1.5% typical
Delivered biomass	$ **3.01**	/MM Btu HHV	7.7	8.57	2.35	0.98	**based on costs below**
Electricity	$ **0.045**	/kWh	3.2	3.57	0.98	0.41	0.04-0.05/kWh typical **industrial** rates
Variable Operating Cost			**12.1**	**13.52**	**3.71**	**1.54**	From plant gate at high pressure
Fixed Operating Cost	**5.0%**	/yr of capital	**6.2**	**6.90**	**1.90**	**0.79**	4-7% typical for refining
Capital Charges	**16%**	/yr of capital	**19.7**	**21.96**	**6.03**	**2.50**	20-25% typical for refining
Total Liquid Hydrogen Costs from Biomass			**38.0**	**42.38**	**11.64**	**4.82**	including return on investment
Carbon Tax	$ **50.00**	/ton	(2.5)	(2.75)	(0.75)	(0.31)	include indirect C from power
CO2 Disposal	$ **10.00**	/ton	2.0	2.27	0.62	0.26	

plant gate still requires distribution

Delivered biomass @ $ 53.02 /bone dry ton (BDT) or $ 3.01 /million Btu LHV based on below:
$ 500 /hectare per yr gross total revenues or $ 200 /acre per yr gross total revenues
10 ton biomass/yr per ha - bone dry basic or 4.0 tons biomass/yr per acre - bone dry
8,000 Btu/lb HHV bone dry and 50% moisture of green biomass
$ 2.08 /mile round trip for typical 25 ton truck hauling green biomass
18 miles round trip haul = $ 1.51 /ton green or $ **3.02 /ton bone dry equivalent transportation**

If waste bio or coproduct lower gross revenue needs but much lower yield/

TABLE E-26
MS Bio-F Seq
Mid-Size Hydrogen via Biomass Gasification Technology Plus CO_2 Capture with Future Optimism

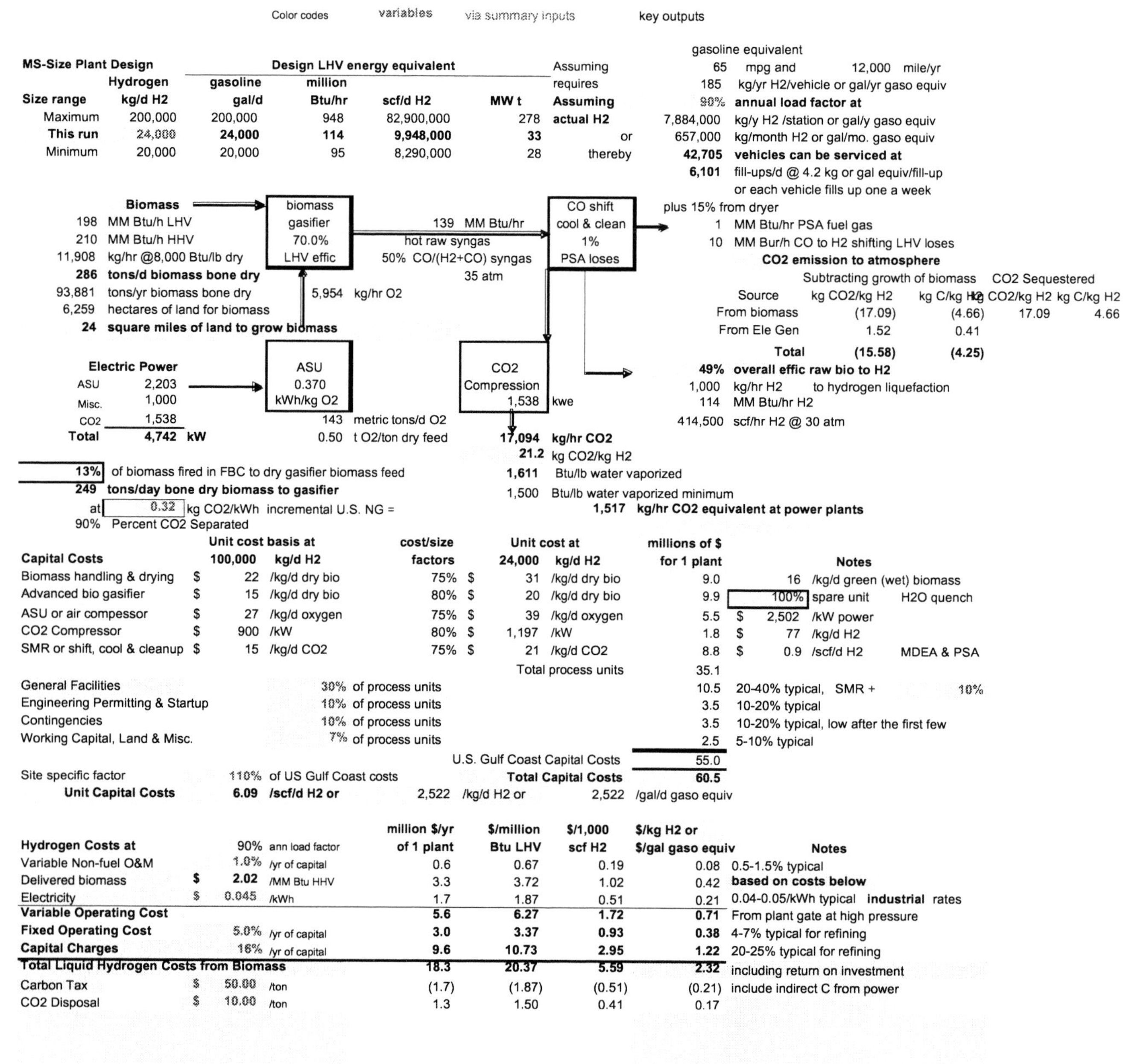

Color codes | variables | via summary inputs | key outputs

MS-Size Plant Design	Hydrogen	gasoline	Design LHV energy equivalent million		
Size range	kg/d H2	gal/d	Btu/hr	scf/d H2	MW t
Maximum	200,000	200,000	948	82,900,000	278
This run	24,000	24,000	114	9,948,000	33
Minimum	20,000	20,000	95	8,290,000	28

Assuming requires | gasoline equivalent 65 mpg and 12,000 mile/yr
185 kg/yr H2/vehicle or gal/yr gaso equiv
Assuming 90% annual load factor at
actual H2 7,884,000 kg/y H2 /station or gal/y gaso equiv
or 657,000 kg/month H2 or gal/mo. gaso equiv
thereby 42,705 vehicles can be serviced at
6,101 fill-ups/d @ 4.2 kg or gal equiv/fill-up
or each vehicle fills up one a week

Biomass
198 MM Btu/h LHV
210 MM Btu/h HHV
11,908 kg/hr @8,000 Btu/lb dry
286 tons/d biomass bone dry
93,881 tons/yr biomass bone dry
6,259 hectares of land for biomass
24 square miles of land to grow biomass

biomass gasifier 70.0% LHV effic
139 MM Btu/hr hot raw syngas
50% CO/(H2+CO) syngas
35 atm
5,954 kg/hr O2

CO shift cool & clean 1% PSA loses
plus 15% from dryer
1 MM Btu/hr PSA fuel gas
10 MM Bur/h CO to H2 shifting LHV loses
CO2 emission to atmosphere

Source	Subtracting growth of biomass kg CO2/kg H2	kg C/kg H2	CO2 Sequestered kg CO2/kg H2	kg C/kg H2
From biomass	(17.09)	(4.66)	17.09	4.66
From Ele Gen	1.52	0.41		
Total	(15.58)	(4.25)		

49% overall effic raw bio to H2
1,000 kg/hr H2 to hydrogen liquefaction
114 MM Btu/hr H2
414,500 scf/hr H2 @ 30 atm

Electric Power	
ASU	2,203
Misc.	1,000
CO2	1,538
Total	4,742 kW

ASU 0.370 kWh/kg O2
143 metric tons/d O2
0.50 t O2/ton dry feed

CO2 Compression 1,538 kwe
17,094 kg/hr CO2
21.2 kg CO2/kg H2
1,611 Btu/lb water vaporized
1,500 Btu/lb water vaporized minimum
1,517 kg/hr CO2 equivalent at power plants

13% of biomass fired in FBC to dry gasifier biomass feed
249 tons/day bone dry biomass to gasifier
at 0.32 kg CO2/kWh incremental U.S. NG =
90% Percent CO2 Separated

Capital Costs	Unit cost basis at 100,000 kg/d H2		cost/size factors	Unit cost at 24,000 kg/d H2		millions of $ for 1 plant	Notes
Biomass handling & drying	$ 22	/kg/d dry bio	75%	$ 31	/kg/d dry bio	9.0	16 /kg/d green (wet) biomass
Advanced bio gasifier	$ 15	/kg/d dry bio	80%	$ 20	/kg/d dry bio	9.9	100% spare unit H2O quench
ASU or air compressor	$ 27	/kg/d oxygen	75%	$ 39	/kg/d oxygen	5.5	$ 2,502 /kW power
CO2 Compressor	$ 900	/kW	80%	$ 1,197	/kW	1.8	$ 77 /kg/d H2
SMR or shift, cool & cleanup	$ 15	/kg/d CO2	75%	$ 21	/kg/d CO2	8.8	$ 0.9 /scf/d H2 MDEA & PSA
				Total process units		35.1	
General Facilities	30%	of process units				10.5	20-40% typical, SMR + 10%
Engineering Permitting & Startup	10%	of process units				3.5	10-20% typical
Contingencies	10%	of process units				3.5	10-20% typical, low after the first few
Working Capital, Land & Misc.	7%	of process units				2.5	5-10% typical
				U.S. Gulf Coast Capital Costs		55.0	
Site specific factor	110%	of US Gulf Coast costs		Total Capital Costs		60.5	
Unit Capital Costs	6.09	/scf/d H2 or	2,522 /kg/d H2 or	2,522	/gal/d gaso equiv		

Hydrogen Costs at	90%	ann load factor	million $/yr of 1 plant	$/million Btu LHV	$/1,000 scf H2	$/kg H2 or $/gal gaso equiv	Notes
Variable Non-fuel O&M	1.0%	/yr of capital	0.6	0.67	0.19	0.08	0.5-1.5% typical
Delivered biomass	$ 2.02	/MM Btu HHV	3.3	3.72	1.02	0.42	based on costs below
Electricity	$ 0.045	/kWh	1.7	1.87	0.51	0.21	0.04-0.05/kWh typical industrial rates
Variable Operating Cost			5.6	6.27	1.72	0.71	From plant gate at high pressure
Fixed Operating Cost	5.0%	/yr of capital	3.0	3.37	0.93	0.38	4-7% typical for refining
Capital Charges	18%	/yr of capital	9.6	10.73	2.95	1.22	20-25% typical for refining
Total Liquid Hydrogen Costs from Biomass			18.3	20.37	5.59	2.32	including return on investment
Carbon Tax	$ 50.00	/ton	(1.7)	(1.87)	(0.51)	(0.21)	include indirect C from power
CO2 Disposal	$ 10.00	/ton	1.3	1.50	0.41	0.17	

notes: Costs are for production at the plant only and do not include distribution costs. Assume no plant storage and compression of hydrogen to 75 atm for distribution.

Delivered biomass @ $ 35.58 /bone dry ton (BDT) or $ **2.02 /million Btu LHV based on below:**
$ 500 /hectare per yr gross total revenues or $ 200 /acre per yr gross total revenues
15 ton biomass/yr per ha - bone dry basic or 6.0 tons biomass/yr per acre - bone dry
8,000 Btu/lb HHV bone dry and 50% moisture of green biomass
$ 2.08 /mile round trip for typical 25 ton truck hauling green biomass
13 miles round trip haul = $ 1.12 /ton green or $ **2.25 /ton bone dry equivalent transportation**

If waste bio or coproduct lower gross revenue needs but much lower yield/ha

TABLE E-27
MS Ele-C
Midsize Hydrogen via Electrolysis of Water with Current Technology

Color codes variables via summary inputs key outputs

MS-Size Plant Design		Design LHV energy equivalent				
	Hydrogen	gasoline	million			
Size range	kg/d H2	gal/d	Btu/hr	scf/d H2	MW t	
Maximum	100,000	100,000	474	41,450,000	139	
This run	24,000	24,000	114	9,948,000	33	
Minimum	10,000	10,000	47	4,145,000	14	

Assuming requires gasoline equivalent 65 mpg and 12,000 mile/yr
185 kg/yr H2/vehicle or gal/yr gaso equiv
Assuming actual H2 90% annual load factor at
7,884,000 kg/y H2 /station or gal/y gaso equiv
or 657,000 kg/month H2 or gal/mo. gaso equiv
thereby 42,705 vehicles can be serviced at
6,101 fill-ups/d @ 4.2 kg or gal equiv/fill-up
or each vehicle fills up one a week

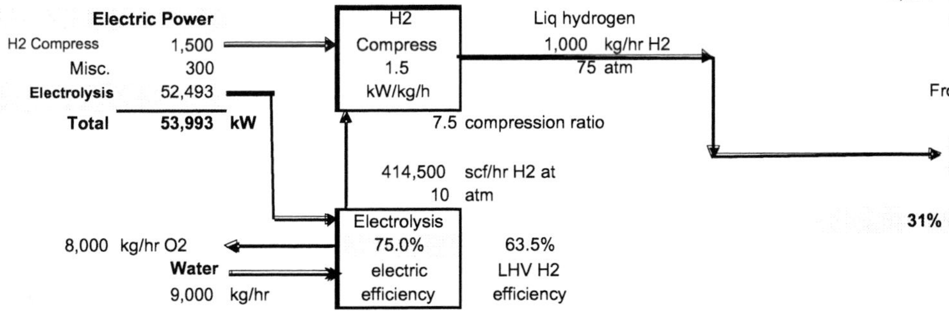

Electric Power	
H2 Compress	1,500
Misc.	300
Electrolysis	52,493
Total	53,993 kW

H2 Compress 1.5 kW/kg/h — Liq hydrogen 1,000 kg/hr H2, 75 atm — 7.5 compression ratio

414,500 scf/hr H2 at 10 atm

Electrolysis 75.0% electric efficiency — 63.5% LHV H2 efficiency

8,000 kg/hr O2; Water 9,000 kg/hr

Source	kg CO2/kg H2	kg C/kg H2
From Ele Gen	17.28	4.71
Total	17.28	4.71

Hydrogen @ 75 atm to hydrogen liquefaction

31% fuel for power to H2 efficiency

theoretical power 39.37 kWh/kg H2 at 100% electric efficiency
actual power 52.49 kWh/kg or 4.73 kWh/Nm3 H2
at 0.32 kg CO2/kWh incremental U.S. NG = 17,278 kg/hr CO2 equivalent at power plants
17.3 kgCO2/kg H2

Capital Costs	Unit cost basis at 100,000 kg/d H2	cost/size factors	Unit cost at 24,000 kg/d H2	millions of $ for 1 plant	Notes
Electrolyser	$ 800 /kW	90%	$ 923 /kW	48.4	$ 4.9 /scf/d H2
H2 Compressor	$ 2,000 /kW	80%	$ 2,661 /kW	4.0	$ 166 /kg/d H2
				52.4	
General Facilities		20% of process units		10.5	20-40% typical
Engineering Permitting & Startup		10% of process units		5.2	10-20% typical
Contingencies		10% of process units		5.2	10-20% typical, low after the first few
Working Capital, Land & Misc.		7% of process units		3.7	5-10% typical
			U.S. Gulf Coast Capital Costs	77.1	
Site specific factor	110% of US Gulf Coast costs		Total Capital Costs	$ 84.8	
Unit Capital Costs of	8.52 /scf/d H2 or	3,532 /kg/d H2 or	3,532 /gal/d gaso equiv		

Hydrogen Costs at	90% ann load factor	million $/yr of 1 plant	$/million Btu LHV	$/1,000 scf H2	$/kg H2 or $/gal gaso equiv	Notes
Non-fuel Variable O&M	1.0% /yr of capital	0.848	0.94	0.26	0.11	0.5-1.5% typical
Oxygen byproduct	$ (10) /ton O2	(0.631)	(0.70)	(0.19)	(0.08)	large amount could create min. value
Electricity	$ 0.045 /kWh	19.16	21.35	5.86	2.43	$0.04-0.05/kWh **industrial** rate
Variable Operating Cost		19.373	21.59	5.93	2.46	
Fixed Operating Cost	5.0% /yr of capital	4.239	4.72	1.30	0.54	4-7% typical for refining
Capital Charges	16% /yr of capital	13.479	15.02	4.12	1.71	20-25% typical for refining
Total Liquid Hydrogen Costs from Electrolysis		37.091	41.34	11.35	4.70	including return on investment
Carbon Tax	$ 50.00 /ton	1.9	2.07	0.57	0.24	include indirect C from power

notes: Costs are for production at the plant only and do not include distribution costs. Assume no plant storage and compression of hydrogen to 75 atm for distribution.

Note: if 12 hr/d at only $ 0.020 /kWh lower off-peak rate and
12 hr/d at $ 0.060 /kWh higher peak rate daily average rate is $ 0.040 /kWh

If only operated during low off-peak rates times would have low ann load factor & need more H2 storage

Assume Hydrogn Systems Electrolysis at 150 psig pressure, Norsk Hydro & Stuard systems are low pressure

Table E-28
MS Ele-F
Midsize Hydrogen via Electrolysis of Water with Future Optimism

Color codes variables via summary inputs key outputs

MS-Size Plant Design	Hydrogen	Design LHV energy equivalent				
		gasoline	million			
Size range	kg/d H2	gal/d	Btu/hr	scf/d H2	MW t	
Maximum	100,000	100,000	474	41,450,000	139	
This run	24,000	**24,000**	**114**	**9,948,000**	**33**	
Minimum	10,000	10,000	47	4,145,000	14	

		gasoline equivalent	
Assuming	65	mpg and	12,000 mile/yr
requires	185	kg/yr H2/vehicle or gal/yr gaso equiv	
Assuming	90%	**annual load factor at**	
actual H2	7,884,000	kg/y H2 /station or gal/y gaso equiv	
or	657,000	kg/month H2 or gal/mo. gaso equiv	
thereby	**42,705**	**vehicles can be serviced at**	
	6,101	fill-ups/d @ 4.2 kg or gal equiv/fill-up	
		or each vehicle fills up one a week	

Electric Power	
H2 Compress	500
Misc.	300
Electrolysis	46,318
Total	**46,818 kW**

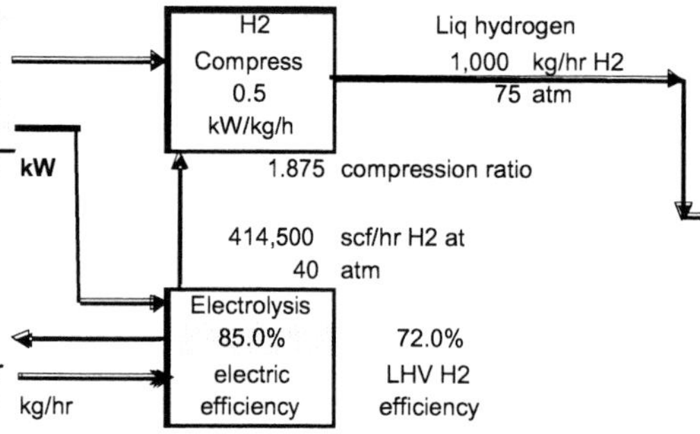

H2 Compress 0.5 kW/kg/h

Liq hydrogen 1,000 kg/hr H2 75 atm

1.875 compression ratio

414,500 scf/hr H2 at 40 atm

Electrolysis 85.0% electric efficiency

72.0% LHV H2 efficiency

8,000 kg/hr O2

Water 9,000 kg/hr

Source	kg CO2/kg H2	kg C/kg H2
From Ele Gen	14.98	4.09
Total	**14.98**	**4.09**

Hydrogen @ 75 atm to hydrogen liquefaction

36% Fuel for power to H2 efficiency

theoretical power 39.37 kWh/kg H2 at 100% electric efficiency

actual power 46.32 kWh/kg or 4.17 kWh/Nm3 H2

at 0.32 kg CO2/kWh incremental U.S. NG = **14,982 kg/hr CO2 equivalent at power plants**

15.0 kgCO2/kg H2

Capital Costs	Unit cost basis at 100,000 kg/d H2	cost/size factors	Unit cost at 24,000 kg/d H2	millions of $ for 1 plant	Notes
Electrolyser	$ 100 /kW	90%	$ 115 /kW	5.3	$ 0.5 /scf/d H2
H2 Compressor	$ 1,000 /kW	80%	$ 1,330 /kW	0.7	$ 28 /kg/d H2
			Total process units	6.0	
General Facilities		20% of process units		1.2	20-40% typical
Engineering Permitting & Startup		10% of process units		0.6	10-20% typical
Contingencies		10% of process units		0.6	10-20% typical, low after the first few
Working Capital, Land & Misc.		7% of process units		0.4	5-10% typical
			U.S. Gulf Coast Capital Costs	8.8	
Site specific factor	110% of US Gulf Coast costs		**Total Capital Costs**	**$ 9.7**	
Unit Capital Costs of	**0.98 /scf/d H2 or**	405 /kg/d H2 or	405 /gal/d gaso equiv		

Hydrogen Costs at	90% ann load factor	million $/yr of 1 plant	$/million Btu LHV	$/1,000 scf H2	$/kg H2 or $/gal gaso equiv	Notes
Non-fuel Variable O&M	1.0% /yr of capital	0.097	0.11	0.03	0.01	0.5-1.5% typical
Oxygen byproduct	$ (10) /ton O2	(0.631)	(0.70)	(0.19)	(0.08)	large amount could create min. value
Electricity	$ 0.045 /kWh	16.610	18.51	5.08	2.11	$0.04-0.05/kWh **industrial** rate
Variable Operating Cost		**16.076**	**17.92**	**4.92**	**2.04**	
Fixed Operating Cost	5.0% /yr of capital	**0.486**	**0.54**	**0.15**	**0.06**	4-7% typical for refining
Capital Charges	16% /yr of capital	**1.545**	**1.72**	**0.47**	**0.20**	20-25% typical for refining
Total Liquid Hydrogen Costs from Electrolysis		**18.107**	**20.18**	**5.54**	**2.30**	including return on investment
Carbon Tax	$ 50.00 /ton	1.6	1.80	0.49	0.20	include indirect C from power

notes: Costs are for production at the plant only and do not include distribution costs. Assume no plant storage and compression of hydrogen to 75 atm for distribution.

Note: if 12 hr/d at only $ 0.020 /kWh lower off-peak rate and
12 hr/d at $ 0.060 /kWh higher peak rate daily average rate is $ 0.040 /kWh

If only operated during low off-peak rates times would have low ann load factor & need more H2 storage

Assume Hydrogn Systems Electrolysis at 150 psig pressure, Norsk Hydro & Stuard systems are low pressure

TABLE E-29
MS Truck-C
Liquid Hydrogen Distribution via Tanker Trucks Based on Current Technology

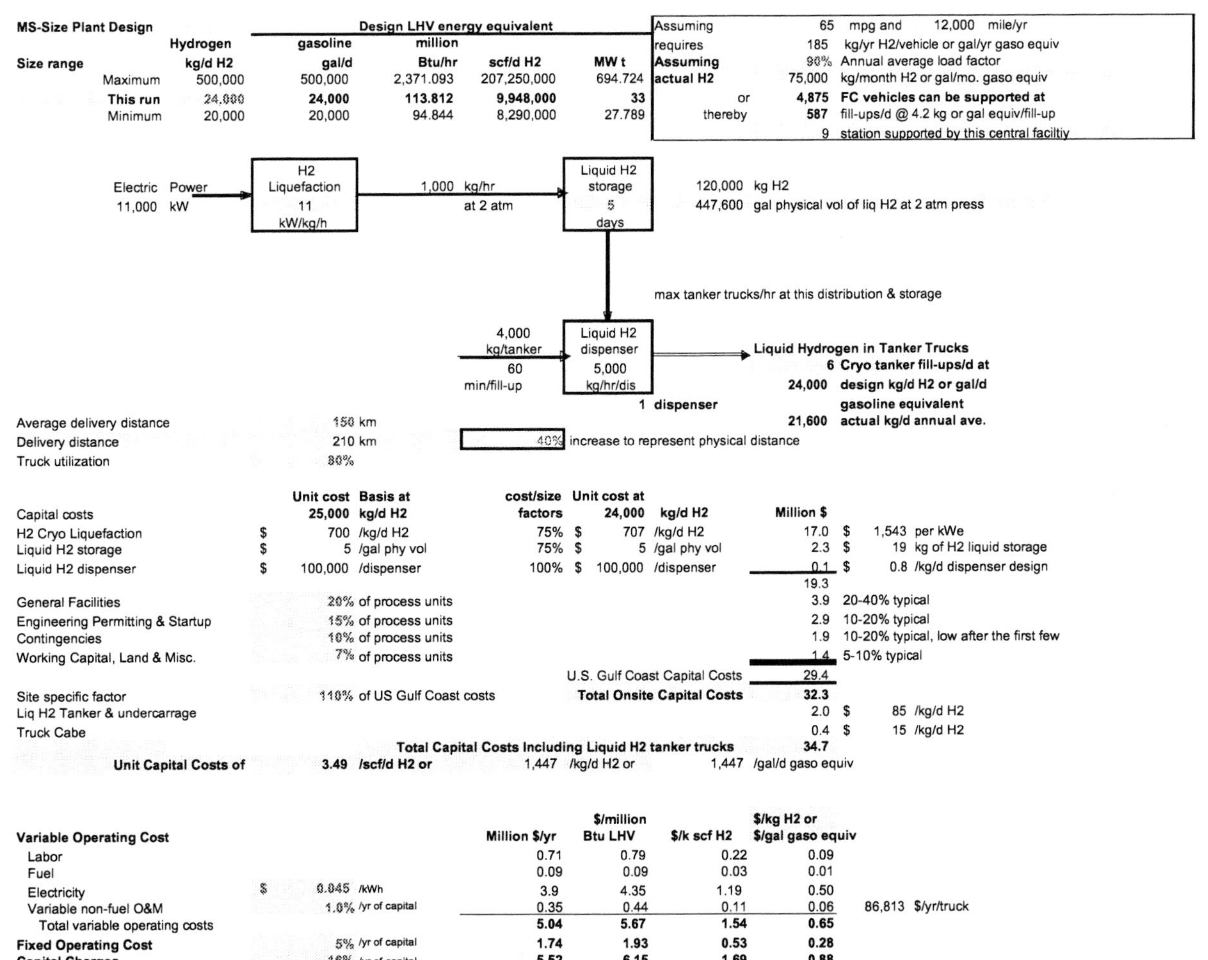

MS-Size Plant Design		Hydrogen	Design LHV energy equivalent gasoline	million		
Size range		kg/d H2	gal/d	Btu/hr	scf/d H2	MW t
	Maximum	500,000	500,000	2,371.093	207,250,000	694.724
	This run	24,000	24,000	113.812	9,948,000	33
	Minimum	20,000	20,000	94.844	8,290,000	27.789

Assuming	65	mpg and 12,000 mile/yr
requires	185	kg/yr H2/vehicle or gal/yr gaso equiv
Assuming	90%	Annual average load factor
actual H2	75,000	kg/month H2 or gal/mo. gaso equiv
or	4,875	FC vehicles can be supported at
thereby	587	fill-ups/d @ 4.2 kg or gal equiv/fill-up
	9	station supported by this central faciltiy

Electric Power 11,000 kW → H2 Liquefaction 11 kW/kg/h → 1,000 kg/hr at 2 atm → Liquid H2 storage 5 days — 120,000 kg H2; 447,600 gal physical vol of liq H2 at 2 atm press

max tanker trucks/hr at this distribution & storage

4,000 kg/tanker, 60 min/fill-up → Liquid H2 dispenser 5,000 kg/hr/dis → Liquid Hydrogen in Tanker Trucks

1 dispenser

6 Cryo tanker fill-ups/d at
24,000 design kg/d H2 or gal/d gasoline equivalent
21,600 actual kg/d annual ave.

Average delivery distance	150	km
Delivery distance	210	km — 40% increase to represent physical distance
Truck utilization	80%	

Capital costs		Unit cost	Basis at 25,000 kg/d H2	cost/size factors		Unit cost at 24,000	kg/d H2	Million $		
H2 Cryo Liquefaction	$	700	/kg/d H2	75%	$	707	/kg/d H2	17.0	$	1,543 per kWe
Liquid H2 storage	$	5	/gal phy vol	75%	$	5	/gal phy vol	2.3	$	19 kg of H2 liquid storage
Liquid H2 dispenser	$	100,000	/dispenser	100%	$	100,000	/dispenser	0.1	$	0.8 /kg/d dispenser design
								19.3		
General Facilities		20%	of process units					3.9		20-40% typical
Engineering Permitting & Startup		15%	of process units					2.9		10-20% typical
Contingencies		10%	of process units					1.9		10-20% typical, low after the first few
Working Capital, Land & Misc.		7%	of process units					1.4		5-10% typical
							U.S. Gulf Coast Capital Costs	29.4		
Site specific factor		110%	of US Gulf Coast costs				Total Onsite Capital Costs	32.3		
Liq H2 Tanker & undercarrage								2.0	$	85 /kg/d H2
Truck Cabe								0.4	$	15 /kg/d H2
							Total Capital Costs Including Liquid H2 tanker trucks	34.7		
Unit Capital Costs of		3.49	/scf/d H2 or			1,447	/kg/d H2 or	1,447		/gal/d gaso equiv

Variable Operating Cost			Million $/yr	$/million Btu LHV	$/k scf H2	$/kg H2 or $/gal gaso equiv	
Labor			0.71	0.79	0.22	0.09	
Fuel			0.09	0.09	0.03	0.01	
Electricity	$ 0.045	/kWh	3.9	4.35	1.19	0.50	
Variable non-fuel O&M	1.0%	/yr of capital	0.35	0.44	0.11	0.06	86,813 $/yr/truck
Total variable operating costs			5.04	5.67	1.54	0.65	
Fixed Operating Cost	5%	/yr of capital	1.74	1.93	0.53	0.28	
Capital Charges	16%	/yr of capital	5.52	6.15	1.69	0.88	
Total operating costs			12.30	13.75	3.76	1.80	

Assumptions

Truck costs			
Tank unit	450,000	$/module	113 $/kg H2 stroage
Undercarrage	60,000	$/trailer	
Cabe	90,000	$/cab	
Truck boil-off rate	0.30	%/day	
Truck capacity	4000	kg/truck	
Fuel economy	6	mpg	
Average speed	50	km/hr	
Load/unload time	4	hr/trip could be lowered with a liquid H2 pump	
Truck availability	24	hr/day	
Hour/driver	12	hr/driver	
Driver wage & benefits	28.75	$/hr	
Fuel price	1	$/gal	
Truck requirement calculations			
Trips per year	1,971		5 trips per day
Total Distance	827,820	km/yr	206,955 km/yr per truck — little high
Time for each trip	8.4	hr/trip	
Trip length	12.4	hr/trip	
Delivered product	7,785,285	kg/yr	
Total delivery time	24,440	hr/yr	
Total driving time	16,556	hr/yr	
Total load/unload time	7,884	hr/yr	
Truck availability	7008	hr/yr	
Truck requirement	4	trucks	
Driver time	3504	hr/yr	
Drivers required	7	persons	
Fuel usage	85,000	gal/yr	
Delivery energy (9.63 kWth/liter diesel)	0.39	kW/Kg H2	

TABLE E-30
MS Truck-F
Liquid Hydrogen Distribution via Tanker Trucks Based on Future Optimism

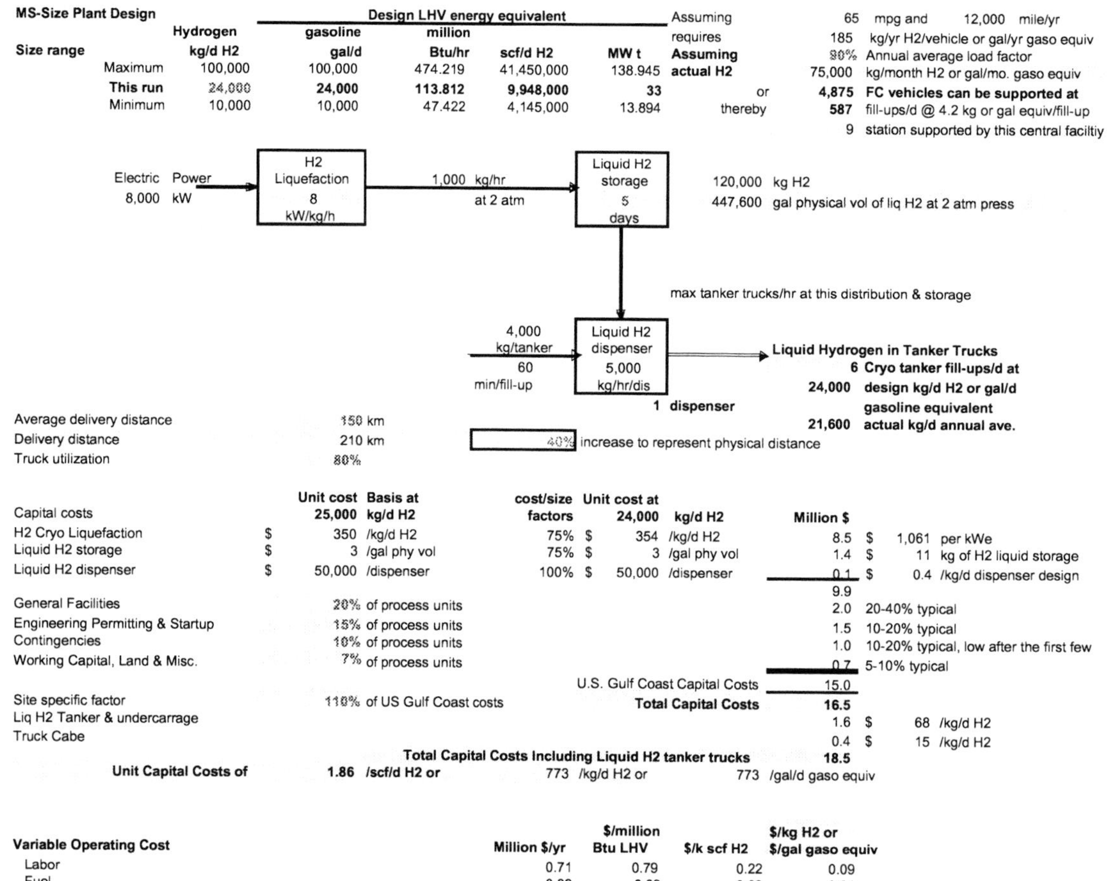

Variable Operating Cost		Million $/yr	$/million Btu LHV	$/k scf H2	$/kg H2 or $/gal gaso equiv	
Labor		0.71	0.79	0.22	0.09	
Fuel		0.09	0.09	0.03	0.01	
Electricity	$ 0.045 /kWh	2.8	3.16	0.87	0.36	
Variable non-fuel O&M	1.0% /yr of capital	0.19	0.23	0.06	0.03	46,352 $/yr/truck
Total variable operating costs		3.81	4.28	1.17	0.49	
Fixed Operating Cost	5% /yr of capital	0.93	1.03	0.28	0.15	
Capital Charges	16% /yr of capital	2.95	3.28	0.90	0.47	
Total operating costs		7.69	8.59	2.35	1.10	

Assumptions

Truck costs		
Tank unit	350,000	$/module 88 $/kg H2 stroage
Undercarrage	60,000	$/trailer
Cabe	90,000	$/cab
Truck boil-off rate	0.30	%/day
Truck capacity	4000	kg/truck
Fuel economy	6	mpg
Average speed	50	km/hr
Load/unload time	4	hr/trip could be lowered with a liquid H2 pump
Truck availability	24	hr/day
Hour/driver	12	hr/driver
Driver wage & benefits	28.75	$/hr
Fuel price	1	$/gal

Truck requirement calculations

Trips per year		1,971	5 trips per day
Total Distance		827,820 km/yr	206,955 km/yr per truck — little high
Time for each trip		8.4 hr/trip	
Trip length		12.4 hr/trip	
Delivered product		7,785,285 kg/yr	
Total delivery time		24,440 hr/yr	
Total driving time		16,556 hr/yr	
Total load/unload time		7,884 hr/yr	
Truck availability		7008 hr/yr	
Truck requirement		4 trucks	
Driver time		3504 hr/yr	
Drivers required		7 persons	
Fuel usage		85,000 gal/yr	
Delivery energy	9.63 kWth/liter diesel	0.39 kW/Kg H2	

TABLE- E-31
MS Disp-C
Liquid-Hydrogen-Based Fueling Stations with Current Technology

MS-Size size hydrogen production	24,000	kg/d design H2 central plant
Annual average load factor	90%	/yr of design

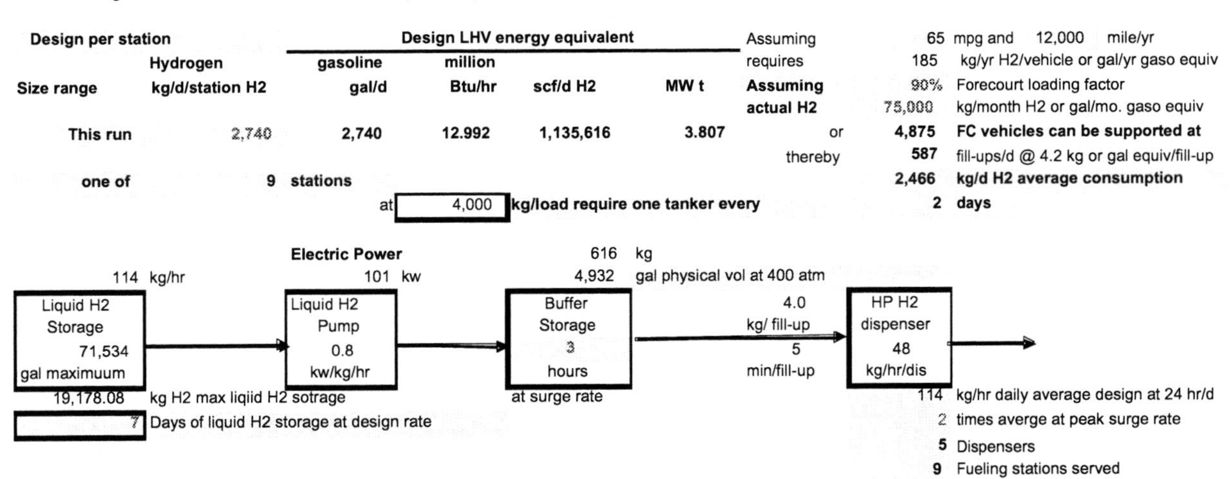

Capital Costs	Unit cost basis at 1,000	kg/d H2	cost/size factors	Unit cost at 2,740	kg/d H2	millions of $	Notes	
Liquid H2 pump/vaporizer	$ 250	/kg/d H2	70%	$ 185	/kg/d H2	0.506	$ 185	/kg/d H2
Liquid H2 storage	$ 10	/gal phy vol	70%	$ 7	/gal phy vol	0.529	$ 28	/kg/d H2
H2 buffer storage	$ 100	/gal phy vol	80%	$ 82	/gal phy vol	0.403	$ 654	/kg/d H2
Liquid H2 dispenser	$ 15,000	/dispenser	100%	$ 15,000	/dispenser	0.075	$ 27	/kg/d dispenser design
					Unit cost	1.513		
General Facilities & permitting	25%	of unit cost				0.378		
Eng. startup & contingencies	10%	of unit cost				0.151		
Contingencies	10%	of unit cost				0.151		
Working Capital, Land & Misc.	7%	of unit cost				0.106		
					Capital Costs	2.300	for 1 of	9 stations
					Total Capital Costs	21	for all	9 stations
Unit Capital Costs of	**2.03**	**/scf/d H2 or**	839	/kg/d H2 or	839	/gal/d gaso equiv		

Hydrogen Costs at	90%	ann load factor	$/yr of 1 station	$/million Btu LHV	$/k scf H2	$/kg H2 or $/gal gaso equiv	
Road tax or (subsidy)	$ -	/gal gaso equiv.	-	-	-	-	can be subsidy like EtOH
Gas Station mark-up	$ -	/gal gaso equiv.	-	-	-	-	if H2 drops total station revenues
Variable Non-fuel O&M	5.0%	/yr of capital	114,990	1.12	0.31	0.13	0.5-1.5 typical many be low here
Electricity	$ 0.070	/kWh	50,400	0.49	0.14	0.06	0.06-0.09 typical commercial rates
Variable Operating Cost			**165,390**	**1.61**	**0.44**	**0.18**	
Fixed Operating Cost	3.0%	/yr of capital	**68,994**	**0.67**	**0.18**	**0.08**	3-5% typical, may be lower here
Capital Charges	14.0%	/yr of capital	**321,973**	**3.14**	**0.86**	**0.36**	20-25% typical for refiners
Fueling Station Cost including return of investment			**556,357**	**5.43**	**1.49**	**0.62**	

Hydrogen Fueling Station Costs Delivery to	9 Stations Million $/yr
Variable Operating Cost	**1.49**
Fixed Operating Cost	**0.62**
Capital Charges	**2.90**
Total Fueling Station Cost	**5.01**

TABLE E-32
MS Disp-F
Liquid-Hydrogen-Based Fueling Stations with Future Optimism

MS-Size size hydrogen production	24,000	kg/d design H2 central plant
Annual average load factor	90%	/yr of design

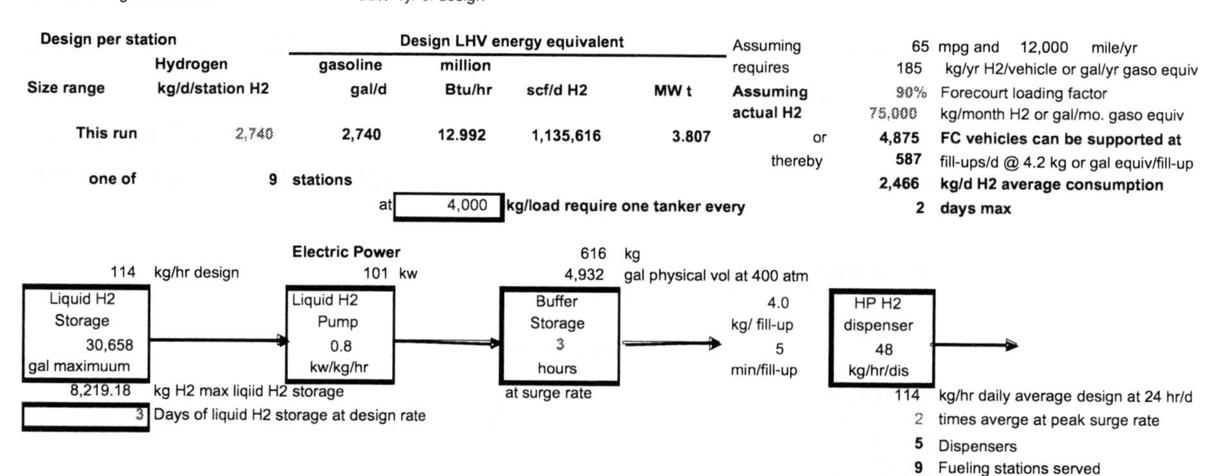

Capital Costs	Unit cost basis at 1,000	kg/d H2	cost/size factors	Unit cost at 2,740	kg/d H2	millions of $	Notes	
Liquid H2 pump/vaporizer	$ 150	/kg/d H2	70%	$ 111	/kg/d H2	0.304	$ 111	/kg/d H2
Liquid H2 storage	$ 5	/gal phy vol	70%	$ 4	/gal phy vol	0.113	$ 14	/kg/d H2
H2 buffer storage	$ 50	/gal phy vol	80%	$ 41	/gal phy vol	0.202	$ 327	/kg/d H2
Liquid H2 dispenser	$ 8,000	/dispenser	100%	$ 8,000	/dispenser	0.040	$ 15	/kg/d dispenser design
					Unit cost	0.659		
General Facilities & permitting	25%	of unit cost				0.165		
Eng. startup & contingencies	10%	of unit cost				0.066		
Contingencies	10%	of unit cost				0.066		
Working Capital, Land & Misc.	7%	of unit cost				0.046		
					Capital Costs	1.001	for 1 of	9 stations
					Total Capital Costs	9	for all	9 stations
Unit Capital Costs of	**0.88**	**/scf/d H2 or**	365	/kg/d H2 or	365	/gal/d gaso equiv		

Hydrogen Costs at	90%	ann load factor	$/yr of 1 station	$/million Btu LHV	$/k scf H2	$/kg H2 or $/gal gaso equiv	
Road tax or (subsidy)	$ -	/gal gaso equiv.	-	-	-	-	can be subsidy like EtOH
Gas Station mark-up	$ -	/gal gaso equiv.	-	-	-	-	if H2 drops total station revenues
Variable Non-fuel O&M	5.0%	/yr of capital	50,052	0.49	0.13	0.06	0.5-1.5 typical many be low here
Electricity	$ 0.070	/kWh	50,400	0.49	0.14	0.06	0.06-0.09 typical commercial rates
Variable Operating Cost			**100,452**	**0.98**	**0.27**	**0.11**	
Fixed Operating Cost	3.0%	/yr of capital	**30,031**	**0.29**	**0.08**	**0.03**	3-5% typical, may be lower here
Capital Charges	14.0%	/yr of capital	**140,146**	**1.37**	**0.38**	**0.16**	20-25% typical for refiners
Fueling Station Cost			**270,630**	**2.64**	**0.73**	**0.30**	

including return of investment

Hydrogen Fueling Station Costs Delivery to	9 Stations Million $/yr
Variable Operating Cost	0.90
Fixed Operating Cost	0.27
Capital Charges	1.26
Total Fueling Station Cost	2.44

Table E-33 begins on page 176.

TABLE E-33
Distributed Plant Summary of Results

Pathway Table E-	Dist NG-C 35	Dist NG-F 36	Dist Elec-C 37	Dist Elec-F 38	Dis NGASE 39
Capital investment, MM	1.85	0.96	2.54	0.57	1.30
Production costs s/kg H_2					
Variable costs					
Road tax or (subsidy)	—	—	—	—	—
Gas Station mark-up	—	—	—	—	—
Feed	1.37	1.17			1.34
Electricity	0.15	0.12	384	3.31	0.40
Non-fuel O&M, %/yr of capital	0.12	0.06	0.16	0.04	0.08
Total variable costs	1.64	1.35	4.00	3.35	1.83
Fixed costs, %/yr of capital	0.23	0.12	0.32	0.07	0.16
Capital charges	1.64	0.85	2.26	0.51	1.15
Total Production Costs	3.51	2.33	6.58	3.93	3.15
Carbon Tax	0.17	0.14	0.24	0.21	0.18
TOTAL H_2 Costs	3.68	2.47	6.82	4.13	3.32
Carbon Dioxide Vented to Atmosphere					
KG Carbon/KG H_2	3.31	2.82	4.79	4.13	355
Direct Use	3.11	2.67			3.05
Indirect Use	0.19	0.15	4.79	4.13	0.50
KG CO_2/KG H_2	12.13	10.34	15.13	15.13	13.03
Direct Use	11.42	9.79			11.19
Indirect Use	0.71	0.55	15.13	15.13	1.85

Dis WI EIe-C 40	Dis WT El-F 41	Di PV El-C 42	Di PV El-F 43	Di WI Gr C 44	Di WT Gr F 45	Di PV-Gr-C 46	Di PV-Gr-F 47
6.86	0.89	9.94	1.43	2.75	0.59	2.74	0.59
—	—	—	—				
—	—	—	—				
3.29	1.90	17.48	4.64	3.67	2.75	6.57	3.58
0.44	0.06	0.63	0.09	0.17	0.04	0.17	0.04
3.73	1.95	18.11	4.73	385	2.78	6.74	3.61
0.87	0.11	1.26	0.18	0.35	0.07	0.35	0.07
6.09	0.79	8.82	1.27	2.44	0.52	2.43	0.52
10.69	2.86	28.19	6.18	6.64	3.38	9.52	4.21
				0.17	0.12	0.19	0.17
10.69	2.86	28.19	6.18	6.81	3.50	9.71	4.37
				3.35	2.48	3.83	3.30
				3.35	2.48	3.83	3.30
				12.28	9.08	14.04	12.11
				12.28	9.08	14.04	12.11

TABLE E-34
Distributed Plant, Onsite Hydrogen Summary of Inputs

Inputs — **Boxed** are the key input variables that can change for specific situations

design basis

Key Variables Inputs		Value	Unit	Notes
Hydrogen Production Inputs				1 kg H2 is the same energy content as 1 gallon of gasoline
Design hydrogen production		480	**kg/d H2 design**	198,960 **scf/d H2** 100 to 10,000 kg/d range for forecourt
Annual average load factor		90%	of design	13,140 kg/month actual or 157,680 kg/yr actual
High pressure H2 storage		3	hr at peak surge rate	"plug & play" 24 hr process unit replacements for high availability
FCV gasoline equiv mileage		65	mpg	28 km/liter **432 kg/d average**
FCV miles per year		12,000	mile/yr	185 kg/yr H2 for each FCV
Capital Cost Buildup Inputs from process unit costs				**All major utilities included as process units**
General Facilities		20%	of process units	20-40% typical, should be low for small "plug & play" units
Engineering, Permitting & Startup		10%	of process units	10-20% typical, assume low eng. of multiple standard designs
Contingencies		10%	of process units	10-20% typical, should be low after the first few
Working Capital, Land & Misc.		5%	of process units	5-10% typical, high land costs for urban onsite
Site specific factor		110%	of US Gulf Coast	90-130% typical; sales tax, labor rates & weather issues
Product Cost Buildup Inputs				
Road tax or (subsidy)		$ -	/gal gasoline equivalent	may need subsidy like EtOH to get it going
Gas Station mark-up		$ -	/gal gasoline equivalent	may be needed if H2 sales drops total station revenues
Non-fuel Variable O&M		1.0%	/yr of capital	0.5-1.5% is typical
Fuels	Natural Gas	$ 6.50	/MM Btu HHV	$4-7/MM Btu typical **commercial** rate, see www.eia.doe.gov
	Electricity	$ 0.070	/kWh	$0.06-.0.09/kWh typical **commercial** rate, see www.eia.doe.gov
	Electricity generation eff	50%		Incremental future efficiency, current average grid is only 30%
	Electricity CO2 emissions	0.32	kg/hr CO2/kWe	Based on NGCC, coal and current grid at 0.75 kg/hr CO2/kWe
	Electricity CO2 emissions	0.32	kg/hr CO2/kWe	Based on NGCC, coal and current grid at 0.75 kg/hr CO2/kWe
Carbon tax		$ 50.00	/tonne	include C from power $ 27.27 /tonne CO2 equivalent
Fixed Operating Cost		2.0%	/yr of capital	4-7% typical for refiners: labor, overhead, insurance, taxes, G&A
Capital Charges		14.0%	/yr of capital	20-25%/yr CC typical for refiners & 14-20%/yr CC for utilities
				20%/yr CC is about 12% IRR DCF on 100% equity where as
				15%/yr CC is about 12% IRR DCF on 50% equity & debt at 7%

TABLE E-35
Dist NG-C
Distributed Size Onsite Hydrogen via Steam Reforming of Natural Gas with Current Technology

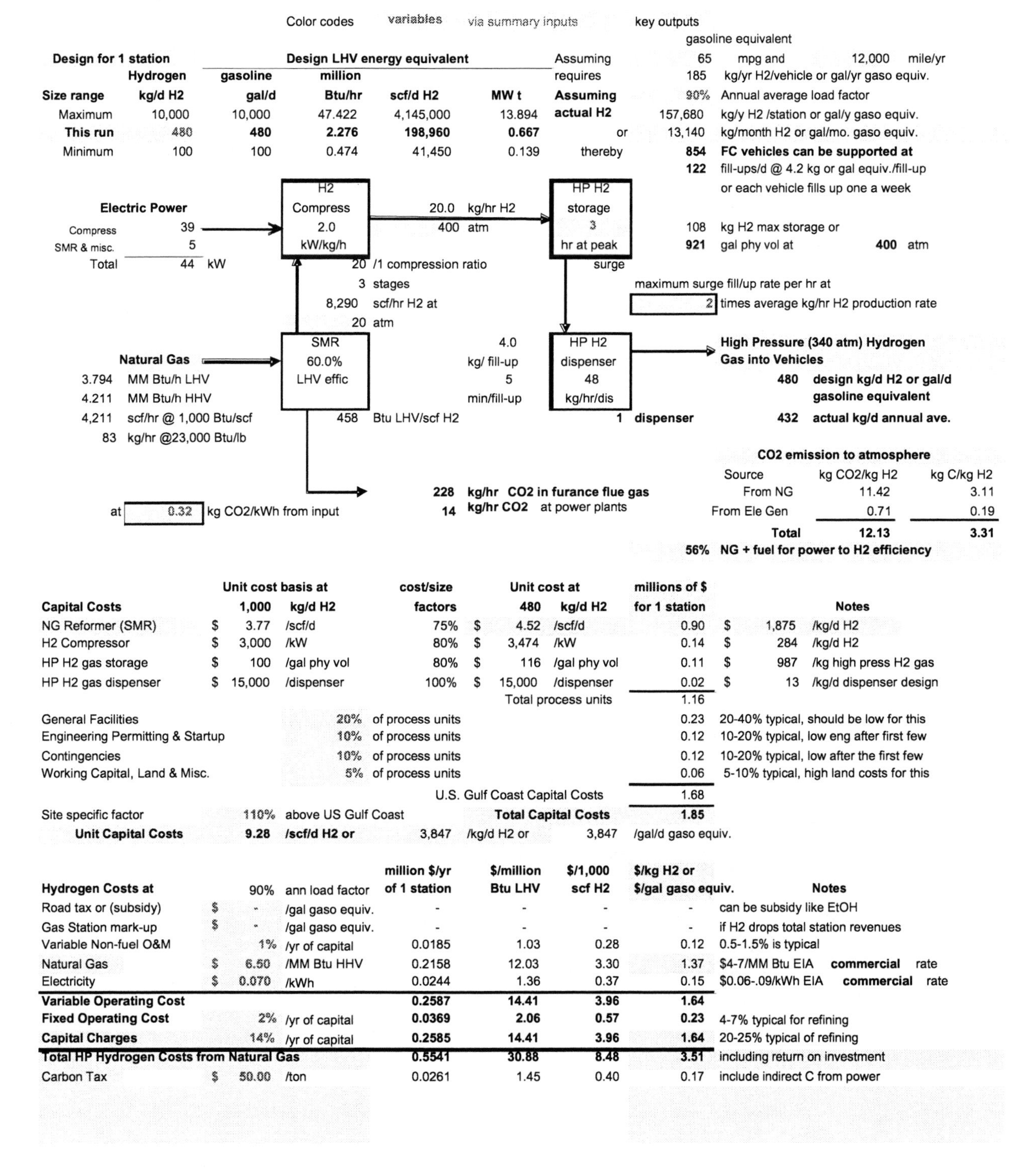

Color codes variables via summary inputs key outputs

Design for 1 station	Hydrogen kg/d H2	gasoline gal/d	Design LHV energy equivalent million Btu/hr	scf/d H2	MW t
Size range					
Maximum	10,000	10,000	47.422	4,145,000	13.894
This run	480	480	2.276	198,960	0.667
Minimum	100	100	0.474	41,450	0.139

Assuming requires; Assuming actual H2 or thereby

gasoline equivalent
65 mpg and 12,000 mile/yr
185 kg/yr H2/vehicle or gal/yr gaso equiv.
90% Annual average load factor
157,680 kg/y H2 /station or gal/y gaso equiv.
13,140 kg/month H2 or gal/mo. gaso equiv.
854 FC vehicles can be supported at
122 fill-ups/d @ 4.2 kg or gal equiv./fill-up
or each vehicle fills up one a week

Electric Power
Compress 39
SMR & misc. 5
Total 44 kW

H2 Compress 2.0 kW/kg/h
20.0 kg/hr H2
400 atm
20 /1 compression ratio
3 stages
8,290 scf/hr H2 at
20 atm

HP H2 storage 3 hr at peak surge
108 kg H2 max storage or
921 gal phy vol at 400 atm
maximum surge fill/up rate per hr at
2 times average kg/hr H2 production rate

Natural Gas
3.794 MM Btu/h LHV
4.211 MM Btu/h HHV
4,211 scf/hr @ 1,000 Btu/scf
83 kg/hr @23,000 Btu/lb

SMR 60.0% LHV effic
458 Btu LHV/scf H2

4.0 kg/ fill-up
5 min/fill-up
HP H2 dispenser 48 kg/hr/dis
1 dispenser

High Pressure (340 atm) Hydrogen Gas into Vehicles
480 design kg/d H2 or gal/d gasoline equivalent
432 actual kg/d annual ave.

228 kg/hr CO2 in furance flue gas
14 kg/hr CO2 at power plants
at 0.32 kg CO2/kWh from input

CO2 emission to atmosphere

Source	kg CO2/kg H2	kg C/kg H2
From NG	11.42	3.11
From Ele Gen	0.71	0.19
Total	**12.13**	**3.31**

56% NG + fuel for power to H2 efficiency

Capital Costs	Unit cost basis at 1,000 kg/d H2		cost/size factors	Unit cost at 480 kg/d H2		millions of $ for 1 station	Notes	
NG Reformer (SMR)	$ 3.77	/scf/d	75%	$ 4.52	/scf/d	0.90	$ 1,875	/kg/d H2
H2 Compressor	$ 3,000	/kW	80%	$ 3,474	/kW	0.14	$ 284	/kg/d H2
HP H2 gas storage	$ 100	/gal phy vol	80%	$ 116	/gal phy vol	0.11	$ 987	/kg high press H2 gas
HP H2 gas dispenser	$ 15,000	/dispenser	100%	$ 15,000	/dispenser	0.02	$ 13	/kg/d dispenser design
				Total process units		1.16		
General Facilities			20%	of process units		0.23	20-40% typical, should be low for this	
Engineering Permitting & Startup			10%	of process units		0.12	10-20% typical, low eng after first few	
Contingencies			10%	of process units		0.12	10-20% typical, low after the first few	
Working Capital, Land & Misc.			5%	of process units		0.06	5-10% typical, high land costs for this	
				U.S. Gulf Coast Capital Costs		1.68		
Site specific factor	110%	above US Gulf Coast		**Total Capital Costs**		**1.85**		
Unit Capital Costs	**9.28**	**/scf/d H2 or**	3,847	/kg/d H2 or	3,847	/gal/d gaso equiv.		

Hydrogen Costs at	90%	ann load factor	million $/yr of 1 station	$/million Btu LHV	$/1,000 scf H2	$/kg H2 or $/gal gaso equiv.	Notes
Road tax or (subsidy)	$ -	/gal gaso equiv.	-	-	-	-	can be subsidy like EtOH
Gas Station mark-up	$ -	/gal gaso equiv.	-	-	-	-	if H2 drops total station revenues
Variable Non-fuel O&M	1%	/yr of capital	0.0185	1.03	0.28	0.12	0.5-1.5% is typical
Natural Gas	$ 6.50	/MM Btu HHV	0.2158	12.03	3.30	1.37	$4-7/MM Btu EIA **commercial** rate
Electricity	$ 0.070	/kWh	0.0244	1.36	0.37	0.15	$0.06-.09/kWh EIA **commercial** rate
Variable Operating Cost			**0.2587**	**14.41**	**3.96**	**1.64**	
Fixed Operating Cost	2%	/yr of capital	**0.0369**	**2.06**	**0.57**	**0.23**	4-7% typical for refining
Capital Charges	14%	/yr of capital	**0.2585**	**14.41**	**3.96**	**1.64**	20-25% typical of refining
Total HP Hydrogen Costs from Natural Gas			**0.5541**	**30.88**	**8.48**	**3.51**	including return on investment
Carbon Tax	$ 50.00	/ton	0.0261	1.45	0.40	0.17	include indirect C from power

note: Assume filling station has existing natural gas pipeline infrastructure, if not more capital or higher NG price

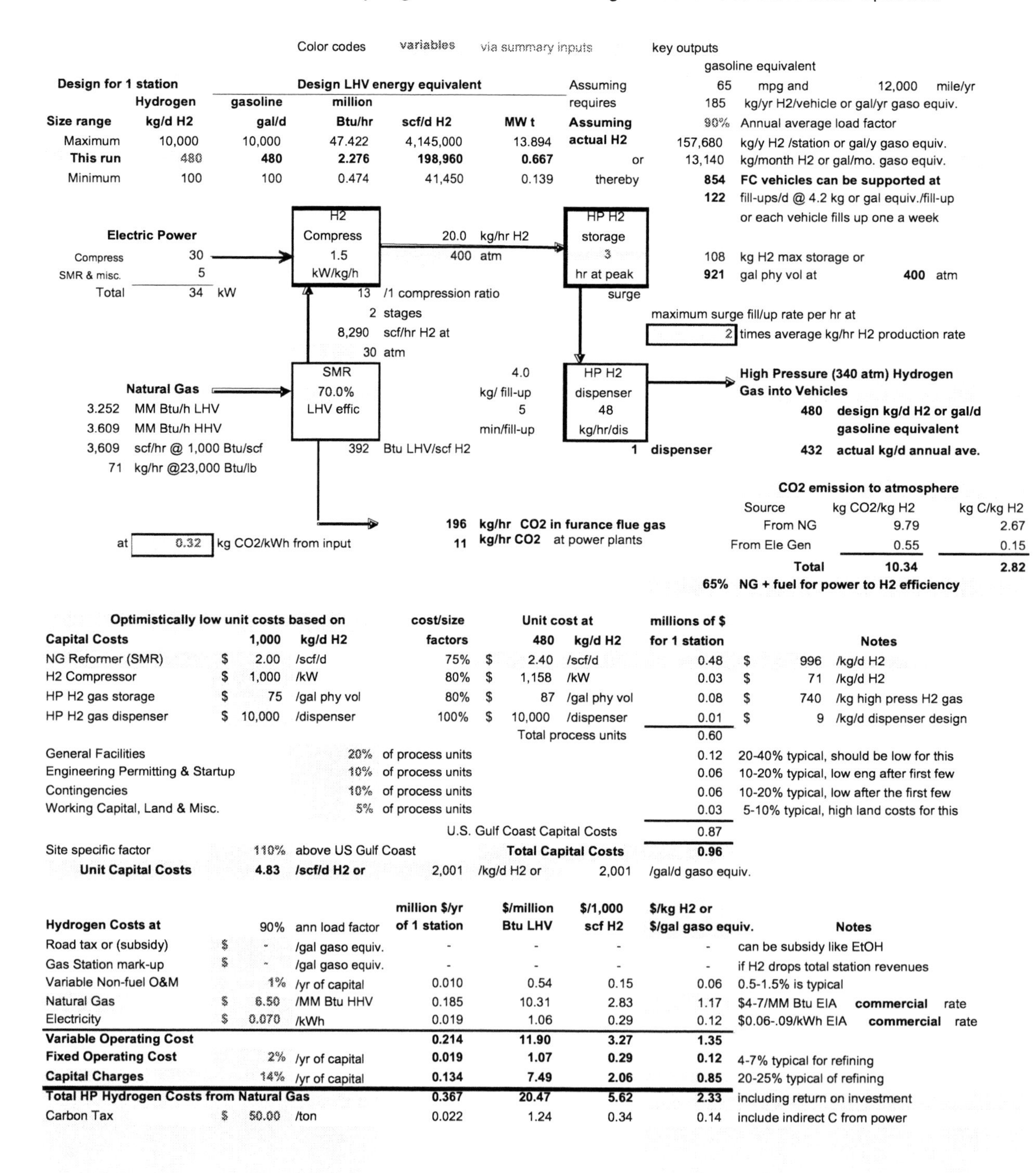

TABLE E-36
Dist NG-F
Distributed Size Onsite Hydrogen via Steam Reforming of Natural Gas with Future Optimism

Color codes variables via summary inputs

Design for 1 station	Hydrogen	gasoline	Design LHV energy equivalent million			Assuming requires
Size range	kg/d H2	gal/d	Btu/hr	scf/d H2	MW t	Assuming actual H2
Maximum	10,000	10,000	47.422	4,145,000	13.894	
This run	480	480	2.276	198,960	0.667	or
Minimum	100	100	0.474	41,450	0.139	thereby

key outputs
gasoline equivalent

65	mpg and 12,000 mile/yr
185	kg/yr H2/vehicle or gal/yr gaso equiv.
90%	Annual average load factor
157,680	kg/y H2 /station or gal/y gaso equiv.
13,140	kg/month H2 or gal/mo. gaso equiv.
854	**FC vehicles can be supported at**
122	fill-ups/d @ 4.2 kg or gal equiv./fill-up
	or each vehicle fills up one a week
108	kg H2 max storage or
921	gal phy vol at **400** atm

Electric Power

Compress	30
SMR & misc.	5
Total	34 kW

H2 Compress 1.5 kW/kg/h — 20.0 kg/hr H2, 400 atm → HP H2 storage 3 hr at peak surge

13 /1 compression ratio
2 stages
8,290 scf/hr H2 at
30 atm

maximum surge fill/up rate per hr at
2 times average kg/hr H2 production rate

Natural Gas → SMR 70.0% LHV effic

3.252	MM Btu/h LHV
3.609	MM Btu/h HHV
3,609	scf/hr @ 1,000 Btu/scf
71	kg/hr @23,000 Btu/lb

392 Btu LHV/scf H2

4.0 kg/ fill-up
5 min/fill-up

HP H2 dispenser 48 kg/hr/dis
1 dispenser

High Pressure (340 atm) Hydrogen Gas into Vehicles
480 design kg/d H2 or gal/d gasoline equivalent
432 actual kg/d annual ave.

196 kg/hr CO2 in furance flue gas
11 kg/hr CO2 at power plants

at 0.32 kg CO2/kWh from input

CO2 emission to atmosphere

Source	kg CO2/kg H2	kg C/kg H2
From NG	9.79	2.67
From Ele Gen	0.55	0.15
Total	**10.34**	**2.82**

65% NG + fuel for power to H2 efficiency

Capital Costs	**Optimistically low unit costs based on 1,000**	**kg/d H2**	**cost/size factors**	**Unit cost at 480**	**kg/d H2**	**millions of $ for 1 station**	**Notes**	
NG Reformer (SMR)	$ 2.00	/scf/d	75%	$ 2.40	/scf/d	0.48	$ 996	/kg/d H2
H2 Compressor	$ 1,000	/kW	80%	$ 1,158	/kW	0.03	$ 71	/kg/d H2
HP H2 gas storage	$ 75	/gal phy vol	80%	$ 87	/gal phy vol	0.08	$ 740	/kg high press H2 gas
HP H2 gas dispenser	$ 10,000	/dispenser	100%	$ 10,000	/dispenser	0.01	$ 9	/kg/d dispenser design
				Total process units		0.60		
General Facilities		20%	of process units			0.12	20-40% typical, should be low for this	
Engineering Permitting & Startup		10%	of process units			0.06	10-20% typical, low eng after first few	
Contingencies		10%	of process units			0.06	10-20% typical, low after the first few	
Working Capital, Land & Misc.		5%	of process units			0.03	5-10% typical, high land costs for this	
				U.S. Gulf Coast Capital Costs		0.87		
Site specific factor	110%	above US Gulf Coast		**Total Capital Costs**		**0.96**		
Unit Capital Costs	**4.83**	**/scf/d H2 or**	2,001	/kg/d H2 or	2,001	/gal/d gaso equiv.		

Hydrogen Costs at	**90%**	**ann load factor**	**million $/yr of 1 station**	**$/million Btu LHV**	**$/1,000 scf H2**	**$/kg H2 or $/gal gaso equiv.**	**Notes**
Road tax or (subsidy)	$ -	/gal gaso equiv.	-	-	-	-	can be subsidy like EtOH
Gas Station mark-up	$ -	/gal gaso equiv.	-	-	-	-	if H2 drops total station revenues
Variable Non-fuel O&M	1%	/yr of capital	0.010	0.54	0.15	0.06	0.5-1.5% is typical
Natural Gas	$ 6.50	/MM Btu HHV	0.185	10.31	2.83	1.17	$4-7/MM Btu EIA **commercial** rate
Electricity	$ 0.070	/kWh	0.019	1.06	0.29	0.12	$0.06-.09/kWh EIA **commercial** rate
Variable Operating Cost			**0.214**	**11.90**	**3.27**	**1.35**	
Fixed Operating Cost	2%	/yr of capital	**0.019**	**1.07**	**0.29**	**0.12**	4-7% typical for refining
Capital Charges	14%	/yr of capital	**0.134**	**7.49**	**2.06**	**0.85**	20-25% typical of refining
Total HP Hydrogen Costs from Natural Gas			**0.367**	**20.47**	**5.62**	**2.33**	including return on investment
Carbon Tax	$ 50.00	/ton	0.022	1.24	0.34	0.14	include indirect C from power

note: Assume filling station has existing natural gas pipeline infrastructure, if not more capital or higher NG price

TABLE E-37
Dist Ele-C
Distributed Size Onsite Hydrogen via Electrolysis of Water with Current Technology

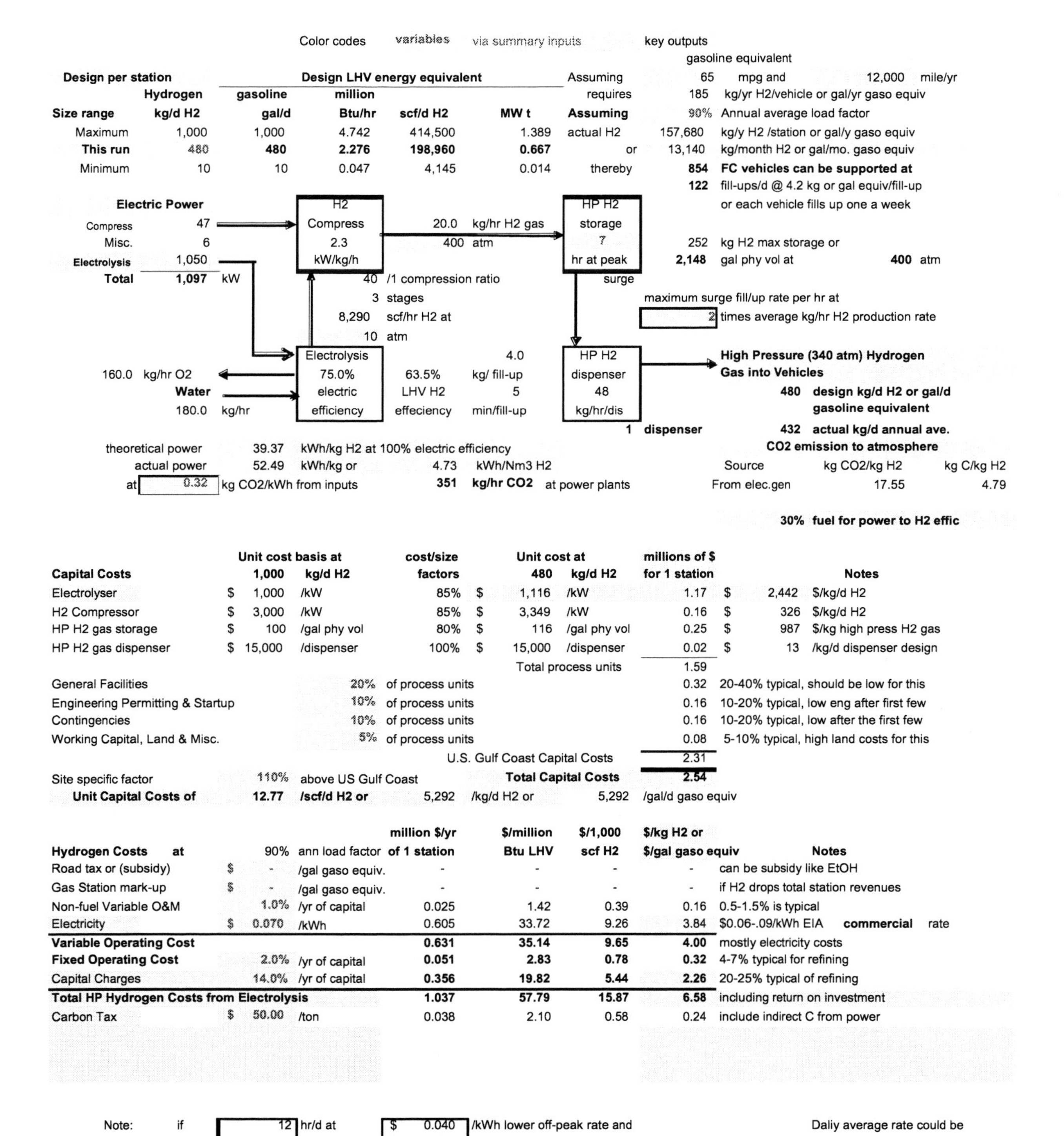

Color codes variables via summary inputs key outputs

Design per station	Hydrogen	gasoline	Design LHV energy equivalent million					gasoline equivalent
Size range	kg/d H2	gal/d	Btu/hr	scf/d H2	MW t	Assuming requires	65 / 185	mpg and 12,000 mile/yr / kg/yr H2/vehicle or gal/yr gaso equiv
Maximum	1,000	1,000	4.742	414,500	1.389	Assuming	90%	Annual average load factor
This run	480	480	2.276	198,960	0.667	actual H2	157,680	kg/y H2 /station or gal/y gaso equiv
Minimum	10	10	0.047	4,145	0.014	or	13,140	kg/month H2 or gal/mo. gaso equiv
						thereby	854	FC vehicles can be supported at
							122	fill-ups/d @ 4.2 kg or gal equiv/fill-up
								or each vehicle fills up one a week

Electric Power	
Compress	47
Misc.	6
Electrolysis	1,050
Total	1,097 kW

H2 Compress 2.3 kW/kg/h — 20.0 kg/hr H2 gas, 400 atm → HP H2 storage 7 hr at peak surge

252 kg H2 max storage or
2,148 gal phy vol at 400 atm

40 /1 compression ratio
3 stages
8,290 scf/hr H2 at
10 atm

maximum surge fill/up rate per hr at
2 times average kg/hr H2 production rate

Electrolysis 75.0% electric efficiency — 63.5% LHV H2 effeciency — 4.0 kg/ fill-up, 5 min/fill-up — HP H2 dispenser 48 kg/hr/dis — 1 dispenser

160.0 kg/hr O2
Water 180.0 kg/hr

High Pressure (340 atm) Hydrogen Gas into Vehicles
480 design kg/d H2 or gal/d gasoline equivalent
432 actual kg/d annual ave.

theoretical power 39.37 kWh/kg H2 at 100% electric efficiency
actual power 52.49 kWh/kg or 4.73 kWh/Nm3 H2
at 0.32 kg CO2/kWh from inputs 351 **kg/hr CO2** at power plants

CO2 emission to atmosphere

Source	kg CO2/kg H2	kg C/kg H2
From elec.gen	17.55	4.79

30% fuel for power to H2 effic

Capital Costs	Unit cost basis at 1,000 kg/d H2		cost/size factors	Unit cost at 480 kg/d H2		millions of $ for 1 station	Notes	
Electrolyser	$ 1,000	/kW	85%	$ 1,116	/kW	1.17	$ 2,442	$/kg/d H2
H2 Compressor	$ 3,000	/kW	85%	$ 3,349	/kW	0.16	$ 326	$/kg/d H2
HP H2 gas storage	$ 100	/gal phy vol	80%	$ 116	/gal phy vol	0.25	$ 987	$/kg high press H2 gas
HP H2 gas dispenser	$ 15,000	/dispenser	100%	$ 15,000	/dispenser	0.02	$ 13	/kg/d dispenser design
				Total process units		1.59		
General Facilities			20%	of process units		0.32	20-40% typical, should be low for this	
Engineering Permitting & Startup			10%	of process units		0.16	10-20% typical, low eng after first few	
Contingencies			10%	of process units		0.16	10-20% typical, low after the first few	
Working Capital, Land & Misc.			5%	of process units		0.08	5-10% typical, high land costs for this	
				U.S. Gulf Coast Capital Costs		2.31		
Site specific factor	110%	above US Gulf Coast		Total Capital Costs		2.54		
Unit Capital Costs of	12.77	/scf/d H2 or	5,292	/kg/d H2 or	5,292	/gal/d gaso equiv		

Hydrogen Costs at	90%	ann load factor	million $/yr of 1 station	$/million Btu LHV	$/1,000 scf H2	$/kg H2 or $/gal gaso equiv	Notes
Road tax or (subsidy)	$ -	/gal gaso equiv.	-	-	-	-	can be subsidy like EtOH
Gas Station mark-up	$ -	/gal gaso equiv.	-	-	-	-	if H2 drops total station revenues
Non-fuel Variable O&M	1.0%	/yr of capital	0.025	1.42	0.39	0.16	0.5-1.5% is typical
Electricity	$ 0.070	/kWh	0.605	33.72	9.26	3.84	$0.06-.09/kWh EIA **commercial** rate
Variable Operating Cost			**0.631**	**35.14**	**9.65**	**4.00**	mostly electricity costs
Fixed Operating Cost	2.0%	/yr of capital	**0.051**	**2.83**	**0.78**	**0.32**	4-7% typical for refining
Capital Charges	14.0%	/yr of capital	**0.356**	**19.82**	**5.44**	**2.26**	20-25% typical of refining
Total HP Hydrogen Costs from Electrolysis			**1.037**	**57.79**	**15.87**	**6.58**	including return on investment
Carbon Tax	$ 50.00	/ton	0.038	2.10	0.58	0.24	include indirect C from power

Note: if 12 hr/d at $ 0.040 /kWh lower off-peak rate and
12 hr/d at $ 0.090 /kWh higher peak rate

Daliy average rate could be $ 0.065 /kWh

If only operated during low off-peak rate times would have low ann load factor & need more expensive H2 storage
Assume Hydrogn Systems Electrolysis at 150 psig pressure, Norsk Hydro & Stuard systems are low pressure
Assumed oxygen recovery for by-product sales with large central plant case, but only minor economic impact

TABLE E-38
Dist Ele-F
Distributed Size Onsite Hydrogen via Electrolysis of Water with Future Optimism

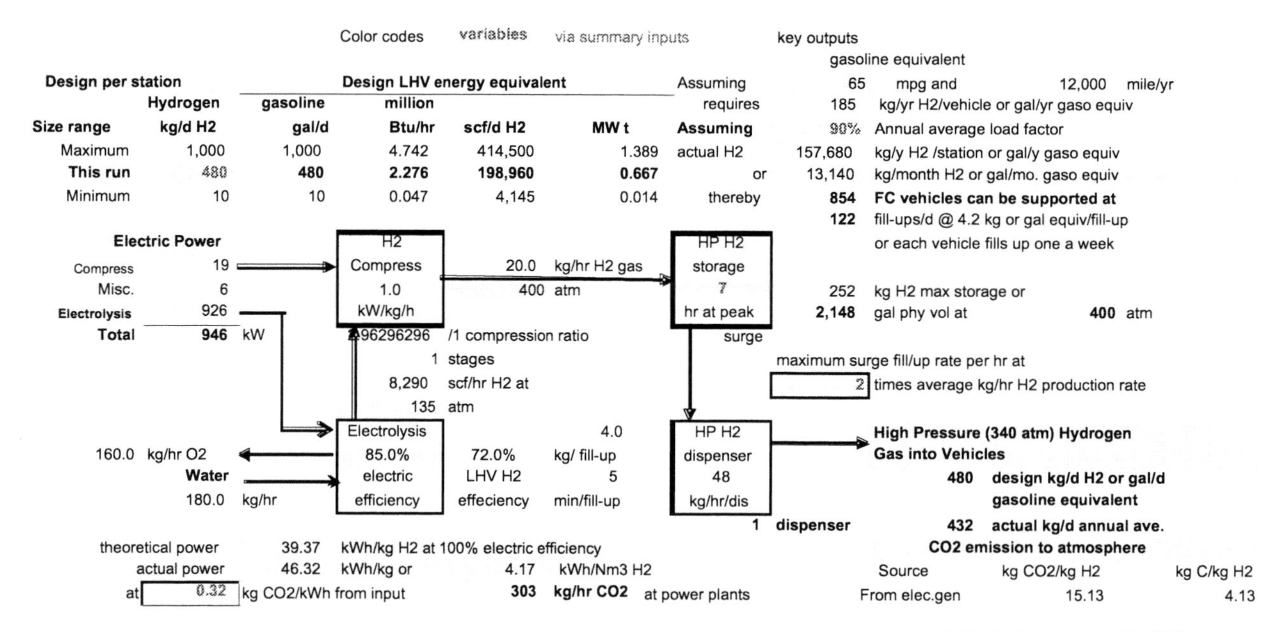

Optimistically low unit costs based on			cost/size		Unit cost at		millions of $		Notes	
Capital Costs	**1,000**	**kg/d H2**	**factors**		**480**	**kg/d H2**	**for 1 station**			
Electrolyser	$ 125	/kW	85%	$	140	/kW	0.13	$	269	$/kg/d H2 less that SMR
H2 Compressor	$ 1,500	/kW	85%	$	1,675	/kW	0.03	$	68	$/kg/d H2
HP H2 gas storage	$ 75	/gal phy vol	80%	$	87	/gal phy vol	0.19	$	740	$/kg high press H2 gas
HP H2 gas dispenser	$ 10,000	/dispenser	100%	$	10,000	/dispenser	0.01	$	9	/kg/d dispenser design
					Total process units		0.36			
General Facilities		20%	of process units				0.07	20-40% typical, should be low for this		
Engineering Permitting & Startup		10%	of process units				0.04	10-20% typical, low eng after first few		
Contingencies		10%	of process units				0.04	10-20% typical, low after the first few		
Working Capital, Land & Misc.		5%	of process units				0.02	5-10% typical, high land costs for this		
					U.S. Gulf Coast Capital Costs		0.52			
Site specific factor	110%	above US Gulf Coast			**Total Capital Costs**		**0.57**			
Unit Capital Costs of	**2.87**	**/scf/d H2 or**	1,191	/kg/d H2 or	1,191	/gal/d gaso equiv				

Hydrogen Costs at			**million $/yr**	**$/million**	**$/1,000**	**$/kg H2 or**	**Notes**
	90%	ann load factor	**of 1 station**	**Btu LHV**	**scf H2**	**$/gal gaso equiv**	
Road tax or (subsidy)	$ -	/gal gaso equiv.	-	-	-	-	can be subsidy like EtOH
Gas Station mark-up	$ -	/gal gaso equiv.	-	-	-	-	if H2 drops total station revenues
Non-fuel Variable O&M	1.0%	/yr of capital	0.006	0.32	0.09	0.04	0.5-1.5% is typical
Electricity	$ 0.070	/kWh	0.522	29.09	7.99	3.31	$0.06-.09/kWh EIA **commercial** rate
Variable Operating Cost			**0.528**	**29.40**	**8.07**	**3.35**	mostly electricity costs
Fixed Operating Cost	2.0%	/yr of capital	**0.011**	**0.64**	**0.17**	**0.07**	4-7% typical for refining
Capital Charges	14.0%	/yr of capital	**0.080**	**4.46**	**1.22**	**0.51**	20-25% typical of refining
Total HP Hydrogen Costs from Electrolysis			**0.619**	**34.50**	**9.47**	**3.93**	including return on investment
Carbon Tax	$ 50.00	/ton	0.033	1.81	0.50	0.21	include indirect C from power

Note: if 12 hr/d at $ 0.040 /kWh lower off-peak rate and — Daliy average rate could be
12 hr/d at $ 0.090 /kWh higher peak rate — $ 0.065 /kWh

If only operated during low off-peak rate times would have low ann load factor & need more expensive H2 storage
Assume high pressure electrolysis cell with drastically lower capital cost of mass production cost & higher efficiency
Assumed oxygen recovery for by-product sales with large central plant case, but only minor economic impact

TABLE E-39
Dist NGASE-F
Distributed Size Onsite Hydrogen via Natural Gas-Assisted Steam Electrolysis of Water with Future Optimism

Color codes **variables** via summary inputs key outputs

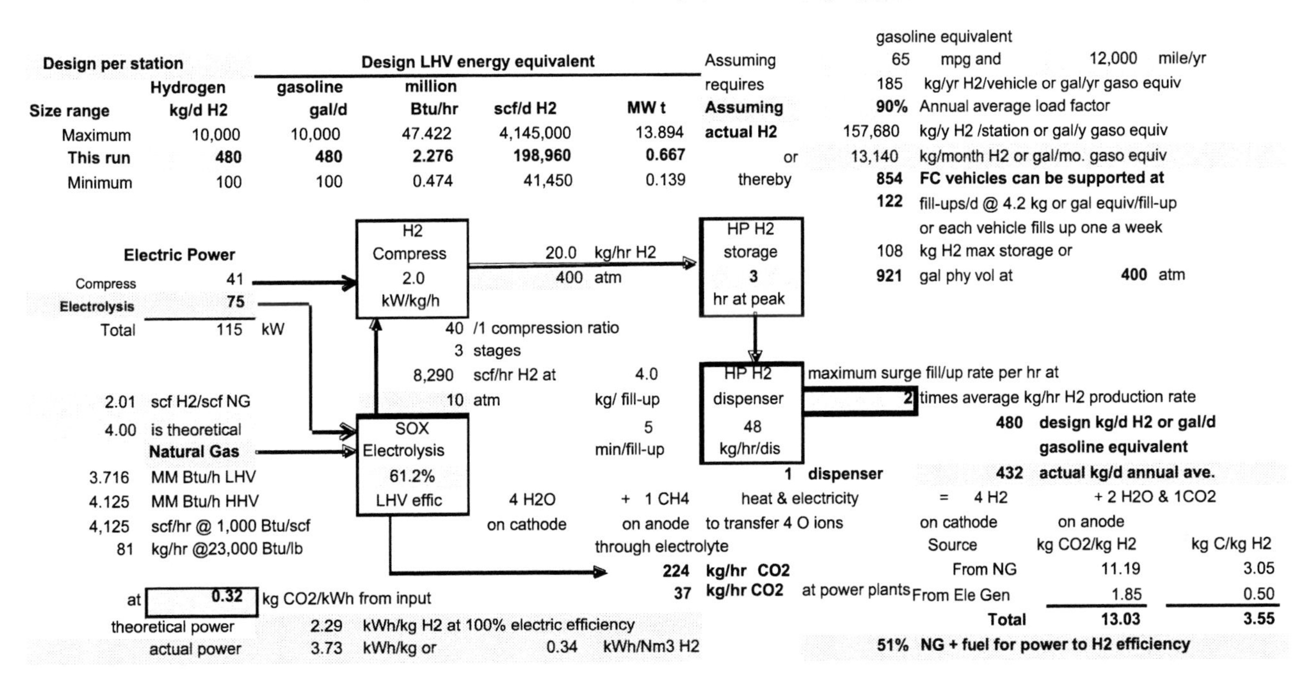

Capital Costs	Unit cost basis at 1,000 kg/d H2		cost/size factors	Unit cost at 480 kg/d H2		millions of $ for 1 station	Notes	
NGASE Electrolysis	$ 3.00	/scf/d	85%	$ 3.35	/scf/d	0.67	1388	$/kg/d H2,twice electro costs
H2 Compressor	$ 1,500	/kW	85%	$ 1,675	/kW	0.07	$ 142	/kg/d H2
HP H2 gas storage	$ 75	/gal phy vol	80%	$ 87	/gal phy vol	0.08	$ 740	$/kg high press H2 gas
HP H2 gas dispenser	$ 10,000	/dispenser	100%	$ 10,000	/dispenser	0.01	$ 9	/kg/d dispenser design
				Total process units		0.81		
General Facilities			20%	of process units		0.16	20-40% typical, should be low for this	
Engineering Permitting & Startup			10%	of process units		0.08	10-20% typical, low eng after first few	
Contingencies			10%	of process units		0.08	10-20% typical, low after the first few	
Working Capital, Land & Misc.			5%	of process units		0.04	5-10% typical, high land costs for this	
				U.S. Gulf Coast Capital Costs		1.18		
Site specific factor	110%	above US Gulf Coast		Total Capital Costs		1.30		
Unit Capital Costs of	**6.53**	**/scf/d H2 or**	2,707	/kg/d H2 or	2,707	/gal/d gaso equiv		

Hydrogen Costs at	90%	ann load factor	million $/yr of 1 station	$/million Btu LHV	$/1,000 scf H2	$/kg H2 or $/gal gaso equiv	Notes		
Road tax or (subsidy)	$ -	/gal gaso equiv.	-	-	-	-	can be subsidy like EtOH		
Gas Station mark-up	$ -	/gal gaso equiv.	-	-	-	-	if H2 drops total station revenues		
Non-fuel Variable O&M	1.0%	/yr of capital	0.013	0.72	0.20	0.08	0.5-1.5% is typical		
Natural Gas	$ 6.50	/MM Btu HHV	0.211	11.78	3.23	1.34	$5-7/MM Btu EIA	commercial	rate
Electricity	$ 0.070	/kWh	0.064	3.55	0.97	0.40	$0.06-.09/kWh EIA	commercial	rate
Variable Operating Cost			**0.288**	**16.05**	**4.41**	**1.83**			
Fixed Operating Cost	**2.0%**	/yr of capital	**0.026**	**1.45**	**0.40**	**0.16**	4-7% typical for refining		
Capital Charges	**14.0%**	/yr of capital	**0.182**	**10.14**	**2.78**	**1.15**	Deprecation, taxes & 20% Return		
Total Hydrogen Cost from Natural Gas SOX			**0.496**	**27.64**	**7.59**	**3.15**	including return on investment		
Carbon Tax	$ **50.00**	/ton	0.028	1.56	0.43	0.18	include indirect C from power		

notes Costs of this solid oxide fuel cell operated as electrolyzer with NG on anode is very speculative
Included as a lower power consumption electrolyze alternative

TABLE E-40
Dist WT Ele-C
Distributed Size Onsite Hydrogen via Wind-Turbine-Based Electrolysis with Current Technology

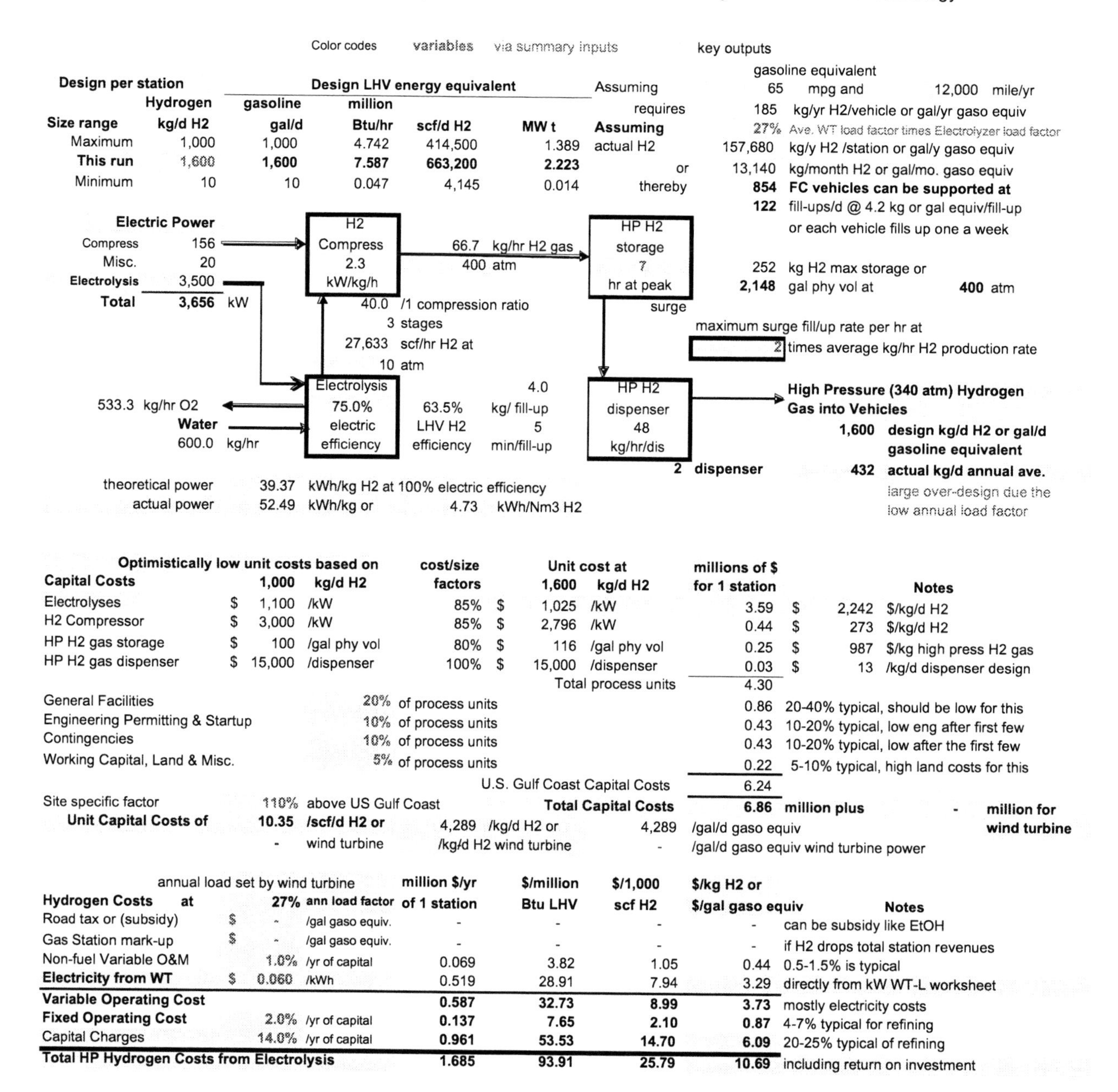

Color codes variables via summary inputs key outputs

Design per station	Hydrogen	gasoline	Design LHV energy equivalent million		
Size range	kg/d H2	gal/d	Btu/hr	scf/d H2	MW t
Maximum	1,000	1,000	4.742	414,500	1.389
This run	1,600	1,600	7.587	663,200	2.223
Minimum	10	10	0.047	4,145	0.014

Assuming requires
Assuming actual H2
or
thereby

gasoline equivalent
65 mpg and 12,000 mile/yr
185 kg/yr H2/vehicle or gal/yr gaso equiv
27% Ave. WT load factor times Electrolyzer load factor
157,680 kg/y H2 /station or gal/y gaso equiv
13,140 kg/month H2 or gal/mo. gaso equiv
854 FC vehicles can be supported at
122 fill-ups/d @ 4.2 kg or gal equiv/fill-up
or each vehicle fills up one a week

Electric Power	
Compress	156
Misc.	20
Electrolysis	3,500
Total	3,656 kW

H2 Compress 2.3 kW/kg/h
66.7 kg/hr H2 gas 400 atm
HP H2 storage 7 hr at peak surge
252 kg H2 max storage or
2,148 gal phy vol at 400 atm
40.0 /1 compression ratio
3 stages
27,633 scf/hr H2 at
10 atm
maximum surge fill/up rate per hr at
2 times average kg/hr H2 production rate

533.3 kg/hr O2
Water 600.0 kg/hr
Electrolysis 75.0% electric efficiency
63.5% LHV H2 efficiency
4.0 kg/ fill-up 5 min/fill-up
HP H2 dispenser 48 kg/hr/dis
2 dispenser

High Pressure (340 atm) Hydrogen Gas into Vehicles
1,600 design kg/d H2 or gal/d gasoline equivalent
432 actual kg/d annual ave.
large over-design due the low annual load factor

theoretical power 39.37 kWh/kg H2 at 100% electric efficiency
actual power 52.49 kWh/kg or 4.73 kWh/Nm3 H2

Capital Costs	Optimistically low unit costs based on 1,000 kg/d H2		cost/size factors	Unit cost at 1,600 kg/d H2		millions of $ for 1 station	Notes	
Electrolyses	$ 1,100	/kW	85%	$ 1,025	/kW	3.59	$ 2,242	$/kg/d H2
H2 Compressor	$ 3,000	/kW	85%	$ 2,796	/kW	0.44	$ 273	$/kg/d H2
HP H2 gas storage	$ 100	/gal phy vol	80%	$ 116	/gal phy vol	0.25	$ 987	$/kg high press H2 gas
HP H2 gas dispenser	$ 15,000	/dispenser	100%	$ 15,000	/dispenser	0.03	$ 13	/kg/d dispenser design
				Total process units		4.30		
General Facilities			20% of process units			0.86	20-40% typical, should be low for this	
Engineering Permitting & Startup			10% of process units			0.43	10-20% typical, low eng after first few	
Contingencies			10% of process units			0.43	10-20% typical, low after the first few	
Working Capital, Land & Misc.			5% of process units			0.22	5-10% typical, high land costs for this	
				U.S. Gulf Coast Capital Costs		6.24		
Site specific factor	110%	above US Gulf Coast		Total Capital Costs		6.86	million plus -	million for wind turbine
Unit Capital Costs of	10.35	/scf/d H2 or	4,289 /kg/d H2 or		4,289	/gal/d gaso equiv		
	-	wind turbine	/kg/d H2 wind turbine		-	/gal/d gaso equiv wind turbine power		

annual load set by wind turbine

Hydrogen Costs at	27% ann load factor	million $/yr of 1 station	$/million Btu LHV	$/1,000 scf H2	$/kg H2 or $/gal gaso equiv	Notes
Road tax or (subsidy)	$ - /gal gaso equiv.	-	-	-	-	can be subsidy like EtOH
Gas Station mark-up	$ - /gal gaso equiv.	-	-	-	-	if H2 drops total station revenues
Non-fuel Variable O&M	1.0% /yr of capital	0.069	3.82	1.05	0.44	0.5-1.5% is typical
Electricity from WT	$ 0.060 /kWh	0.519	28.91	7.94	3.29	directly from kW WT-L worksheet
Variable Operating Cost		**0.587**	**32.73**	**8.99**	**3.73**	mostly electricity costs
Fixed Operating Cost	2.0% /yr of capital	**0.137**	**7.65**	**2.10**	**0.87**	4-7% typical for refining
Capital Charges	14.0% /yr of capital	**0.961**	**53.53**	**14.70**	**6.09**	20-25% typical of refining
Total HP Hydrogen Costs from Electrolysis		**1.685**	**93.91**	**25.79**	**10.69**	including return on investment

Assume high pressure electrolysis cell with drastically lower capital cost of mass production cost & higher efficiency
Assume **no** transmission or distributions cost for the wind turbine power which usually adds $0.02-.04/kWh
Assumed oxygen recovery for by-product sales with large central plant case, but only minor economic impact

TABLE E-41
Dist WT Ele-F
Distributed Size Onsite Hydrogen via Wind-Turbine-Based Electrolysis with Future Optimism

Color codes variables via summary inputs

Design per station			Design LHV energy equivalent			Assuming
	Hydrogen	gasoline	million			requires
Size range	kg/d H2	gal/d	Btu/hr	scf/d H2	MW t	Assuming
Maximum	1,000	1,000	4.742	414,500	1.389	actual H2
This run	1,200	1,200	5.691	497,400	1.667	or
Minimum	10	10	0.047	4,145	0.014	thereby

key outputs

gasoline equivalent	
65	mpg and 12,000 mile/yr
185	kg/yr H2/vehicle or gal/yr gaso equiv
36%	Ave. WT load factor times Electrolyzer load factor
157,680	kg/y H2 /station or gal/y gaso equiv
13,140	kg/month H2 or gal/mo. gaso equiv
854	**FC vehicles can be supported at**
122	fill-ups/d @ 4.2 kg or gal equiv/fill-up
	or each vehicle fills up one a week
252	kg H2 max storage or
2,148	gal phy vol at **400** atm

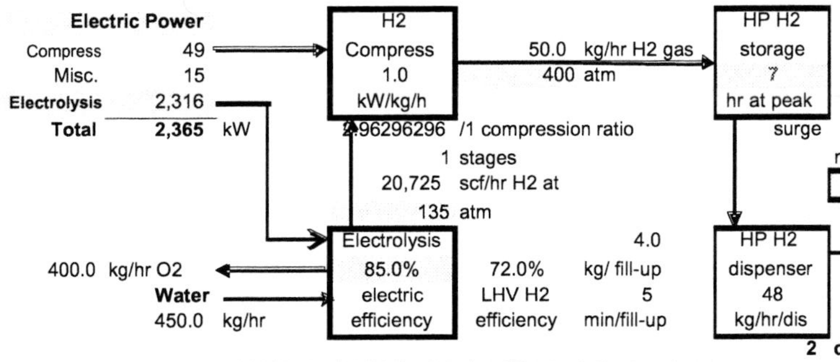

Electric Power		
Compress	49	
Misc.	15	
Electrolysis	2,316	
Total	2,365	kW

H2 Compress 1.0 kW/kg/h — 50.0 kg/hr H2 gas, 400 atm — HP H2 storage 7 hr at peak surge

2.96296296 /1 compression ratio
1 stages
20,725 scf/hr H2 at
135 atm

maximum surge fill/up rate per hr at
2 times average kg/hr H2 production rate

400.0 kg/hr O2 ← Electrolysis 85.0% electric efficiency | 72.0% LHV H2 efficiency | 4.0 kg/ fill-up, 5 min/fill-up | HP H2 dispenser 48 kg/hr/dis | **2** dispenser

Water → 450.0 kg/hr

High Pressure (340 atm) Hydrogen Gas into Vehicles
1,200 **design kg/d H2 or gal/d gasoline equivalent**
432 **actual kg/d annual ave.**
large over-design due the low annual load factor

theoretical power 39.37 kWh/kg H2 at 100% electric efficiency
actual power 46.32 kWh/kg or 4.17 kWh/Nm3 H2

Capital Costs	Optimistically low unit costs based on 1,000 kg/d H2		cost/size factors	Unit cost at 1,200 kg/d H2		millions of $ for 1 station	Notes		
Electrolyses	$ 125	/kW	85%	$ 122	/kW	0.28	$ 235	$/kg/d H2	less that SMR
H2 Compressor	$ 1,500	/kW	85%	$ 1,460	/kW	0.07	$ 59	$/kg/d H2	
HP H2 gas storage	$ 75	/gal phy vol	80%	$ 87	/gal phy vol	0.19	$ 740	$/kg high press H2 gas	
HP H2 gas dispenser	$ 10,000	/dispenser	100%	$ 10,000	/dispenser	0.02	$ 9	/kg/d dispenser design	
				Total process units		0.56			
General Facilities	20%	of process units				0.11	20-40% typical, should be low for this		
Engineering Permitting & Startup	10%	of process units				0.06	10-20% typical, low eng after first few		
Contingencies	10%	of process units				0.06	10-20% typical, low after the first few		
Working Capital, Land & Misc.	5%	of process units				0.03	5-10% typical, high land costs for this		
				U.S. Gulf Coast Capital Costs		0.81			
Site specific factor	110%	above US Gulf Coast		**Total Capital Costs**		**0.89**	**million plus**		**million for wind turbine**
Unit Capital Costs of	**1.79**	**/scf/d H2 or**		743 /kg/d H2 or		743	/gal/d gaso equiv		
	-	**wind turbine**		/kg/d H2 wind turbine		-	/gal/d gaso equiv wind turbine power		

annual load set by wind turbine

Hydrogen Costs at	36%	ann load factor	million $/yr of 1 station	$/million Btu LHV	$/1,000 scf H2	$/kg H2 or $/gal gaso equiv	Notes
Road tax or (subsidy)	$ -	/gal gaso equiv.	-	-	-	-	can be subsidy like EtOH
Gas Station mark-up	$ -	/gal gaso equiv.	-	-	-	-	if H2 drops total station revenues
Non-fuel Variable O&M	1.0%	/yr of capital	0.009	0.50	0.14	0.06	0.5-1.5% is typical
Electricity from WT	$ 0.040	/kWh	0.299	16.68	4.58	1.90	directly from kW WT-L worksheet
Variable Operating Cost			**0.308**	**17.18**	**4.72**	**1.95**	mostly electricity costs
Fixed Operating Cost	2.0%	/yr of capital	**0.018**	**0.99**	**0.27**	**0.11**	4-7% typical for refining
Capital Charges	14.0%	/yr of capital	**0.125**	**6.96**	**1.91**	**0.79**	20-25% typical of refining
Total HP Hydrogen Costs from Electrolysis			**0.451**	**25.13**	**6.90**	**2.86**	27% of relative high

Assume high pressure electrolysis cell with drastically lower capital cost of mass production cost & higher efficiency
Assume **no** transmission or distributions cost for the wind turbine power which usually adds $0.02-.04/kWh
Assumed oxygen recovery for by-product sales with large central plant case, but only minor economic impact

TABLE E-42
Dist PV Ele-C
Distributed Size Onsite Hydrogen via PV Solar-Based Electrolysis with Current Technology

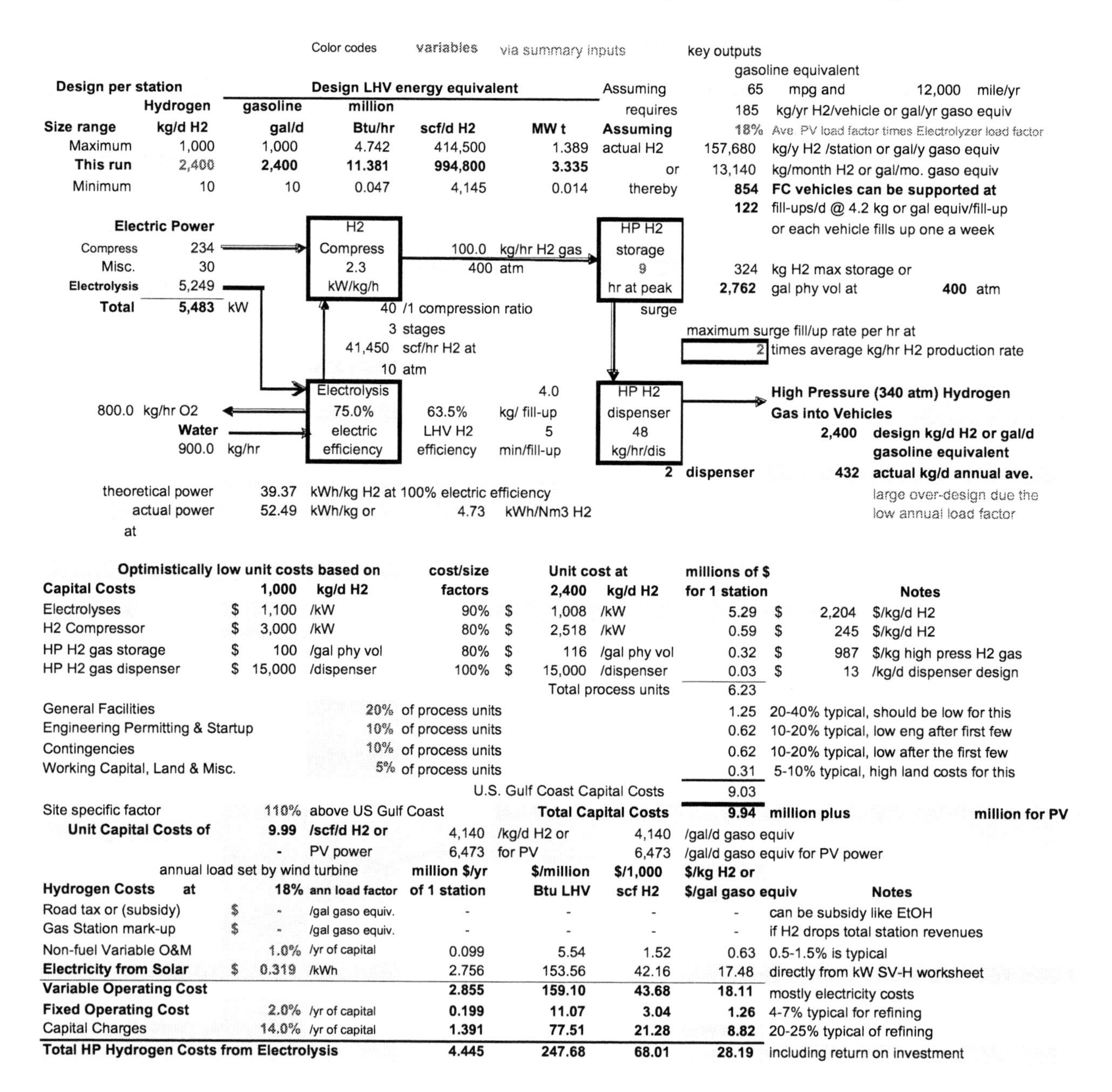

Color codes | variables | via summary inputs | key outputs

Design per station	Hydrogen	gasoline	Design LHV energy equivalent million			Assuming
Size range	kg/d H2	gal/d	Btu/hr	scf/d H2	MW t	requires Assuming actual H2
Maximum	1,000	1,000	4.742	414,500	1.389	
This run	2,400	2,400	11.381	994,800	3.335	or
Minimum	10	10	0.047	4,145	0.014	thereby

gasoline equivalent
65 mpg and 12,000 mile/yr
185 kg/yr H2/vehicle or gal/yr gaso equiv
18% Ave PV load factor times Electrolyzer load factor
157,680 kg/y H2 /station or gal/y gaso equiv
13,140 kg/month H2 or gal/mo. gaso equiv
854 FC vehicles can be supported at
122 fill-ups/d @ 4.2 kg or gal equiv/fill-up
or each vehicle fills up one a week

Electric Power		
Compress	234	
Misc.	30	
Electrolysis	5,249	
Total	**5,483**	kW

H2 Compress 2.3 kW/kg/h — 100.0 kg/hr H2 gas, 400 atm — HP H2 storage 9 hr at peak surge

40 /1 compression ratio
3 stages
41,450 scf/hr H2 at
10 atm

324 kg H2 max storage or
2,762 gal phy vol at **400** atm

maximum surge fill/up rate per hr at
2 times average kg/hr H2 production rate

800.0 kg/hr O2
Water 900.0 kg/hr

Electrolysis 75.0% electric efficiency — 63.5% LHV H2 efficiency — 4.0 kg/ fill-up, 5 min/fill-up — HP H2 dispenser 48 kg/hr/dis — 2 dispenser

High Pressure (340 atm) Hydrogen Gas into Vehicles
2,400 design kg/d H2 or gal/d gasoline equivalent
432 actual kg/d annual ave.
large over-design due the low annual load factor

theoretical power 39.37 kWh/kg H2 at 100% electric efficiency
actual power at 52.49 kWh/kg or 4.73 kWh/Nm3 H2

Optimistically low unit costs based on **Capital Costs**	**1,000 kg/d H2**	cost/size factors	Unit cost at **2,400 kg/d H2**	millions of $ for 1 station	**Notes**
Electrolyses	$ 1,100 /kW	90%	$ 1,008 /kW	5.29	$ 2,204 $/kg/d H2
H2 Compressor	$ 3,000 /kW	80%	$ 2,518 /kW	0.59	$ 245 $/kg/d H2
HP H2 gas storage	$ 100 /gal phy vol	80%	$ 116 /gal phy vol	0.32	$ 987 $/kg high press H2 gas
HP H2 gas dispenser	$ 15,000 /dispenser	100%	$ 15,000 /dispenser	0.03	$ 13 /kg/d dispenser design
			Total process units	6.23	
General Facilities		20% of process units		1.25	20-40% typical, should be low for this
Engineering Permitting & Startup		10% of process units		0.62	10-20% typical, low eng after first few
Contingencies		10% of process units		0.62	10-20% typical, low after the first few
Working Capital, Land & Misc.		5% of process units		0.31	5-10% typical, high land costs for this
			U.S. Gulf Coast Capital Costs	9.03	
Site specific factor	110% above US Gulf Coast		**Total Capital Costs**	**9.94**	**million plus million for PV**
Unit Capital Costs of	**9.99 /scf/d H2 or**	4,140 /kg/d H2 or	4,140	/gal/d gaso equiv	
	- PV power	6,473 for PV	6,473	/gal/d gaso equiv for PV power	

annual load set by wind turbine

Hydrogen Costs at	**18% ann load factor**	**million $/yr of 1 station**	**$/million Btu LHV**	**$/1,000 scf H2**	**$/kg H2 or $/gal gaso equiv**	**Notes**
Road tax or (subsidy)	$ - /gal gaso equiv.	-	-	-	-	can be subsidy like EtOH
Gas Station mark-up	$ - /gal gaso equiv.	-	-	-	-	if H2 drops total station revenues
Non-fuel Variable O&M	1.0% /yr of capital	0.099	5.54	1.52	0.63	0.5-1.5% is typical
Electricity from Solar	$ 0.319 /kWh	2.756	153.56	42.16	17.48	directly from kW SV-H worksheet
Variable Operating Cost		**2.855**	**159.10**	**43.68**	**18.11**	mostly electricity costs
Fixed Operating Cost	2.0% /yr of capital	**0.199**	**11.07**	**3.04**	**1.26**	4-7% typical for refining
Capital Charges	14.0% /yr of capital	**1.391**	**77.51**	**21.28**	**8.82**	20-25% typical of refining
Total HP Hydrogen Costs from Electrolysis		**4.445**	**247.68**	**68.01**	**28.19**	including return on investment

Assume high pressure electrolysis cell with drastically lower capital cost of mass production cost & higher efficiency
Assume no transmission or distributions cost for the wind turbine power which usually adds $0.02-.04/kWh
Assumed oxygen recovery for by-product sales with large central plant case, but only minor economic impact

TABLE E-43
Dist PV Ele-F
Distributed Size Onsite Hydrogen via PV Solar-Based Electrolysis with Future Optimism

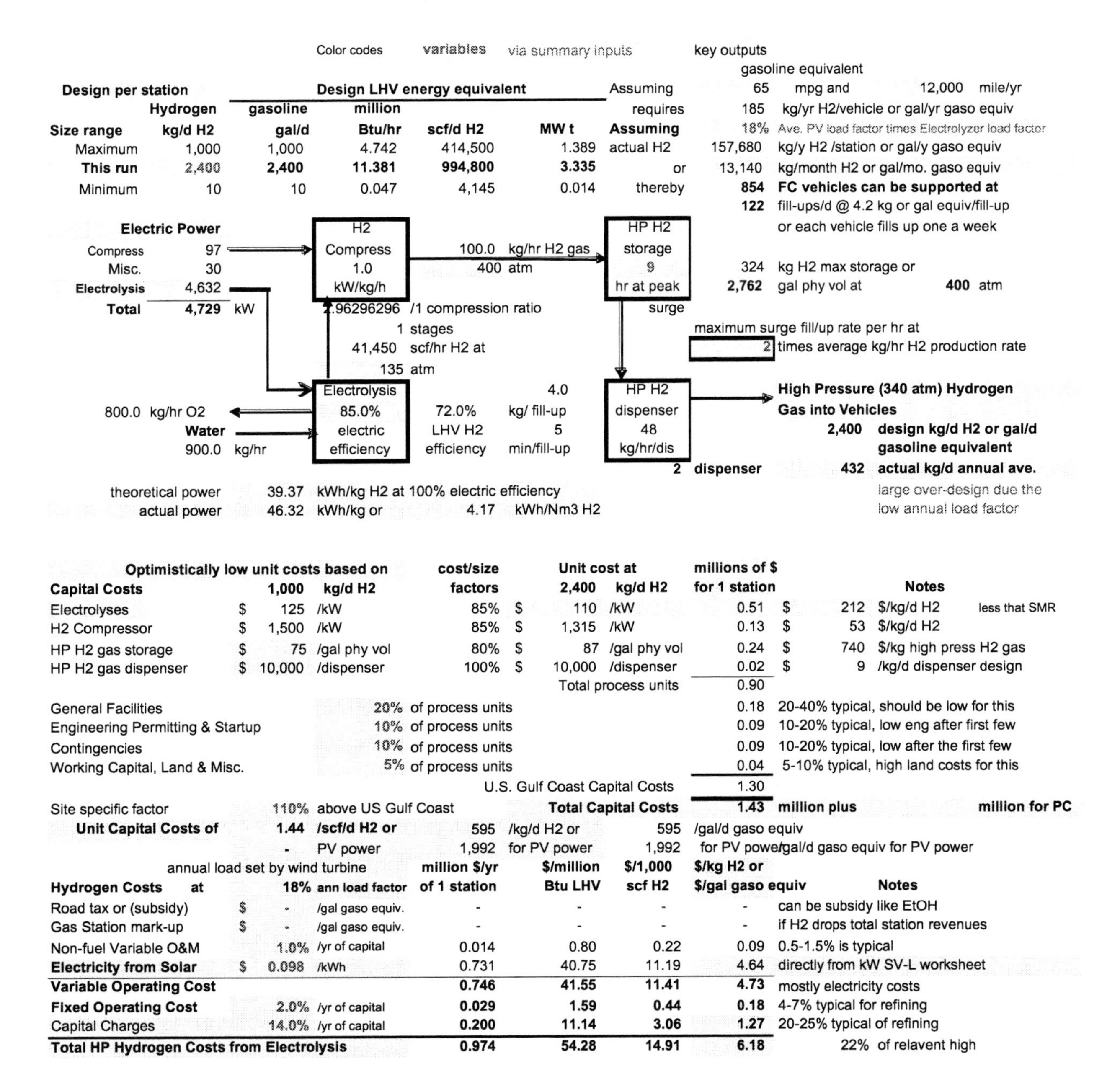

Color codes variables via summary inputs

Design per station	Hydrogen	Design LHV energy equivalent			
Size range	kg/d H2	gasoline gal/d	million Btu/hr	scf/d H2	MW t
Maximum	1,000	1,000	4.742	414,500	1.389
This run	2,400	2,400	11.381	994,800	3.335
Minimum	10	10	0.047	4,145	0.014

Assuming requires / Assuming actual H2 or thereby

key outputs
gasoline equivalent
65 mpg and 12,000 mile/yr
185 kg/yr H2/vehicle or gal/yr gaso equiv
18% Ave. PV load factor times Electrolyzer load factor
157,680 kg/y H2 /station or gal/y gaso equiv
13,140 kg/month H2 or gal/mo. gaso equiv
854 FC vehicles can be supported at
122 fill-ups/d @ 4.2 kg or gal equiv/fill-up
or each vehicle fills up one a week
324 kg H2 max storage or
2,762 gal phy vol at 400 atm
maximum surge fill/up rate per hr at
2 times average kg/hr H2 production rate

Electric Power
Compress 97
Misc. 30
Electrolysis 4,632
Total 4,729 kW

H2 Compress 1.0 kW/kg/h
100.0 kg/hr H2 gas
400 atm
HP H2 storage 9 hr at peak surge
2.96296296 /1 compression ratio
1 stages
41,450 scf/hr H2 at
135 atm

800.0 kg/hr O2
Water 900.0 kg/hr
Electrolysis 85.0% electric efficiency
72.0% LHV H2 efficiency
4.0 kg/ fill-up
5 min/fill-up
HP H2 dispenser 48 kg/hr/dis
2 dispenser

High Pressure (340 atm) Hydrogen Gas into Vehicles
2,400 design kg/d H2 or gal/d gasoline equivalent
432 actual kg/d annual ave.
large over-design due the low annual load factor

theoretical power 39.37 kWh/kg H2 at 100% electric efficiency
actual power 46.32 kWh/kg or 4.17 kWh/Nm3 H2

Optimistically low unit costs based on		cost/size	Unit cost at	millions of $		
Capital Costs	**1,000 kg/d H2**	**factors**	**2,400 kg/d H2**	**for 1 station**	**Notes**	
Electrolyses	$ 125 /kW	85%	$ 110 /kW	0.51	$ 212 $/kg/d H2	less that SMR
H2 Compressor	$ 1,500 /kW	85%	$ 1,315 /kW	0.13	$ 53 $/kg/d H2	
HP H2 gas storage	$ 75 /gal phy vol	80%	$ 87 /gal phy vol	0.24	$ 740 $/kg high press H2 gas	
HP H2 gas dispenser	$ 10,000 /dispenser	100%	$ 10,000 /dispenser	0.02	$ 9 /kg/d dispenser design	
			Total process units	0.90		
General Facilities	20% of process units			0.18	20-40% typical, should be low for this	
Engineering Permitting & Startup	10% of process units			0.09	10-20% typical, low eng after first few	
Contingencies	10% of process units			0.09	10-20% typical, low after the first few	
Working Capital, Land & Misc.	5% of process units			0.04	5-10% typical, high land costs for this	
			U.S. Gulf Coast Capital Costs	1.30		
Site specific factor	110% above US Gulf Coast		**Total Capital Costs**	**1.43**	**million plus**	**million for PC**

Unit Capital Costs of **1.44 /scf/d H2 or** 595 /kg/d H2 or 595 /gal/d gaso equiv
\- PV power 1,992 for PV power 1,992 for PV powe/gal/d gaso equiv for PV power

annual load set by wind turbine

Hydrogen Costs at	**18% ann load factor**	**million $/yr of 1 station**	**$/million Btu LHV**	**$/1,000 scf H2**	**$/kg H2 or $/gal gaso equiv**	**Notes**
Road tax or (subsidy)	$ - /gal gaso equiv.	-	-	-	-	can be subsidy like EtOH
Gas Station mark-up	$ - /gal gaso equiv.	-	-	-	-	if H2 drops total station revenues
Non-fuel Variable O&M	1.0% /yr of capital	0.014	0.80	0.22	0.09	0.5-1.5% is typical
Electricity from Solar	$ 0.098 /kWh	0.731	40.75	11.19	4.64	directly from kW SV-L worksheet
Variable Operating Cost		**0.746**	**41.55**	**11.41**	**4.73**	mostly electricity costs
Fixed Operating Cost	2.0% /yr of capital	**0.029**	**1.59**	**0.44**	**0.18**	4-7% typical for refining
Capital Charges	14.0% /yr of capital	**0.200**	**11.14**	**3.06**	**1.27**	20-25% typical of refining
Total HP Hydrogen Costs from Electrolysis		**0.974**	**54.28**	**14.91**	**6.18**	22% of relavent high

Assume high pressure electrolysis cell with drastically lower capital cost of mass production cost & higher efficiency
Assume no transmission or distributions cost for the wind turbine power which usually adds $0.02-.04/kWh
Assumed oxygen recovery for by-product sales with large central plant case, but only minor economic impact

Table E-44
Dist WT Gr Ele-C
Distributed Size Onsite Hydrogen via Wind Turbine/Grid Hybrid-Based Electrolysis with Current Costs

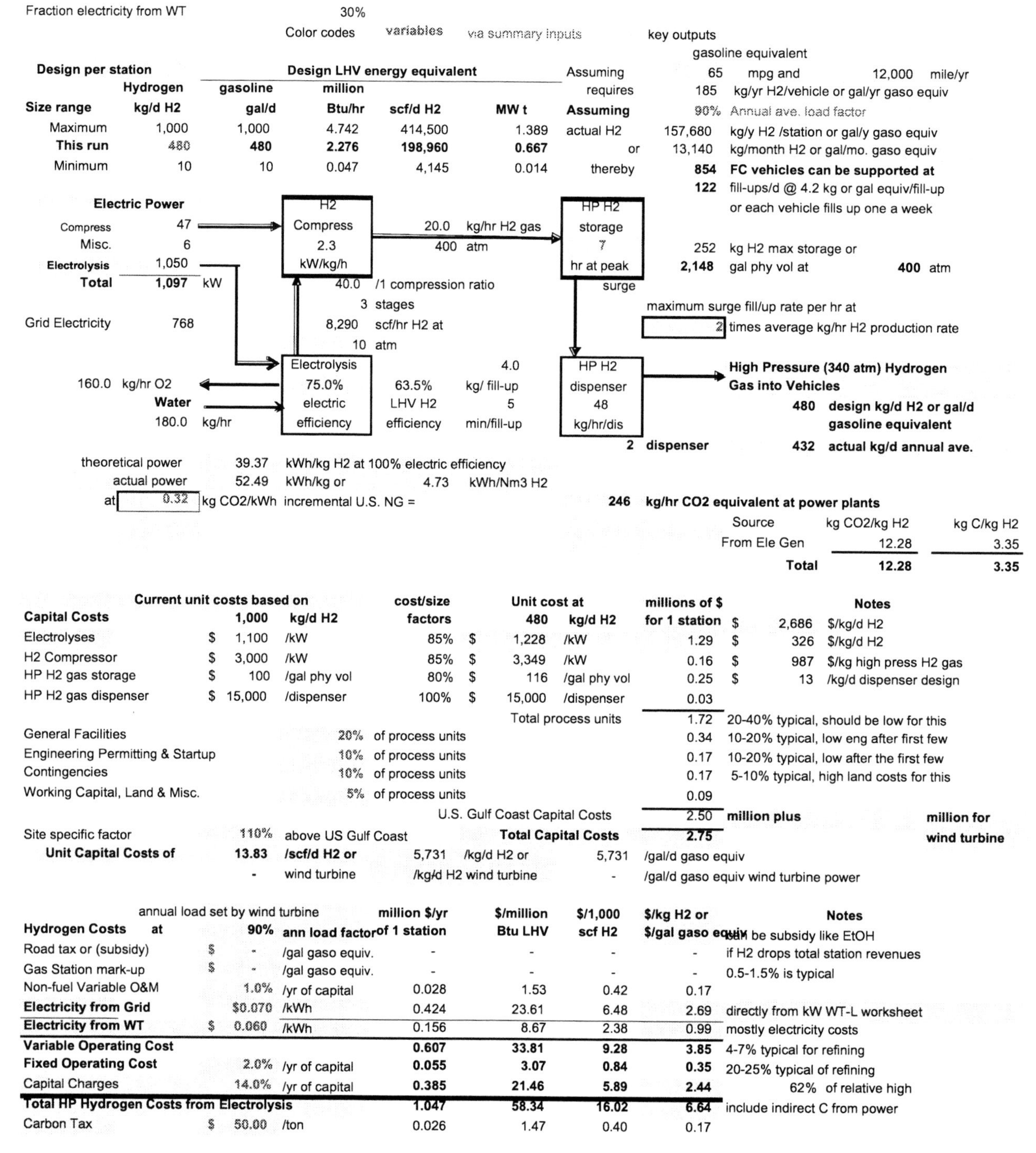

Fraction electricity from WT 30%
Color codes variables via summary inputs

Design per station	Hydrogen	gasoline	Design LHV energy equivalent million		
Size range	kg/d H2	gal/d	Btu/hr	scf/d H2	MW t
Maximum	1,000	1,000	4.742	414,500	1.389
This run	480	480	2.276	198,960	0.667
Minimum	10	10	0.047	4,145	0.014

Assuming requires
Assuming actual H2 or thereby

key outputs
gasoline equivalent
65 mpg and 12,000 mile/yr
185 kg/yr H2/vehicle or gal/yr gaso equiv
90% Annual ave. load factor
157,680 kg/y H2 /station or gal/y gaso equiv
13,140 kg/month H2 or gal/mo. gaso equiv
854 FC vehicles can be supported at
122 fill-ups/d @ 4.2 kg or gal equiv/fill-up or each vehicle fills up one a week

Electric Power	
Compress	47
Misc.	6
Electrolysis	1,050
Total	**1,097** kW

Grid Electricity 768

H2 Compress 2.3 kW/kg/h
20.0 kg/hr H2 gas
400 atm
40.0 /1 compression ratio
3 stages
8,290 scf/hr H2 at
10 atm

HP H2 storage 7 hr at peak surge
252 kg H2 max storage or
2,148 gal phy vol at 400 atm
maximum surge fill/up rate per hr at
2 times average kg/hr H2 production rate

160.0 kg/hr O2
Water 180.0 kg/hr
Electrolysis 75.0% electric efficiency
63.5% LHV H2 efficiency
4.0 kg/ fill-up
5 min/fill-up
HP H2 dispenser 48 kg/hr/dis
2 dispenser

High Pressure (340 atm) Hydrogen Gas into Vehicles
480 design kg/d H2 or gal/d gasoline equivalent
432 actual kg/d annual ave.

theoretical power 39.37 kWh/kg H2 at 100% electric efficiency
actual power 52.49 kWh/kg or 4.73 kWh/Nm3 H2
at 0.32 kg CO2/kWh incremental U.S. NG =

246 kg/hr CO2 equivalent at power plants

Source	kg CO2/kg H2	kg C/kg H2
From Ele Gen	12.28	3.35
Total	**12.28**	**3.35**

Capital Costs	Current unit costs based on 1,000 kg/d H2	cost/size factors	Unit cost at 480 kg/d H2	millions of $ for 1 station	Notes
Electrolyses	$ 1,100 /kW	85%	$ 1,228 /kW	1.29	$ 2,686 $/kg/d H2
H2 Compressor	$ 3,000 /kW	85%	$ 3,349 /kW	0.16	$ 326 $/kg/d H2
HP H2 gas storage	$ 100 /gal phy vol	80%	$ 116 /gal phy vol	0.25	$ 987 $/kg high press H2 gas
HP H2 gas dispenser	$ 15,000 /dispenser	100%	$ 15,000 /dispenser	0.03	$ 13 /kg/d dispenser design
			Total process units	1.72	20-40% typical, should be low for this
General Facilities		20% of process units		0.34	10-20% typical, low eng after first few
Engineering Permitting & Startup		10% of process units		0.17	10-20% typical, low after the first few
Contingencies		10% of process units		0.17	5-10% typical, high land costs for this
Working Capital, Land & Misc.		5% of process units		0.09	
			U.S. Gulf Coast Capital Costs	2.50	million plus million for
Site specific factor	110% above US Gulf Coast		Total Capital Costs	2.75	wind turbine
Unit Capital Costs of	13.83 /scf/d H2 or	5,731 /kg/d H2 or	5,731	/gal/d gaso equiv	
	- wind turbine	/kg/d H2 wind turbine	-	/gal/d gaso equiv wind turbine power	

annual load set by wind turbine

Hydrogen Costs at	90% ann load factor	million $/yr of 1 station	$/million Btu LHV	$/1,000 scf H2	$/kg H2 or $/gal gaso equiv	Notes
Road tax or (subsidy)	$ - /gal gaso equiv.	-	-	-	-	can be subsidy like EtOH if H2 drops total station revenues
Gas Station mark-up	$ - /gal gaso equiv.	-	-	-	-	0.5-1.5% is typical
Non-fuel Variable O&M	1.0% /yr of capital	0.028	1.53	0.42	0.17	
Electricity from Grid	$0.070 /kWh	0.424	23.61	6.48	2.69	directly from kW WT-L worksheet
Electricity from WT	$ 0.060 /kWh	0.156	8.67	2.38	0.99	mostly electricity costs
Variable Operating Cost		**0.607**	**33.81**	**9.28**	**3.85**	4-7% typical for refining
Fixed Operating Cost	2.0% /yr of capital	**0.055**	**3.07**	**0.84**	**0.35**	20-25% typical of refining
Capital Charges	14.0% /yr of capital	**0.385**	**21.46**	**5.89**	**2.44**	62% of relative high
Total HP Hydrogen Costs from Electrolysis		**1.047**	**58.34**	**16.02**	**6.64**	include indirect C from power
Carbon Tax	$ 50.00 /ton	0.026	1.47	0.40	0.17	

Assume high pressure electrolysis cell with drastically lower capital cost of mass production cost & higher efficiency
Assume no transmission or distributions cost for the wind turbine power which usually adds $0.02-.04/kWh
Assumed oxygen recovery for by-product sales with large central plant case, but only minor economic impact

TABLE E-45
Dist WT- Gr Ele-F
Distributed Size Onsite Hydrogen via Wind Turbine/Grid Hybrid-Based Electrolysis with Future Optimism

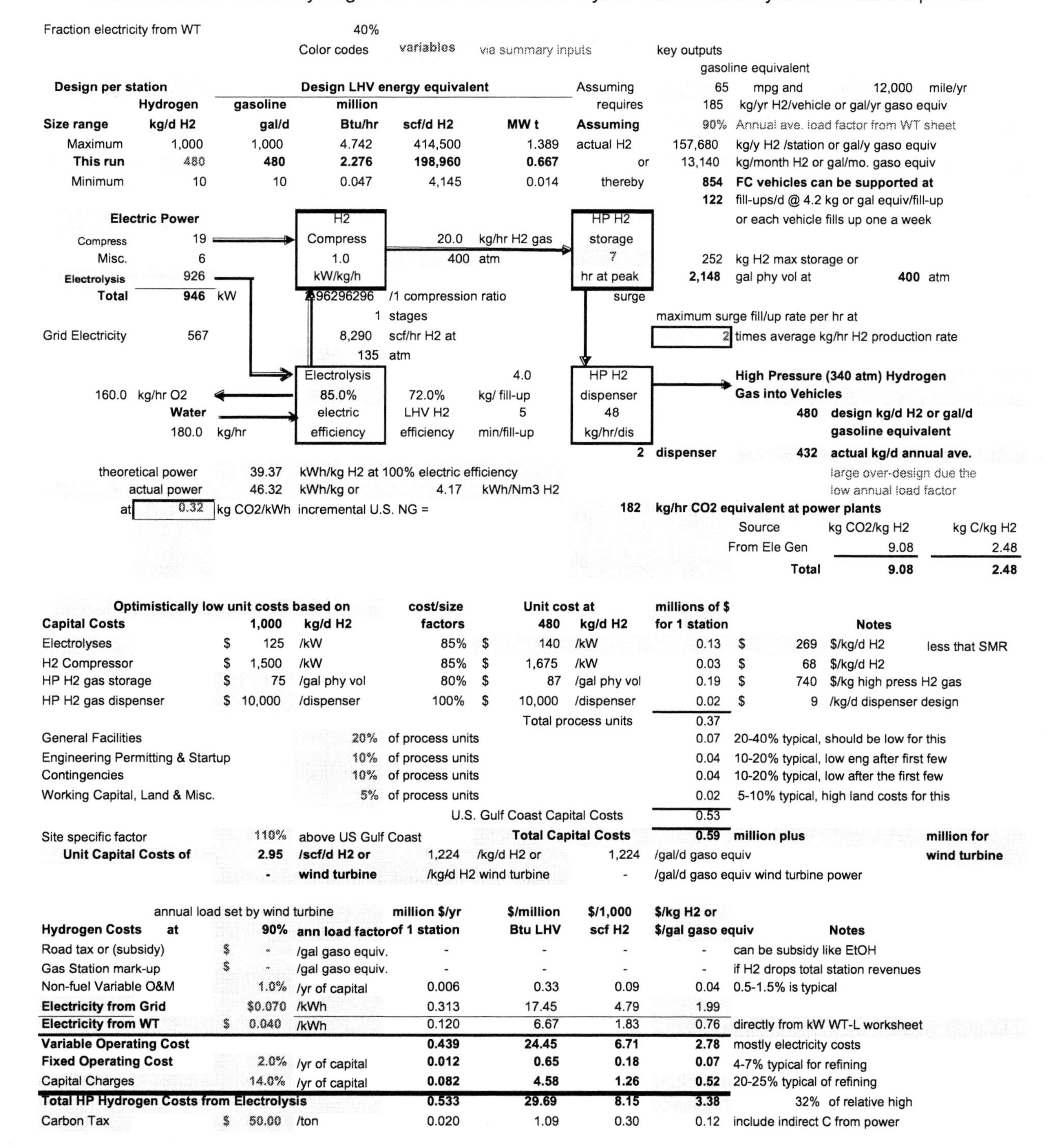

Fraction electricity from WT 40%

Color codes variables via summary inputs

Design per station	Hydrogen	gasoline	Design LHV energy equivalent million		
Size range	kg/d H2	gal/d	Btu/hr	scf/d H2	MW t
Maximum	1,000	1,000	4.742	414,500	1.389
This run	480	480	2.276	198,960	0.667
Minimum	10	10	0.047	4,145	0.014

key outputs

gasoline equivalent

Assuming requires 65 mpg and 12,000 mile/yr

185 kg/yr H2/vehicle or gal/yr gaso equiv

Assuming 90% Annual ave. load factor from WT sheet

actual H2 157,680 kg/y H2 /station or gal/y gaso equiv

or 13,140 kg/month H2 or gal/mo. gaso equiv

thereby 854 FC vehicles can be supported at

122 fill-ups/d @ 4.2 kg or gal equiv/fill-up

or each vehicle fills up one a week

Electric Power	
Compress	19
Misc.	6
Electrolysis	926
Total	946 kW

Grid Electricity 567

H2 Compress 1.0 kW/kg/h

2.96296296 /1 compression ratio

1 stages

8,290 scf/hr H2 at 135 atm

20.0 kg/hr H2 gas 400 atm

HP H2 storage 7 hr at peak surge

252 kg H2 max storage or

2,148 gal phy vol at 400 atm

maximum surge fill/up rate per hr at

2 times average kg/hr H2 production rate

160.0 kg/hr O2

Water 180.0 kg/hr

Electrolysis 85.0% electric efficiency

72.0% LHV H2 efficiency

4.0 kg/ fill-up 5 min/fill-up

HP H2 dispenser 48 kg/hr/dis

2 dispenser

High Pressure (340 atm) Hydrogen Gas into Vehicles

480 design kg/d H2 or gal/d gasoline equivalent

432 actual kg/d annual ave.

large over-design due the low annual load factor

theoretical power 39.37 kWh/kg H2 at 100% electric efficiency

actual power 46.32 kWh/kg or 4.17 kWh/Nm3 H2

at 0.32 kg CO2/kWh incremental U.S. NG =

182 kg/hr CO2 equivalent at power plants

Source	kg CO2/kg H2	kg C/kg H2
From Ele Gen	9.08	2.48
Total	9.08	2.48

Capital Costs	Optimistically low unit costs based on 1,000 kg/d H2		cost/size factors	Unit cost at 480 kg/d H2		millions of $ for 1 station	Notes		
Electrolyses	$ 125	/kW	85%	$ 140	/kW	0.13	$ 269	$/kg/d H2	less that SMR
H2 Compressor	$ 1,500	/kW	85%	$ 1,675	/kW	0.03	$ 68	$/kg/d H2	
HP H2 gas storage	$ 75	/gal phy vol	80%	$ 87	/gal phy vol	0.19	$ 740	$/kg high press H2 gas	
HP H2 gas dispenser	$ 10,000	/dispenser	100%	$ 10,000	/dispenser	0.02	$ 9	/kg/d dispenser design	
				Total process units		0.37			
General Facilities			20%	of process units		0.07	20-40% typical, should be low for this		
Engineering Permitting & Startup			10%	of process units		0.04	10-20% typical, low eng after first few		
Contingencies			10%	of process units		0.04	10-20% typical, low after the first few		
Working Capital, Land & Misc.			5%	of process units		0.02	5-10% typical, high land costs for this		
				U.S. Gulf Coast Capital Costs		0.53			
Site specific factor	110%	above US Gulf Coast		Total Capital Costs		0.59	million plus		million for wind turbine
Unit Capital Costs of	2.95	/scf/d H2 or	1,224	/kg/d H2 or	1,224	/gal/d gaso equiv			
	-	wind turbine		/kg/d H2 wind turbine	-	/gal/d gaso equiv wind turbine power			

annual load set by wind turbine

Hydrogen Costs at	90%	ann load factor	million $/yr of 1 station	$/million Btu LHV	$/1,000 scf H2	$/kg H2 or $/gal gaso equiv	Notes
Road tax or (subsidy)	$ -	/gal gaso equiv.	-	-	-	-	can be subsidy like EtOH
Gas Station mark-up	$ -	/gal gaso equiv.	-	-	-	-	if H2 drops total station revenues
Non-fuel Variable O&M	1.0%	/yr of capital	0.006	0.33	0.09	0.04	0.5-1.5% is typical
Electricity from Grid	$0.070	/kWh	0.313	17.45	4.79	1.99	
Electricity from WT	$ 0.040	/kWh	0.120	6.67	1.83	0.76	directly from kW WT-L worksheet
Variable Operating Cost			0.439	24.45	6.71	2.78	mostly electricity costs
Fixed Operating Cost	2.0%	/yr of capital	0.012	0.65	0.18	0.07	4-7% typical for refining
Capital Charges	14.0%	/yr of capital	0.082	4.58	1.26	0.52	20-25% typical of refining
Total HP Hydrogen Costs from Electrolysis			0.533	29.69	8.15	3.38	32% of relative high
Carbon Tax	$ 50.00	/ton	0.020	1.09	0.30	0.12	include indirect C from power

Assume high pressure electrolysis cell with drastically lower capital cost of mass production cost & higher efficiency

Assume no transmission or distributions cost for the wind turbine power which usually adds $0.02-.04/kWh

Assumed oxygen recovery for by-product sales with large central plant case, but only minor economic impact

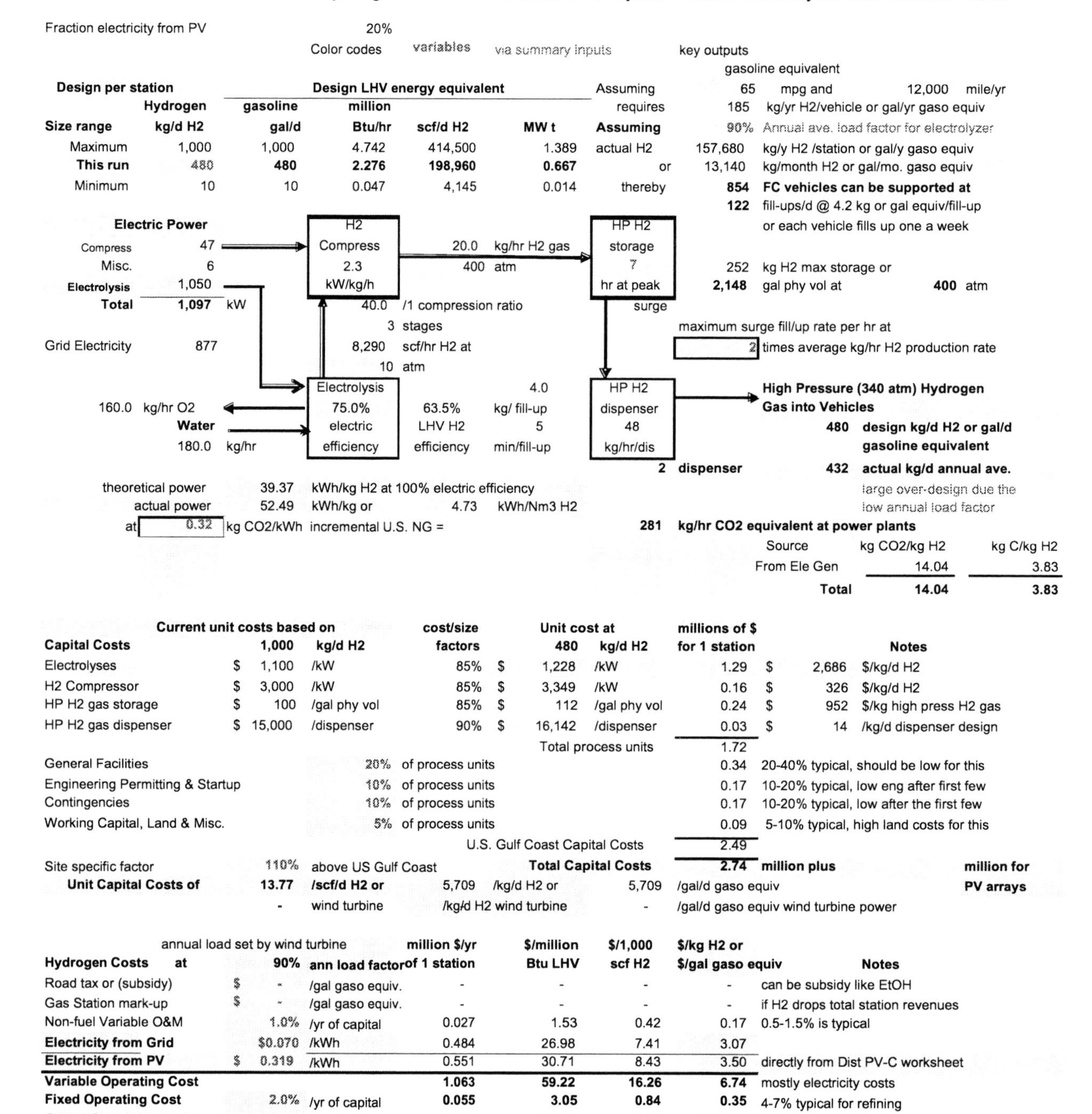

Table E-46
Dist PV-Gr Ele-C
Distributed Size Onsite Hydrogen via Photovoltaics/Grid Hybrid-Based Electrolysis with Current Costs

Fraction electricity from PV 20%

Color codes: variables — via summary inputs — key outputs

Design per station	Hydrogen	gasoline	Design LHV energy equivalent million		
Size range	kg/d H2	gal/d	Btu/hr	scf/d H2	MW t
Maximum	1,000	1,000	4.742	414,500	1.389
This run	480	480	2.276	198,960	0.667
Minimum	10	10	0.047	4,145	0.014

Assuming requires — gasoline equivalent: 65 mpg and 12,000 mile/yr
185 kg/yr H2/vehicle or gal/yr gaso equiv
Assuming 90% Annual ave. load factor for electrolyzer
actual H2 157,680 kg/y H2 /station or gal/y gaso equiv
or 13,140 kg/month H2 or gal/mo. gaso equiv
thereby **854 FC vehicles can be supported at**
122 fill-ups/d @ 4.2 kg or gal equiv/fill-up
or each vehicle fills up one a week

Electric Power	
Compress	47
Misc.	6
Electrolysis	1,050
Total	1,097 kW

Grid Electricity 877

H2 Compress: 2.3 kW/kg/h; 20.0 kg/hr H2 gas; 400 atm
40.0 /1 compression ratio
3 stages
8,290 scf/hr H2 at 10 atm

HP H2 storage: 7 hr at peak surge
252 kg H2 max storage or
2,148 gal phy vol at 400 atm

maximum surge fill/up rate per hr at 2 times average kg/hr H2 production rate

160.0 kg/hr O2
Water 180.0 kg/hr
Electrolysis: 75.0% electric efficiency; 63.5% LHV H2 efficiency
4.0 kg/ fill-up; 5 min/fill-up
HP H2 dispenser: 48 kg/hr/dis; 2 dispenser

High Pressure (340 atm) Hydrogen Gas into Vehicles
480 **design kg/d H2 or gal/d gasoline equivalent**
432 **actual kg/d annual ave.**
large over-design due the low annual load factor

theoretical power 39.37 kWh/kg H2 at 100% electric efficiency
actual power 52.49 kWh/kg or 4.73 kWh/Nm3 H2
at 0.32 kg CO2/kWh incremental U.S. NG =
281 **kg/hr CO2 equivalent at power plants**

Source	kg CO2/kg H2	kg C/kg H2
From Ele Gen	14.04	3.83
Total	**14.04**	**3.83**

Capital Costs	Current unit costs based on 1,000 kg/d H2		cost/size factors	Unit cost at 480 kg/d H2		millions of $ for 1 station	Notes	
Electrolyses	$ 1,100	/kW	85%	$ 1,228	/kW	1.29	$ 2,686	$/kg/d H2
H2 Compressor	$ 3,000	/kW	85%	$ 3,349	/kW	0.16	$ 326	$/kg/d H2
HP H2 gas storage	$ 100	/gal phy vol	85%	$ 112	/gal phy vol	0.24	$ 952	$/kg high press H2 gas
HP H2 gas dispenser	$ 15,000	/dispenser	90%	$ 16,142	/dispenser	0.03	$ 14	/kg/d dispenser design
				Total process units		1.72		
General Facilities	20%	of process units				0.34	20-40% typical, should be low for this	
Engineering Permitting & Startup	10%	of process units				0.17	10-20% typical, low eng after first few	
Contingencies	10%	of process units				0.17	10-20% typical, low after the first few	
Working Capital, Land & Misc.	5%	of process units				0.09	5-10% typical, high land costs for this	
				U.S. Gulf Coast Capital Costs		2.49		
Site specific factor	110%	above US Gulf Coast		Total Capital Costs		2.74	million plus	million for PV arrays
Unit Capital Costs of	13.77	/scf/d H2 or	5,709	/kg/d H2 or	5,709	/gal/d gaso equiv		
	-	wind turbine		/kg/d H2 wind turbine	-	/gal/d gaso equiv wind turbine power		

annual load set by wind turbine

Hydrogen Costs at	90%	ann load factor	million $/yr of 1 station	$/million Btu LHV	$/1,000 scf H2	$/kg H2 or $/gal gaso equiv	Notes
Road tax or (subsidy)	$ -	/gal gaso equiv.	-	-	-	-	can be subsidy like EtOH
Gas Station mark-up	$ -	/gal gaso equiv.	-	-	-	-	if H2 drops total station revenues
Non-fuel Variable O&M	1.0%	/yr of capital	0.027	1.53	0.42	0.17	0.5-1.5% is typical
Electricity from Grid	$0.070	/kWh	0.484	26.98	7.41	3.07	
Electricity from PV	$ 0.319	/kWh	0.551	30.71	8.43	3.50	directly from Dist PV-C worksheet
Variable Operating Cost			**1.063**	**59.22**	**16.26**	**6.74**	mostly electricity costs
Fixed Operating Cost	2.0%	/yr of capital	**0.055**	**3.05**	**0.84**	**0.35**	4-7% typical for refining
Capital Charges	14.0%	/yr of capital	**0.384**	**21.38**	**5.87**	**2.43**	20-25% typical of refining
Total HP Hydrogen Costs from Electrolysis			**1.501**	**83.65**	**22.97**	**9.52**	89% of relative high
Carbon Tax	$ 50.00	/ton	0.030	1.68	0.46	0.19	include indirect C from power

Assume high pressure electrolysis cell with drastically lower capital cost of mass production cost & higher efficiency
Assume no transmission or distributions cost for the wind turbine power
Assumed oxygen recovery for by-product sales with large central plant case, but only minor economic impact

TABLE E-47
Dist PV-Gr Ele-F
Distributed Size Onsite Hydrogen via PV/Grid Hybrid-Based Electrolysis with Future Optimism

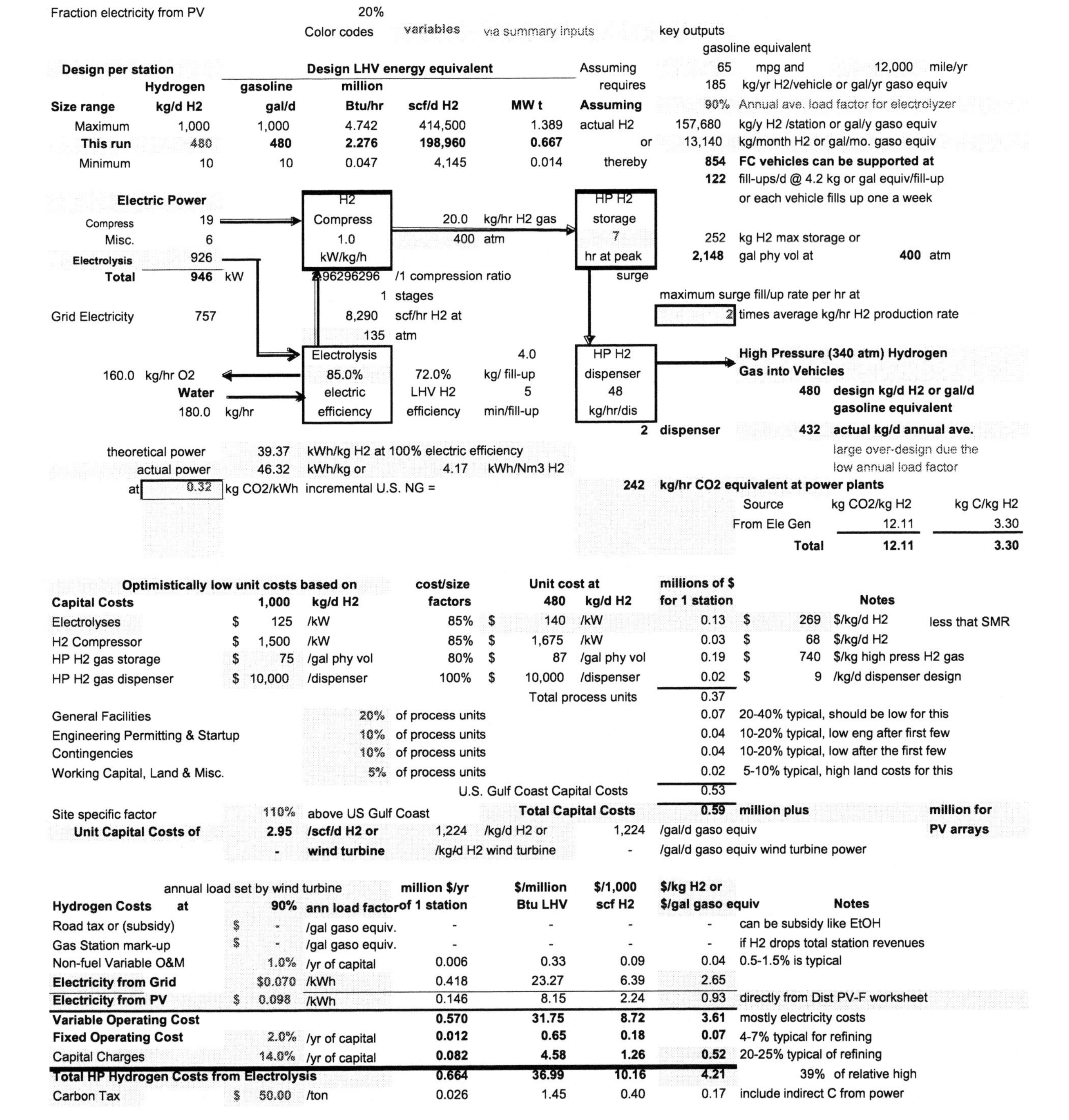

Fraction electricity from PV 20%

Color codes variables via summary inputs key outputs

Design per station			Design LHV energy equivalent		
	Hydrogen	gasoline	million		
Size range	kg/d H2	gal/d	Btu/hr	scf/d H2	MW t
Maximum	1,000	1,000	4.742	414,500	1.389
This run	480	**480**	**2.276**	**198,960**	**0.667**
Minimum	10	10	0.047	4,145	0.014

gasoline equivalent

Assuming 65 mpg and 12,000 mile/yr

requires 185 kg/yr H2/vehicle or gal/yr gaso equiv

Assuming 90% Annual ave. load factor for electrolyzer

actual H2 157,680 kg/y H2 /station or gal/y gaso equiv

or 13,140 kg/month H2 or gal/mo. gaso equiv

thereby **854 FC vehicles can be supported at**

122 fill-ups/d @ 4.2 kg or gal equiv/fill-up

or each vehicle fills up one a week

Electric Power

Compress	19
Misc.	6
Electrolysis	926
Total	**946** kW

Grid Electricity 757

H2 Compress 1.0 kW/kg/h

20.0 kg/hr H2 gas 400 atm

HP H2 storage 7 hr at peak surge

252 kg H2 max storage or **2,148** gal phy vol at **400** atm

2.96296296 /1 compression ratio

1 stages

8,290 scf/hr H2 at 135 atm

maximum surge fill/up rate per hr at 2 times average kg/hr H2 production rate

Electrolysis 85.0% electric efficiency

72.0% LHV H2 efficiency

4.0 kg/ fill-up 5 min/fill-up

HP H2 dispenser 48 kg/hr/dis

2 dispenser

160.0 kg/hr O2

Water 180.0 kg/hr

High Pressure (340 atm) Hydrogen Gas into Vehicles

480 design kg/d H2 or gal/d gasoline equivalent

432 actual kg/d annual ave.

large over-design due the low annual load factor

theoretical power 39.37 kWh/kg H2 at 100% electric efficiency

actual power 46.32 kWh/kg or 4.17 kWh/Nm3 H2

at 0.32 kg CO2/kWh incremental U.S. NG = **242 kg/hr CO2 equivalent at power plants**

Source	kg CO2/kg H2	kg C/kg H2
From Ele Gen	12.11	3.30
Total	**12.11**	**3.30**

Optimistically low unit costs based on			**cost/size**		**Unit cost at**		**millions of $**			
Capital Costs	**1,000**	**kg/d H2**	**factors**		**480**	**kg/d H2**	**for 1 station**		**Notes**	
Electrolyses	$ 125	/kW	85%	$	140	/kW	0.13	$	269	$/kg/d H2 less that SMR
H2 Compressor	$ 1,500	/kW	85%	$	1,675	/kW	0.03	$	68	$/kg/d H2
HP H2 gas storage	$ 75	/gal phy vol	80%	$	87	/gal phy vol	0.19	$	740	$/kg high press H2 gas
HP H2 gas dispenser	$ 10,000	/dispenser	100%	$	10,000	/dispenser	0.02	$	9	/kg/d dispenser design
					Total process units		0.37			
General Facilities	20%	of process units					0.07	20-40% typical, should be low for this		
Engineering Permitting & Startup	10%	of process units					0.04	10-20% typical, low eng after first few		
Contingencies	10%	of process units					0.04	10-20% typical, low after the first few		
Working Capital, Land & Misc.	5%	of process units					0.02	5-10% typical, high land costs for this		
					U.S. Gulf Coast Capital Costs		0.53			
Site specific factor	110%	above US Gulf Coast			**Total Capital Costs**		**0.59**	**million plus**		**million for PV arrays**
Unit Capital Costs of	**2.95**	**/scf/d H2 or**	1,224	/kg/d H2 or	1,224		/gal/d gaso equiv			
	-	**wind turbine**	/kg/d H2 wind turbine		-		/gal/d gaso equiv wind turbine power			

annual load set by wind turbine

Hydrogen Costs at	**90%**	**ann load factor**	**million $/yr of 1 station**	**$/million Btu LHV**	**$/1,000 scf H2**	**$/kg H2 or $/gal gaso equiv**	**Notes**
Road tax or (subsidy)	$ -	/gal gaso equiv.	-	-	-	-	can be subsidy like EtOH
Gas Station mark-up	$ -	/gal gaso equiv.	-	-	-	-	if H2 drops total station revenues
Non-fuel Variable O&M	1.0%	/yr of capital	0.006	0.33	0.09	0.04	0.5-1.5% is typical
Electricity from Grid	$0.070	/kWh	0.418	23.27	6.39	2.65	
Electricity from PV	$ 0.098	/kWh	0.146	8.15	2.24	0.93	directly from Dist PV-F worksheet
Variable Operating Cost			**0.570**	**31.75**	**8.72**	**3.61**	mostly electricity costs
Fixed Operating Cost	2.0%	/yr of capital	**0.012**	**0.65**	**0.18**	**0.07**	4-7% typical for refining
Capital Charges	14.0%	/yr of capital	**0.082**	**4.58**	**1.26**	**0.52**	20-25% typical of refining
Total HP Hydrogen Costs from Electrolysis			**0.664**	**36.99**	**10.16**	**4.21**	39% of relative high
Carbon Tax	$ 50.00	/ton	0.026	1.45	0.40	0.17	include indirect C from power

Assume high pressure electrolysis cell with drastically lower capital cost of mass production cost & higher efficiency

Assume no transmission or distributions cost for the PV power which usually adds $0.02-.04/kWh

Assumed oxygen recovery for by-product sales with large central plant case, but only minor economic impact

TABLE E-48
Dist PV-C
Photovoltatic Solar Power Generation Economics for Current Technology

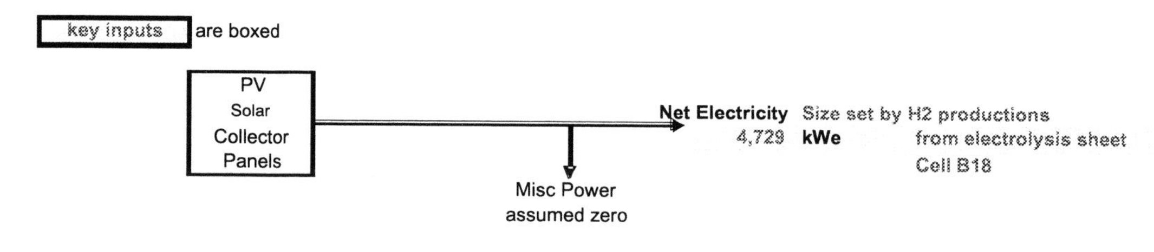

Solar intensity	1 kW/m2	at Standard Test Conditions (i.e. best possible) sunny day at noon	
PV efficiency	15%	half theoretical efficiency - crystalline silicon, thin film only 6%	
Maximum PV rating	0.15 kWp-dc/m2	assume no degrading with time or higher summer temperatures	
Required peak capacity	4,729 kWp-dc	sunny day at noon	
area of PV panels	31,527 m2 or	**339,233 ft2 or**	**7.79 acres or 3.15 hectares**

Capital Costs	**Optimistically low unit costs based on 1,000 kWe**	**cost/size factors**	**Unit cost at 4,729 kW**	**$ Millions**	**$/kW net**	
PV panels installed	$ 2,500 /kW	85%	$ 1,980 /kW	9.36	1,980	28 $/ft2 or 297 $/m2 PV panel
Power conditioning	100 /kW	85%	$ 79 /kW	0.37	79	
	below from input summary		Subtotal of process units	9.74	2,059	
General Facilities	20% of process units			1.95	412	
Engineering Permitting & Startup	10% of process units			0.97	206	
Contingencies	10% of process units			0.97	206	
Working Capital, Land & Misc.	5% of process units			0.49	103	
			U.S. Gulf Coast Capital Costs	14.12	2,986	
Site specific factor	110% above US Gulf Coast		**Total Capital Costs**	**15.53**	**3,285**	**$/kW design**
					6,473	$/kg/d H2 design capacity
					15.62	$/scf/d H2 design capacity

Electricity Cost	**Inputs for summary**	**US dollars $ MM/yr**	**$/kWh**	
	20% **ann. capacity factor**			Sunny location like CA
Capital charges	14% of capital per yr	2.175	0.262	
Fixed O&M	2%	0.311	0.037	
Variable O&M	1%	0.155	0.019	
Solar Subsidies	$ - /kWh	-	-	zero for fair comparison
Solar PV power costs		**2.641**	**0.319**	**0.000 mt CO2/MWh**
Assumed transmission and distribution costs			-	
Total delivered Solar PV power costs			**0.319**	**$/kWh**

Note: Solar PV subside is equivalent to $ - /tonne CO2 avoided assuming solar power replaces NGCC at 0.35 mt CO2/MWh

TABLE E-49
Dist PV-F
Photovoltaic Solar Power Generation Economics of Future Optimism

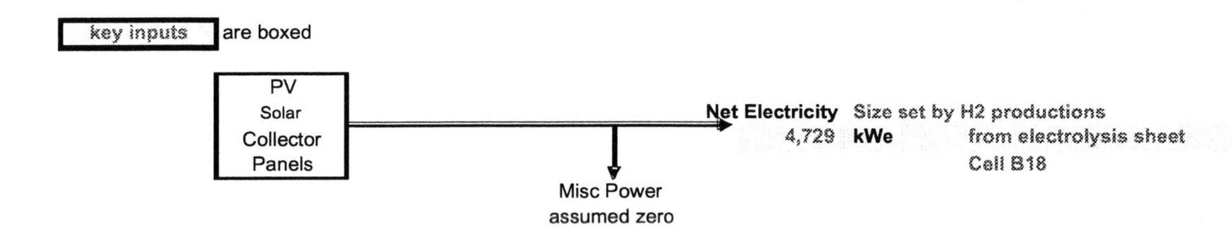

Solar intensity	1	kW/m2	at Standard Test Conditions (i.e. best possible) sunny day at noon					
PV efficiency	20%							
Maximum PV rating	0.2	kWp-dc/m2	assume no degrading with time or higher summer temperatures					
Required peak capacity	4,729	kWp-dc	sunny day at noon					
area of PV panels	23,645	m2 or	**254,425**	**ft2 or**	**5.84**	**acres or**	**2.36**	**hectares**

	Optimistically low unit costs based on		cost/size	Unit cost at				
Capital Costs	**1,000**	**kWe**	**factors**	**4,729 kW**	**$ Millions**	**$/kW net**	11	$/ft2 or
PV panels installed	$ 750	/kW	85%	$ 594 /kW	2.81	594	119	$/m2 PV panel
Power conditioning	50	/kW	85%	$ 40 /kW	0.19	40		
	below from input summary			Subtotal of process units	3.00	634		
General Facilities	20%	of process units			0.60	127		
Engineering Permitting & Startup	10%	of process units			0.30	63		
Contingencies	10%	of process units			0.30	63		
Working Capital, Land & Misc.	5%	of process units			0.15	32		
				U.S. Gulf Coast Capital Costs	4.35	919		
Site specific factor	110%	above US Gulf Coast		**Total Capital Costs**	**4.78**	**1,011**	**$/kW design**	
						1,992	$/kg/d H2 design capacity	
						4.80	$/scf/d H2 design capacity	

	Inputs for summary		**US dollars**			
Electricity Cost	20%	**ann. capacity factor**	**$ MM/yr**	**$/kWh**	Sunny location like CA	
Capital charges	14%	of capital per yr	0.669	0.081		
Fixed O&M	2%		0.096	0.012		
Variable O&M	1%		0.048	0.006		
Solar Subsidies	$ -	/kWh	-	-	zero for fair comparison	
Solar PV power costs at wind turbine			**0.813**	**0.098**	**0.000**	**mt CO2/MWh**
Assumed transmission and distribution costs				-		
Total delivered Solar PV power costs				**0.098**	**$/kWh**	

Note: Solar PV subside is equivalent to **$ -** /tonne CO2 avoided assuming the wind turbine power replaces NGCC at 0.35 mt CO2/MWh

Appendix F

U.S. Energy Systems

INTRODUCTION

This appendix briefly describes some of the major characteristics of the current and evolving energy system. These characteristics will be important factors in determining the competitiveness of hydrogen and the feasibility of introducing it without disrupting the system. For example, natural gas is the most likely feedstock for producing hydrogen, at least at first, but resource constraints may limit growth.

ENERGY SUPPLY SYSTEMS

Petroleum

Petroleum is used for a variety of purposes in the domestic economy, the most important being for transportation, especially automobiles, trucks, and airplanes. The United States was self-sufficient in petroleum roughly 50 years ago, after which demand outpaced domestic production and the United States began to import growing volumes of petroleum from abroad. Today, the United States depends on foreign sources for about 55 percent of its petroleum needs, and this fraction is continually increasing (EIA, 2003).

Petroleum is extracted from underground reservoirs via oil wells. It is then moved by pipeline or ship to a refinery where it is transformed into finished products, including various grades of gasoline, diesel fuels, jet fuels, home heating oil, bunker fuels, lubricants, and chemical feedstocks. Most finished fuels are moved by product pipelines to distribution terminals. From the terminals, fuels are moved to gasoline stations, truck stops, airports, and so on, where most end users obtain their fuels.

The U.S. petroleum system is massive. It includes 161 oil refineries; 2,000 oil storage terminals; approximately 220,000 miles of crude oil and oil products lines; and more than 175,000 gasoline service stations (NRC, 2002). The lifetimes of this capital stock vary considerably. Crude and product pipelines, refineries, and terminals have extremely long lifetimes (many decades), because they are in a constant state of repair, upgrade, and partial replacement.

In addition to the supply system, there is a vast array of end-use equipment dependent on petroleum products. Automobiles and light trucks, the end users most likely to be converted to hydrogen, have a lifetime of about 15 years. Thus, gasoline-burning vehicles produced in 2020 will still require a fuel supply in 2035 and later.

One of the materials extensively used in petroleum refining and chemical plants is hydrogen. A significant infrastructure currently exists for the manufacture, distribution, and use of hydrogen. Table F-1 provides information on the current U.S. hydrogen system and compares that system with the current U.S. gasoline system to provide some perspective on the challenge of substituting hydrogen for gasoline as an automobile fuel (EIA, 2003). Most hydrogen is produced onsite by the user. Nevertheless, this production (about 8 billion kg/yr) is equal to the hydrogen that would be required to fuel the light-duty vehicles postulated to be operating in 2028 (see Figure 6-3 in Chapter 6 of this report). Thus, the United States has substantial experience with producing and handling hydrogen.

Petroleum, natural gas, and coal are responsible for the vast majority of world energy production. These fuels are finite resources, and they are being depleted. Figure F-1 shows the world's estimated remaining fossil fuel resources.

For decades, various analysts have predicted petroleum resource constraints. U.S. production peaked in the 1970s, but international production has so far shown no signs of faltering. At some point, resource constraints will prevent supply from keeping up with demand, and prices will increase. This report makes no attempt to determine when this might happen, and assumes a price of $30/bbl for petroleum from now until 2050 (see Chapters 4 and 5, and Appendix E). However, the committee notes that constraints might emerge within the time frame of this study. Two EIA production scenarios are shown in Figure F-2 (EIA, 2000). These projections include both conventional petroleum and unconventional petroleum resources, such as tar sands and heavy oils.

TABLE F-1 Some Perspective on the Size of the Current Hydrogen and Gasoline Production and Distribution Systems in the United States

Characteristic	Hydrogen System	Gasoline System
Production	9 million tons per year	150 million tons per year of gasoline equivalent
Pipeline capacity	<700 miles	~200,000 miles of petroleum and product
Distribution stations	<15	>175,000
Delivery trucks	~19 2,600 psi hydrogen trucks to deliver the equivalent of one gasoline tank truck	

SOURCE: EIA (2002).

Natural Gas

Approximately 24 percent of U.S. energy demand in 2001 was satisfied by natural gas. Over 90 percent was produced domestically, with almost all of the rest imported from Canada (EIA, 2003). Natural gas is found in underground reservoirs, often in combination with petroleum, and is accessed by wells similar to oil wells. Once it is brought to the surface, natural gas is processed to remove impurities, and a mercaptan is added to provide the "rotten egg" smell that facilitates easy detection of leaks. Natural gas is then compressed and transmitted to storage facilities and end users.

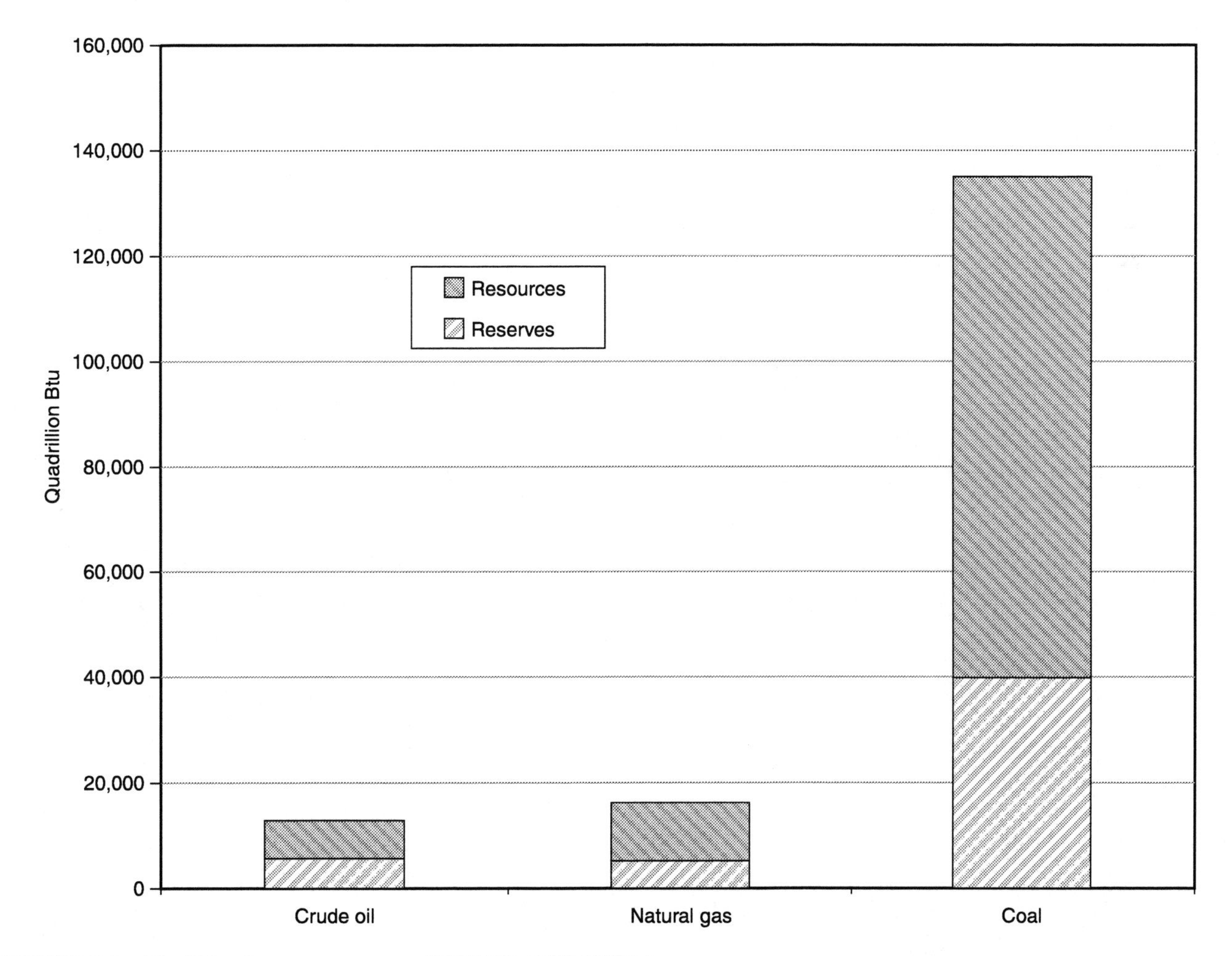

FIGURE F-1 World fossil energy resources. SOURCE: IPCC (2001b).

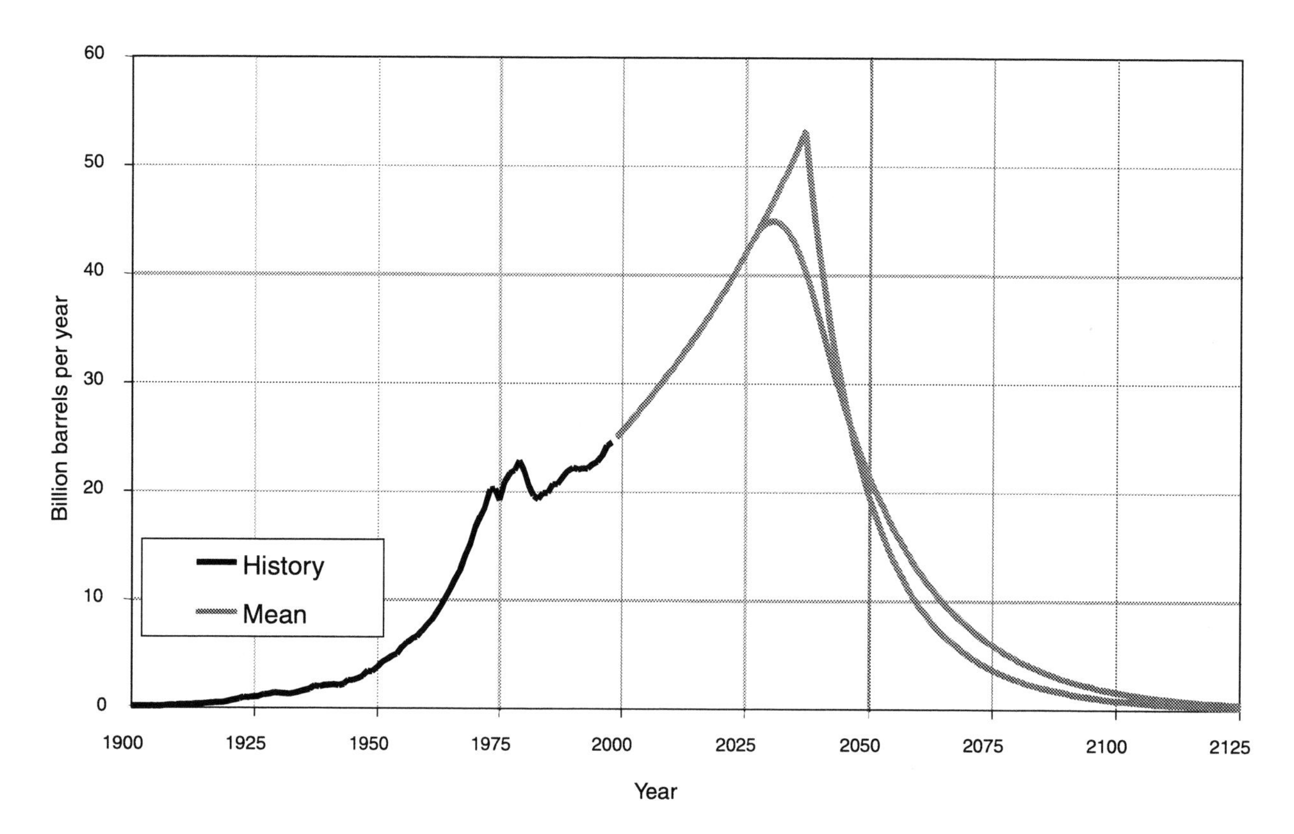

FIGURE F-2 Annual production scenarios for the mean resource estimate showing sharp and rounded peaks, 1900–2125. Growth rate leading to either peak is 2 percent. Sharp peak occurs in 2037 followed by decline at reserve to production ratio of 10. Rounded peak occurs in 2030 followed by decline at 5 percent. U.S. volumes were added to the USGS (2000) foreign volumes estimate to obtain a world total of 3,000 billion barrels (mean value) of ultimately recoverable resources. SOURCE: EIA (2000).

The U.S. natural gas system includes 726 natural-gas-processing plants; 410 underground gas storage fields; 254,000 miles of gas transmission lines; and 980,000 miles of local distribution pipelines (NPC, 2001). Natural gas is used extensively in homes and buildings for heating, and it is an important feedstock for chemical plants, fertilizer production, and industrial processes. Finally, its most rapidly growing use is in the production of electrical power, where it has become the fuel of choice for the majority of new electric power generation plants.

Natural gas resources are as much a cause of concern as petroleum. While much natural gas is still available in this country (especially in Alaska, if a pipeline can be built to access it), and significant new reserves may still be discovered, consumption already is rising rapidly. Furthermore, gas fields tend to deplete faster than oil fields, and advanced techniques such as enhanced oil recovery do not apply to gas. In addition, gas is more difficult to import from overseas. The main technique is liquefying it at very low temperatures, shipping it via insulated tanker ships, and reheating it at the port of entry, a technique which is also vulnerable to terrorism. Thus it is not clear that natural gas will be more than an interim source of hydrogen.

Coal

U.S. coal reserves total about 270 billion short tons, approximately 25 percent of total world reserves. Annual U.S. consumption is just over 1 billion short tons, giving a reserve life of approximately 275 years at today's level of use. The reserves are sufficient to warrant consideration of coal as a primary feedstock for future hydrogen production.

In 2002, the top four coal-producing states were Wyoming (373 million short tons [st]), West Virginia (150 million st), Kentucky (124 million st) and Pennsylvania (68 million st) (National Mining Association, 2003). Many other states also have significant resources. Coal can be shipped long distances by train at low cost. Thus, coal can be considered as an option for primary feedstock in all regions of the United States.

In 2002, approximately 92 percent of all coal produced in the United States—that is, about 982 million st—was used to generate electric power, representing about 50 percent of the total power produced (EIA, 2003). The remaining 8 percent of consumption occurs in coke plants, other industrial plants (including combined heat and power applications), and residential and commercial uses.

On a worldwide basis, approximately 34 percent of all power is generated with coal, which is expected to fall to 31 percent by 2025. Outside the United States and Europe, the most common use of coal is in the steel industry and for steam and direct heating in industrial applications (e.g., chemical, cement, and pulp and paper industries). China, India, and Russia are other large users of coal.

Nuclear Power

About 20 percent of the nation's electricity is produced by nuclear power plants, which consume uranium. Low-cost uranium for nuclear reactors is currently very plentiful in the United States and elsewhere in the world. Very few nuclear reactors are being built these days, so little exploration for uranium has occurred in recent decades. However, a nuclear revival, whether for electricity or hydrogen, would spur uranium prospecting and might cause uranium prices to escalate in the long term.

Current known, recoverable world resources of uranium are approximately 3.1 million tons, estimated to be sufficient for about 50 years at current levels of consumption. A doubling of price from present levels is projected to create a 10-fold increase in these resources. Moving from current nuclear power technology to breeder reactors is estimated to increase uranium utilization another 60-fold (World Nuclear Association, 2002). Breeder reactors, however, would aggravate some of the issues now associated with the nuclear industry, including those surrounding safety and nuclear proliferation, while possibly reducing the waste disposal problem.

It is clear that there is an enormous supply of available fuel for use in nuclear power plants. However, the future of nuclear power in the United States is by no means clear. Current problems with the further use of nuclear power in the United States include economics—costs for new nuclear power plants are above current market acceptability—and public acceptance, which may have moderated in recent years but remains to be tested.

THE ELECTRIC POWER SYSTEM

Electric power is produced from a variety of fuels and energy sources. In 2001, coal was responsible for over 50 percent of U.S. electric power, nuclear for roughly 20 percent, natural gas for 17 percent, hydroelectric for 8 percent, and others for the remaining 5 percent (EIA, 2003). The electric power system includes approximately 10,400 generating stations, with a total installed capacity of 786 gigawatts (EIA, 2001). In addition, there is a significant distributed electric power generation capacity of about 70 GW from about 10 million generators, which operate for widely varying periods of time each year (see Chapter 3 in this report).

Nearly 160,000 miles of high-voltage electrical transmission lines in the United States carry power from power stations to load centers (Edison Electric Institute, 2002). In addition, distribution lines carry the power from substations to end users. The electrical power system is fundamentally different from the liquid or gaseous fuel supply systems, which involve fluid flows that are relatively easy to direct and control. Electric power flow, which is dictated by complex physics principles, can often be difficult to control.

New transmission lines are increasingly difficult to build, largely because of public opposition. This appears to have been a contributing factor behind the widespread blackout in August 2003. The transmission system is being used for purposes for which it was not originally designed, and upgrades are not keeping pace with the increasing loads on it. Unless this situation is corrected, it may hamper the use of electrolyzers in distributed hydrogen generation facilities. Building pipelines to carry hydrogen may encounter some of the same siting problems. Distributed power systems, using small generating plants (probably burning natural gas) close to hydrogen load centers may help to overcome transmission constraints but may also increase vulnerability to natural gas disruptions.

Electricity is very expensive to store, so it is generated as needed. Hydrogen is somewhat easier to store and, as discussed elsewhere in this report, hydrogen could be used in conjunction with the electric system as backup storage, so that hydrogen would be generated at times of ample power in a reversible fuel cell and reconverted as needed (see Chapter 8 and Appendix G in this report).

Appendix G

Hydrogen Production Technologies: Additional Discussion

This appendix discusses in more detail the technologies that can be used to produce hydrogen and which are addressed in Chapter 8. Cost analyses for them are presented in Chapter 5. In this appendix, the committee addresses the following technologies: (1) reforming of natural gas to hydrogen, (2) conversion of coal to hydrogen, (3) nuclear energy to produce hydrogen, (4) electrolysis, (5) wind energy to produce hydrogen, (6) production of hydrogen from biomass, and (7) production of hydrogen from solar energy. The following major sections—one for each of the technologies—include a brief description of the current technology; possible improvements for future technology; refer to Chapter 5 and Appendix E (which presents spreadsheet data from the committee's cost analyses), where applicable, for the current and possible future costs, CO_2 emissions, and energy efficiencies; note the potential advantages and disadvantages of using the technology for hydrogen production; and comment on the Department of Energy's (DOE's) research, development, and demonstration (RD&D) plan for hydrogen.

In general, in developing estimates about future possible technologies, the committee systematically adopted an optimistic posture. The estimates are meant to represent what possibly could be achieved with concerted research and development (R&D). But the committee is not predicting that the requisite R&D will be pursued, nor is it predicting that these technical advances necessarily will be achieved, even with a concerted R&D program. Estimates were made of what might be achieved with appropriate R&D.

The state of development referred to as "possible future" technologies is based on technological improvements that may be achieved if the appropriate research and development are successful. These improvements are not guaranteed; rather, they may be the result of successful R&D programs. And they may require significant technological breakthroughs. Generally, these possible future technologies are available at a significantly lower cost than are the "current technologies" using the same feedstocks.

HYDROGEN FROM NATURAL GAS

Compared with other fossil fuels, natural gas is a cost-effective feed for making hydrogen, in part because it is widely available, is easy to handle, and has a high hydrogen-to-carbon ratio, which minimizes the formation of by-product CO_2. However, as pointed out elsewhere in this report, natural gas is already imported as liquefied natural gas (LNG)[1] into the United States today, and imports are projected to increase. Thus, increased use of natural gas for a hydrogen economy would only increase imports further. As a result, the committee considers natural gas to be a transitional fuel for distributed generation units, not a long-range fuel for central station plants for the hydrogen economy.

Production Techniques

The primary ways in which natural gas, mostly methane, is converted to hydrogen involve reaction with either steam (steam reforming), oxygen (partial oxidation), or both in sequence (autothermal reforming). The overall reactions are shown below:

$$CH_4 + 2H_2O \rightarrow CO_2 + 4H_2$$
$$CH_4 + O_2 \rightarrow CO_2 + 2H_2$$

In practice, gas mixtures containing carbon monoxide (CO) as well as carbon dioxide (CO_2) and unconverted methane (CH_4) are produced and require further processing. The reaction of CO with steam (water-gas shift) over a catalyst produces additional hydrogen and CO_2, and after purification, high-purity hydrogen (H_2) is recovered. In most cases,

[1]Importation of large amounts of LNG would require major investments to provide LNG marine terminals and related infrastructure. These would be potential targets for terrorist attacks, which would threaten the security of LNG supplies.

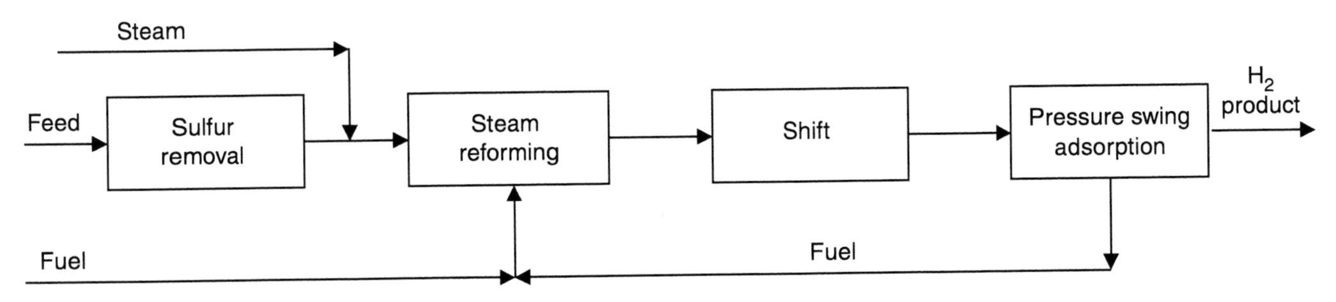

FIGURE G-1 Schematic representation of the steam methane reforming process.

CO_2 is vented to the atmosphere today, but there are options for capturing it for subsequent sequestration.

Worldwide production of hydrogen is about 41 million tons per year (ORNL, 2003). Since over 80 percent of this production is accomplished by steam methane reforming (SMR), this method is discussed first.

Steam Methane Reforming

Steam methane reforming involves four basic steps (see Figure G-1). Natural gas is first catalytically treated with hydrogen to remove sulfur compounds. It is then reformed by mixing it with steam and passing it over a nickel-on-alumina catalyst, making CO and hydrogen. This step is followed by catalytic water-gas shift to convert the CO to hydrogen and CO_2. Finally, the hydrogen gas is purified with pressure swing adsorption (PSA). The reject stream from PSA forms a portion of the fuel that is burned in the reformer to supply the needed heat energy. Therefore, CO_2 contained in the PSA reject gas is currently vented with the flue gas. If the CO_2 were to be sequestered, a separations process would be added to capture it.

The reforming reactions are as follows:

$$CH_4 + H_2O \rightarrow CO + 3H_2$$
$$CO + H_2O \rightarrow CO_2 + H_2 \text{ (water-gas-shift reaction)}$$
$$\text{Overall: } CH_4 + 2\,H_2O \rightarrow CO_2 + 4H_2$$

The reaction of natural gas with steam to form CO and H_2 requires a large amount of heat (206 kJ/mol methane). In current commercial practice, this heat is added using fired furnaces containing tubular reactors filled with catalyst.

Partial Oxidation

Partial oxidation (POX) of natural gas with oxygen is carried out in a high-pressure, refractory-lined reactor. The ratio of oxygen to carbon is carefully controlled to maximize the yield of CO and H_2 while maintaining an acceptable level of CO_2 and residual methane and minimizing the formation of soot. Downstream equipment is provided to remove the large amount of heat generated by the oxidation reaction, shift the CO to H_2, remove CO_2, which could be sequestered, and purify the hydrogen product. Of course, this process requires a source of oxygen, which is usually provided by including an air separation plant. Alternatively, air can be used instead of oxygen and product hydrogen recovered from nitrogen and other gases using palladium diffusion. POX can also be carried out in the presence of an oxidation catalyst, and in this case is called catalytic partial oxidation.

Autothermal Reforming

As already indicated, SMR is highly endothermic, and tubular reactors are used commercially to achieve the heat input required. When oxygen and steam are used in the conversion and are combined with SMR in autothemal reforming (ATR), the heat input required can be achieved by the partial combustion of methane. The reformer consists of a ceramic-lined reactor with a combustion zone and a subsequent fixed-bed catalytic SMR zone. Heat generated in the combustion zone is directly transferred to the catalytic zone by the flowing reaction gas mixture, thus providing the heat needed for the endothermic reforming reaction. As will be discussed, ATR is used today primarily for very large conversion units. There are several other design concepts that combine direct oxygen injection and catalytic conversion, including secondary reforming.

It has been suggested that methane conversion to hydrogen and elemental carbon might also be an attractive route, but the committee believes that this is unlikely. Such an approach would generate a large amount of carbon byproduct,[2] and less than 60 percent of the combined heats of combustion of the hydrogen and carbon products is associated with the hydrogen. For this approach to become a viable alternative, uses for large amounts of carbon must be found.

Natural Gas Conversion Today

Steam methane reforming is widely used worldwide to generate both synthesis gas and hydrogen. The gas produced is

[2]On a stoichiometric basis, 3 kg C would be made per kilogram of hydrogen.

used to make chemicals such as ammonia and methanol, to refine petroleum, metals, and electronic materials, and to process food components. More than 32 million tons per year (t/yr) H_2 (80 million kg/day) are produced using natural gas SMR. Hydrogen is also made today using partial oxidation and ATR.

The vast commercial experience based on this manufacturing capacity has led to many improvements in the technology, reducing costs and increasing efficiency. Perhaps the most important element is the tubular reactor in which the SMR reaction takes place. Progress has led to higher tube wall temperatures, better control of carbon formation, and feedstock flexibility.[3] This progress in turn has led to lower steam-to-carbon ratios and improved efficiency. The water-gas-shift unit has also been improved, and now one-step shift can be employed to replace the former two-step operation at different temperatures. Finally, purification of the hydrogen product has been simplified by using PSA to remove methane, carbon oxides, and trace impurities in a single step. While designs today do not generally include CO_2 capture, technology is currently available to accomplish this. Using a commercial selective absorption process, CO_2 could be recovered for subsequent sequestration.

Progress has also been made in designing and building larger SMR plants. Currently, single-train commercial plants of up to 480,000 kg H_2 per day (200 million standard cubic feet per day [scf/d]) are being built, and even larger plants can be constructed using multiple trains. Units as small as 300 kg/day are also being built.[4] In many cases, the units built are one of a kind, with specific features to meet the requirements of a site, application, or customer. At least one company is fabricating commercial SMR hydrogen plants as small as 300 kg/day using components of fixed design, one of the elements of mass production.[5]

Partial oxidation utilizing natural gas is fully developed and used commercially. In most cases today, commercial units use feeds of lower value than natural gas, such as coal, coke, petroleum residues, or other by-products, because of economics. However, natural gas is a preferred feed for POX from a technical standpoint and can be used to generate hydrogen where competitive.

Oxygen-blown ATR with natural gas is used today in very large units that generate a mixture of CO and H_2 for the Fischer-Tropsch process or methanol synthesis. This is attractive in part because the units can produce the hydrogen-to-carbon monoxide ratio needed in the synthesis step. Since the heat of reaction is added by combustion with oxygen, the catalyst can be incorporated as a fixed bed that can be scaled up to achieve further benefits of larger plant size in both the ATR and the oxygen plant that is required. ATR also offers benefits when CO_2 capture is included. This is because the optimum separation technology for this design recovers CO_2 at 3 atmospheres (atm), thus reducing the cost of compression to pipeline pressure (75 atm).

In summary, all three processes (SMR, POX, and ATR) are mature technologies today for the conversion of natural gas to hydrogen. SMR is less costly except in very large units, where ATR has an advantage. SMR is also somewhat more efficient when the energy for air separation is included. POX has the advantage of being applicable to lower-quality feeds such as petroleum coke, but this is not directly relevant to natural gas conversion.

Future Natural Gas Conversion Plants

Given the current interest in possibilities for a hydrogen economy and the current commercial need for hydrogen, significant effort is being focused on improving natural gas conversion to hydrogen. Improved catalysts and materials of construction, process simplification, new separations processes, and reactor concepts that could improve the integration of steam reforming and partial oxidation are being investigated. Catalytic partial oxidation is also under consideration. Since steam reforming and partial oxidation are mature technologies, the primary opportunities for improvement involve developing designs for specific applications that are cost-effective and efficient.

Several thousand distributed generators will be needed for the hydrogen economy, and it should be possible to lower the cost of these generators significantly through mass production of a generation "appliance." Such appliances may be further improved by tailoring the design to the fueling application. For example, hydrogen would likely be stored at roughly 400 atm, and to the extent that the conversion reactor pressure can be increased, hydrogen compression costs would be reduced and efficiency improved. For distributed generators incorporating POX or ATR, suitable cost-effective methods for hydrogen purification need to be developed. Alternatively, in such cases there are potentially attractive opportunities to recover the oxygen needed with membranes and thus to lower the cost.

Other concepts are also in the exploratory research stage. These involve new or modified ways of providing the endothermic heat of steam reforming or utilizing the heat of reaction in partial oxidation.

New, lower-cost designs for distributed generation probably can be advanced to the commercial prototype stage in the next 5 to 7 years. Some of these improvements could be applicable to large plants.

Economics

The committee undertook cost studies as described elsewhere (in Chapter 5 and Appendix E) to identify the areas

[3]J.R. Rostrup-Nielsen, Haldor Topsoe, "Methane Conversion," presentation to the committee, April 25, 2003.

[4]Personal communication from Dale Simbeck, SFA Pacific, to committee member Robert Epperly, April 30, 2003.

[5]Dennis Norton, Hydro-Chem, "Hydro-Chem," presentation to the committee, June 11, 2003.

TABLE G-1 Economics of Conversion of Natural Gas to Hydrogen

	Plant Size (kilograms of hydrogen per stream day [SD]) and Case					
	1,200,000[a]		24,000[b]		480[c]	
	Current	Possible Future	Current	Possible Future	Current	Possible Future
Investment (no sequestration), \$/kg/SD	411	297	897	713	3847	2001[d]
Investment (with sequestration), \$/kg/SD[e]	520	355	1219	961	—	—
Total H_2 cost (no sequestration), \$/kg	1.03[f]	0.92[f]	1.38[f]	1.21[f]	3.51[g]	2.33[g]
Total H_2 cost (with sequestration), \$/kg[e]	1.22[f]	1.02[f]	1.67[f]	1.46[f]	—	—
CO_2 emissions (no sequestration), kg/kg H_2	9.22	8.75	9.83	9.12	12.1	10.3
CO_2 emissions (with sequestration), kg/kg H_2	1.53	1.30	1.71	1.53	—	—
Overall thermal efficiency (no sequestration), %[h]	72.3[a]	77.9[a]	46.1	53.1	55.5	65.2
Overall thermal efficiency (with sequestration), %[e, h]	61.1	68.2	43.4	49.0	—	—

[a]Includes compression of product hydrogen to pipeline pressure of 75 atm.
[b]Includes liquefaction of H_2 prior to transport.
[c]Includes compression of H_2 to 400 atm for storage/fueling vehicles.
[d]Includes estimated benefits of mass production.
[e]Includes capture and compression of CO_2 to 135 atm for pipeline transport to sequestration site.
[f]Based on natural gas at \$4.50/million Btu.
[g]Based on natural gas at \$6.50/million Btu.
[h]Based on lower heating values for natural gas and hydrogen; includes hydrogen generation, purification, and compression, and energy imported from offsite, as well as distribution and dispensing.

that could have the greatest impact on the introduction of hydrogen fuel. For hydrogen production from natural gas, plant sizes of 1,200,000 kg per stream day (kg/SD), 24,000 kg/SD, and 480 kg/SD were studied (see Table G-1).[6] For each plant size, a current case representing what can be done today with modern technology and a future case representing what might be possible in the future were included. The possible future case for the 480 kg/SD plant includes the estimated benefits of mass production. For the two larger plants, options were included to capture CO_2 and to compress it to pipeline pressure (75 atm) for sequestration offsite. Capture was not included for the smallest plant, since the cost for collection of CO_2 from distributed plants was considered to be prohibitive, in that forecourt sequestration of CO_2 added \$4.40/kg H_2 to the cost (DiPietro, 1997).

As shown in Table G-1, current investments vary with plant size, from \$411 to \$3847/kg/SD as size is decreased from 1.2 million to 480 kg/SD. While improved technology visualized in the possible future cases lowers investment by 20 to 48 percent, plant size has a more pronounced effect (see Figure G-2). For the two larger plants, CO_2 capture increases investment by 22 to 35 percent.

As illustrated in Figure G-3, hydrogen cost[7] in the largest plant with no CO_2 capture is \$1.03/kg of hydrogen with current technology and \$0.92/kg with future technology. This cost increases to \$1.38/kg and \$1.21/kg in a midsize plant, and to \$3.51/kg and \$2.33/kg in the smallest plant. CO_2 capture adds 11 to 21 percent, depending on the case. Table G-1 shows overall thermal efficiency[8] for the largest plant to be 72.3 to 77.9 percent without CO_2 capture (for current and possible future technology, respectively), and 61.1 to 68.2 percent with CO_2 capture (for current and possible future technology, respectively). Efficiency for the smallest plant is 55.5 to 65.2 percent.[9] Without capture, the CO_2 emissions are 8.8 to 12.1 kg CO_2 per kilogram hydrogen. Capture lowers these emissions to 1.3 to 1.7 kg CO_2 per kilogram of

[6]All plant capacities are in kilograms of hydrogen per stream day.

[7]Hydrogen costs are based on a natural gas price of \$4.50/million Btu for the two larger plants and \$6.50/million Btu for the smallest one.

[8]Based on lower heating values of natural gas and hydrogen; includes production.

[9]The thermal efficiencies for the midsize plant are 43.4 and 46.1 percent with current technology (with and without CO_2 capture, respectively) and 49.0 and 53.1 percent with possible future technology (with and without CO_2 capture, respectively). These numbers are lower than might be expected, because it is assumed that hydrogen from these plants would be delivered to fueling stations as a liquid. These cases include the liquefaction of hydrogen.

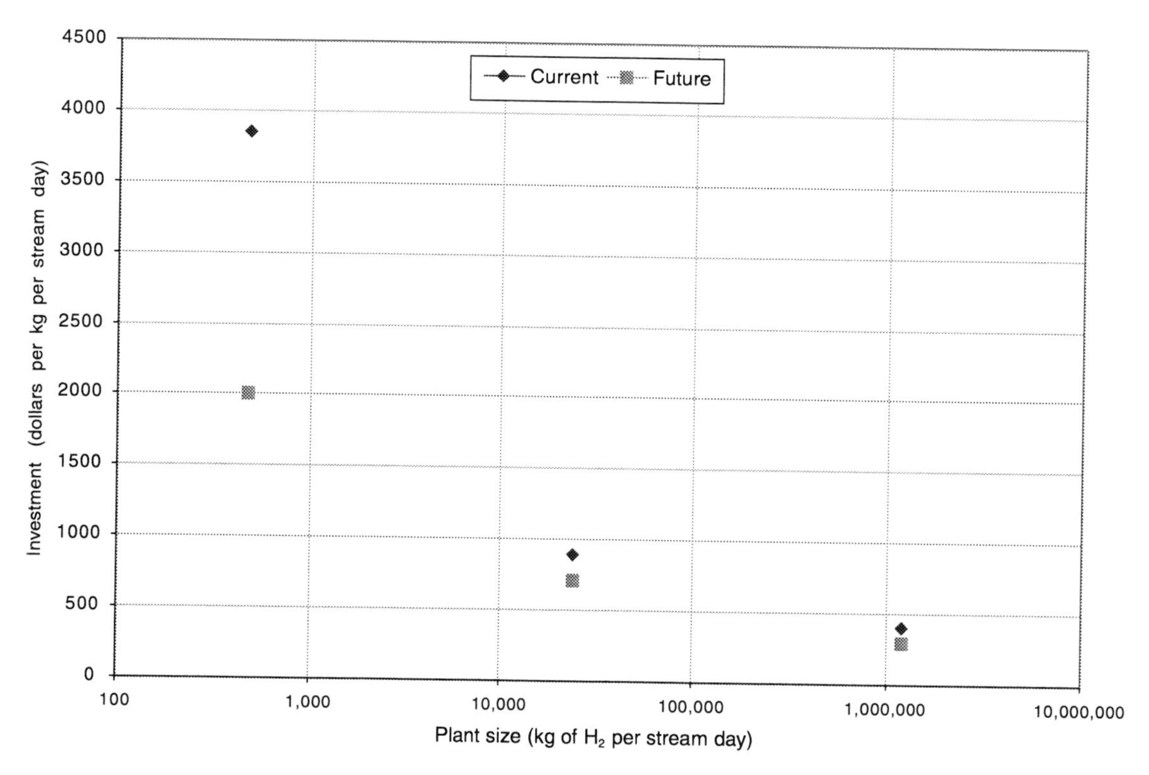

FIGURE G-2 Estimated investment costs for current and possible future hydrogen plants (with no carbon sequestration) of three sizes.

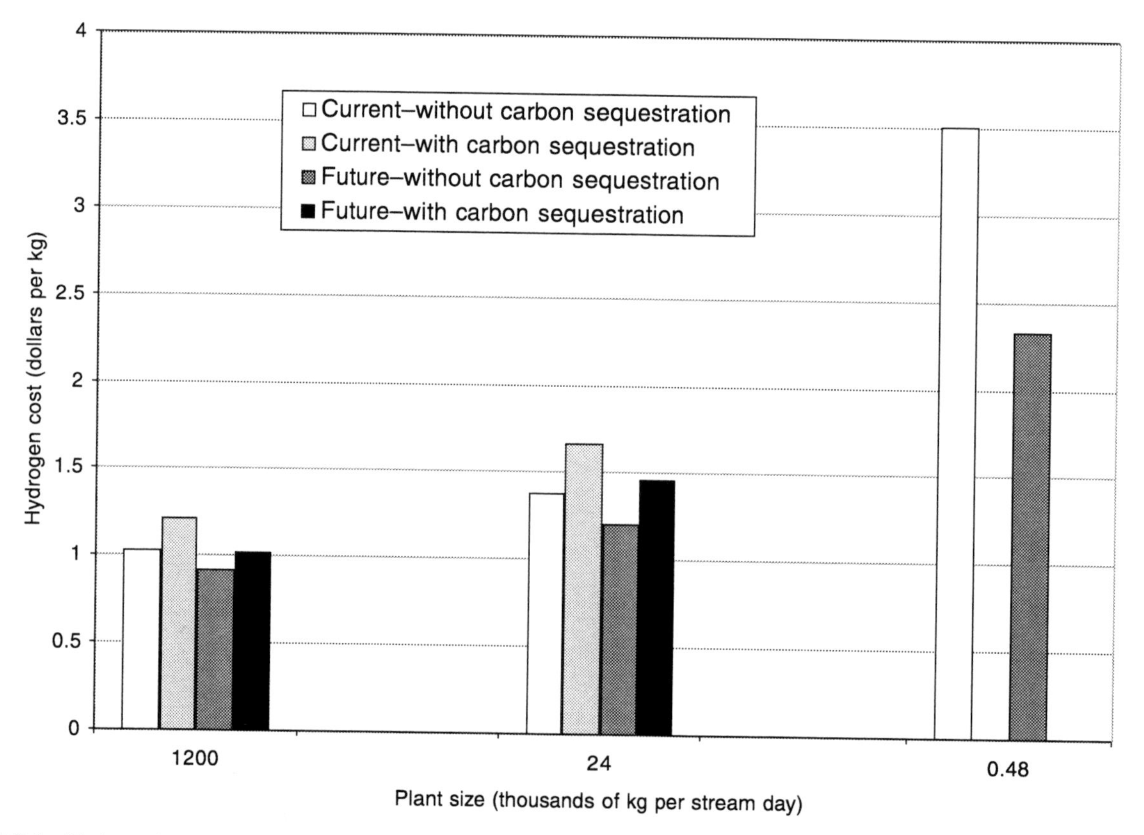

FIGURE G-3 Estimated costs for conversion of natural gas to hydrogen in plants of three sizes, current and possible future cases, with and without sequestration of CO_2.

hydrogen. Emissions and thermal efficiency estimates include the effects of generating the required electricity offsite in state-of-the-art power generation facilities with 65 percent efficiency with 0.32 kg CO_2/kWh of electricity.

The DOE states that its goal by 2010 is to reduce the cost of the distributed production of hydrogen from natural gas and/or liquid fuels to $1.50/kg (delivered, untaxed, without sequestration) at the pump, based on a natural gas price of $4/million Btu (DOE, 2003a). The committee's analysis indicates that this goal will be very difficult to achieve for the distributed-size hydrogen plants and will likely require additional time. The possible future case for distributed generation, which already incorporates the estimated benefits of mass production of SMR units, yields a hydrogen cost of $1.88/kg with $4/million Btu of natural gas. Achievement of the DOE goal would require additional thermal efficiency improvements and investment reductions. The goal could be met if, for example, the SMR thermal efficiency were further increased from 70 to 80 percent (excluding the compression of product hydrogen to storage pressure) and the SMR investment was cut by 35 percent, assuming that the benefits of mass production have been appropriately included. The committee did not study the likelihood of achieving these additional improvements. It is also important to note that the committee's cost estimates are based on the assumption that distributed generators operate throughout the year at 90 percent of design capacity. As a consequence, units would have to operate at or near design capacity 24 hours per day, or else the actual cost of hydrogen from such units would be higher than calculated.[10] Achieving a 90 percent capacity factor would require careful integration of the design rate of the hydrogen generator, hourly demand variations at fueling stations, and onsite storage capability.

The committee believes that there is considerable uncertainty regarding the future cost of hydrogen from small hydrogen plants. This uncertainty is further increased by the need for high reliability and safe operation with infrequent attention from relatively unskilled operators (i.e., customers and station attendants). In the committee's view, the DOE program should address these issues on a priority basis, as discussed below.

Hydrogen cost using steam methane reforming is sensitive to the price of natural gas, as shown in Figure G-4. Based on current technology cases, an increase in natural gas price from $2.50 to $6.50/million Btu increases hydrogen cost by 97 percent in a 1.2 million kg/SD plant and by 68 percent in a 24,000 kg/SD unit. For the 480 kg/SD unit, an increase from $4.50 to $8.50/million Btu raises hydrogen cost by 28 percent. These numbers highlight the importance of focusing research on improving efficiency in addition to reducing investment.

[10]Based on the committee's model, a reduction of on-stream time from 90 to 70 percent would increase the cost of hydrogen in a 480 kg/SD unit by 11 to 15 percent.

TABLE G-2 U.S. Natural Gas Consumption and Methane Emissions from Operations, 1990 and 2000

Consumption/Emissions	1990	2000
Natural gas consumption (Tcf)[a]	18.7	22.6
Methane emissions (Gg)[b]	5772	5541

[a]See EPA (2002).
[b]U.S. Department of Energy, Energy Information Sheets, "Natural Gas Consumption," May 12, 2003, Washington, D.C.

Other Environmental Impacts

Natural gas is lost to the atmosphere during the production, processing, transmission, storage, and distribution of hydrogen. Since methane, the major component of natural gas, has a global warming potential of 23,[11] this matter deserves discussion.

Methane is produced primarily in biological systems through the natural decomposition of organic waste. Methane emissions include those from the cultivation of agricultural land and the decomposition of animal wastes. The Environmental Protection Agency (EPA) estimates that 70 percent of methane emissions result from human activities and the balance from natural processes.[12] Less than 20 percent of total global emissions of methane are related to fossil fuels, including natural gas operations (IPCC, 1995). The EPA reports that 19 percent of the anthropogenic emissions of methane in 2000 came from natural gas operations, and 25 percent of that came from distribution of natural gas within cities, primarily to individual users (EPA, 2002).

Perhaps the most compelling statistic is that between 1990 and 2000, methane emissions from natural gas operations decreased even though natural gas consumption increased (Table G-2). Clearly, improvements are being made to reduce losses from natural gas operations. For example, the EPA says that a voluntary program with industry, the Natural Gas STAR Program,[13] has reduced methane emissions by 216 billion cubic feet (Bcf) since its inception in 1993.

As already pointed out, the advent of hydrogen-powered cars would increase natural gas consumption significantly.

[11]The Intergovernmental Panel on Climate Change (IPCC) has defined global warming potential as follows: "An index, describing the radiative characteristics of well mixed greenhouse gases, that represents the combined effect of the differing times these gases remain in the atmosphere and their relative effectiveness in absorbing outgoing infrared radiation. This index approximates the time-integrated warming effect of a unit mass of a given greenhouse gas in today's atmosphere, relative to that of carbon dioxide" (IPCC, 2001a).

[12]See Environmental Protection Agency (EPA), "Current and Future Methane Emissions from Natural Sources." Available online at http://www.epa.gov/ghginfo/reports/curr.htm. Accessed December 10, 2003.

[13]Information on the U.S. Environmental Protection Agency's STAR Program is available online at http://www.epa.gov.gasstar/. Accessed November 15, 2003.

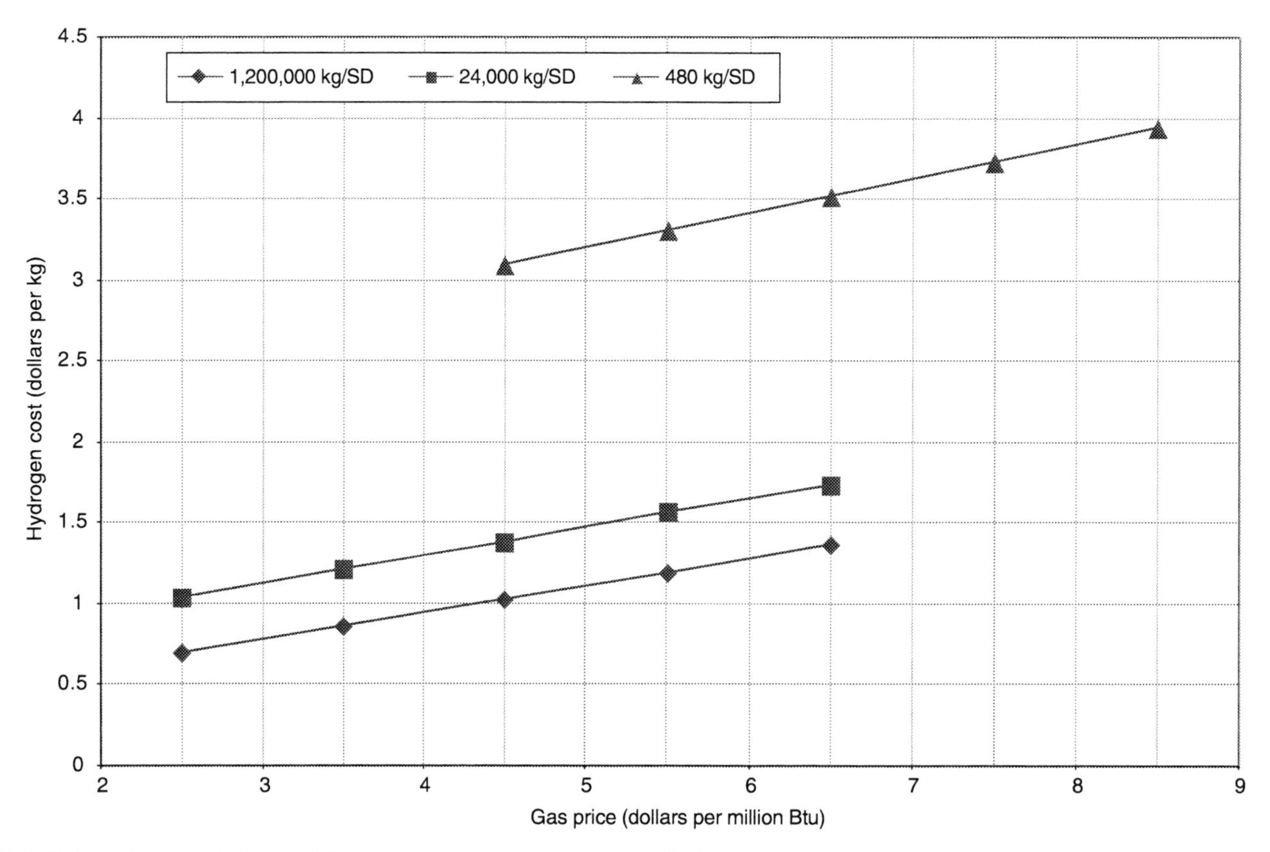

FIGURE G-4 Estimated effects of the price of natural gas on the cost of hydrogen at plants of three sizes using steam methane reforming. Costs based on current technology. NOTE: SD = stream day.

However, this increase would not necessarily increase losses from the natural gas system.

Advantages and Disadvantages

There are several advantages to generating hydrogen from natural gas. Feedstock availability is quite widespread, since an extensive pipeline distribution system for natural gas already exists in the United States and natural gas is available in most populated areas of the country. Further, there is extensive commercial experience, and natural-gas-to-hydrogen conversion technology is widely used commercially throughout the world and is at an advanced stage of optimization in large plants. If centralized, large-scale natural gas conversion plants are built, CO_2 can be captured for subsequent sequestration, although its separation and capture are probably not economically feasible with small, distributed hydrogen generators. Furthermore, the committee believes that small-scale reformers at fueling stations are one of the technologies most likely to be implemented in the transition period if policies are put in place to stimulate a transition to hydrogen for light-duty vehicles.

The primary disadvantages of using natural gas are that it is a nonrenewable, limited resource, and increasing amounts are projected to be imported in the future to meet U.S. market needs—which runs counter to the DOE's goal of improving national security. Also, natural gas prices are volatile and are very sensitive to seasonal demand. Over the past 12 months, for example, the price has varied from $2.70 to more than $9.50/million Btu,[14] and there has been an upward trend in the U.S. wellhead gas price since 1998. This variability becomes even more important given that SMR economics are sensitive to natural gas price.

Research Needs and the Department of Energy Program

Distributed generation of hydrogen from natural gas in fueling facilities could be the lowest-cost option for hydrogen production during the transition. However, the future cost of this option is uncertain, given the technical and engineering uncertainties and special requirements that demand priority attention in the DOE program, as it is advanced by contract research organizations.

Distributed generation of hydrogen as envisioned has never before been achieved because of two particular requirements: (1) the mass production of the thousands of generating units, incorporating the latest technology improve-

[14]See NYMEX Henry-Hub NATURAL GAS PRICE, available online at http://www.oilnergy.com/1gnymex.htm#year. Accessed December 10, 2003.

ments, needed to meet demand, minimize cost, and improve efficiency; and (2) unit designs and operating procedures that ensure the reliable and safe operation of these appliances with only periodic surveillance by relatively unskilled personnel (station attendants and consumers). Currently, there is a market for such units in the merchant industrial sector, which accounts for about 12 percent of the total hydrogen market in the United States (ORNL, 2003). It is clear that the DOE must provide the impetus for the program.

In contrast, centralized generation of hydrogen in one-of-a-kind, medium-sized and large plants is widely practiced, and as a result there is extensive commercial experience in this area. Given the commercial market for hydrogen, the committee believes that suppliers will continue to search for ways to improve the technology and make it even more competitive for medium- and large-scale plants.

Publications from the DOE hydrogen program indicate that the program on distributed generation will include demonstration of a "low-cost, small-footprint plant" (DOE, 2003a, b). However, it is not clear whether the program gives priority to distributed generation or includes an effort to demonstrate the benefits of and specific designs for mass production in the specified time frame of the program. The needed designs would involve concomitant engineering that would create designs for manufacturing engineering, to guide research and to prepare for mass production of the appliance, and would also develop a system design for a typical fueling facility, including the generation appliance, compression, high-pressure storage incorporating the latest storage technology, and dispensers. With today's technology, such ancillary systems cost 30 percent as much as the reformer. The committee believes that these costs can be reduced by over 50 percent and that efficiency can be improved through system integration and the incorporation of the latest technology. Compression and high-pressure storage are examples of systems in which significant improvements are expected.

The DOE hydrogen program is positioned to stimulate the development of newer concepts, such as membrane separation coupled with chemical conversion, and this seems appropriate to the committee. However, most of the effort in this area appears directed toward POX and ATR. The committee believes that SMR could be the preferred process for this application, and that it should also be pursued in parallel with the effort involving POX and ATR.

HYDROGEN FROM COAL

This section presents the basics of making hydrogen from coal in large, central station plants. The viability of this option is contingent on demand for hydrogen large enough to support an associated distribution system, a large resource base, competitive uses of coal, the environmental impacts of production and transportation, and the technologies and the associated costs for converting coal into hydrogen.

Many of the issues and technologies associated with making hydrogen from coal are similar to those of making power from coal. These subjects are closely linked and should be considered in concert—particularly with respect to clean coal technologies. These technologies will be required for making hydrogen, and they also offer the best opportunity for low-cost, high-efficiency, and low-emission power production. The lowest-cost hydrogen coal plants are likely to be ones that coproduce power and hydrogen.[15]

Coal is a viable option for making hydrogen in large, central station plants when the demand for hydrogen becomes sufficient to support an associated, large distribution system. The United States has enough coal to make hydrogen far into the future. A substantial coal infrastructure already exists, commercial technologies for converting coal to hydrogen are available from several licensors, the cost of hydrogen from coal is among the lowest available, and technology improvements are identified that should reach future DOE cost targets. The major consideration is that, because of the high carbon content in coal, the CO_2 emissions from making hydrogen from coal are larger than those from any other conversion technology for making hydrogen. This underscores the need to develop carbon sequestration techniques that can handle very large amounts of CO_2 before the widespread implementation of coal to make hydrogen should occur.

Coal Transportation

If coal is to be a major source for future hydrogen production, the infrastructure for delivering it to the future hydrogen plants will need to be expanded enough to handle these future requirements. Based on the assumptions used by the committee, the current production and delivery infrastructure capacity would need to be increased by 11 percent to meet the 2030 hydrogen demand, and by 57 percent to meet the 2050 hydrogen demand. Coal is a viable option for making hydrogen in large, central station plants when the demand for hydrogen becomes large enough to support an associated transport, storage, and distribution system.

Most bulk coal transportation is by rail, with trucks used for local transport. For reasons of economics, most of the world's coal consumption is in power plants located nearby coal mines, which minimizes the necessity for long-distance transportation. More than 60 percent of the coal used for power generation worldwide is consumed within 50 km of the mine site. In the United States, the average distance that coal is shipped by rail is farther, at about 800 miles. That distance has increased in recent years owing to the move toward greater use of coals with lower sulfur content (found mainly in the West) to meet sulfur oxide emissions standards in plants located mainly in the South and the East. As coal is currently shipped over great distances in the United States,

[15]David Gray and Glen Tomlinson, Mitretek Systems, "Hydrogen from Coal," presentation to the committee, April 24, 2003.

delivery to broad geographic areas should not be a barrier to the use of coal to make hydrogen for at least the next 30 years, since demand will not be much different from current trends.

Environmental Impacts of Coal Consumption and Transportation

Using more coal to produce hydrogen will have a number of environmental consequences. Coal mining itself causes numerous environmental issues, ranging from widespread land disturbance, soil erosion, dust, biodiversity impacts, waste piles, and so forth, to subsidence and abandoned mine workings. Once coal has been extracted, it needs to be moved from the mine to the power plant or other place of use.

The main pollutants resulting from conventional combustion of coal are sulfur oxides (SO_x), nitrogen oxides (NO_x), particulates, CO_2, and mercury (Hg). SO_x is dealt with through lower-sulfur-content coal as well as flue gas desulfurization (FGD). Approximately 30 percent of U.S. coal power generating equipment had some sort of FGD or SO_x reduction technology at the end of 1999, according to data gathered by DOE's Energy Information Administration.[16] Newer processes for power generation, such as integrated gasification combined cycle power generation, which involves a conversion rather than a combustion process, is more effective at reducing criteria pollutants than existing pollution control technologies are (East-West Center, 2000).

Potentially the most significant future issue for coal combustion is CO_2 emissions, since on a net energy basis coal combustion produces 80 percent more CO_2 than the combustion of natural gas does, and 20 percent more than does residual fuel oil, which is the most widely used other fuel for power generation (EIA [2001], Table B1). Likewise, the CO_2 emissions associated with making hydrogen from coal will be larger than those for making hydrogen from natural gas. Using currently available technology, the CO_2 emissions are about 19 kg CO_2 per kilogram of hydrogen produced, compared with approximately 10 kg CO_2 per kilogram of hydrogen manufactured from natural gas.

Atmospheric emissions from coal-fired generating plants are of concern to various bodies—national (criteria pollutants [CO, particulates,[17] O_3, NO_2, SO_2, and Pb], are defined and regulated by the EPA under the National Ambient Air Quality Standards) and international (greenhouse gases, considered under the UN Framework Convention on Climate Change, are mainly CO_2, CH_4, N_2O, hydrofluorocarbons, perfluorocarbons, and SF_6). Since the 1970s, the U.S. electricity industry has made considerable progress in reducing SO_2, NO_2, and particulate emissions, despite a large increase in coal consumption, through the use of FGD, filtration, electrostatic precipitators, and selective catalytic reduction (SCR). To the extent that new emission control technologies can be applied to existing plants and that new generating technologies can be used, further progress is expected in overall emissions reductions (Ness et al., 1999).

Current Coal Technologies

Conventional coal-fired power generation uses a combustion boiler that heats water to make steam, which is used to drive an expansion steam turbine and generator. Various designs of coal combustion boilers exist, the most modern and efficient of which use pulverized coal and produce supercritical (high-pressure/high-temperature) steam. Overall efficiencies are typically in the 36 to 40 percent range. Although a staple for power generation for decades, this conventional combustion technique is not suitable for making hydrogen. Hydrogen-making technologies employ a conversion process rather than a combustion process. These conversion processes, such as gasification, are suitable for making power and/or hydrogen.

Clean Coal Technologies

Clean coal technologies use alternative ways of converting coal so as to reduce plant emissions and increase plant thermal efficiency, leading to an overall cost of electricity that is lower than the cost for electricity from conventional plants. Systems under development include low-emission boiler systems (LEBSs), high-performance power systems (HIPPSs), integrated gasification combined cycle (IGCC), and pressurized fluidized-bed combustion (PFBC) (Ness et al., 1999). The goal is to attain thermal efficiencies in the 55 to 60 percent range (higher heating value [HHV]) (Ness et al., 1999). With the exception of the IGCC systems, all of the others rely on increasingly sophisticated emissions control systems; IGCC uses a different conversion system to reduce emissions at the outset. It is this gasification technology that is best suited to making hydrogen from coal.

Gasification Technology

Gasification systems typically involve partial oxidation of the coal with oxygen and steam in a high-temperature and elevated-pressure reactor. The short-duration reaction proceeds in a highly reducing atmosphere that creates a synthesis gas, a mix of predominantly CO and H_2 with some steam and CO_2. This syngas can be further shifted to increase H_2 yield. The gas can be cleaned in conventional ways to recover elemental sulfur (or make sulfuric acid), and a high-concentration CO_2 stream can be easily isolated and sent for

[16]Energy Information Administration, Form EIA-767, "Steam-Electric Plant Operation and Design Report"; Form EIA-860A, "Annual Electric Generator Report-Utility"; and Form EIA-860B, "Annual Electric Generator Report—Non-utility."

[17]Two sizes are considered criteria pollutants, PM_{10} and $PM_{2.5}$. The 2.5 mm particles result from combustion; the larger, 10 mm particulates typically take the form of airborne dust. Both can penetrate the lungs and are known to cause long-term damage resulting in respiratory and bronchial diseases.

disposal. The use of high temperature and pressure and oxygen minimizes NO_x production. The slag and ash that is drawn off from the bottom of the reactor encapsulate heavy metals in an inert, vitreous material, which currently is used for road fill. The high temperature also eliminates any production of organic materials, and more than 90 percent of the mercury is removed in syngas processing. Syngas produced from current gasification plants is used in a variety of applications, often with multiple applications from a single facility. These applications include syngas used as feedstock for chemicals and fertilizers, syngas converted to hydrogen used for hydro-processing in refineries, production, generation of electricity by burning the syngas in a gas turbine, and additional heat recovery steam generation using a combined cycle configuration.

There are currently at least 111 operating gasification plants running on a variety of feedstocks. These include residual oils from refining crude oil, petroleum coke, and to a lesser extent, coal. The syngas that is generated has typically been used for subsequent chemicals manufacture; making power from IGCC systems is a more recent innovation, successfully demonstrated in the mid-1980s and commercially operated since the mid-1990s. Gasification is, therefore, a well-proven commercial process technology, and several companies offer licenses for its use.

Oxygen-Blown Versus Air-Blown Gasification

Gasification plants exist that use either air-blown or oxygen-blown designs. Air-blown designs save the capital cost and operating expense of air separation units, but the dilution of the combustion products with nitrogen makes the separation of CO_2, in particular, a much more expensive exercise. In addition, the extra inert nitrogen volume going through the plant increases vessel sizes significantly and increases the cost of downstream equipment. Oxygen-blown designs do not introduce the additional nitrogen, so once the sulfur compounds have been removed from the syngas, what is left is a high-purity stream of CO_2 that can be more easily and cheaply separated. Because of the need to consider CO_2 capture and sequestration for future hydrogen generation plants, only oxygen-blown designs are feasible for consideration.

Estimated Costs of Hydrogen Production and Carbon Dioxide Emissions

Most gasification plants produce syngas for chemical production, and often for steam. IGCC plants then burn the syngas to produce power. The flexibility to polygenerate multiple products to suit a given situation is one of the strengths of the gasification system. Thus, relatively few gasification plants are dedicated to producing hydrogen only (or indeed any other single product). The future large-scale hydrogen generation plant will likely also generate some amounts of power because of the advantages provided through polygeneration. It is necessary therefore to preface any remarks concerning the costs of producing only hydrogen or the costs of sequestering CO_2 with this caveat.

All of the technology needed to produce hydrogen from coal is commercially proven and in operation today, and designs already exist for hydrogen and power coproduction facilities. However, technology advances currently in development will continue to drive down the costs and increase the efficiency of these facilities. Hydrogen-from-coal plants combine a number of technologies including oxygen supply, gasification, CO shift, sulfur removal, and gas turbine technologies. All of these technology areas have advances under development that will significantly improve the plant's capital and operating costs and thermal efficiency. Examples of these pending technology advances include Ion Transport Membrane (ITM) technology for air separation (oxygen supply); advances in gasifier technology (feedstock preparation, conversion, availability); warm gas cleanup; advanced gas turbines for both syngas and hydrogen; CO_2 capture technology advances; new, lower-cost sulfur-removal technology; and slag-handling improvements.

It is estimated that today a gasification plant producing hydrogen only would be able to deliver hydrogen to the plant gate at a cost of about \$0.96/kg H_2 with no CO_2 sequestration. If CO_2 capture were also required, it would cost \$1.03/kg H_2. This pricing reflects costs for producing hydrogen from very large, central station plants at which hydrogen will be distributed through pipelines. In these plants a single gasifier can produce more than 100 million scf/day H_2. It is envisioned that a typical installation would include two to three gasifiers.

The economics of making hydrogen from coal is somewhat different from that for making it from other fossil fuels, in that the capital costs needed per kilogram of produced hydrogen are larger for coal plants, but the raw material costs per kilogram of produced hydrogen are lower. Coal is inexpensive, but the coal gasification plant is expensive. If the coal price is changed by 25 percent, the hydrogen cost is changed by only \$0.05/kg. If the cost of the plant is changed by 25 percent, the hydrogen cost is changed by \$0.16/kg. This should lead to a very stable cost of hydrogen production that can be lowered through future improvements in technology.

In addition to the CO_2 produced from making the electricity consumed in producing hydrogen, CO_2 emissions result from the carbon in the coal. The emissions depend on the type and quality of coal, but for typical Western coal with 2 percent sulfur and 12,000 Btu/dry lb, approximately 18.8 kg CO_2 are emitted per kilogram of hydrogen produced. With a CO_2 capture system in place, it is estimated that this figure could be reduced by as much as 80 to 90 percent, the exact amount depending on capital efficiency and cost-benefit analysis. Although the economics of hydrogen production from coal does vary somewhat with the quality of coal being gasified, essentially any coal can be gasified to produce hydrogen. Coals with ash content greater than 30 percent are already being gasified. The main effects of coal-quality variance on hydrogen production are the amount of by-products produced (primarily slag and elemental sulfur) and the capital cost, which would be affected mostly by the amount of additional inert material in the coal that has to be handled. For a gasification plant producing maximum hydrogen from coal, the variance in potential feed coal quality is estimated to produce a variance of less than 15 percent in the amount of CO_2 generated per ton of hydrogen produced. The lower-quality coals generate lower amounts of CO_2 per ton of hydrogen. Other effects of coal quality are less significant.

Research and Development Needs

In terms of its stage of development, coal gasification is a less mature commercial process than coal combustion processes and other hydrogen generation processes using other fossil fuels, especially in the aspects of capturing CO_2 and providing flexibility in hydrogen and electricity production. In that sense the potential for improvement through technology development is significant. The main issues are capital cost and reliability (the latter is usually addressed through including standby equipment). Both are major reasons why IGCC technology has not been widely adopted for power generation, which is a very competitive business. The flexibility to vary between hydrogen production and power production will cost extra capital, which has to be recovered.

For the commercial processes available from several different licensors, the R&D needs should be directed at capital cost reduction, standardization of plant design and execution concept, gas cooler designs, process integration, oxygen plant optimization, and acid gas removal technology. The potential efficiency and capital cost improvements in these areas could combine to lower the overall cost of hydrogen from coal by about 10 to 15 percent from today's costs. Since many parts of the coal-to-hydrogen process are the same as for coal-to-power processes, similar improvements in power costs from IGCC should be possible. These areas are improvements to existing technology, so they should be able to be achieved in the near term.

The potential also exists for new technologies to make larger improvements in the efficiency and cost of making hydrogen from coal. For new gasification technologies, the best opportunities for R&D appear to be for new reactor designs (entrained bed gasification), improved gas separation (hot gas separation), and purification techniques. These technologies, and the concept of integrating them with one another, are in very early development phases and will require longer-term development to verify the true potential and to reach commercial readiness. Recent studies have indicated that the combined potential of these new technologies could lower the cost of making hydrogen from coal by about 25 percent.

Future Costs

Evolutionary improvements in current technology can lower the cost of hydrogen from coal from the estimated $0.96/kg to about $0.90/kg. The evolution of future costs will be a function of the number of units constructed over time, since each subsequent plant gives an additional opportunity to apply the experience derived from prior plants, as well as economies of scale for process unit production.

The introduction of new technologies can lower costs even further. New gasification technologies along with new syngas cleanup and separation technologies hold potential for further improving efficiencies and lowering the costs of producing hydrogen to about $0.71/kg (see Chapter 5 and Appendix E). Separating and capturing CO_2 will increase these costs to $0.77/kg.

Department of Energy Programs for Coal to Hydrogen

The DOE programs for making hydrogen from coal reside in the Office of Fossil Energy and are related to programs to make electricity from coal. The overall goal of the Hydrogen from Coal Program is to have an operational, zero-emissions, coal-fueled facility in 2015 that coproduces hydrogen and electricity with 60 percent overall efficiency (DOE, 2003c). Major milestones for reaching this goal include these:

- *2006*—Advanced hydrogen separation technology, including membranes tolerant of trace contaminants, identified;
- *2011*—Hydrogen modules for coal gasification combined-cycle coproduction facility demonstrated; and
- *2015*— Zero-emission, coal-based plant producing hydrogen and electric power (with sequestration) that reduces the cost of hydrogen by 25 percent compared with the cost at current coal-based plants demonstrated.

To reach these milestones, R&D activities within the Hydrogen from Coal Program are focused on the development of novel processes that include these:

- Advanced water-gas-shift reactors using sulfur-tolerant catalysts,
- Novel membranes for hydrogen separation from CO_2,
- Technology concepts that combine hydrogen separation and water-gas shift, and
- Fewer-step designs to separate impurities from hydrogen.

Associated coal gasification R&D programs in which success is dependent on efficiency improvements and lower cost include these:

- Advanced ITM technology for oxygen separation from air,
- Advanced cleaning of raw synthesis gas,

- Improvements in gasifier design, and
- CO_2 capture and sequestration technology.

Summary

The United States has enough coal to make all of the hydrogen that the economy will need for a very long time, a substantial coal infrastructure already exists, commercial technologies for converting coal to hydrogen are available from several licensors, the cost of hydrogen from coal is among the lowest available, and technology improvements are identified to reach the future DOE cost targets. As such, coal is a viable option for making hydrogen in large, central station plants when the demand for hydrogen becomes large enough to support an associated distribution system.

The key to the efficient and clean manufacture of hydrogen from coal is to gasify the coal first, to produce a synthesis gas—a mixture of hydrogen and CO—and then to further process the CO with water to produce additional hydrogen and CO_2.

Combinations of coal gasifiers and gas cleanup processes have been built, tested, and used to produce electric power in the integrated gasification combined cycle (IGCC) process. While IGCC power plants have been built and operated on a commercial scale, further process improvements to lower costs and to improve reliability are both possible and desirable. Accordingly, a number of years ago the DOE initiated a related R&D program called Vision 21, which is up and running and has been reviewed by the National Research Council, most recently in early 2003 (NRC, 2003b). Major aspects of this program will be applicable to making hydrogen from coal and will lead to more efficient and lower-cost hydrogen production designs.

Making hydrogen from coal produces a large amount of CO_2 as a by-product. At present, the United States does not restrict the emissions of CO_2 from any sources, but it is possible that such restrictions might be invoked in the future. Because of the possible effects of CO_2 on global climate change, the government has accelerated R&D aimed at reducing or eliminating CO_2 emissions from energy-producing systems, one of these being coal-fueled systems. A part of the Department of Energy's hydrogen program is aimed at developing safe and economic methods of sequestering CO_2 in a variety of underground geologic formations. Indeed, a sequestration R&D program was initiated in the department's Office of Fossil Energy a number of years ago and is now supported at a significant level. The new coal-based power systems being developed under the DOE's Vision 21 program are aimed at coupling power plant with sequestration systems.

Beyond the Vision 21 program, the DOE recently announced its intention to proceed with a large, coal-to-electricity-and-hydrogen verification plant with coupled sequestration. This plant, called FutureGen, is now in the early stages of detailed planning. In addition to demonstrating coproduction of electricity and hydrogen with sequestration, the system is intended to act as a large-scale testbed for innovative new technologies aimed at reducing systems costs.

HYDROGEN FROM NUCLEAR ENERGY

Nuclear Power Technology Today

The United States derived about 20 percent of its electricity from nuclear energy in 2002 (EIA, 2003). While no nuclear power plants have been ordered in the United States since 1975, the orders prior to that date resulted in the 103 power reactors operating today. With a total capacity of nearly 100 gigawatts electric (GWe), they constitute about 13 percent of the installed U.S. electric generation capacity. Since their operating costs are relatively low, the existing plants tend to be part of the base load for their owner companies, and their output has been increasing since the late 1980s. The current U.S. plants use water as the coolant and neutron moderator, and they rely on the steam Rankine cycle as the thermal-to-electrical power conversion cycle. Nearly 65 percent of these light-water reactors (LWRs) are of the pressurized-water reactor (PWR) type, and 35 percent are of the boiling-water reactor (BWR) type. The LWR technology has dominated the reactor market and constitutes about 80 percent of the nearly 440 operating plants in the world today. Different technologies have been deployed in Great Britain, which depends mostly on gas-cooled reactors (GCRs) and advanced gas-cooled reactors (AGRs) cooled by CO_2 but using an indirect steam power cycle. Canada, India, and a few other countries operate heavy-water reactors (HWRs), also with an indirect steam power cycle.

Other reactor technology options were tested in several countries. These include helium-cooled high-temperature gas-cooled reactors (HTGRs) and sodium-cooled fast reactors (SFRs). However, the operation of these plants did not spur wider market penetration. The HTGR has a significant technical base due to the experience gained from power plants in the United States and Germany and the more recent, smaller test reactors in Japan and China. Coupling gas-cooled reactors to a direct or indirect gas turbine Brayton power cycle can yield thermal efficiencies much higher than the 33 percent of current LWRs. However, there is no experience in gas-turbine powered nuclear plants, since the U.S. and German HTGR plants use an indirect steam Rankine cycle for electricity production.

While the LWR technology dominates the global nuclear energy market, the fuel cycle technology has not had similar unanimity. The United States, Spain, Sweden, and several other countries opted for a once-through uranium-based fuel cycle, in which the used fuel is destined for a geologic repository for highly radioactive waste, after a storage period of a decade or more. France, Germany, Japan, and Russia, among other countries, have preferred to extract and recycle the fissile material in the spent fuel to increase the energy derived from the fuel and to reduce the volume and toxicity of the waste that will be disposed of in geologic repositories. The U.S. approach is less costly while the supply of inexpensive uranium lasts—which at the current rate of consumption should be for at least 50 to 100 years. When additional fuel material is needed, chemical reprocessing of the used fuel to recycle it can extend the fuel availability to thousands of years—that is when fast reactors, such as the sodium-cooled reactors, would become desirable.

Nuclear Power Technology in the Future

In the past 20 years, several advanced versions of the LWR, collectively called advanced LWRs (ALWRs), have been designed, but only one type has been built: the advanced boiling-water reactor (ABWR), which was built in Japan. These reactors are generally known today as Generation III reactors. Some of these designs have been certified as safe by the Nuclear Regulatory Commission (USNRC), but no orders have materialized for them in the United States. New versions of light-water reactors are now under review for safety certification by the USNRC.

Two versions of the HTGR are being designed by international consortia. One is led by the South African utility, ESKOM, with a direct helium gas turbine power cycle. This reactor builds on the German experience with circulating graphite pebbles containing ceramic-coated oxide fuel microparticles. The fuel is designed to be robust for the temperature range of operation and accidents. The ability of the microparticle fuel to reach very high burnup induced another consortium (Framatome ANP, General Atomics [GA], and Russian collaborators) to design a plutonium consumption reactor. In this case, the microparticles will be housed in stationary graphite blocks, typical of the earlier GA-designed HTGRs in the United States.

In 2002, several reactor concepts were selected by an international team representing 10 countries as promising technologies that should be further explored for availability beyond 2025; these technologies are collectively known as the Generation IV reactors. The goals for the new reactor systems are to improve the economics, safety, waste characteristics, and security of the reactors and the fuel cycle. The emphasis in the development was given to the following six concepts:

1. Very high temperature reactor (VHTR), a version of the HTGR;
2. Supercritical water reactor (SCWR), with a direct power cycle;
3. Fast gas-cooled reactor (FGR), with a direct helium or CO_2 gas turbine power cycle;
4. Heavy metal (lead alloy)-cooled reactor (HMCR), with an indirect power cycle;
5. Sodium-cooled reactor (SCR), with an indirect steam power cycle; and
6. Molten salt-cooled reactor (MSR), with a fluid fuel and an indirect power cycle.

It is notable that all Generation IV reactors aim to operate at higher coolant temperatures than those of the LWRs, thereby increasing the efficiency of thermal-to-electrical-energy conversion. The main characteristics of some of the reactors mentioned here and others are given in Table G-3.

Proposed Technologies for Hydrogen Production

Hydrogen can be produced using current reactor technology for electricity for electrolysis (water splitting). Potentially more efficient hydrogen production may be attained by significantly raising the water temperature before splitting the molecules using thermochemistry or electrolysis. Such approaches require temperatures in the range of 700°C to 1000°C. Current water-cooled reactors produce temperatures under 350°C, and cannot be used for such purposes. However, the coolants from several advanced reactor concepts do reach such high temperatures and may be coupled to thermochemical plants (Brown et al., 2003; Doctor et al., 2002; and Forsberg, 2003). A recent experimental helium-cooled reactor at the Japan Atomic Energy Research Institute (JAERI) was built specifically with the goal of hydrogen production. Its desired coolant maximum temperature is 900°C. It started operation in 1999 and is still undergoing testing of its fuels and its operations at lower temperatures.

Another possibility for producing hydrogen is the use of nuclear heat to provide the energy needed for heating in the steam methane reforming (SMR) process, as suggested recently by the Electric Power Research Institute (EPRI) (Sandell, 2003). That too requires high temperatures, above 700°C, for efficiency. Therefore, it must be coupled to a high-temperature reactor. This process reduces but does not eliminate the CO_2 emissions associated with conventional SMR. It also reduces the amount of natural gas required for hydrogen production.

TABLE G-3 Nuclear Reactor Options and Their Power Cycle Efficiency

Current and Advanced Reactor Type	T_{outlet} (°C)	η_{th} (%)
Current light-water reactor (LWR)	280–320	32–34
Advanced light-water reactor (ALWR)	285–330	32–35
Supercritical water-cooled reactor (SCWR)[a]	400–600	38–45
He high-temperature graphite reactor (HTGR)	850–950	42–48
Supercritical CO_2 advanced gas reactor (S-AGR)	650–750	46–51
Molten salt-cooled reactor (AHTR)[b]	750–1000	NE
Heavy metal (lead alloy)-cooled reactor (HMCR)[a]	540–650	NE

NOTE: NE = not evaluated.
[a]One of the Generation IV reactors.
[b]The fuel resembles that of an HTGR but with a salt coolant.

The various options for nuclear hydrogen production are given in Table G-4. The basic chemistry, projected efficiency, established experience, and other related issues for each technology option are now briefly addressed.

High-Temperature Electrolysis of Steam

The electrical energy demand in the electrolysis process decreases with increasing water (or steam) temperature. While the demand for heat energy is increased, the decrease in the electrical energy demand improves the overall thermal-to-hydrogen heat conversion efficiency. Higher temperatures also help lower the cathodic and anodic overvoltages. Therefore, it is possible to increase the current density at higher temperatures, which yields a significant increase in the process efficiency. Thus, the high-temperature electrolysis of stream (HTES) is advantageous from both thermodynamic and kinetic standpoints. The electrodes of the HTES unit can be made of ceramic materials, which avoids corrosion problems, though hydrogen embrittlement might still be a problem for electrode durability. High-pressure operation would also be preferable, in order to reduce the size of the chemical units and transmission lines.

The HTES process is potentially advantageous when coupled to high-efficiency power cycles and can consequently yield high overall thermal-to-hydrogen energy efficiency. The efficiency of hydrogen production via coupling of HTES to either of two high-temperature nuclear reactors is given in Figure G-5 (Yildiz and Kazimi, 2003). One reactor is the gas turbine modular high-temperature reactor (GT-MHR) (LaBar, 2002). The second is an advanced gas-cooled reactor (AGR) coupled to a direct supercritical CO_2 power cycle. The cycle was originally proposed for fast reactors (Dostal et al., 2002). The supercritical AGR (S-AGR), also referred to as the S-CO_2, necessitates upgrading the AGR design pressure from the current 4 megapascals (MPa) to about 20 MPa, which has not been attempted before in a concrete containment. A reference HTES design called HOTELLY (high-operating-temperature electrolysis) (Doenitz et al., 1988) is chosen as the basis for this example.

Implementation of the GT-MHR-HTES at the temperature of 850°C for the near term appears possible, while achieving temperatures of 950°C and higher might be expected for the years 2025 and beyond. Similarly, for the S-AGR-HTES, the near-term and far-term goals may be 650°C

TABLE G-4 An Overview of Nuclear Hydrogen Production Options

	Approach			
	Electrolysis		Thermochemistry	
Feature	Water	High-Temperature Steam	Methane Reforming	Water Splitting
Required temperature (°C)	>0	>300 for LWR >600 for S-AGR	>700	>850 for SI cycle >600 for Cu-Cl cycle
Efficiency (%) of chemical process	75–80	85–90	70–80	>45, depending on temperature
Efficiency (%) coupled to LWR	27	30	Not feasible	Not feasible
Efficiency (%) coupled to HTGR, AHTR, or S-AGR	Below 40	40–60, depending on temperature	>70	40–60, depending on cycle and temperature
Advantages	Proven technology with LWRs	Can be coupled to reactors operating at intermediate temperatures	Proven chemistry	Eliminates CO_2 emissions
	Eliminates CO_2 emissions	Eliminates CO_2 emissions	40% reduction in CO_2 emissions	
Disadvantages	Low efficiency	Requires high-temperature reactors	CO_2 emissions are not eliminated	Aggressive chemistry
		Also requires development of durable HTES units	Depends on methane prices	Requires development

NOTE: LWR = light-water reactor; S-AGR = supercritical CO_2 advanced gas reactor; S-I = sulfur-iodine; Cu-Cl = copper-chlorine; HTGR = high-temperature gas-cooled reactor; AHTR = advanced high-temperature reactor; HTES = high-temperature electrolysis of steam.

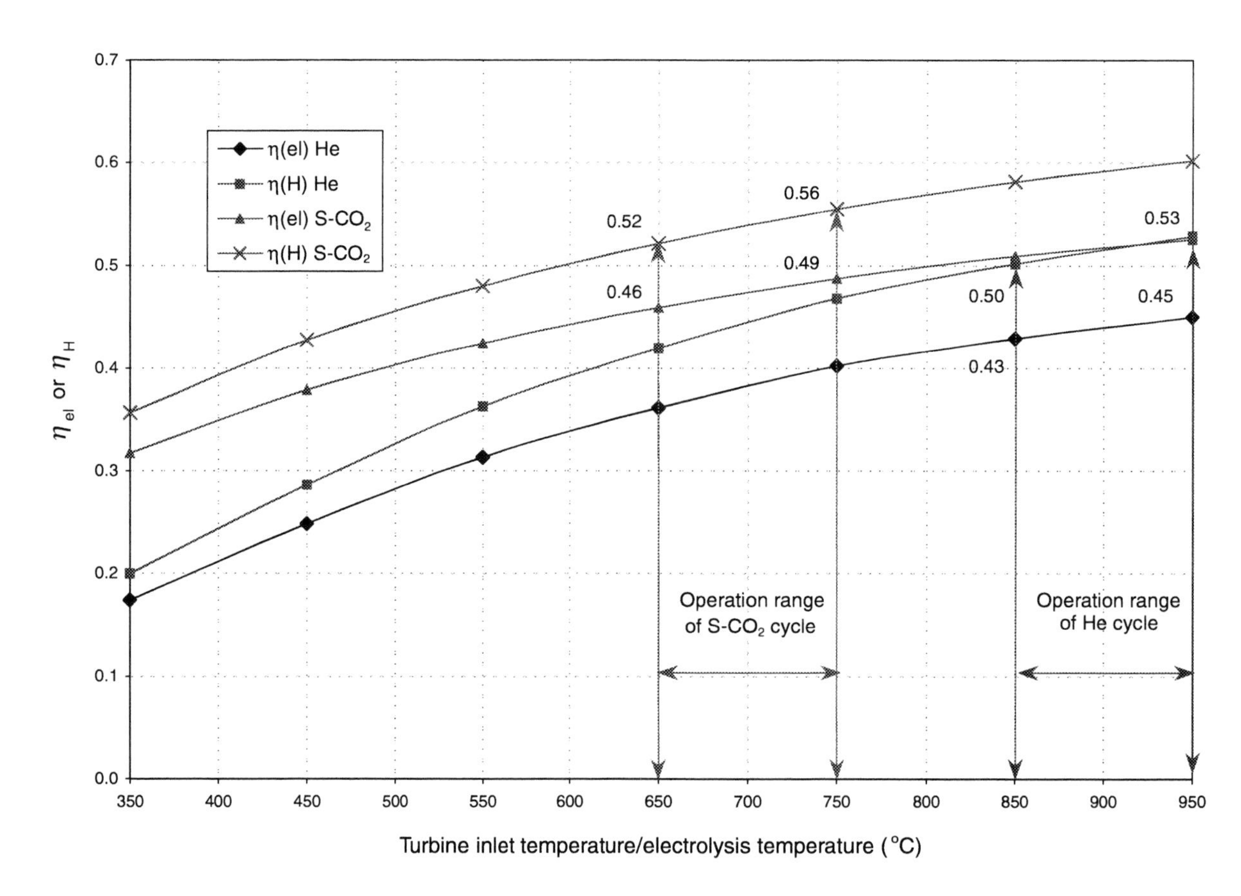

FIGURE G-5 Power cycle net efficiency (η_{el}) and thermal-to-hydrogen efficiency (η_H) for the gas turbine modular helium reactor (He) high-temperature electrolysis of steam (HTES) and the supercritical CO_2 (S-CO_2) advanced gas-cooled reactor HTES technologies. SOURCE: Yildiz and Kazimi (2003).

and 750°C, respectively. The thermal energy (MJ) needed to produce 1 kg H_2 is presented in Figure G-6.[18]

Nuclear reactors coupled to HTES are capital-intensive technologies, due to both the nuclear plant and the electrolysis plant. The development of economical and durable HTES unit materials, which can be similar to those of the solid oxide fuel cell materials, can contribute to cost reduction. The development of improved HTES units with low electrode overvoltage at lower temperatures can enable their use with lower-temperature and thereby lower-cost nuclear plants. Improved HTES cell designs are currently being investigated at Lawrence Livermore National Laboratory (Pham, 2000) and Idaho National Engineering and Environmental Laboratory (Herring, 2002). In addition, attaining high power cycle efficiency at the nuclear plant with relatively low temperatures can contribute to cost reduction. Finally, development of economic high-temperature radiation-resistant graphite or ceramic-coated graphite materials for the nuclear plant is needed.

Thermochemical Water Splitting

A recent screening of several hundred possible reactions (Besenbruch et al., 2000) has identified two candidate thermochemical cycles that have the highest commercialization potential, with high efficiency and practical applicability to nuclear heat sources. These are the sulfur-iodine (SI) and calcium-bromine-iron (Ca-Br) cycles. The S-I cycle is being investigated by General Atomics and JAERI. The Ca-Br cycle, which is sometimes called UT-3 to honor its origin at the University of Tokyo, is being investigated by JAERI. Argonne National Laboratory (ANL) is currently working on achieving thermochemical water-splitting processes at lower temperatures than the SI and Ca-Br cycles. ANL has identified the copper-chlorine (Cu-Cl) thermochemical cycle for this purpose (Doctor et al., 2002).

Sulfur-Iodine Cycle and Other Sulfur Cycles The SI cycle has been proposed in several forms. (The SI cycle and other

[18] η_{th} (power cycle thermal efficiency) is taken from Dostal et al. (2002), with the adjustment of 9 percent reduction for the He cycle and 3 percent reduction for the S-CO_2 cycle in finding the η_{el}, to reflect the heat losses due to component cooling and leakage.

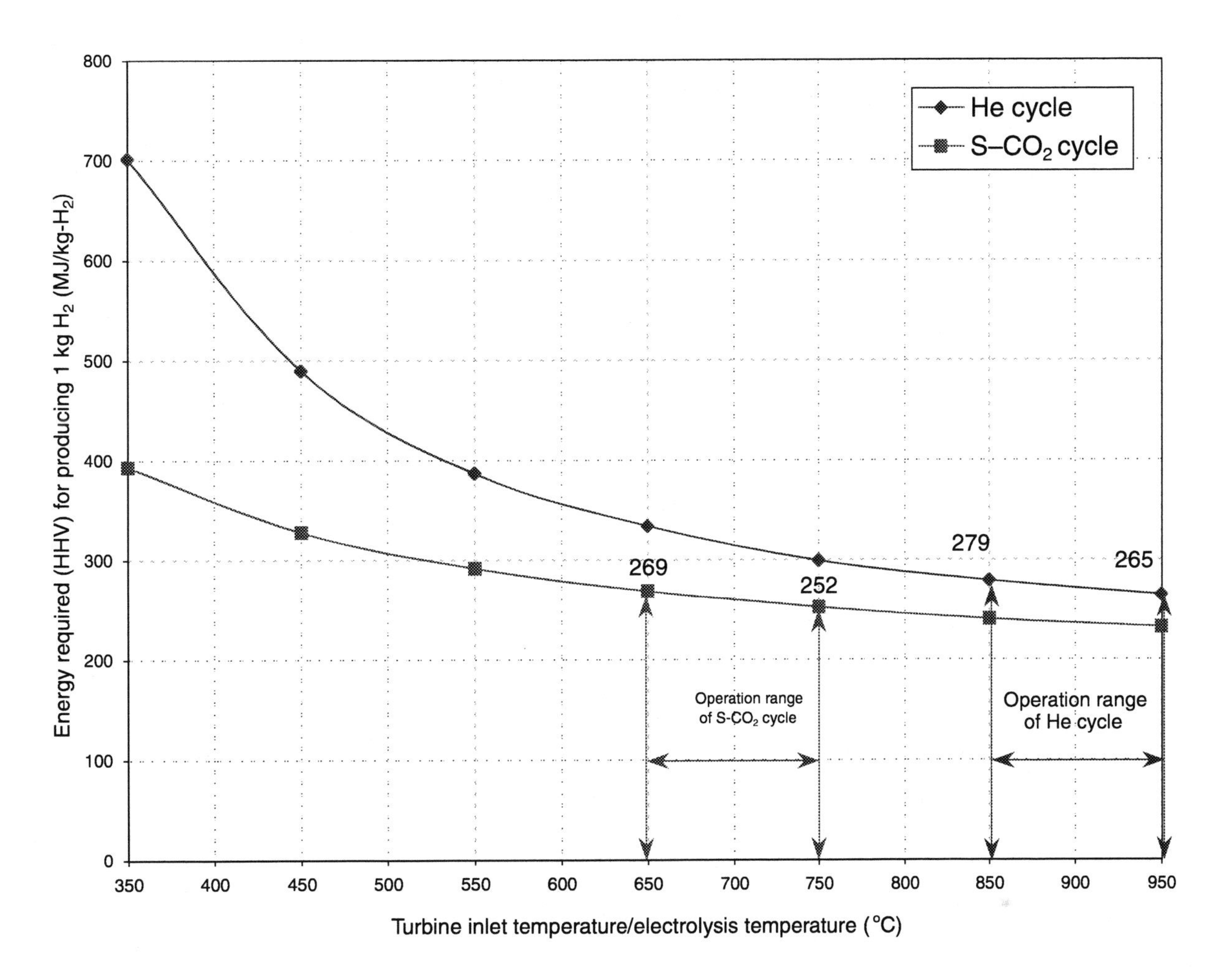

FIGURE G-6 The energy needs for hydrogen production by the gas turbine modular helium reactor (He cycle) high-temperature electrolysis of steam (HTES) and the supercritical CO_2 (S-CO_2 cycle) advanced gas-cooled reactor HTES technologies. NOTE: HHV = higher heating value.

sulfur cycles are depicted schematically in Figure G-7.) The most promising form consists of the following three chemical reactions, which yield the dissociation of water (Brown et al., 2003):

$$I_2 + SO_2 + 2H_2O \rightarrow 2HI + H_2SO_4 \quad (120°C)$$
$$H_2SO_4 \rightarrow SO_2 + H_2O + {}^1/_2O_2 \quad (830°C–900°C)$$
$$2HI \rightarrow I_2 + H_2 \quad (300°C–450°C)$$

A hybrid sulfur-based process does not require iodine and has the same high-temperature step as sulfur iodine but a single electrochemical low-temperature step that forms sulfuric acid. That electrolysis step makes sulfuric acid at very low voltage (power). The low-voltage electrolysis step (low power compared with electrolysis of water) may allow much larger scale-up of the electrochemical cells. (High-voltage systems have high internal heat generation rates that often limit the scale-up of a single cell.) The efficiency of this process is about the same as that of the SI process, but is influenced by the efficiency of the electrical power cycle. It is one of only four processes for which a fully integrated process has been demonstrated in a hood. It is the only process for which a full conceptual design report for a full-scale facility has been developed. Lastly, like the SI process, it has the potential for major improvements.

The SI cycle requires high operating temperatures but offers the opportunity for high-efficiency conversion of heat to hydrogen energy, η_H, as shown in Figure G-8. The SI cycle can be coupled to the modular high-temperature reactor (MHR) (a version of the HTGR) (LaBar, 2002). This reactor consists of 600 megawatt-thermal (MWth) modules, which are cooled by helium gas, with high coolant exit temperatures that can provide the necessary heat to the SI reactions. The coupling of the MHR and SI cycle, MHR-SI, provides a large-scale, centralized production of hydrogen.

The MHR-SI is a capital-intensive technology. Future cost reduction can be achieved from high efficiency by devising materials that can withstand higher temperatures. Reactor materials that are temperature-, irradiation- and corrosion-resistant would be needed. Also, possible reduc-

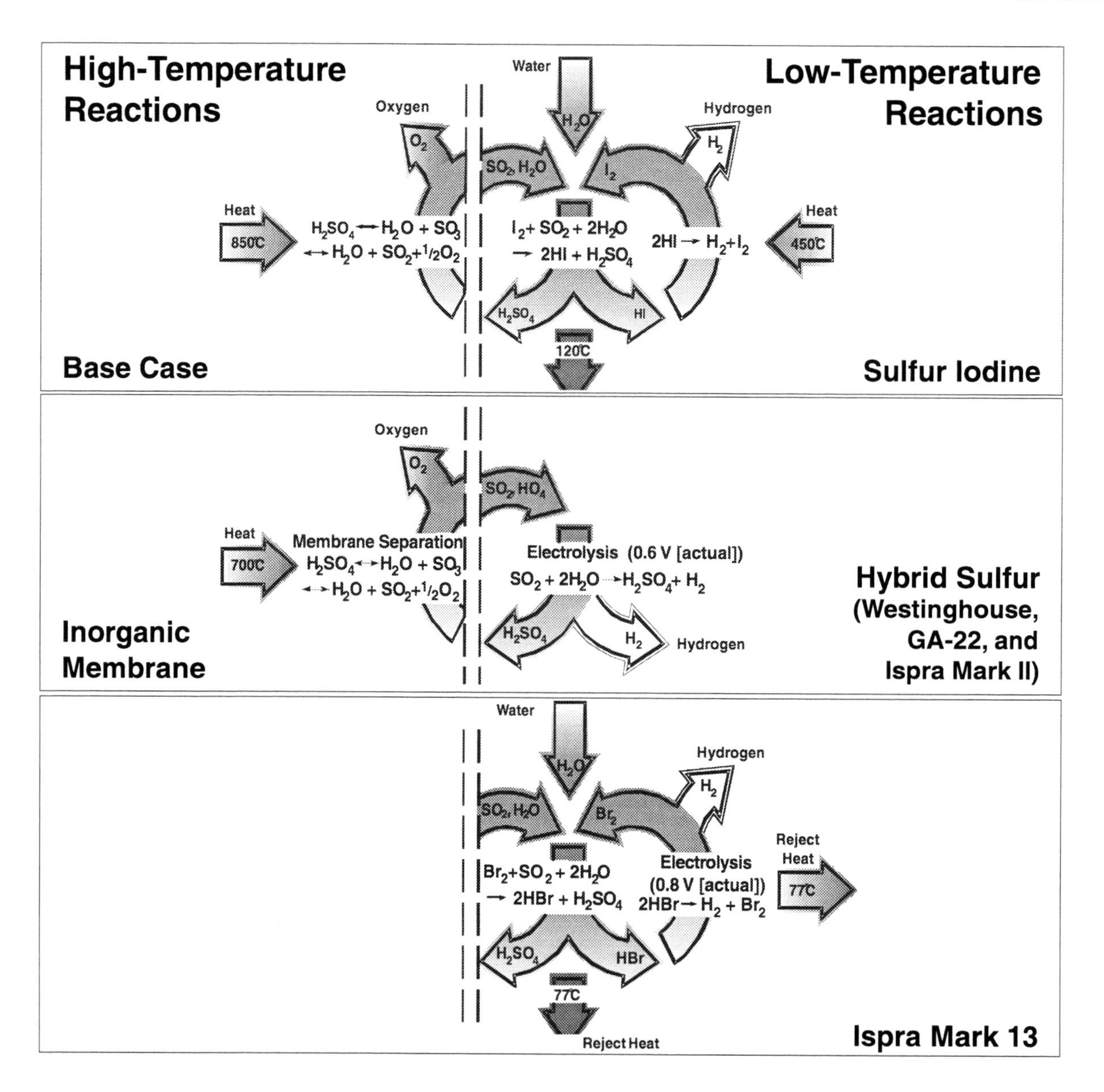

FIGURE G-7 Depiction of the most promising sulfur thermochemical cycles for water splitting. Courtesy of Charles Forsberg, Oak Ridge National Laboratory.

tion in the capital cost may result from improved catalytic materials and higher hydrogen production capacity in each facility.

Calcium-Bromine-Iron Cycle The calcium-bromine-iron (Ca-Br, or UT-3) cycle involves solid-gas interactions that may facilitate the reagent-product separations, as opposed to the all-fluid interactions in the SI cycle, but it will introduce the problems of solids handling, support, and attrition. This process is formed of the following reactions (Doctor et al., 2002):

$$CaBr_2 + H_2O \rightarrow CaO + 2HBr \quad (730°C)$$
$$CaO + Br_2 \rightarrow CaBr_2 + {}^1/_2O_2 \quad (550°C)$$
$$Fe_3O_4 + 8HBr \rightarrow 3FeBr_2 + 4H_2O + Br_2 \quad (220°C)$$
$$3FeBr_2 + 4H_2O \rightarrow Fe_3O_4 + 6HBr + H_2 \quad (650°C)$$

The thermodynamics of these reactions have been found favorable. However, the hydrogen production efficiency of the process is limited to about 40 percent, owing to the melting point of Ca-Br_2 at 760°C (Schultz et al., 2002).

Other Cycles Argonne National Laboratory's Chemical Engineering Division is studying other cycles like the copper-chlorine thermochemical cycle. The energy efficiency of the process is projected to be 40 to 45 percent (ANL, 2003). This work is currently being investigated only by ANL, at a bench-scale R&D level, and no pilot demonstra-

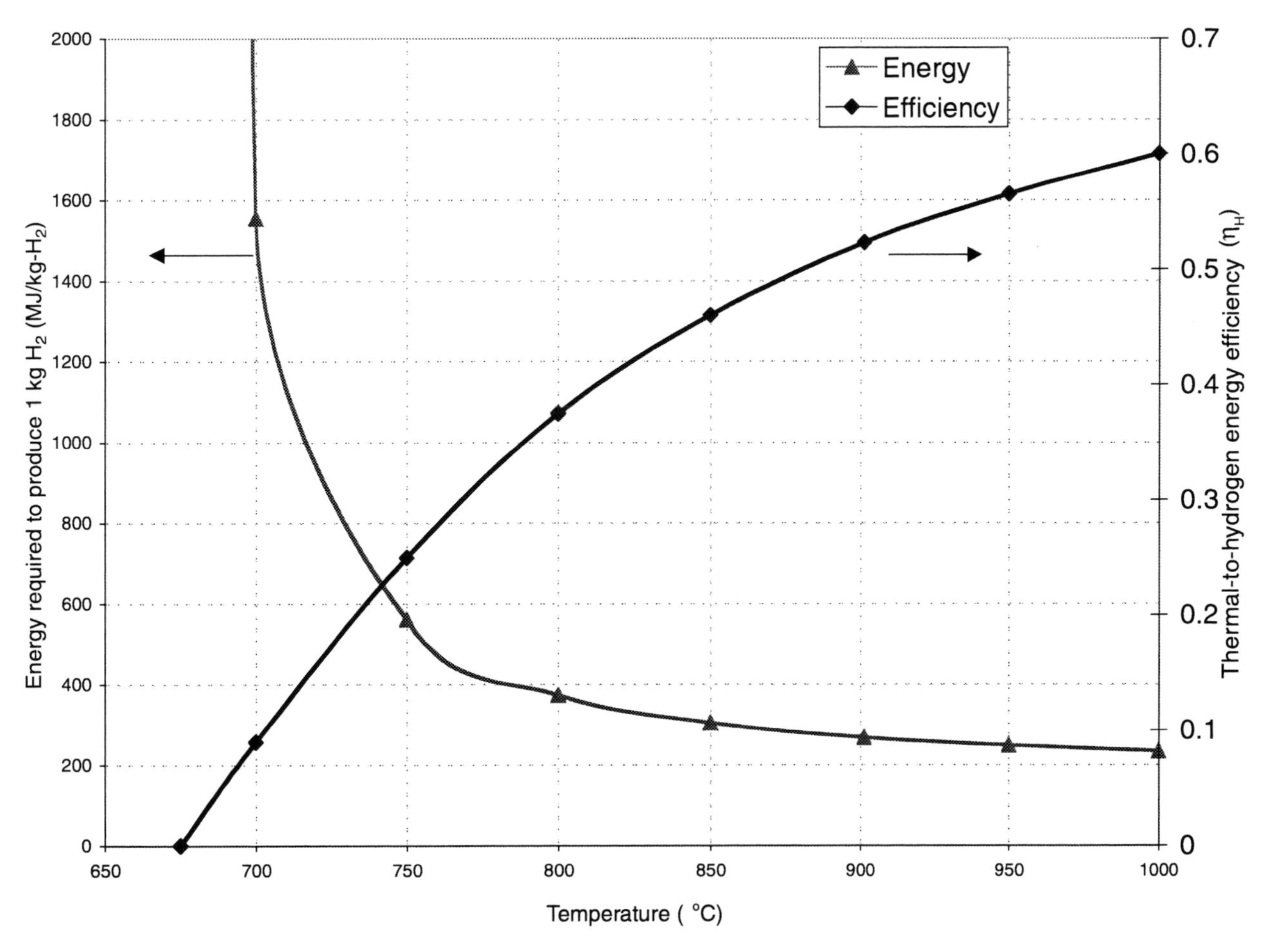

FIGURE G-8 Estimated thermal-to-hydrogen efficiency (η_H) of the sulfur-iodine (SI) process and thermal energy required to produce a kilogram of hydrogen from the modular high-temperature reactor-SI technology. SOURCE: Brown et al. (2003).

tions have been undertaken. One of the main advantages of this process is that construction materials and corrosion-resistance are more tractable at 500°C than at higher temperatures. Another advantage is that, owing to its relatively low operating temperature, it can become compatible with several current and advanced nuclear reactor technologies.

Steam Methane Reforming

Steam methane reforming (SMR) is currently the main commercial technology for hydrogen production in the United States. The SMR process requires high temperature, and the most common means of providing the heat for the process is through the burning of natural gas in the reforming furnaces, as described in the section "Hydrogen from Natural Gas," earlier in this appendix.

The SMR process can be coupled to a high-temperature helium-cooled reactor, such as the MHR. The MHR can function as the heat source operating at about 850°C, to replace the natural gas burning. The high operating temperature can enable the process to take place at about 80 percent efficiency. This approach (which might be called N [nuclear]-SMR) reduces the CO_2 emissions to the atmosphere by large quantities. Elimination of the natural-gas-burning furnace in this process reduces the CH_4 consumption by about 40 percent (Spath et al., 2000), which is parallel to the amount of CO_2 emission reduction.

Cost of Nuclear Hydrogen Production Plants

The cost of hydrogen produced by electricity generated from existing nuclear power through water electrolysis is equivalent to using the electricity supplied by the grid for hydrogen production. Today this cost is about a factor of 3 higher than what is achievable by conventional SMR, with natural gas prices at $4.5/million Btu, even when the cost of hydrogen distribution is taken into account. The improved power-cycle efficiencies of the advanced nuclear power plants may bring this cost differential in the future down to a factor of 1.5.

The cost of hydrogen production using the MHR-SMR option is dependent on the cost of natural gas feedstock to the reforming process. However, the cost from MHR-SMR is less sensitive to the cost of natural gas than is conventional

SMR. Hydrogen production cost estimates by EPRI (Sandell et al., 2003) and by the Massachusetts Institute of Technology (Yildiz and Kazim, 2003) indicate that this approach may be competitive with conventional SMR if the natural gas prices go above $6/million Btu. However, their analyses did not include any taxes or fees for CO_2 production.

The cost of hydrogen production by the *n*th-of-a-kind of MHR using the SI process was assessed by Brown et al. (2003). The authors considered the cost of producing 800 t H_2 per day using heat from four units of 600 MWth, each producing a coolant at 850ºC and having an overall efficiency of 42 percent. Starting with an overnight cost of $470/MWth for the nuclear electric plant, adding a heat exchanger, and replacing the electric generation capacity with a thermochemical plant, the total plant capital cost was found to be about $750/MWth. (A recent review of the costs of nuclear power at recent plants—built in the past 10 years in Korea, Finland, and Japan—finds the overnight costs of plants to be in the range of $530 to $800/MWth [Deutch and Moniz, 2003].) The cost of running the MHR nuclear plant is estimated to be $93.9 million per year and the hydrogen plant to be $50.7 million per year. This resulted in the cost of hydrogen production being about $1.50/kg. However, it is possible to argue that future developments could facilitate reaching higher efficiency in the conversion of the nuclear thermal energy into hydrogen production. Furthermore, larger numbers of units in one place could lead to lower costs; thus, larger plants could be associated with lower plant and operating costs. Using optimistic assumptions about advances in nuclear plant construction and thermochemical plant efficiency, the cost of a 1200 t/day MHR-SI hydrogen plant may be assumed to reach a level of $600/MWth as the technology matures. Including the usual contingency and permitting costs could add about one-third to this cost, thus leading to an effective plant cost estimate of $800/MWth and, assuming a 3-year construction time, the hydrogen production cost would be about $1.60/kg.

Advantages of Nuclear Energy Use for Hydrogen Production

- *Long-term domestic source.* Nuclear fuel will be available for a long time in the future, both domestically and worldwide. Its price is not subject to global geopolitical pressures.
- *Carbon implications.* If nuclear energy is used in the short term as the heat source in the SMR process, the result would be to reduce CO_2 emissions by nearly 40 percent. If one of the water-splitting processes is used, whether via a thermochemical process or an electrolysis approach, there will be no CO or CO_2 emissions.
- *Efficiency of the overall process.* In comparison with several other sources of hydrogen, the capability of attaining overall thermal-to-hydrogen energy efficiency in excess of 50 percent values by future technologies (e.g., the N-SMR, HTES, SI, and possibly other paths) is one of the advantages of nuclear energy use in hydrogen production. The higher the temperatures that can be achieved for the reactors, the higher their efficiencies.
- *Environmental implication.* There are no polluting emissions, or toxic gas, or particulate releases due to nuclear energy use for water splitting as the means for hydrogen production. N-SMR will have CO_2 emissions. The water-splitting processes coupled to high-temperature reactors assume complete recycling of all reactants. The volume of waste from the nuclear reactor cycle, while highly radioactive, is confined to small quantities compared with that from several other sources of energy, but it will have high levels of concentrated radioactivity.

Disadvantages of Nuclear Energy Use for Hydrogen Production

- *Efficiency of the conventional electrolysis process.* Even though it is a proven and clean technology, the low efficiency of low-temperature electrolysis makes the process uneconomic.
- *Capital cost.* Both the new nuclear reactor plants and the hydrogen plants coupled to the nuclear plants are capital-intensive. While the operating costs will be low owing to the expected high thermal efficiencies, the economics of the whole process may be disadvantageous. Capital and life-cycle costs remain high, and plant designs are in need of simplification. Enabling shorter periods of construction and increased factory-based manufacturing of components will also reduce the cost of the plants.
- *Nuclear waste.* The nuclear waste disposal scheme remains to be finalized. The Yucca Mountain project in Nevada has made good advances recently, and when licensed it can provide a destination for the spent fuel accumulating at the plant sites. The development of a closed fuel cycle that involves the extraction and use of the fissile contents from the irradiated fuel would reduce the long-lived radioactivity associated with the waste to be sent to the repository.
- *Proliferation.* Nuclear-fuel-cycle operations leave open the possibility of improper access to fissile material through theft or diversion. Proliferation can be addressed through near-term measures designed to improve the proliferation-resistance of current nuclear reactor operations and through long-term research to explore proliferation-resistant designs (PCAST, 1999).
- *Public concerns and permitting needs.* There is a public perception that nuclear energy and its emissions during normal operations increase radiation risks. There is also some fear of widespread devastation in case of accidents. These concerns would be reduced by the continued safe operation of existing plants and increased safety margins in new plants. In addition, the recent concerns about terrorism may add to the public fear of nuclear plants. The concerns of the public have led in the past to prolonged permitting periods for nuclear plants. Thus, the permitting of commercial

nuclear energy may pose a barrier to any expansion of this technology.

Research and Development Needs for Economic Hydrogen Production Using Nuclear Energy

1. A high priority should be given to the development of high-temperature reactors that can provide coolants at temperatures higher than 800°C. This objective seems most readily achievable using the helium-cooled gas reactor technology of HTGRs. The ability of the reactor's structural materials to operate for a long time at temperatures between 800°C and 1000°C needs to be established. The R&D program should include the following:

- Qualification of particle fuel materials to operate at the desired high temperatures,
- Qualification of the irradiation properties of graphite and other structural materials at the desired range of temperatures, and
- Operation and control of the helium power cycle at very high temperatures.

2. The efficiency of thermochemical schemes to accomplish water splitting without any CO_2 emissions should be examined at a laboratory scale for the promising cycles, such as the SI cycles. Materials compatibility issues and catalysts to enhance the reaction at lower temperatures should be pursued. Reasonably-sized demonstration plants using the integrated process should be pursued in a few years for the most promising scheme.

3. Development of the high-temperature steam electrolysis process should be pursued. The issues of materials durability, reduction of overvoltages, effects of the operating pressure, and separation of gas products in an efficient and safe manner should be investigated.

4. Development of a supercritical CO_2 cycle should be given a high priority. It can be directly used with a CO_2-cooled reactor such as the AGR, or indirectly used with the other reactors such as an HTGR. It can be the bottoming cycle for a high-temperature reactor, whose coolant would supply heat at higher temperatures to a thermochemical plant. Demonstration of the thermal conversion efficiency for a moderate-size turbine and compressor (in the MWe range) is needed to validate the cycle thermodynamics.

5. The safety issues of coupling the nuclear island to the hydrogen-producing chemical island need to be examined in order to establish the guidelines necessary for avoiding accident propagation from one island to the other. Such guidelines would be needed even if the first application of nuclear hydrogen production was based on the nuclear-assisted SMR approach.

Comments on the Department of Energy Program

The DOE nuclear hydrogen program is being pursued in two streams: one for reactor technology development and one for the chemical processes. The DOE's total program for reactor development was not reviewed by the committee, but it is understood that the DOE is pursuing the development of several versions of high-temperature reactors and is giving priority to the gas-cooled reactor options. This priority is compatible with the reactors suitable for hydrogen production. However, the molten salt-cooled graphite reactor may be a variant missing in the DOE program.

The DOE R&D program related to hydrogen from nuclear energy includes the chemical processes as well as the high-temperature electrolysis path. A balanced approach is wise in order to benefit from high-efficiency electricity generation at lower temperatures than appear to be required for the thermochemical processes. A systems analysis of the electrolysis approach is needed in order to determine the impact of the more efficient distributed generation capability. The electrolyzer units can use materials similar to those of fuel cells that operate at high temperatures, and a synergistic materials program may be possible. Finally, both electrolysis and thermochemistry are potentially applicable to the use of solar energy for hydrogen production.

The overall size of the hydrogen plant R&D appears modest at this point, seeking $2 million in new funds under the Hydrogen Fuel Initiative in FY 2004, in addition to the $2 million being spent through the Nuclear Energy Research Initiative. That level might be appropriate for laboratory-level investigations of one option. Covering several options in both the thermochemistry and high-temperature electrolysis properly requires a level of $10 million to $20 million per year for 4 or 5 years before reaching a conclusion on the best approach for a large (100 MWth) demonstration plant based on the most promising option.

The current R&D portfolio does not allow for "out of the box" thinking. It needs to encourage exploratory basic research involving other approaches, such as methods of enhancing hydrogen production by radiolysis or photolysis in properly designed radiation sources.

Summary

Hydrogen can be produced from current nuclear reactors using electrolysis of water. More efficient hydrogen production may be attained by thermochemical splitting of water or electrolysis of high-temperature steam. Another possibility is the use of nuclear energy as the source of heat for steam methane reforming (SMR). The water-splitting approach releases no CO_2. Efficient water-splitting processes and nuclear-SMR all require temperatures well above 700°C. Current water-cooled reactors produce temperatures under 350°C, and cannot be used for efficient hydrogen production. Advanced reactors, such as gas-cooled reactors, involve coolants that can achieve the required high temperatures.

As indicated, the DOE's total program for reactor development was not reviewed by the committee, but it is understood to include high-temperature reactors, with focus on gas-cooled reactor options. This priority is compatible with the reactors suitable for hydrogen production. The DOE R&D program on the chemical processes for nuclear hydrogen production appears to favor thermochemical processing over the high-temperature electrolysis path. A more balanced approach would be wiser in order to make use of potentially high efficiency electricity generation at lower temperatures than are required for thermochemical processes. Furthermore, the electrolyzer units can use materials similar to those for fuel cells that operate at high temperatures, and a synergistic materials program may be possible.

The research budget for the hydrogen technology part of the Department of Energy's nuclear hydrogen program is at the level of $4 million for FY 2004, which appears to be modest. The examination of several options for promising cycles, including the process kinetics, the material's ability to withstand the aggressive chemistry and temperatures, the separation of fluids, and the overall efficiency of the systems involved, requires a significantly higher level of funding for a few years, until the most promising process is selected for demonstration. Advances made in the thermochemical cycles or in high-temperature electrolysis are of benefit to hydrogen production using other fuel sources, such as solar energy. A portfolio of advancing near-term technologies needs to be maintained while innovative approaches are being examined.

The research portfolio should also include safety aspects of integrating the nuclear reactor with the chemical plant for hydrogen production. This aspect of the program is an important ingredient in establishing guidelines for the designs to avoid the potential for accident propagation. The involvement of industry in assessing the practicality and cost of the technology that might be selected for development in order to ensure the highest economic potential should be emphasized.

HYDROGEN FROM ELECTROLYSIS

Two basic options exist for producing hydrogen. One way is to separate the hydrogen from hydrocarbons through processes referred to as reforming or fuel processing. The second way to make hydrogen is from water, using the process of electrolysis to dissociate water into its separate hydrogen and oxygen constituents. Electrolysis technologies that have been in use for decades both dissociate water and capture oxygen and/or hydrogen, primarily to meet industrial chemical needs. Electrolysis has also played a critical role in life support (oxygen replenishment) in space and submarine applications over the past several decades.

Importance of Electrolysis

Making hydrogen through electrolysis generally consumes considerably more energy per unit of hydrogen produced than does making hydrogen from hydrocarbons. Nonetheless, electrolysis is of interest as a potential source of hydrogen energy for several reasons. First, water (and the hydrogen it contains) is more abundant than hydrocarbons are. Depletion and geopolitical concerns for water are in general far less serious than are those for hydrocarbons. Further, there are geographical regions in the nation and around the world where hydrocarbons (especially natural gas, the predominant source of hydrogen reformation) are simply not available; hydrogen from water may be the only practical means of providing hydrogen in such settings.

Second, the net energy costs of making hydrogen through electrolysis must be viewed in an economic context. Electrolysis can be a means of converting low-cost Btus (e.g., coal) into much-higher-value Btus if the result is to replace gasoline or other transport fuels.

Third (and as is discussed further in the analysis that follows), electrolysis is seen as a potentially cost-effective means of producing hydrogen on a distributed scale and at costs appropriate to meet the challenges of supplying the hydrogen needs of the early generations of fuel cell vehicles. Electrolyzers are compact and can realistically be situated at existing fueling stations.

Fourth, electrolysis presents a path to hydrogen production from renewably generated electrical power. From an energy perspective, electrolysis is literally a way to transform electricity into fuel. Electrolysis is thus the means of linking renewably generated power to transport fuels markets. Currently, renewable solar, wind, and hydro power, by themselves, produce only electricity.

And finally, electrolyzers operating in tandem with power-generating devices (including fuel cells) present a new architecture for markets related to distributed energy storage. Various electrolyzer makers are developing products that can make hydrogen when primary electricity is available, and then store and use that hydrogen for subsequent regeneration into electricity as needed. For example, several firms are involved with developing backup power devices that operate in the 1 to 20 kilowatt (kW) range for up to 24 hours, well beyond the capability of conventional batteries. This same concept is being applied directly to renewable sources, creating the means to produce power-on-demand from inherently intermittent renewables. And finally, electrolysis may play a role in regenerative braking on vehicles. Electrolyzers and hydrogen have the appropriate scale and functionality to become part of the distributed generation marketplace as the cost of electrolyzers comes down over time.

Technology Options

Current electrolysis technologies fall into two basic categories: (1) solid polymer (which provides for a solid electrolyte) and (2) liquid electrolyte, most commonly potassium hydroxide (KOH). In both technologies, water is introduced into the reaction environment and subjected to an electrical current that causes dissociation; the resulting hydrogen and oxygen atoms are then put through an ionic transfer mechanism that causes the hydrogen and oxygen to accumulate in separate physical streams.

Solid polymer, or proton exchange, membranes were developed at General Electric and other companies in the 1950s and 1960s to support the U.S. space program. A proton exchange membrane (PEM) electrolyzer is literally a PEM fuel cell operating in reverse mode. When water is introduced to the PEM electrolyzer cell, hydrogen ions are drawn into and through the membrane, where they recombine with electrons to form hydrogen atoms. Oxygen gas remains behind in the water. As this water is recirculated, oxygen accumulates in a separation tank and can then be removed from the system. Hydrogen gas is separately channeled from the cell stack and captured.

Liquid electrolyte systems typically utilize a caustic solution to perform the functions analogous to those of a PEM electrolyzer. In such systems, oxygen ions migrate through the electrolytic material, leaving hydrogen gas dissolved in the water stream. This hydrogen is readily extracted from the water when directed into a separating chamber.

KOH systems have historically been used in larger-scale applications than PEM systems. Electrolyzer Corporation of Canada (now Stuart Energy) and the electrolyzer division of Norsk Hydro have built relatively large plants (100 kg/hour and larger) to meet fertilizer production needs in locations around the globe where natural gas is not available to provide hydrogen for the process.

The all-inclusive costs of hydrogen from PEM and KOH systems today are roughly comparable. Reaction efficiency tends to be higher for KOH systems because the ionic resistance of the liquid electrolyte is lower than the resistance of current PEM membranes. But the reaction efficiency advantage of KOH systems over PEM systems is offset by higher purification and compression requirements, especially at small scale (1 to 5 kg/hour).

Today's Electrolysis Markets

Chemical and Niche Energy Applications

Electrolyzers are today commercially viable only in selected industrial gas applications (excepting various noncommercial military and aerospace applications). Commercial applications include the previously mentioned remote fertilizer market in which natural gas feedstock is not available. The other major commercial market for electrolysis today is the distributed, or "merchant," industrial hydrogen market. This merchant market involves hydrogen delivered by truck in various containers. Large containers are referred to as tube trailers. An industrial gas company will deliver a full tube trailer to a customer and take the empty trailer back for refilling. Customers with smaller-scale requirements are served by cylinders that are delivered by truck and literally installed by hand.

In general, the smaller the quantities of hydrogen required by a customer, the higher will be the all-inclusive delivered cost. Tube trailer customers (e.g., semiconductor, glass, or specialty metals manufacturers) pay in the range of $3.00/100 scf, or about $12/kg. Cylinder customers (e.g., laboratories, research facilities, and smaller manufacturing concerns) pay at least twice the tube trailer price. The value of hydrogen in distributed chemical markets today is much higher than the value of hydrogen if it were to be used as fuel. The price of hydrogen will need to be in the $2.00/kg range to compete with conventional fuels for transportation.

It will take significant cost-reduction and efficiency improvements for electrolytic hydrogen to compete in vehicle fueling markets. Nonetheless, a number of stationary energy-related applications for electrolytic hydrogen are beginning to materialize. These smaller but higher-value energy applications merit the DOE's attention and support as a means of advancing the practical development of hydrogen from electrolysis for future, larger-scale fueling markets.

Off-Grid Renewables Applications

Power-on-demand from inherently intermittent renewables is another interesting application for electrolysis. Off-grid, renewable-based systems need electricity at night or when the wind doesn't blow. The value difference between electricity when available and when needed is often great enough to merit the utilization of batteries to fill this gap. In circumstances in which the amount and duration of stored energy becomes relatively large in relation to battery functionality, an electrolyzer-hydrogen regenerative system may prove a lower-cost solution, ultimately enabling greater use of renewables for meeting off-grid energy needs.

Current Electrolyzer Technology and Fueling Costs

The cost of hydrogen from electrolysis is dominated by two factors: (1) the cost of electricity and (2) capital-cost recovery for the system. A third cost factor—operation and maintenance expenses (O&M)—adds perhaps 3 to 5 percent to total annual costs. The electrochemical efficiency of the unit, coupled with the price of electricity, determine the variable cost. The total capital cost of the electrolyzer unit, including compression, storage, and dispensing equipment, is the basis of fixed-cost recovery.

Electrochemical Efficiency

Proton exchange membranes, whether operating in electrolysis mode or fuel cell mode, have the property of higher efficiency at lower current density. There is a 1:1 relationship in electrolysis between the rate of hydrogen production and current applied to the system.

The energy required in the theoretical efficiency limit of any water electrolysis process is 39.4 kWh per kilogram. PEM electrolyzers operating at low current density can approach this efficiency limit. However, the quantities of hydrogen produced at low current density are small, resulting in very high capital costs per unit of hydrogen produced. As shown in Figure G-9, cell stack efficiencies drop to 75 percent when current densities rise into the range of 1000 amps per square foot (ASF). As previously stated, the electrochemical efficiency of KOH systems is higher over a broader range of current densities, but this higher reaction efficiency is offset at least in part by higher compression and purification costs, as well as by higher costs associated with managing the liquid electrolyte itself.

The committee believes that current technology is capable of producing an electrolyzer-based fueling facility having the capacity to produce 480 kg/day, or 20 kg/hour. This plant would be capable of fueling 120 cars per day, assuming an average purchase of 4 kg per car. A plant of this scale would of necessity today be a KOH system, but with additional development, PEM technology should be capable of providing systems of comparable scale.

Electrolyzer systems of this scale should be capable of operating with an overall efficiency of 63.5 percent lower heating value [LHV], including all parasitic loads other than compression. The electrolyzer is assumed to be able to generate hydrogen at an internal pressure in the 150 psi range; supplementary compression will be required to raise the pressure to automotive fueling pressures in the 7000 psi (400 atm) range. The electrical requirement associated with compression is assumed at 2.3 kW/kg/hour, adding about 5 percent to the plant's electrical consumption and bringing overall efficiency down to about 59 percent.

Equipment Costs

Regarding capital cost recovery, the cost of the 480 kg/day system, excluding compression and dispensing, is assumed at $1000/kW input. The cost of the complete fueling system is summarized in Table G-5.

The total cost of a system at this scale would be about $2.5 million. It is anticipated that electrolysis technology scales with an 85 percent factor, so smaller-scale systems, with somewhat higher unit costs, are entirely feasible. For example, a facility with half the fueling capability (60 cars per day) would cost about $1.25 million, plus a 15 percent scaling factor. The scalability of electrolysis is one of the important factors relating to its likely use in early-stage fuel cell vehicle adoption. The electrochemical efficiency of electrolysis is essentially independent of scale.

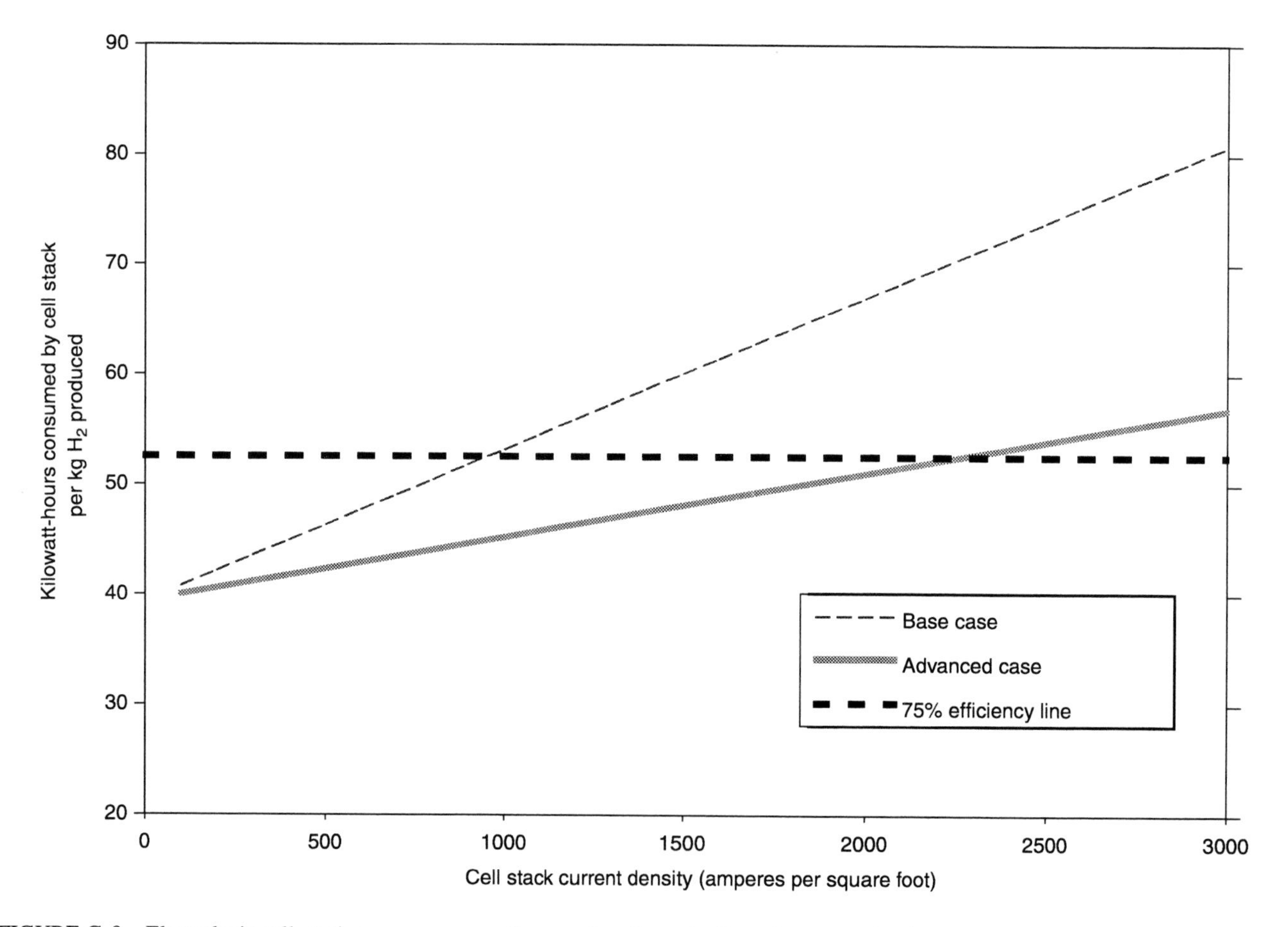

FIGURE G-9 Electrolysis cell stack energy consumption as a function of cell stack current density.

TABLE G-5 Capital Costs of Current Electrolysis Fueler Producing 480 Kilograms of Hydrogen per Day

	Unit Cost ($)	Total Cost ($ millions)
Electrolyzer unit	1,000/kW	1.17
Hydrogen compressor	3,000/kW	0.16
Hydrogen storage	100/gal.	0.24
Hydrogen dispenser	15,000/unit	0.02
Total process units		1.59
General facilities	20%	0.32
Engineering, permitting, start-up	10%	0.16
Contingencies	10%	0.16
Working capital and miscellaneous	5%	0.08
Total capital		2.31
Siting factor (110% of Gulf Coast)		0.23
Total		2.54

NOTE: See Table E-37 in Appendix E in this report.

All-Inclusive Cost of Hydrogen Fuel from Electrolysis

The total cost of electrolytic hydrogen from currently available technology is summarized in Table G-6. This table assumes a 14 percent capital cost-recovery factor, and presents the total cost (variable, capital, and O&M) associated with the assumed fueling facility. The delivered cost of grid electricity is assumed at 7 cents/kWh. Total costs are in the range of $6.50/kg.

Future Electrolysis Technology Enhancements

Among the research priorities that can improve the efficiency and/or reduce the cost of future electrolysis fueling devices are the following:

Efficiency-Enhancing Objectives

1. *Reducing the ionic resistance of the membrane.* New membranes will be thinner and will incorporate improved ion-conducting formulations that lower the resistance of the membrane and cause more of the electrical energy delivered to the membrane to be translated into hydrogen chemical energy and less into heat. In alkaline (KOH) systems, ionic resistance tends to be less than in proton exchange membrane systems, but KOH systems tend to have more complex materials handling and pressurization regimes.

2. *Reducing other (parasitic) system energy losses.* A variety of parasitic loads, such as power conditioning, can be reduced through system redesign and optimization. Power conditioning is one area of efficiency loss; current systems lose as much as 10 percent electrical efficiency with currently available inverters. These losses will be reduced by half or more with new inverters redesigned to meet the specific needs of electrolyzers. Power supply companies will need to see enough market assurance before those redesigns will be forthcoming.

TABLE G-6 All-Inclusive Cost of Hydrogen from Current Electrolysis Fueling Technology

	Cost per Year per Station ($ million)	Cost per Kilogram ($)
Nonfuel variable operation and maintenance (1% of capital)	0.025	0.16
Electricity (7 cents/kWh)	0.605	3.84
Variable operating costs	0.630	4.00
Fixed operating costs (2%/year of capital)	0.051	0.32
Capital charges (14%/year of capital)	0.354	2.24
Total cost	1.035	6.56

Other cost reductions can come from optimizing an array of components and the overall operating system. Volume manufacturing and pricing are also important cost factors.

In calling out the efficiency costs of alternating current/direct current (ac/dc) power conversion, one advantage of renewable power becomes worthy of note. Renewables generate dc power that can be applied to the dc-using electrolyzer cell stack without inversion. This incremental efficiency advantage associated with renewables may become material as the cost of power from renewables continues to drop.

3. *Reducing current density.* Conversion efficiencies are a function of electric current density, so the substitution of more electrolyte or more cell surface area has the impact of reducing overall power requirements per unit of hydrogen produced. Improved catalyst deposition technology will also lower the amount and cost of materials per unit of hydrogen production. Operating system redesign for optimization is another area of cost reduction opportunity.

Technology advances will be required to get to efficiencies beyond the current level. One area that promises to improve efficiency is higher temperature, which has the effect of lowering the ionic resistance within the cell environment.

4. *Higher temperatures.* PEM technologies typically operate at low temperatures (below 100°C) because of membrane durability limitations. Higher-temperature proton exchange membranes are in development; these should be able to tolerate significantly higher temperatures and thereby deliver higher efficiencies.

Cost Reduction Objectives

The committee believes that PEM electrolysis is subject to the same basic cost reduction drivers as those for fuel cells. Cost breakthroughs in (1) catalyst formulation and loading, (2) bipolar plate/flow field, (3) membrane expense and durability, (4) volume manufacturing of subsystems and modules by third parties, (5) overall design simplifications, and (6) scale economies (within limits) all promise to lower the cost per unit of production. The committee finds it plausible that electrolyzer capital costs can fall by a factor of 8—from $1000 per kW in the near term to $125 per kW over the next 15 to 20 years, contingent on similar cost reductions occurring in fuel cells. This reduction seems attainable when considered against the claims by fuel cell developers that they can bring the cost of fuel cells to $50/kW from today's nearly $5000/kW prices.

Advanced Future Electrolysis Technologies

The committee was presented with the view that technologies beyond PEM may offer higher overall efficiency by going to significantly higher temperatures and design concepts. Solid oxide fuel cell technology operates at much higher temperatures than PEM technology does, and so it may be a source of advanced electrolyzer performance going forward. Efficiencies moving toward 95 percent may be possible with solid oxide. But solid oxide systems operating at 500°C to 1000°C are probably at least 5 and perhaps 10 years in the future.

Solid oxide systems, because of their thermal management needs, may be confined to systems of significantly larger scale than PEM systems. Solid oxide electrolyzers may be scalable down to gas station duty, but that remains to be proven. Clearly, PEM systems can scale appropriately for distributed refueling duty.

Electrolysis/Oxidation Hybrids Still further advances in electrolysis technology, such as have been conceived at Lawrence Livermore National Laboratory, involve solid oxide electrolyzer/hydrocarbon hybrids. The hybrid concept involves enhancing the efficiency of the already-high-temperature electrolysis process by using the oxidation of natural gas as a means of intensifying the migration of oxygen ions through the electrolyte and thereby reducing the effective amount of electric energy required to transport the oxygen ion. The concept appears to offer the potential for significantly improved net electrochemical efficiency. However, the concept relies on a number of technical breakthroughs in harnessing solid oxide technology and ultimately requires a separate stream of methane or another combustible fuel supply in addition to water and electricity.

Future Electrolytic Hydrogen Fuel Costs

The committee's assessment of electrolysis improvements focused on PEM-based technologies rather than on advanced concepts. The effect is to offer a view of futures that are based on today's technology and do not rely on new technological breakthroughs that, should they occur, would only enhance the cost and performance picture.

Overall, improvements in electrolyzer performance will come from three advancements: (1) improved electrochemical efficiency—efficiency gains from 63.5 percent system efficiency to 75 percent system efficiency (LHV) could be attainable; (2) system costs—as stated above, the system capital costs may be reduced by a factor of 8, from $1000/kW to $125/kW, driven largely by the same cost factors that must be addressed by fuel cell developers if there is to be any meaningful penetration by fuel cells into the transportation marketplace; and (3) compressor performance and cost are seen to be improving as a result of a variety of emerging hydrogen energy alternatives, all of which depend on taking hydrogen to significantly higher energy densities than can today be attained with only hydrogen compression.

The resulting impact of technology development on the future cost of hydrogen from electrolysis is summarized in Table G-7. Variable costs (electricity) fall as a result of improved electrochemical efficiency. The biggest change comes from the large drop in capital costs, which translates directly into lower capital cost per unit of production. This, along with lower compression costs, results in reduced all-inclusive costs of hydrogen from $6.58/kg using current technology to $3.94/kg as a result of future improvements.

Sensitivity to Electricity Costs

Figure G-10 illustrates the considerable sensitivity of the cost of hydrogen from electrolysis to the price of input electricity. Each 1 cent reduction in the price of electricity reduces the cost of electrolytic hydrogen fuel by 53 cents/kg, or more than 8 percent per penny. Effective utilization of electrolysis as a fueling option will involve the cooperation of utilities and rate-making bodies.

Environmental Impacts of Electrolysis

The environmental impact of the use of electrolysis to produce hydrogen depends on the source of electricity. The

TABLE G-7 Cost of Hydrogen from Future Electrolysis Fueling Technology

Capital Cost	Unit Cost ($)	Cost per Station ($ million)
Electrolyzer	125/kW	0.13
Compressor	1,500/kW	0.03
Storage	75/gal	0.19
Dispenser	10,000/unit	0.01
Other		0.17
Total capital (with a 1.1 siting factor)		0.57
Cost		**$/kg**
Nonfuel variable cost		0.04
Electricity		3.31
Fixed operating costs		0.07
Capital charges		0.51
Total		3.93

NOTE: See Table E-38 in Appendix E in this report.

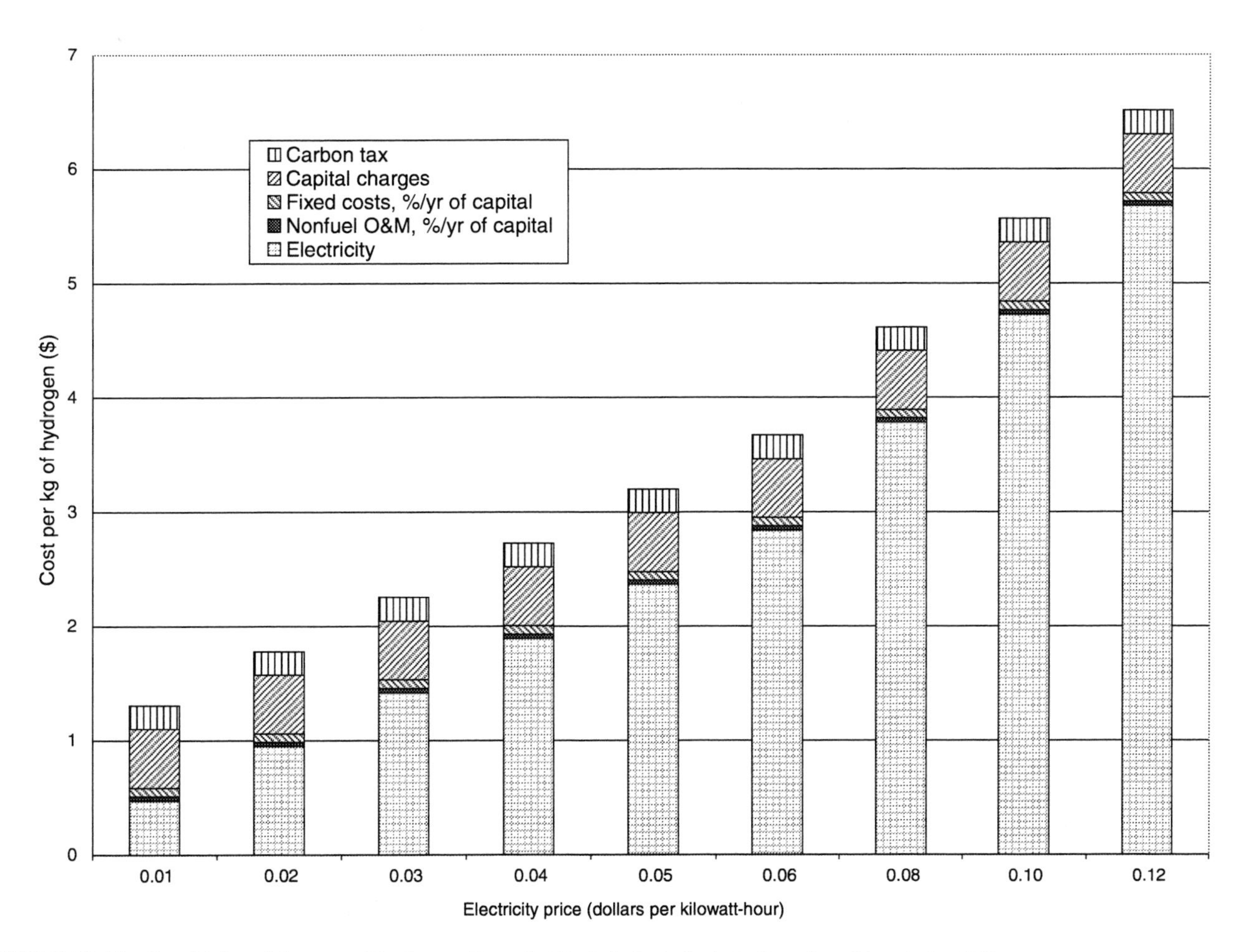

FIGURE G-10 Sensitivity of the cost of hydrogen from distributed electrolysis to the price of input electricity.

electrolysis process produces little if any CO_2 or other greenhouse gas emissions per se. Electrolyzers contain no combustion devices, and the only input to the process other than electricity is pure water.

However, there does exist a relationship between emissions and electrolysis. Any pollution associated with electricity consumed by the electrolyzer needs to be taken into account. As stated previously, one fundamental appeal of electrolysis is that it creates a path for converting renewable power into fuel. But the low capacity factors of renewables (other than geothermal and hydro power) make an all-renewables case very difficult on an economic basis. Electricity from nuclear plants is also non-emitting on a greenhouse gas emissions basis, but the outlook for additional nuclear plants is uncertain at best.

Power from the grid is assumed to derive from the grid's average generating mix. With today's grid mix, about 17.6 kg CO_2 are emitted per kilogram of hydrogen. As the portfolio of energy resources utilized to supply electric power evolves, the amount of CO_2 emitted to produce 1 kg H_2 could either increase or decrease.

Electrolysis as an Early-Stage Transitional Hydrogen Fuel Source

Electrolysis may be particularly well suited to meeting the early-stage fueling needs of a fuel cell vehicle market. Electrolyzers scale down reasonably well; the efficiency of the electrolysis reaction is independent of the size of the cell or cell stacks involved. And the balance of plant costs in an electrolyzer are also fairly scalable.

The compact size of electrolyzers makes them suitable to be placed at or near existing fueling stations. And finally, electrolyzers can utilize existing water and electricity infrastructures to a considerable extent, obviating the need for a new pipeline or surface hydrogen transport infrastructure that would be required of larger, central station hydrogen production technologies.

Summary

Electrolytic hydrogen production is an existing technology that serves a high-value industrial chemicals-based market today. The key to adapting this technology to meet energy-related applications in the future is cost reduction and performance enhancement. The Department of Energy has already identified several technology objectives relating to electrolytic hydrogen production.

Hydrogen can be made from renewable sources, enabling a perfectly sustainable energy path. The falling cost of renewable energy resources and the improving cost and efficiency outlook for electrolysis contribute to the prospect that renewably sourced electrolytic hydrogen may be competitive with other hydrogen supply in at least some instances.

Electrolyzers typically operate from grid-quality power, so a new variety of power control and conditioning equipment needs to be developed in order for electrolyzers to operate efficiently from renewable sources. The prospect exists for good efficiency in converting renewable power to hydrogen, insofar as electrolyzers require direct current and renewables generate direct current, so there are no losses associated with ac/dc conversion.

HYDROGEN PRODUCED FROM WIND ENERGY

Introduction

The production of hydrogen from renewable energy sources is often stated as the long-term goal of a mature hydrogen economy (Turner, 1999). As such the development of cost-effective renewable technologies should clearly be a priority in the hydrogen program, especially since considerable progress is required before these technologies reach the levels of productivity and economic viability needed to compete effectively with the traditional alternatives. Thus, basic renewables research needs to be expanded and the development of renewable hydrogen production systems accelerated.

Of all the renewables currently on the drawing boards, in the near to medium term, wind arguably has the highest potential as an excellent source for producing pollution-free hydrogen, using the electricity generated by the wind turbines to electrolyze water into hydrogen and oxygen. The issues for its successful development and deployment are threefold: (1) further reducing the cost of wind turbine technology and the cost of the electricity generated by wind, (2) reducing the cost of electrolyzers, and (3) optimizing the wind turbine-electrolyzer with hydrogen storage system. This section discusses current costs and projections for future costs of electricity produced by wind energy and then looks at the cost of producing hydrogen using an integrated wind turbine-electrolyzer system. (Discussion of electrolyzer technology is presented in the section "Hydrogen from Electrolysis.") This section focuses on wind energy systems that would be deployed on a distributed scale.

Status of Wind Energy in the World Today

While wind energy has been one of humanity's primary energy sources for transporting goods, milling grain, and pumping water for several millennia, its use as an energy source began to decline as industrialization took place in Europe and then in America. The decline was at first gradual as the use of petroleum and coal, both cheaper and more reliable energy sources, became widespread, and then it fell more sharply as power transmission lines were extended into most rural areas of industrialized countries. The oil crises of the 1970s, however, triggered renewed interest in wind energy technology for grid-connected electricity production, water pumping, and power supply in remote areas, promoting the industry's rebirth. By 2002, grid-connected wind power in operation surpassed 31,000 MW worldwide (see Figure G-11).

In the early 1980s, the United States accounted for 95 percent of the world's installed wind energy capacity (see Figure G-11). The U.S. share has since dropped to 15 percent in 2002. Other countries dramatically increased their capacity starting in the mid-1990s, while the U.S. capacity essentially stagnated until 1999, when more than 600 MW in new capacity were installed in a rush to beat an expiring production tax credit for utility-scale projects. This credit has since been extended through December 31, 2003. In 2001 and 2002, the total installed wind capacity doubled in the United States, and in 2003 it was expected to increase another 25 percent, to more than 6000 MW, with installations of 1400 to 1600 MW of new wind power (AWEA, 2003).

The decline in the U.S. capacity world share can be explained by a combination of economic factors and changes in government-sponsored support programs that impeded the development of new capacity. The U.S. wind industry was born in 1981 in the aftermath of the world oil crises of 1973–1974 and 1978–1979. Wind energy was not cost-competitive with fossil fuel energy, but federal legislation guaranteed a market for wind-generated power and offered generous tax credits to developers of wind energy. However, 1986 marked the beginning of the slowdown in U.S. wind energy development. The availability of relatively cheap oil and natural gas and improvements in gas generating technology, coupled with the expiration of federal tax credits at the end of 1985, meant that wind energy remained significantly more costly than fossil fuels. The tax credit incentives had been more effective in building capacity than in maintaining productivity, and as a consequence electricity generation from wind did not grow as rapidly as initially anticipated. This trend appears to have reversed itself in the past 5 years, with more than a 22 percent annual increase in installed generating capacity since 1998, despite the recent problems permeating the electric utility industry. This recent growth, coupled with progressive state policies—30 states have installed wind capacity—the continuing extension of the federal wind energy production tax credit, and maturing wind turbine technology, appears to have signaled a rebirth for the industry in the United States.

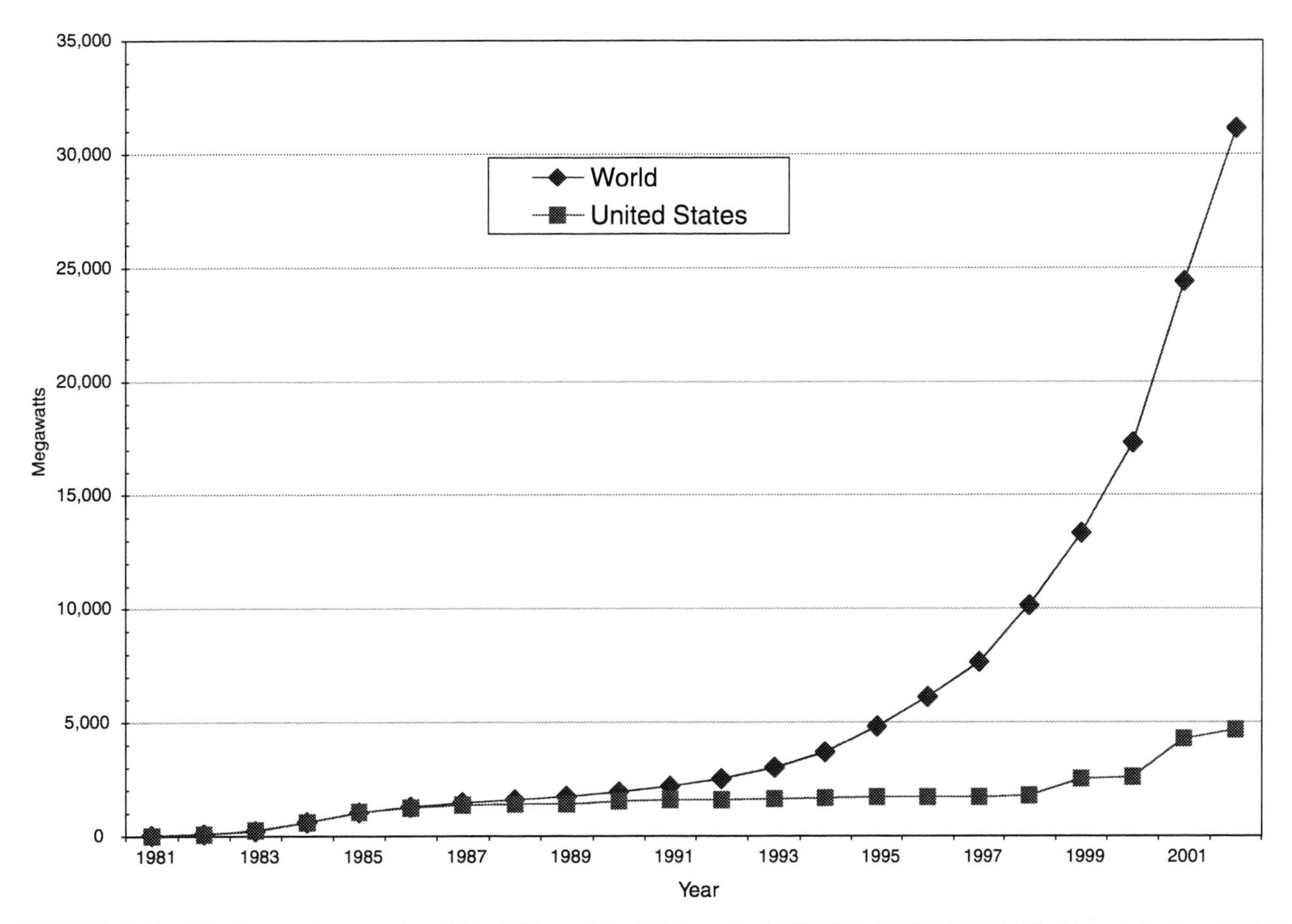

FIGURE G-11 Wind generating capacity, 1981–2002, world and U.S. totals. SOURCES: AWEA (2003), Worldwatch Institute (1999), and WEA (2000).

Potential for Wind Energy: Technical and Resource Availability

The main technical parameter determining the economic success of a wind turbine system is its annual energy output, which in turn is determined by parameters such as average wind speed, statistical wind speed distribution, distribution of occurring wind directions, turbulence intensities, and roughness of the surrounding terrain. Of these, the most important and sensitive parameter is the wind speed, which increases exponentially with height above the ground; the power in the wind is proportional to the third power of the momentary wind speed. As accurate meteorological measurements and wind energy maps (as shown in Figure G-12) become more commonly available, wind project developers can more reliably assess the long-term economic performance of wind farms.

Estimates show that U.S. wind resources could provide more than 10 trillion kWh (Deyette et al., 2003; Elliott and Schwartz, 1993), which includes land areas with wind class 3 or above (corresponds to wind speeds greater than 7 meters per second [m/s] [15.7 mph] at a height of 50 m), within 20 miles of existing transmission lines, and excludes all urban and environmentally sensitive areas. This is over 4 times the total electricity currently generated in the United States. In the DOE's *Hydrogen Posture Plan* (DOE, 2003a), wind availability is estimated to be 3250 GW, equivalent to the above value for a capacity factor of 35 percent. In 2002, installed wind capacity was about 5 GW generating 12.16 billion kWh, corresponding to a capacity factor of 29 percent (EIA, 2003).

There has been a gradual growth of the unit size of commercial machines since the mid-1970s. In the mid-1970s the typical size of a wind turbine was 30 kW. By 1998, the largest units installed had a capacity of 1.65 MW, while turbines with an installed power of 2 MW have now been introduced into the market with over 3 MW machines being developed. The trend toward larger machines is driven by the demand side of the market to utilize economies of scale and to reduce visual impact on the landscape per unit of installed power, and by the expectation that the offshore potential will be growing.

Recent technical advances have made wind turbines more controllable and grid-compatible and have reduced the number of components, making them more reliable and robust. The technology is likely to continue to improve. Such im-

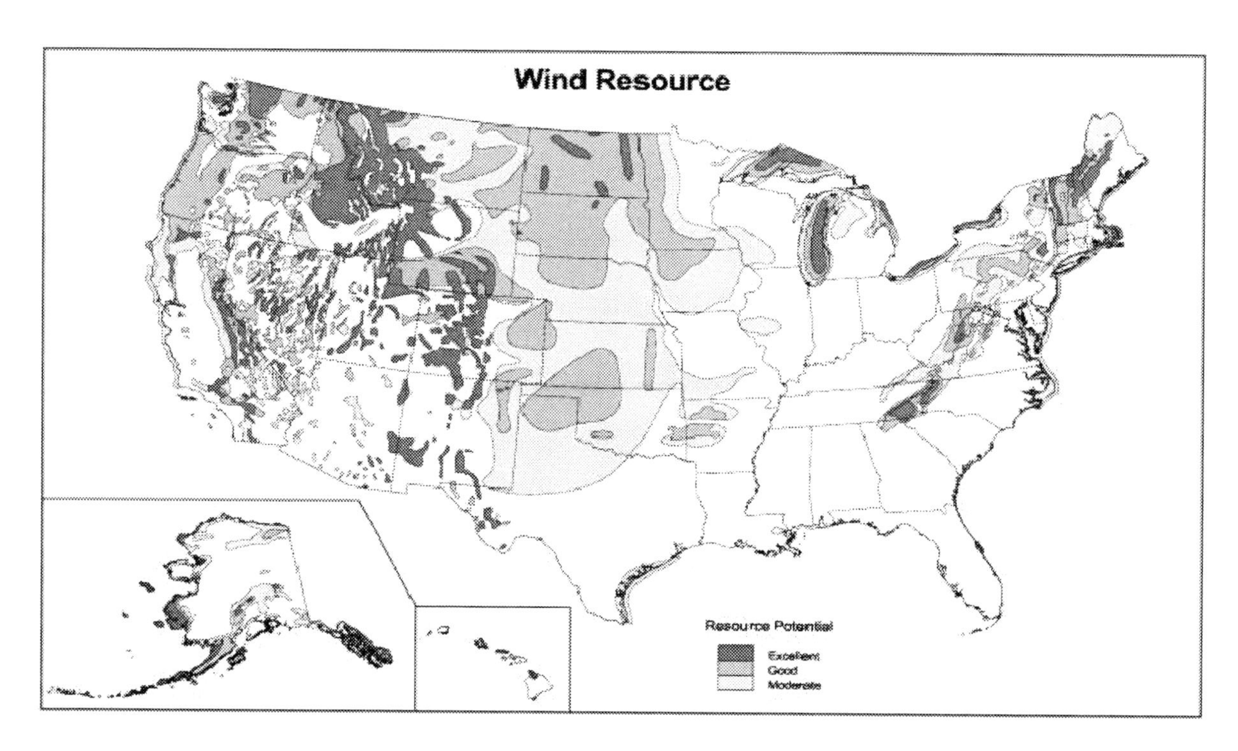

FIGURE G-12 Hydrogen from wind power availability. SOURCE: U.S. Department of Energy, National Renewable Energy Laboratory.

provements include enhanced performance at variable wind speeds, thereby capturing the maximum amount of wind according to local wind conditions, and better grid-compatibility. These advancements can occur through better turbine design and optimization of rotor blades, more efficient power electronic controls and drive trains, and better materials. Furthermore, economies of scale and automated production may also continue to reduce costs (Corey et al., 1999).

Economics of Wind Energy

Larger turbines, more efficient manufacturing, and careful siting of wind machines have brought the installed capital cost of wind turbines down from more than $2500/kW in the early 1980s to less then $1000/kW today at the best wind sites. However, on-stream capacity factor for wind is generally in the range of 30 to 40 percent, which raises the effective cost. While this decrease is due primarily to improvements in wind turbine technology, it is also a result of the general increase in wind farm size, which benefits from economies of scale, as fixed costs can be spread over a larger generating capacity. As a result, wind energy is currently one of the most cost-competitive renewable energy technologies, and in some places it is beginning to compete with new fossil fuel generation (Reeves, 2003).

Worldwide, the cost of generating electricity from wind has fallen by more than 80 percent, from about 38 cents/kWh in the early 1980s to a current range for good wind sites located across the United States of 4 to 7 cents/kWh,[19] with average capacity factors of close to 30 percent. The current federal production tax credit of 1.8 cents/kWh for wind-generated electricity lowers this cost to below 3 cents/kWh at the best wind sites. This is an order-of-magnitude decrease in cost in two decades. Analysts generally forecast that costs will continue to drop significantly as the technology improves further and the market grows around the world (Corey et al., 1999), though some do not (for example, the EIA). In the committee's analysis, for possible future technologies it is assumed that the cost of electricity generated using wind turbines decreases to 4 cents/kWh (including transmission costs). This assumption is based on a wind tur-

[19]Cost of electricity (COE) estimates from the National Renewable Energy Laboratory (NREL), Lawrence Berkeley National Laboratory (LBNL), Northern Power, and GE regarding the current cost of wind-generated electricity excluding the federal production tax credit (PTC) subsidy of 1.8 cents/kWh. NREL: Personal communication with Lee Fingersh: 3.2 to 5 cents/kWh today, depending on location; August 2003. See the web site http://www.eere.energy.gov/wind/web.html. Accessed December 10, 2003.

LBNL: Personal communication with Ryan Wiser: Wind prices are about 4.3 to 5.3 cents/kWh throughout the Midwest, 5.8 cents/kWh in the Mid-Atlantic, around 5.8 to 6.8 cents/kWh in California, and perhaps 4.3 to 5.8 cents in the Northwest; August 2003.

Northern Power: 4 to 6 cents/kWh for wind farms greater than 50 MW located at good wind sites, while for one or two turbines located at a marginal wind site, prices can be as high as 8 to 12 cents/kWh or higher. Dan Reicher, Northern Power Systems, "Hydrogen: Opportunities and Challenges," presentation to the committee, June 2003.

bine capital cost of \$500/kW, total capital costs of \$745/kW, and a capacity factor of 40 percent.[20] The expectation is that wind turbine design will be refined and economies of scale will accrue. While these values can be considered optimistic (e.g., by the EIA), others predict even lower values, given successful technology advancement and supportive policy conditions (Bailie et al., 2003; Corey et al., 1999; WEA, 2000). In the future, cost reduction can occur with multiple advancements: further improvements in turbine design and optimization of rotor blades, more-efficient power controls and drive trains, and improvements in materials. The improvements in materials are expected to facilitate increased turbine height, leading to better access to the higher-energy wind resources available at these greater heights. The desire of new U.S. vendors to participate in wind energy markets will increase competition, leading to an overall optimization and lower cost of the wind turbine system.

Wind technology does not have fuel requirements as do coal, gas, and petroleum generating technologies. However, both the equipment costs and the costs of accommodating special characteristics such as intermittence, resource variability, competing demands for land use, and transmission and distribution availability can add substantially to the costs of generating electricity from wind. For wind resources to be useful for electricity generation and/or hydrogen production, the site must (1) have sufficiently powerful winds, (2) be located near existing distribution networks, and (3) be economically competitive with respect to alternative energy sources. While the technical potential of wind power to fulfill the need for energy services is substantial, the economic potential of wind energy will remain dependent on the cost of wind turbine systems as well as the economics of alternative options.

Hydrogen Production by Electrolysis from Wind Power

Hydrogen production from wind power and electrolysis is a particularly interesting proposition since, as just discussed, among renewable sources, wind power is economically the most competitive, with electricity prices at 4 to 5 cents/kWh at the best wind sites (without subsidies). This means that wind power can generate hydrogen at lower costs than those for any of the other renewable options available today.

In the committee's analysis, it considered wind deployed on a distributed scale, thus bypassing the extra costs and requirements of hydrogen distribution. Since hydrogen from wind energy can be produced close to where it will be used, there is a clear role for it to play in the early years of hydrogen infrastructure development, especially as the committee believes that a hydrogen economy is most likely, at least initially, to develop in a distributed manner.

For distributed wind-electrolysis-hydrogen generation systems, it is estimated that by using today's technologies hydrogen can be produced at good wind sites (class 4 and above) without a production tax credit for approximately \$6.64/kg H_2, using grid electricity as backup for when the wind is not blowing. The committee's analysis considers a system that uses the grid as backup to alleviate the capital underutilization of the electrolyzer with a wind capacity factor of 30 percent. It assumes an average cost of electricity generated by wind of 6 cents/kWh (including transmission costs), while the cost of grid electricity is pegged at 7 cents/kWh, a typical commercial rate. This hybrid hydrogen production system has pros and cons. It reduces the cost of producing the hydrogen, which without grid backup would otherwise be \$10.69/kg H_2, but it also incurs CO_2 emissions from what would otherwise be an emission-free hydrogen production system. The CO_2 emissions are a product of using grid electricity; they are 3.35 kg C per kilogram of hydrogen.

In the future the wind-electrolysis-hydrogen system could be substantially optimized. The wind turbine technology could improve, reducing the cost of electricity to 4 cents/kWh with an increased capacity factor of 40 percent, as discussed previously, and the electrolyzer could also come down substantially in cost and could increase in efficiency (see the discussion in the section "Hydrogen from Electrolysis"). The combination of the increase in capacity factor and the reduction in the capital cost of the electrolyzer and cost of wind-generated electricity results in eliminating the need for using grid electricity (price still pegged at 7 cents/kWh) as a backup. The wind machines and the electrolyzer are assumed to be made large enough that sufficient hydrogen can be generated during the 40 percent of the time that the wind turbines are assumed to provide electricity. Due to the assumed reductions in the cost of the electrolyzer and the cost of wind-turbine-generated electricity, this option is now less costly than using a smaller electrolyzer and purchasing grid-supplied electricity when the wind turbine is not generating electricity. Hydrogen produced in this manner from wind with no grid backup is estimated to cost \$2.85/kg H_2, while for the alternative system with grid backup it is \$3.38/kg H_2. Furthermore, there is now the added advantage of a hydrogen production system that is CO_2-emission free. The results of the committee's analysis are summarized in Table G-8.

Wind-electrolysis-hydrogen production systems are currently far from optimized. For example, the design of wind turbines has to date been geared toward electricity production, not hydrogen. To optimize for better hydrogen production, integrated power control systems between the wind turbine and electrolyzer need to be analyzed, as should hydrogen storage tailored to the wind turbine design. Furthermore, there is the potential to design a system that can coproduce electricity and hydrogen from wind. Under the right circumstances this could be more cost-effective and

[20]This is an average value. Sites in the Great Plains, for example, could have higher capacity factors. The committee decided against using ranges for technology performance parameters in its analysis.

TABLE G-8 Results from Analysis Calculating Cost and Emissions of Hydrogen Production from Wind Energy

	Current Technology		Future Technology	
	With Grid Backup	No Grid Backup	With Grid Backup	No Grid Backup
Average cost of electricity (cents/kWh)	6	6	4	4
Wind turbine capacity factor (%)	30	30	40	40
Hydrogen ($/kg)	6.64	10.69	3.38	2.86
Carbon emissions (kg C/kg H_2)	3.35	0	2.48	0

provide broader system utility, thereby facilitating wind hydrogen system deployment (Fingersh, 2003).

Electricity systems have evolved so that they can now deliver power to consumers with high efficiency through a highly integrated system that aggregates supply and demand. Wind power benefits from this level of aggregation in that system. Numerous utility studies have indicated that wind can readily be absorbed into an integrated network until the wind capacity accounts for about 20 percent of maximum demand. Beyond this, changes to operational practice would likely be needed. Practical experience, as wind penetrates to higher levels, will continue to provide a better understanding of these system integration issues. The degree to which grid compatibility and integration play into future hydrogen production from wind needs to be better understood.

Advantages and Disadvantages

There are obvious environmental advantages to hydrogen produced from wind power. It does not generate solid, radioactive, or hazardous wastes; it does not require water; and it is essentially emission free, producing no CO_2 or criteria pollutants, such as NO_x and SO_2. In addition, it is a domestic source of energy. Thus, it addresses the main concerns that are motivating the current drive toward a hydrogen economy—environmental quality and energy security. But wind power is not problem free.

Environmental Issues

Wind energy, although considered an environmentally sound energy option, does have several negative environmental aspects connected to its use. These include acoustic noise, visual impact on the landscape, impact on bird life, shadows caused by the rotors, and electromagnetic interference influencing the reception of radio, TV, and radar signals. In practice, the noise and visual impacts appear to cause the most problems for siting projects. Noise issues have been reduced by progress in aero-acoustic research providing design tools and blade configurations that have successfully made blades considerably quieter. With careful siting, the impact on bird life appears to be a relatively minor problem. Avoiding habitats of endangered species and major migration routes in the siting of wind farms can for the most part eliminate this problem.

A growing and often intractable problem involves land use issues, particularly the "not in my backyard" phenomenon (i.e., NIMBY). In densely populated countries where the best sites on land are occupied, there is increasing public resistance, making it impossible to realize projects at acceptable cost. This is one of the main reasons that countries such as Denmark and the Netherlands are concentrating on offshore projects, despite the fact that technically and economically they are expected to be less favorable than good land sites are. In countries such as the United Kingdom and Sweden, offshore projects are being planned not because of scarcity of suitable land sites but because preserving the landscape is such an important national value—though there is now also growing resistance to offshore wind projects for the same reason, as seen for a recently proposed wind project off Cape Cod in the United States.

Technical Issues

Wind energy has some technical advantages, in addition to being both a clean and secure energy source, as compared with conventional fossil fuel generation and even some other renewable energy sources. First, it is modular: that is, the generating capacity of wind farms can easily be expanded, since new turbines can be quickly manufactured and installed; this is not the case for either coal-fired or nuclear power plants. Furthermore, a repair to one wind turbine does not affect the power production of all the others. Second, the energy generated by wind turbines can pay for the materials used to make them in as little time as 3 to 4 months for good wind sites (AWEA, 2003).

Despite these advantages, wind's biggest drawback continues to be its intermittence and mismatch with demand, an issue both for electricity generation and hydrogen production (Johansson, 1993). The best wind site locations are often not in close proximity to populations with the greatest energy needs, as in the U.S. Midwest; this problem makes such sites potentially impractical for onsite hydrogen production, owing to the high costs of storage and long-distance hydrogen distribution. On the other hand, if hydrogen storage and distribution were to become more cost-effective, potentially large quantities of relatively cheap hydrogen could be produced at remote, high-quality wind sites and distributed around the country.

Conclusions

Wind energy has some very clear advantages as a source of hydrogen. It fulfills the two main motivations that are propelling the current push toward a hydrogen economy, namely, reducing CO_2 emissions and reducing the need for hydrocarbon imports. In addition, it is the most affordable renewable technology deployed today, with expectations that costs will continue to decline. Since renewable technologies effectively address two of the major public benefits of a move to a hydrogen energy system, and wind energy is the closest to practical utilization with the technical potential to produce a sizable percentage of future hydrogen, it deserves continued, focused attention in the DOE's hydrogen program.

Although wind technology is the most commercially developed of the renewable technologies, it still faces many barriers to deployment as a hydrogen production system. There is a need to develop optimized wind-to-hydrogen systems. Partnerships with industry are essential in identifying the R&D needed to help advance these systems to the next level.

Department of Energy's Multi-Year Research, Development, and Demonstration Plan

There is little mention of hydrogen production from wind throughout the entire June 2003 draft of "Hydrogen, Fuel Cells and Infrastructure Technologies Program: Multi-Year Research, Development and Demonstration Plan" (DOE, 2003b) or in the July 2003 *Hydrogen Posture Plan: An Integrated Research, Development, and Demonstration Plan* (DOE, 2003a). An RD&D plan for hydrogen production from wind power needs to be developed and integrated into the overall hydrogen strategic RD&D plan.

Summary

Energy security and environmental quality, including reduction of CO_2 emissions, are strong factors motivating a hydrogen economy. These goals can both be fulfilled by wind-hydrogen systems. Thus, wind has the potential to play a significant role in a future hydrogen economy, both during the transition and in the long term. Since wind is currently the renewable technology that is most developed and lowest cost, wind-electrolysis-hydrogen systems merit serious attention.

Wind-electrolysis-hydrogen systems have yet to be fully optimized. There are integration opportunities and issues with respect to wind machines and electrolyzers and hydrogen storage that need to be explored. For example, coproduction of electricity and hydrogen can potentially reduce costs and increase the function of the wind-hydrogen system. This could facilitate the development of wind energy systems that are more cost-effective and have broader utility, thereby assisting their development and deployment.

HYDROGEN PRODUCTION FROM BIOMASS AND BY PHOTOBIOLOGICAL PROCESSES

Two basic avenues for molecular hydrogen production by biological processes are currently being considered: (1) via photosynthetically produced biomass followed by subsequent thermochemical processing, and (2) via direct photobiological processes without biomass as intermediate. The first process is well known and intensely researched, while the second is still in the early research stage. These processes have in common the capturing and conversion of solar energy into chemical energy mediated by photosynthetic processes. In both cases, solar energy serves ultimately as the primary energy source for the production of molecular hydrogen by biological processes. In contrast to processes using fossil fuels as primary energy sources, biological processes do not involve net production of CO_2.

Efficiency of Photosynthetic Biomass Production

In photosynthesis as carried out by plants, cyanobacteria, and microalgae, solar energy is converted into biomass in commonly occurring ecosystems at an overall thermodynamic efficiency of about 0.4 percent (see Figure G-13; Hall and Rao, 1999). This low efficiency is due to the molecular properties of the photosynthetic and biochemical machinery, as well as to the ecological and physical-chemical properties of the environment. Of the incident light energy, only about 50 percent is photosynthetically useful. This light energy is used at an efficiency of about 70 percent by the photosynthetic reaction center and is converted into chemical energy, which is converted further into glucose as the primary CO_2 fixation end product at an efficiency of about 30 percent. Of this energy, about 40 percent is lost due to dark respiration. Because of the photo inhibition effect and the nonoptimal conditions in nature, a further significant loss in efficiency is observed when growing plants in natural ecosystems. Therefore, the energy content of common biomass collected from natural ecosystems contains only on the order of 0.4 percent of the primary incoming energy (see Figure G-13). Although higher yields (in the 1 to 5 percent range) have been reported for some crops (e.g., sugarcane), the theoretical maximal efficiency is about 11 percent.

Generally, two types of biomass resources can be considered in the discussion on renewable energy feedstock: (1) primary biomass, such as energy crops, including switchgrass, poplar, and willow, and (2) biomass residues (primary when derived from wood or processed agricultural biomass; secondary when derived from food or fiber processing by-products, or animal waste; and tertiary when derived from urban residues).[21]

Biomass Availability

Today about 4 percent of total energy use in the United States is based on the use of biomass, mainly in the form of

[21]M.K. Mann and R.P. Overend, National Renewable Energy Laboratory, "Hydrogen from Biomass: Prospective Resources, Technologies, and Economics," presentation to the committee, January 22, 2003.

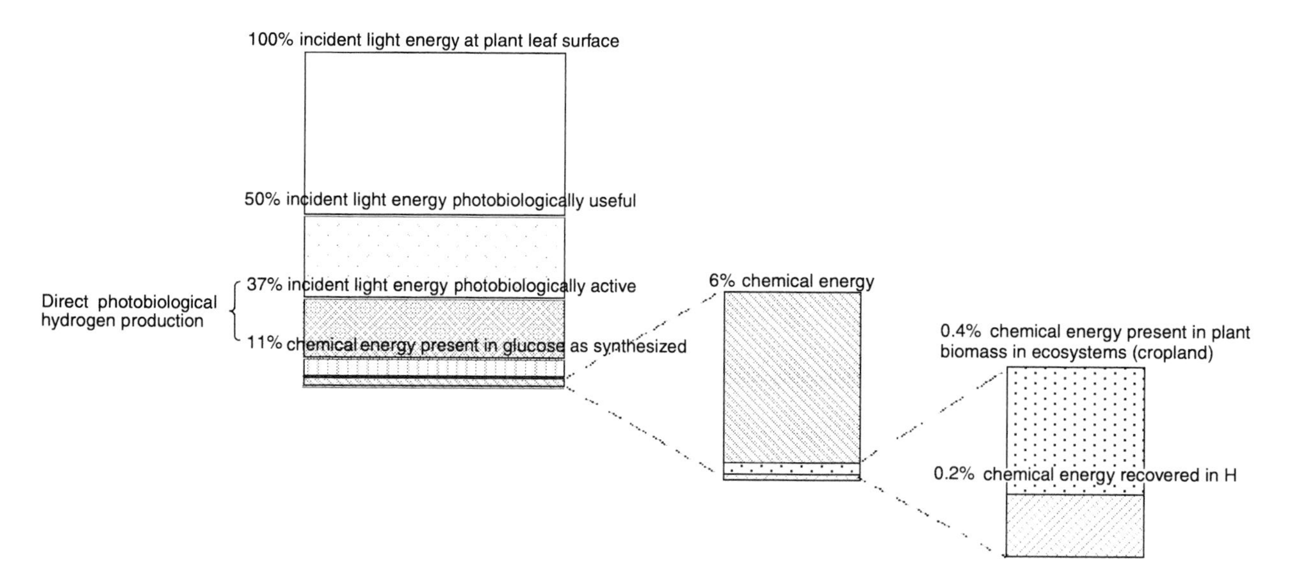

FIGURE G-13 Efficiency of biological conversion of solar energy (adapted from Hall and Rao, 1999).

forest residues. At a cost of \$30 to \$40/t, available biomass can be estimated to be between 220 million and 335 million dry tons per year.[22] This biomass consists mainly of urban residues, sludge, energy crop, and wood and agricultural residues. A significant fraction of this biomass, especially forest residues, is already used by industry or in other competing processes, such as energy generation directly. However, if all of this theoretically available biomass could be converted to hydrogen, the annually available amount would be on the order of 17 million to 26 million t H_2. As Figure 6-3 indicates, in an all-fuel-cell-vehicle scenario in the year 2050, 112 million t H_2 would be required annually. Considering this demand and the competing demands for other uses of biomass, the currently available biomass is insufficient to satisfy the entire demand in a hydrogen economy, and new sources for biomass production would need to be considered.

Primary biomass in the form of energy crops is expected to have the quantitatively most significant impact on hydrogen production for use as transportation fuel by 2050.[23] Estimates of energy that can potentially be derived from energy crops to produce biomass by 2050 range between 45 and 250 exajoules (EJ) per year. Bioenergy crops are currently not produced as dedicated bioenergy feedstock in the United States. Therefore, crop yields, management practices, and associated costs are based on agricultural models rather than on empirical data (Milne et al., 2002; de la Torre Ugarte et al., 2003; Walsh et al., 2000).

Land Use for Additional Biomass Production

In the most aggressive scenario for a hydrogen economy as considered in Chapter 6, a land area between 280,000 and 650,000 square miles is required to grow energy crops in order to support 100 percent of a hydrogen economy. The magnitude for this demand on land becomes apparent when comparing these numbers with the currently used cropland area of 545,000 square miles in the United States. Consequently, bioenergy crop production would require a significant redistribution of the land currently dedicated to food crop production and/or the development of a new land source from the U.S. Department of Agriculture's (USDA's) Conservation Reserve Program (CRP).

Although bioenergy crops can be grown in all regions of the United States, regional variability in productivity, rainfall conditions, and management practices limit energy crop farming to states in the Midwest, South, Southeast, and East (see Figure G-14) (Milne et al., 2002; de la Torre Ugarte et al., 2003; Walsh et al., 2000). Considering all cropland used for agriculture, as well as cropland in the CRP, in pasture and idle cropland, de la Torre Ugarte et al. (2003) considered two management scenarios for profitable bioenergy crop production: one to achieve high biomass production (production management scenario, or PMS), and another to achieve high levels of wildlife diversity (wildlife management scenario, or WMS). The production management scenario would annually produce about 188 million tons of dry biomass, which would be equivalent to 15 million tons of H_2, requiring 41.8 million acres of cropland, of which about

[22] Mark Pastor, Department of Energy, "DOE's Hydrogen Feedstock Strategy," presentation to the committee, June 2003; Roxanne Danz, Department of Energy, Office of Energy Efficiency and Renewable Energy, "Hydrogen from Biomass," presentation to the committee, December 2, 2002.

[23] M.K. Mann and R.P. Overend, National Renewable Energy Laboratory, "Hydrogen from Biomass: Prospective Resources, Technologies, and Economics," presentation to the committee, January 22, 2003.

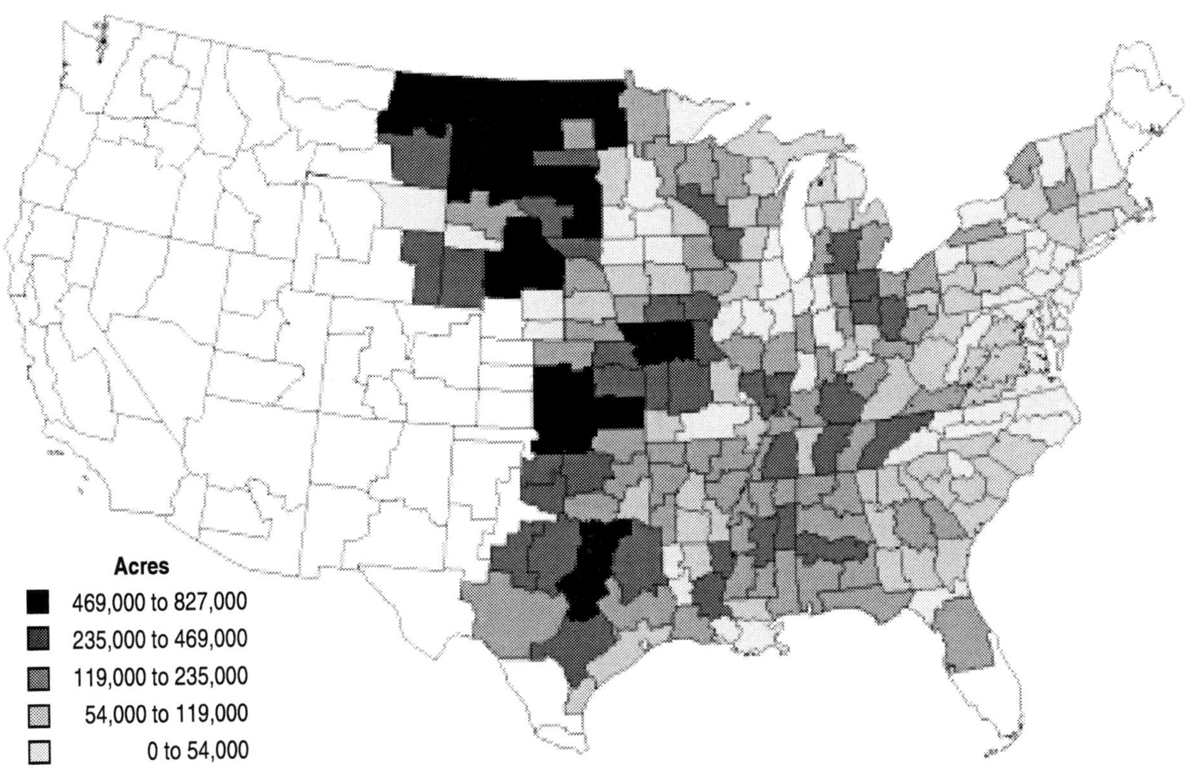

FIGURE G-14 Geographic distribution of projected bioenergy crop plantings on all acres in 2008 in the production management scenario (after Walsh et al., 2000).

56 percent would be from currently used cropland, 30 percent from the CRP, and 13 percent from idle cropland and pasture. The crop would be exclusively switchgrass. In the wildlife management scenario, 96 million dry tons (dt) of biomass (equivalent to 7.6 million t H_2) would be produced on 19.4 million acres of cropland, of which about 53 percent would be from currently used cropland, 42 percent from the CRP, and 4 percent from idle cropland and pasture. Land from the CRP would become a significant source for farming biomass crops. The CRP sets aside environmentally sensitive acres under 10- to 15-year contracts. Appropriate management practices must be developed before CRP lands are used. Environmental ramifications of various management practices must be examined to ensure that there is no substantial loss of environmental benefits, including biodiversity and soil and water quality. It is conceivable that a farming scenario alternating between agricultural crops and bioenergy crops on existing agricultural and CRP lands could be developed; however, those unproven cases were not considered in this analysis.

Biomass Cost

Bioenergy crop production is considered profitable at $40/dt, and could compete with currently grown agricultural crops (TIAX LLC, 2003; Milne et al., 2002). Based on assumed yields, management practices, and input costs, switchgrass is the least-expensive bioenergy crop to produce on a per dry ton basis. Production costs (farm gate costs) for switchgrass are estimated to range from $30/dt to $40/dt, depending on the management scenarios (WMS versus PMS) (de la Torre Ugarte et al., 2003). Adding processing and delivery costs would result in an approximate delivered biomass price on the order of $40 to $50/dt, respectively. Using these feedstock costs as well as current and projected gasifier efficiencies (50 percent versus 70 percent) in the committee's analysis, the future costs per kilogram of hydrogen produced from biomass and delivered at the vehicle is about $3.60 (scenario MS Bio-F; see Figure 5-4 in Chapter 5). In this scenario, a reduction in biomass cost was assumed to be achieved by increasing the crop yield per hectare by 50 percent, which presents significant technical challenges.

The profitability of bioenergy crop farming will vary with given field and soil types (Milne et al., 2002). Notably, the price per dry ton of bioenergy crop is predicted to increase with the total biomass produced. A shift of cropland use from traditional agricultural crops to bioenergy crops will also result in higher prices for traditional crops. Because of land ownership, management, and crop establishment, biomass production by energy crop production will be more expensive than using residue biomass. Also, regional variation in the availability of residue biomass, such as in woody areas in the northeastern United States, could make hydrogen production from biomass competitive in such regions in the short term. However, such operations would be restricted to selected regions in the United States, and, in a long-term sustainable scenario, would require biomass production at the same rate as its consumption. The committee considered it to be unlikely that such localized operations would contribute significantly to the nation's H_2 supply. Therefore, such cases were not considered further in this analysis, nor were fertilizer costs and the energy required to produce, harvest, and transport biomass.

Environmental Impact of Biomass Used for Hydrogen Production

In the overall process of biomass production and gasification, no net CO_2 is generated, except for the CO_2 released from fossil fuels used for (1) harvesting and transportation of biomass, (2) operation of the gasification systems, and (3) electricity, as well as for (4) production and delivery of fertilizers in an advanced biomass system. Biomass handling alone is estimated to consume about 25 percent of the total capital costs of operation of a midsize biomass gasification plant. Furthermore, biomass production requires, in addition to land (see above), about 1000 to 3000 t of water per ton of biomass, as well as nutrients in the form of nitrogen (ammonia), phosphorus (phosphate), sulfur, and trace metals. Profitable future hydrogen production from biomass will require energy crops with increased growth yields, which translates into increased need for fertilizers, energy for production of fertilizers, and potentially water. As is the case with the production of food crops, erosion, nutrient depletion of the soil, and altered water use practices could result in potentially significant environmental impacts as a consequence of farming activities. These effects need to be carefully considered.

Technologies for Hydrogen Production from Biomass

Current Technologies

Current technologies for converting biomass into molecular hydrogen include gasification/pyrolysis of biomass coupled to subsequent steam reformation[24] (Milne et al., 2002; Spath et al., 2000). The main conversion processes are (1) indirectly heated gasification, (2) oxygen-blown gasification, and (3) pyrolysis, as well as (4) biological gasification (anaerobic fermentation). Biomass gasification has been demonstrated at a scale of 100 tons of biomass per day.[25] Only a small, 10 kg/day of H_2 pilot biomass plant is in operation, and no empirical data on the operation, performance, and economics of a full-scale biomass-to-hydrogen plant are available.[26] The thermodynamic efficiencies of these processes are currently around 50 percent. Considering the low energy content of biomass, between 0.2 percent and 0.4 percent of the total available solar energy is converted to molecular hydrogen.

Biomass gasifiers are designed to operate at low pressure and are limited to midsize-scale operations, owing to the heterogeneity of biomass, its localized production, and the relatively high costs of gathering and transporting biomass. Therefore, current biomass gasification plants are associated inherently with unit capital costs that are at least five times as high as those for coal gasification (see Figure 5-2 in Chapter 5) and operate at lower efficiency.

Coproduction (biorefinery) of, for example, phenolic adhesives, polymers, waxes, and other products with hydrogen production from biomass, is being discussed in the context of plant designs to improve the overall economics of biomass-to-hydrogen conversion[27] (Milne et al., 2002). The technical and economic viability of such coproduction plants is unproven and was not considered in this analysis.

Several major technical challenges of biomass gasification/pyrolysis exist and include variable efficiencies, tar production, and catalyst attrition[28] (Milne et al., 2002). Moisture content as well as the relative composition and heterogeneity of biomass can result in significant deactivation of the catalyst. Recent fundamental research has identified a new, potentially inexpensive class of catalysts for aqueous-phase reforming of biomass-derived polyalcohols (Huber et al., 2003). In contrast to residue biomass, the use of bioenergy crops as biomass for gasification is advantageous, as its composition and moisture content are predictable, and the gasification process can be optimized for the corresponding crop.

Using anaerobic fermentation to convert biomass into hydrogen, a maximum of about 67 percent of the energy content (e.g., of glucose) can be recovered in hydrogen theoretically (calculated after Thauer et al., 1977). Considering the currently known fermentation pathways, a practical efficiency of biomass conversion to hydrogen by fermentation is between 15 and 33 percent (4 mol H_2/mol glucose), although this is only possible at low hydrogen partial pressure. However, more efficient fermentation pathways could be conceived and would require significant bioengineering ef-

[24]M.K. Mann and R.P. Overend, National Renewable Energy Laboratory, "Hydrogen from Biomass: Prospective Resources, Technologies, and Economics," presentation to the committee, January 22, 2003.

[25]Roxanne Danz, Department of Energy, Office of Energy Efficiency and Renewable Energy, "Hydrogen from Biomass," presentation to the committee, December 2, 2002.

[26]M.K. Mann and R.P. Overend, National Renewable Energy Laboratory, "Hydrogen from Biomass: Prospective Resources, Technologies, and Economics," presentation to the committee, January 22, 2003.

[27]M.K. Mann and R.P. Overend, National Renewable Energy Laboratory, "Hydrogen from Biomass: Prospective Resources, Technologies, and Economics," presentation to the committee, January 22, 2003.

[28]M.K. Mann and R.P. Overend, National Renewable Energy Laboratory, "Hydrogen from Biomass: Prospective Resources, Technologies, and Economics," presentation to the committee, January 22, 2003.

forts. These values compare with a biomass gasification efficiency of around 50 percent. The impurity of the hydrogen from biomass may be of concern, as fuel cell operations require relatively high-grade quality.

Economic Analysis of Current and Future Biomass-to-Hydrogen Conversion

In the past, and through funding support by the DOE, the process of biomass gasification has received most attention. Gasification technology using biomass, typically wood residues as feedstock, was adapted from coal gasification, and a few small-scale prototypes of biomass gasification plants have been built. Thus, the committee considered only the economics of biomass gasification. However, no midsize gasification facility exists to date that converts biomass to hydrogen, and no empirical data are available on the operation, performance, and economics of a midsize biomass-to-hydrogen plant, as assumed in the economic model. The assumptions made for the committee's analysis of current technology consist of modular combinations adopted from existing technical units for coal gasification (shell gasifier, air separation unit, traditional shift), without considering the variability in chemical composition and moisture content of typical biomass. An overall gasification efficiency of 50 percent is assumed. Furthermore, the committee assumed a scenario in which 100 percent of the H_2 demand would need to be met by biomass-derived hydrogen, acknowledging that in a possible future scenario, a mix of different primary energy sources is more likely. As the relative proportion of such mixes of primary feedstock is unknown, the committee considered the simplified case.

Estimation of the economics of future technology for biomass-to-hydrogen conversion using gasification is more problematic and much more uncertain because of the necessary extrapolations. The committee made the following assumptions for a midsize plant: (1) advanced biomass gasifiers can be developed and will use newly developed technology, such as fluidized catalytic cracking; (2) biomass gasifiers can be modified to produce a CO and H_2 syngas, as does coal gasification; (3) biomass gasification will operate at an overall efficiency of about 70 percent; and (4) through genetic engineering and other breeding methods, the growth yield of switchgrass can be increased by 50 percent. The committee also assumed that the future biomass is derived from bioenergy crops at a price of $50/dt, as opposed to coming from less expensive biomass residues, although it is possible that a mixture of bioenergy crops and residues could be used for future gasifications. With these assumptions, the current price per kilogram of hydrogen delivered at the vehicle of $7.04 (see Figure 5-2) could be reduced in the future to about $3.60. As can be see in Figure 5-4, two factors contribute to the high price: the high capital charges for gasification and the high biomass costs.

Photobiological Hydrogen Production

In recent years, fundamental research on hydrogen production by photosynthetic organisms has received significant attention. In photosynthesis, water is oxidized photobiologically to molecular oxygen and hydrogen in order to satisfy the organism's need to build biomass from CO_2. This notion has prompted the idea of reengineering this process to release those equivalents as molecular hydrogen directly. Such direct production of molecular hydrogen is probably thermodynamically the most efficient use of solar energy in biological hydrogen production (theoretically about 10 percent to 30 percent), because it circumvents inefficiencies in the biochemical steps involved in biomass production, as well as those involved in biomass conversion to hydrogen (see Figure G-13). The photosynthetic formation of molecular hydrogen from water is thermodynamically feasible even at high hydrogen partial pressure. However, such biological capability does not occur in any known organism; thus, it will require substantial metabolic engineering using new approaches in molecular biotechnology. In a variation of this approach, electron flow from the photosynthetic reaction center could be coupled to nitrogenase, which also releases H_2. Another mode of hydrogen production, discussed in context of photosynthetic H_2 production, is dark fermentation mediated by photosynthetic microorganisms. In all cases, the reducing equivalents for producing hydrogen are derived from water, which is abundant and inexpensive. It is unclear to what extent the DOE is providing substantial funds for such research.

Technologies Competing for Land Surface Area

Since the primary energy of all biological processes for hydrogen production is renewable solar energy, all other technologies using solar energy, including photovoltaic and other newer processes, such as thin-film technology, are competing for (land) surface area. Wind energy is indirectly solar energy. Currently, the solar-to-electrical conversion efficiency of newer photoelectric processes is 15 to 18 percent, compared with 0.4 percent for bulk biomass formation, and about 10 percent, potentially, for direct hydrogen production by photosynthetic organisms. Because solar energy harvesting technologies are competing for land use among each other and with other societal activities, such as farming, housing, and recreation, the overall efficiency of a solar energy conversion process will be a key determinant for its economic viability.

Advantages and Disadvantages of Using Biomass and Photobiological Systems for Molecular Hydrogen Production

Hydrogen production from biomass is an attractive technology, as the primary energy is solar (i.e., "renewable"), with no net CO_2 being released (except for transport). Notably, when coupled to CO_2 capture and sequestration on a larger technical scale, this technology might be the most important means to achieve a net reduction of atmospheric CO_2 (see Chapter 6, Figures 6-9 and 6-10). Furthermore, different forms of biomass (bioenergy crops, residues including municipal waste, etc.) could be used in different combinations.

The current concept of biomass-to-hydrogen conversion has several limitations. Biomass conversion to hydrogen is intrinsically inefficient, and only a small percentage of solar energy is converted into hydrogen. Moreover, in order to contribute significantly to a hydrogen economy, the quantity of biomass that needs to be available necessitates the farming of bioenergy crops. However, bioenergy crops obtained by farming will be intrinsically expensive. Residue biomass is less expensive but more variable and heterogeneous in composition, thus making the gasification process less efficient. In addition, significant costs are associated with the collection and transportation of dispersed, low-energy-density bioenergy crops and residues. Most importantly, large-scale biomass production also would pose significant demand on land, nutrient supply, water, and the associated energy for increased biomass production. The environmental impact of significant energy crop farming is unclear, but it can be assumed to be similar to that in crop farming and include soil erosion, significant water and fertilizer demand, eutrophication of downstream waters, and impact on biological diversity. Biomass production is also sensitive to seasonal variability as well as to vagaries of weather and to diseases, with significant demands regarding the storage of biomass in order to compensate for the anticipated fluctuations. The public acceptance of growing and using potentially genetically engineered, high-yield energy crops is also unclear. In addition, competing uses of biomass for purposes other than hydrogen production will also control the price of biomass. Overall, it appears that hydrogen production from farmed and agriculture-type biomass by gasification/pyrolysis will only be marginally economical and competitive.

Biomass gasification could play a significant role in meeting the DOE's goal of greenhouse gas mitigation. It is likely that both in the transition phase to a hydrogen economy and in the steady state, a significant fraction of hydrogen might be derived from domestically abundant coal. In co-firing applications with coal, biomass can provide up to 15 percent of the total energy input of the fuel mixture. The DOE could address greenhouse gas mitigation by co-firing biomass with coal to offset the losses of carbon dioxide to the atmosphere that are inherent in coal combustion processes (even with the best-engineered capture and storage of carbon). Since growth of biomass fixes atmospheric carbon, its combustion leads to no net addition of atmospheric CO_2 even if vented. Thus, co-firing of biomass with coal in an efficient coal gasification process, affording the opportunity for capture and storage of CO_2, could lead to a net reduction of atmospheric CO_2. The co-firing fuel mixture, being dilute in biomass, places lower demands on biomass feedstock. Thus cheaper, though less plentiful, biomass residue could supplant bioenergy crops as feedstock. Using residue biomass would also have a much less significant impact on the environment than would farming of bioenergy crops.

Photobiological hydrogen production is a significantly more efficient process and requires nutrients to a lesser extent than does biomass-to-hydrogen conversion. The objective is to engineer a (micro)organism that catalyzes the light-mediated cleavage of water with the concomitant production of hydrogen at high rates and high thermodynamic efficiency. This process does not take place in naturally occurring organisms at an appreciable rate or scale. While this approach has much potential, there are also major challenges. Substantial bioengineering efforts have to be undertaken to engineer microorganisms with a robust metabolic pathway, including improved kinetics for hydrogen production and efficiencies in light energy conversion and hydrogen production, before a pilot-scale photobiological system could be evaluated. This requires long-term, fundamental research at a significant funding level. Also, inexpensive, large-scale reactor systems need to be designed that minimize the susceptibility of the reactor system to biological contamination. In addition, the public perception of the use and possible concerns over the potential "escape" of genetically engineered microorganisms need to be addressed.

The Department of Energy's Research and Development Program

According to the June 2003 draft of "Hydrogen, Fuel Cells and Infrastructure Technologies Program: Multi-Year Research, Development and Demonstration Plan" (DOE, 2003b), DOE's Office of Energy Efficiency and Renewable Energy program has set technical targets for the years 2005, 2010, and 2015 to reduce costs for biomass gasification/pyrolysis and subsequent steam reforming. Specific goals include the reduction of costs for (1) biomass feedstock, (2) gasification operation (including efficiency), (3) steam reforming, and (4) hydrogen gas purification. Although no specific budget amounts were reported (except at a very high level of aggregation), major funding for R&D of hydrogen production from biomass is apparently in improving gasification/pyrolysis processes. The goals are quite ambitious. The committee's economic analysis (Chapter 5) shows that gasification and the availability of large quantities of inexpensive biomass are major economic barriers for hydrogen derived from biomass. Although listed in the draft report, the EERE program seems to support photobiological hydrogen research, but specific funding levels are unclear. The DOE's R&D targets for increasing the utilization efficiency of absorbed light and hydrogen production are extremely ambitious, and it is unclear how realistic they are. It appears that if such molecular projects are funded, they are for small amounts.

Summary

The committee's analysis indicates the following:

- Considering the assumptions for future technology, biomass-to-hydrogen conversion is unlikely to produce hydrogen at a competitive price, even when compared with hydrogen generated from distributed natural gas.
- The environmental impact of growing significant quantities of biomass as energy crops, including engineered, high-yield crops, will most likely place significant strains on natural resources, including water, soil, land availability, and biodiversity.
- Because of the inherently high cost for collecting and transporting biomass, a biomass gasification plant will be limited in size, will not make full use of the economics of scale, and will be limited to certain geographic regions in the United States.
- Biomass-to-hydrogen conversion is a thermodynamically inefficient path for using solar energy.
- The use of biomass (residues), when co-fired (e.g., with coal) and coupled to subsequent carbon sequestration, might be an important technical option for achieving zero emission and, potentially, a net reduction of atmospheric CO_2.
- Photobiological hydrogen production is a theoretically more efficient process, but significant fundamental molecular research is needed to identify and improve the limiting factors in order to evaluate fully this approach for hydrogen production.

HYDROGEN FROM SOLAR ENERGY

Introduction

It has been estimated that solar energy has the potential of meeting the energy demand of the human race well into the future.[29] One of the methods of recovering solar energy is through the use of photovoltaic (PV) cells. Upon illumination with sunlight, PV cells generate electric energy. Commercial PV modules are available for a wide range of applications. However, they represent a miniscule contribution to U.S. electric power production. The current cost of electricity from a PV module is 6 to 10 times the cost of electricity from coal or natural gas. Therefore, if PV electricity were to be used to make hydrogen, the cost would be significantly higher than if fossil fuels were used. The key for solar energy to be used on a large scale for electricity or hydrogen production is cost reduction. This would require a number of advancements in the current technology.

Current State of Technology

Approximately 85 percent of the current commercial PV modules are based on single-crystal or polycrystalline silicon. The single-crystal or polycrystalline silicon cells are generally of the dimension of 10 to 15 centimeter (cm) (Archer and Hill, 2001). They are either circular or rectangular. In a module, a number of cells are soldered together. Each cell is capable of providing a maximum output of 0.6 volt (V), with the total module output approaching 20 V. The output current of each cell in bright sunlight is generally in the range of to 2 to 5 amps. The single-crystal silicon cells are made from wafers obtained by continuous wire sawing of single-crystal ingots grown by the Czochralski process. Similarly, a large portion of the polycrystalline silicon cells are made from ingots obtained by directional solidification of silicon within a mold. The wafer thickness is generally in the range of 250 to 400 microns. It is worth noting that nearly half of the silicon is wasted as "kerf" loss during cutting. Polycrystalline silicon cells are also made from silicon sheet or ribbon grown by other techniques (Archer and Hill, 2001). This process avoids the cost associated with cutting silicon ingots into wafers. The silicon wafers or ribbons are then further processed to develop *p-n* junctions and wire contacts. The array of cells is laminated using glass and transparent polymer, called ethylvinylacetate (EVA), to provide the final PV module. The modules are known to have long lifetime (10- to 25-year warranty from manufacturers). The current technology gives about 18 percent cell efficiency and 15 percent module efficiency.[30]

A second type of PV technology is based on deposition of thin films. PV cells are prepared by deposition of amorphous as well as microcrystalline silicon from a variety of techniques, including plasma-enhanced chemical vapor deposition, hot wire chemical vapor deposition, and so on. Polycrystalline thin-film compounds based on group II-VI of the periodic table, such as cadmium telluride (CdTe), and group I-III-VI ternary mixtures such as copper-indium-diselenide (CIS), have been used to make thin-film solar cells (Ullal et al., 2002). The thickness of deposited layers is much less than 1 micron. As compared with crystalline silicon solar cells, the thin-film technology potentially has a number of significant advantages in manufacturing: (1) lower consumption of materials; (2) fewer processing steps; (3) automation of processing steps; (4) integrated, monolithic circuit design leading to elimination of the assembly of individual solar cells into final modules; and (5) fast roll-to-roll deposition (Wieting, 2002). It has been estimated that for crystalline silicon solar cells, the complete process involves more than two dozen separate steps to prepare and process ingots, wafers, cells, and circuit assemblies before a module is complete (Wieting, 2002). On the other hand, thin-film module

[29] Nathan Lewis, California Institute of Technology, "Hydrogen Production from Solar Energy," presentation to the committee, April 25, 2003.

[30] The efficiency in this section is defined at 25°C under 1000 W/m^2 of sunlight intensity with the standard global air mass 1.5 spectral distribution. Thus, 15 percent module efficiency refers to peak watt efficiency (W_p) and implies that 15 percent of the incident sunlight energy is converted to electricity.

production requires only half as many process steps, with simplified materials handling.

Thin-film technology appears to hold greater promise for cost reduction, which has led to research by several laboratories over the past two or three decades. Some of the results in efficiency improvement of small laboratory research-size cells, typically of the size of 1 cm^2, are shown in Figure G-15. Research cell efficiencies as high as 21.5 percent for copper-indium (gallium)-diselenide (CIGS) are reported (Ullal et al., 2002). Similarly, high efficiency of 16.5 percent has been reported for CdTe research cells. Amorphous silicon is deposited by using silane (SiH_4) and hydrogen mixtures. In laboratory-scale cells of amorphous silicon, the highest efficiencies obtained are about 12 percent.

One big challenge for thin-film solar cells is to overcome the large drop in efficiency from the laboratory-scale cell to that of a real module. For example, commercial modules of CdTe and CIGS have efficiencies in the range of 7 percent to 12 percent (as compared with laboratory-scale cell efficiencies of 16.5 percent and 21.5 percent). Similarly, commercial amorphous silicon modules have efficiencies less than 10 percent (Shah et al., 1999). The drop in efficiency as cell size is increased is substantial. Attempts are being made to increase the efficiency of amorphous and microcrystalline silicon cells by making dual and triple junction cells (Yang et al., 1997). This change leads to multiple layers, each having a different optimum band gap. However, the deposition of multiple layers increases the processing steps and therefore the cost. A final note is that amorphous silicon modules, when exposed to sunlight, undergo light-induced degradation, operating thereafter at a lower, stabilized efficiency (Shah et al. 1999; Staebler and Wronski, 1977).

In spite of its promise, the thin-film technology has been unable to reduce the cost of solar modules, owing to low deposition rates that have led to low capital utilization of expensive machines. The yields and throughputs have been low. These plants need better inline controls. In recent times, owing to manufacturing problems, some corporations have shut down their thin-film manufacturing facilities. Clearly, easier and faster deposition techniques leading to reproducible results are needed. Also, deposition techniques that would not result in a substantial drop in efficiency from laboratory scale to module scale are required.

Today there is no one clear "winner technology." More than a dozen firms produce solar modules. Even the largest of these firms do not have world-class, large-scale produc-

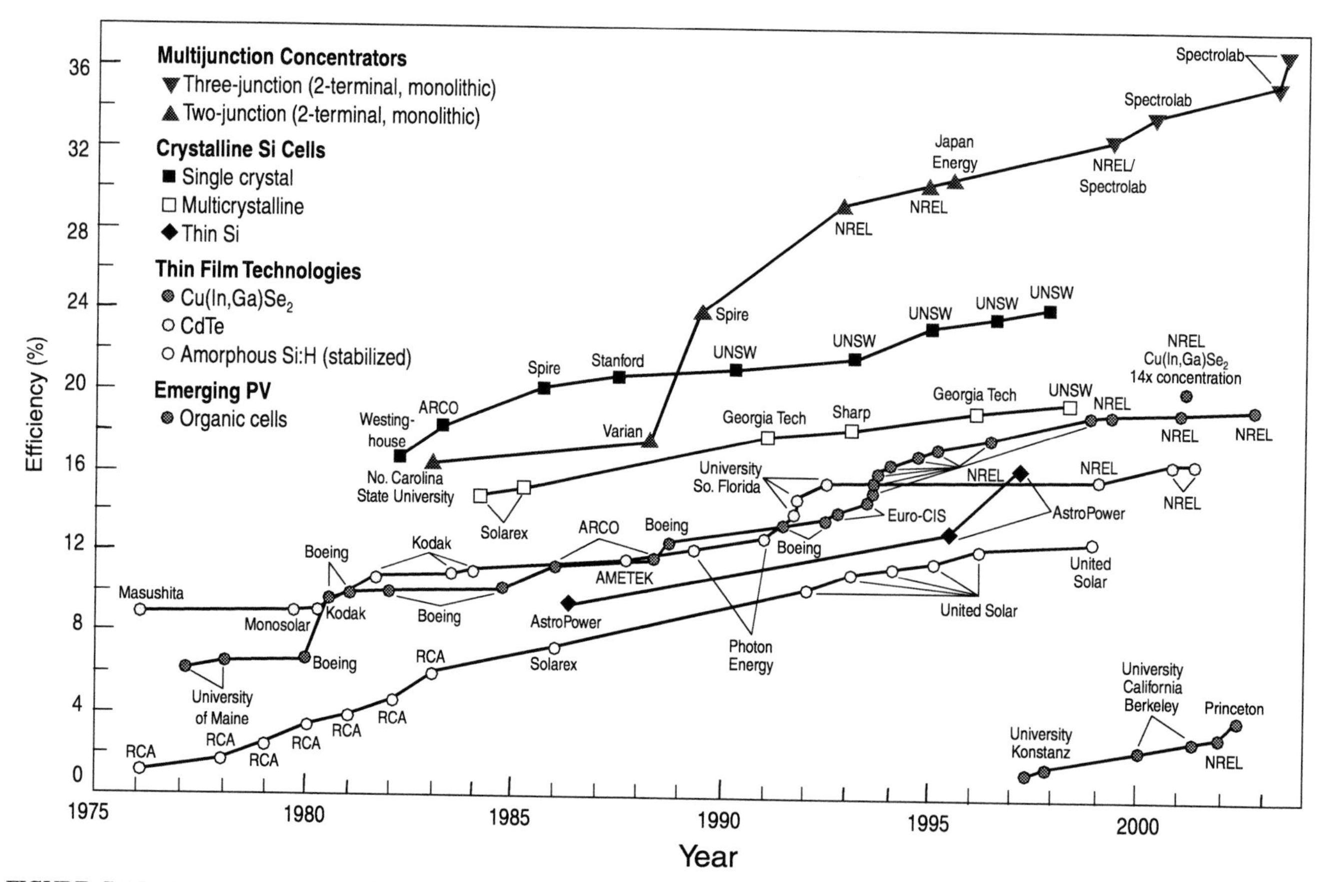

FIGURE G-15 Best research cell efficiencies for multijunction concentrator, thin-film, crystalline silicon, and emerging photovoltaic technologies. SOURCE: National Renewable Energy Laboratory.

tion facilities (greater than 100 MW_p worth of solar modules per year). This size limitation does not allow the economy-of-scale benefits for the solar cell production. Many companies use multiple technologies. The current cost of solar modules is in the range of \$3 to \$6 per peak watt (W_p). For solar cells to be competitive with the conventional technologies for electricity production, the module cost must come down below \$1/W_p. Table G-9 provides cost estimates of producing electricity as well as hydrogen calculated by the committee. In the current scenario, with a favorable, installed cost of about \$3.285/W_p, the electricity cost is estimated to be about \$0.319/kWh (scenario Dist PV-C of Chapter 5). For a futuristic case with all of the expected technology and production advances, the anticipated installed cost of \$1.011/W_p provides electricity cost of \$0.098/kWh (scenario Dist PV-F, Table E-49 in Appendix E). While this target is attractive for electricity generation, it does not produce hydrogen at a competitive cost.

Energy is consumed in the manufacture of solar modules. It has been estimated by NREL that for a crystalline silicon module, the payback period of energy is about 4 years. For an amorphous silicon module this period is currently about 2 years, with the expectation that it will eventually be less than 1 year.

Future Technology

Photovoltaic Cells

Various developments are likely to improve the economic competitiveness of solar technology, especially for thin-film technology. The current research on microcrystalline silicon deposition techniques is leading to higher efficiencies. Techniques leading to higher deposition rates at moderate pressures are being developed (Schroeder, 2003). Better barrier materials to eliminate moisture ingress in the thin-film modules will prolong the module life span. Robust deposition techniques will increase the yield from a given type of equipment. Inline detection and control methods will help to reduce the cost. Some of this advancement will require creative tools and methods.

The committee believes that installed costs of roughly \$1/W_p are attainable. Material costs are quite low, but substrate material, expensive coating equipment, low utilization of equipment, and labor-intensive technology lead to high overall costs. It is expected that in the next decade or two, improvements in these areas have a potential to bring the cost much below \$1/W_p. World-class plants with economies of scale will further contribute to the lowering of cost. For crystalline-silicon-wafer-based technology, the raw material costs by themselves are almost \$1/W_p. However, improvements in operating efficiency, the cost of raw materials, and reduced usage of certain materials are expected to bring overall cost in the neighborhood of \$1/W_p.

A concept that has been proposed is the dye-sensitized solar cell, also known as the Grätzel cell (O'Regan and Grätzel, 1991). A dye is incorporated in a porous inorganic matrix such as TiO_2, and a liquid electrolyte is used for positive charge transport. Photons are absorbed by the dye, and electrons are injected from the dye into *n*-type titania nanoparticles. The nanoparticles of titania are fused together and carry electrons to a conducting electrode. The dye gets its electron from the electrolyte, and the positive ion of the electrolyte moves to the other electrode (Grätzel, 2001). This type of cell has a potential to be low-cost. However, the current efficiencies are quite low, and the stability of the cell in sunlight is very poor. Research is needed to improve performance at both fronts.

Another area of intense research is that on the integration of organic and inorganic materials at the nanometer scale into hybrid solar cells. The current advancement in conductive polymers and the use of such polymers in electronic devices and displays provides the impetus for optimism. The nano-sized particles or rods of the suitable inorganic materials are embedded in the conductive organic polymer matrix. Once again, the research is in the early phase and the current efficiencies are quite low. However, the production of solar cells based either solely on conductive polymers or on hybrids with inorganic materials has a large potential to provide low-cost solar cells. It is hoped that one would be able to cast thin-film solar cells of such materials at a high speed, resulting in low cost.

Regarding production costs, all of the technologies discussed so far convert solar energy into electricity and use the electricity to generate hydrogen through the electrolysis of water. Since PV cells produce dc currents, the electric power can be directly used for electrolysis. As discussed in the section above on electrolyzers, considerable cost reductions are anticipated, which will lower the cost of hydrogen from solar cells. These cost reductions will be particularly valuable for solar cell electricity because the low usage factor associated with PV modules also contributes to the low usage of electrolyzers. This contributes heavily to the cost of hydro-

TABLE G-9 Estimated Cost of Hydrogen Production for Solar Cases

Case	Installed Cost (\$/kW)	Electricity Cost (\$/kWh)	Hydrogen Cost with Electrolyzer (\$/kg)
Current (Dist PV-C)	3285	0.319	28.19 (Dist PV Ele-C)
Future (Dist PV-F)	1011	0.098	6.18 (Dist PV Ele-F)

NOTE: See Appendix E for definition of the symbols for the solar technology cases. See also Tables E-48 and E-49 of Appendix E.

gen produced. For example, in the committee's analysis of costs discussed in Chapter 5 (summarized in Table G-9), for the future optimistic case the cost of hydrogen is calculated to be \$6.18/kg (Dist PV Ele-F). For this case, the cost of the installed PV panels, including all of the general facilities, is estimated to be \$1.011/W_p, and is used in conjunction with an electrolyzer that is assumed to take advantage of all of the advancements made in the fuel cell. The PV part is responsible for \$4.64/kg, and the electrolyzer is \$1.54/kg. Compared with this, the cost of hydrogen from a future central coal plant at the dispensing station is estimated to be \$1.63/kg with carbon tax (CS Coal-F). This cost implies that for a PV-electrolyzer to compete in the future with a coal plant, either the cost of PV modules must be reduced by an order of magnitude or the electrolyzer cost must drop substantially from \$125/kW. A factor contributing to this is the low utilization of the electrolyzer capital. It has been proposed to use electricity from the grid to run the electrolyzer when solar electricity is unavailable. This use increases the time on-stream for the electrolyzer; however, in the long term, for solar to play a dominant role in the hydrogen economy, it cannot rely on power from the grid to supplement equipment utilization. Therefore, while electricity at \$0.098/kWh from a PV module can be quite attractive for distributive applications where electricity is directly used, its use in conjunction with electrolysis to produce hydrogen is certainly not competitive with the projected cost of hydrogen from coal.

Direct Production

Research is being done to create photoelectrochemical cells for the direct production of hydrogen (Grätzel, 2001).[31] In this method, light is converted to electrical and chemical energy. The technical challenge stems from the fact that energy from two photons is needed to split one water molecule. A solid inorganic oxide electrode is used to absorb photons and provide oxygen and electrons. The electrons flow through an external circuit to a metal electrode, and hydrogen is liberated at this electrode. The candidate inorganic oxides are $SrTiO_3$, $KTaO_3$, TiO_2, SnO_2, and Fe_2O_3. When successful, such a method holds promise of directly providing low-cost hydrogen from solar energy.

Regarding production costs, it seems that a photoelectrochemical device in which all of the functions of photon absorption and water splitting are combined in the same equipment may have better potential for hydrogen production at reasonable costs. However, it is instructive to do a quick "back of the envelope" analysis for the acceptable cost by such a system. It is assumed that cost per peak watt for a photoelectrochemical device is the same as that for the possible future PV modules (see Table E-48 of Appendix E.) It is further assumed that this energy is recovered as hydrogen rather than as electricity. Therefore, a recovery of 39.4 kWh translates into a kilogram of hydrogen. This implies that 4729 kWe worth of solar plant in the Dist PV-F spreadsheet will produce about 576 kg/day of hydrogen (assuming an annual capacity factor of 20 percent). At the total cost of \$0.813 million per year, this gives \$3.87/kg of hydrogen! This cost is still too high when compared with that of hydrogen from coal or natural gas plants. It implies that photoelectrochemical devices should recover hydrogen at an energy equivalent of \$0.4 to \$0.5/W_p. This cost challenge is similar to that for electricity production from the solar cells.

Advantages and Disadvantages of Solar Energy

Solar energy holds the promise of being inexhaustible. If harnessed, it can meet all of the energy needed in the foreseeable future. It is clean and environmentally friendly. It converts solar energy into hydrogen without the emission of any greenhouse gas. Because of its distributed nature of power production, it contributes to the national security.

There are certain challenges associated with the use of solar energy. The intermittent nature of sunshine, on both a daily and a seasonal basis, presents a number of challenges. A backup system, or a storage system for electricity/hydrogen, is needed for the periods when sunshine is not available and power demand exists. Furthermore, this intermittent availability means that four to six times more solar modules have to be installed than the peak watt rating would dictate. This intermittency also implies that a significant decrease in the module cost is required. Another challenge is to ensure that no toxic materials are discharged during the fabrication of solar cells and over the complete life cycle of the cell. Such questions have been raised in the context of cadmium-containing solar cells, and public perception in such cases will play a key role.

Challenges and Research and Development Needs

Large-scale use of solar energy for hydrogen economy will require research and development efforts on multiple fronts. In the short term, there is a need to reduce the cost of thin-film solar cells. This reduction will require the development of silicon deposition techniques that are robust and provide high throughput rates. New deposition techniques at moderate pressures with microcrystalline silicon structures for higher efficiencies are needed. Inline detection and control and the development of better roll-to-roll coating processes can lead to reductions in the manufacturing costs. Increased automation will also contribute to the decreased cost. Issues related to a large decrease in efficiency from small laboratory samples to the module level should be addressed. In the short run, thin-film deposition methods can potentially gain from a fresh look at the overall process from the laboratory scale to the manufacturing scale. The research in

[31]Nathan Lewis, California Institute of Technology, "Hydrogen Production from Solar Energy," presentation to the committee, April 25, 2003.

this area is expensive. Some additional centers for such research in academia with industrial alliances could be beneficial. It will be necessary to collect multifunctional teams from different engineering disciplines for such studies.

In the midterm to long term, organic-polymer-based solar cells hold promise for mass production at low cost. They have an appeal for being cast as thin films at very high speeds using known polymer film casting techniques. Currently, the efficiency of such a system is quite low (in the neighborhood of 3 to 4 percent or lower), and stability in sunlight is poor. However, owing to the tremendous development in conducting polymers and other electronics-related applications, it is anticipated that research in such an area has a high potential for success.

Similarly, the search for a stable dye material and better electrolyte material in dye-sensitized cells (Grätzel cells) has a potential to lead to lower-cost solar cells. There is a need to increase the stable efficiency of such cells; a stable efficiency of about 10 percent could be quite useful.

In the long run, the success of directly splitting water molecules by using photons is quite attractive. Research in this area could be very fruitful.

Department of Energy Programs for Solar Energy to Hydrogen

The current DOE target for photoelectrochemical hydrogen production in 2015 is \$5/kg H_2 at the plant gate. Even if this target is met, solar hydrogen is unlikely to be competitive. Therefore, beyond 2015 a much more aggressive cost target for hydrogen production by photoelectrochemical methods is needed.

Since photoelectrochemical hydrogen production is in an embryonic stage, a parallel effort to reduce the cost of electricity production from PV modules must be made. A substantial reduction in PV module cost (lower than \$1/W_p), coupled with a similar reduction in electrolyzer costs (much below \$125/kW at a reasonably high efficiency of about 70 percent based on lower heating value), could provide hydrogen at reasonable cost. In the long run, considering the environmental issues associated with fossil fuels and considering the limitless supply of solar energy, this has a potential to be quite attractive. This option will be especially attractive if advances in battery technology are unable to substantially increase the electricity storage density (based on mass of battery) and greatly reduce the cost of batteries. Therefore, it is recommended that thin-film technologies and other emerging PV technologies that hold the promise for cost reduction be aggressively pursued. As stated earlier, it means that more efficient and robust methods for thin-film coating must be developed. Organic-polymer-based solar cells should also be funded. There is tremendous development underway in conducting polymers for light-emitting diodes and other display technologies. The potential of these materials for solar cell PVs must be actively explored.

Summary

All of the current methods and the projected technologies of producing hydrogen from solar energy are much more expensive (greater than a factor of 3) when compared with hydrogen production from coal or natural gas plants. This is due partly to the lower annual utilization factor of about 20 percent (as compared with say, wind, at 30 to 40 percent). This high cost puts enormous pressure on the need to reduce the cost of a solar energy recovery device. While an expected future installed module cost of about \$1/W_p is very attractive for electricity generation and deserves a strong research effort in its own right, this cost fails to provide hydrogen at a competitive value. The raw material cost for crystalline silicon-wafer-based technologies is a large fraction of the \$1/W_p value and is therefore less likely to provide hydrogen economically. On the other hand, thin-film technologies do not use much raw material in thin films themselves but require tremendous progress in the deposition technology. There is a need for a robust deposition method that would have a potential to reduce cost much below \$1/W_p. Emerging polymer-based technologies have a potential to provide low-cost devices to harness solar energy. It is apparent that there is no one method of harnessing solar energy that is clearly preferable. However, it appears possible that new concepts may emerge that would be competitive. The benefits of such developments would be very substantial.

In the future, as the cost of the fuel cell approaches \$50 per kilowatt, the cost of an electrolytic cell to electrolyze water is also expected to approach a low number (about \$125/kW). With such low-cost electrolyzer units, the electricity cost of about \$0.02 to \$0.03/kWh is expected to result in a competitive hydrogen cost. It is also estimated that for a photoelectrochemical method to compete, its cost must approach \$0.04 to \$0.05/kWh. The order-of-magnitude reductions in cost for both hydrogen processes are similar.

Appendix H

Useful Conversions and Thermodynamic Properties

TABLE H-1 Conversion Factors

metric ton (tonne) = 1000 kg = 1.1023 short tons
Btu = 1055 J
quad = 10^{15} Btu = 1.055 EJ
liter = 0.2642 gallons U.S.
cubic meter (m^3) = 35.31 cubic feet
conversions for hydrogen:
1 million scf/day = 2.65 short tons/day
1 kg = 11.13 N-m^3 (0 degrees Celsius and 1 atmosphere)
1 kg = 415.6 scf (60 degrees Fahrenheit and 1 atmosphere)

NOTE: scf = standard cubic feet; Btu = British thermal unit; EJ = exajoule = 10^{18} joules; N-m^3 = normal cubic meter; kg = kilogram.

TABLE H-2 Thermodynamic Properties of Chemicals of Interest

Parameter	Value
Hydrogen HHV (ΔH)	–286 kJ/mol
Hydrogen LHV (ΔH)	–242 kJ/mol
Methane gross heat of combustion HHV (ΔH_c)	–891 kJ/mol
Energy content of 1 kg hydrogen	141.9 MJ (HHV) = 39.4 kWh 120.1 MJ (LHV) = 33.3 kWh
of 1 N-m^3 hydrogen	12.7 MJ (HHV)
of 1 pound of hydrogen	64.4 MJ (HHV) = 61.0 kBtu
of 1 gallon gasoline	121.3 MJ (LHV); 115,000 Btu (LHV)

NOTE: HHV = higher heating value; LHV = lower heating value; ΔH = enthalpy; J = joule; Btu = British thermal unit; M = million; k = thousand; mol = mole; N-m^3 = normal cubic meter; kWh = kilowatt hour. SOURCE: NIST (2003), except DOE (2003f) for gasoline data.